Modern Birkhäuser Classics

Many of the original research and survey monographs in pure and applied mathematics published by Birkhäuser in recent decades have been groundbreaking and have come to be regarded as foundational to the subject. Through the MBC Series, a select number of these modern classics, entirely uncorrected, are being re-released in paperback (and as eBooks) to ensure that these treasures remain accessible to new generations of students, scholars, and researchers.

Introduction to Quantum Groups

George Lusztig

Reprint of the 1994 Edition

Birkhäuser

George Lusztig
Department of Mathematics
MIT
Cambridge, MA 02139

ISBN 978-0-8176-4716-2 e-ISBN 978-0-8176-4717-9
DOI 10.1007/978-0-8176-4717-9
Springer New York Dordrecht Heidelberg London

Printed on acid-free paper

www.birkhauser-science.com

George Lusztig

Introduction to Quantum Groups

Birkhäuser
Boston • Basel • Berlin

George Lusztig
Department of Mathematics
MIT
Cambridge, MA 02139

Library of Congress Cataloging In-Publication Data

Lusztig, George, 1946-
Introduction to quantum groups / George Lusztig.
p. cm. -- (Progress in mathematics ; v. 110)
Includes bibliographical references and index.
ISBN 0-8176-3712-5 (acid free). -- ISBN 3-7643-3712-5 (acid free)
1. Quantum groups. 2. Mathematical physics. I. Title
II. Series: Progress in mathematics (Boston, Mass.) ; vol. 110
QC 20.7.G76L88 1993 93-7800
530.1'5255--dc20 CIP

Printed on acid-free paper

Second printing 1994

Birkhäuser

ISBN 0-8176-3712-5
ISBN 3-7643-3712-5
Typeset by the Author in AMSTEX.
Printed and bound by Quinn-Woodbine, Woodbine, NJ.
Printed in the U.S.A.

9 8 7 6 5 4 3 2

Contents

Preface

According to Drinfeld, a quantum group is the same as a Hopf algebra. This includes as special cases, the algebra of regular functions on an algebraic group and the enveloping algebra of a semisimple Lie algebra. The quantum groups discussed in this book are the quantized enveloping algebras introduced by Drinfeld and Jimbo in 1985, or variations thereof.

Although such quantum groups appeared in connection with problems in statistical mechanics and are closely related to conformal field theory and knot theory, we will regard them purely as a new development in Lie theory. Their place in Lie theory is as follows. Among Lie groups and Lie algebras (whose theory was initiated by Lie more than a hundred years ago) the most important and interesting ones are the semisimple ones. They were classified by E. Cartan and Killing around 1890 and are quite central in today's mathematics. The work of Chevalley in the 1950s showed that semisimple groups can be defined over arbitrary fields (including finite ones) and even over integers.

Although semisimple Lie algebras cannot be deformed in a non-trivial way, the work of Drinfeld and Jimbo showed that their enveloping (Hopf) algebras admit a rather interesting deformation depending on a parameter v. These are the quantized enveloping algebras of Drinfeld and Jimbo. The classical enveloping algebras could be obtained from them for $v \to 1$.

Subsequent work showed that the algebras of Drinfeld and Jimbo have a natural form over $\mathbf{Z}[v, v^{-1}]$; this specializes for $v = 1$ to the Kostant $\mathbf{Z}$-form of the classical enveloping algebras. On the other hand, it can be specialized to v equal to a root of 1, giving some new objects which include quantum versions of the semisimple groups over fields of positive characteristic.

In addition to extending the range of the theory of semisimple groups from $\mathbf{Z}$ to $\mathbf{Z}[v, v^{-1}]$, the theory of quantum groups has led to a new, extremely rigid structure, in which the objects of the theory are provided with canonical bases with rather remarkable properties; in particular, in the simply laced case, the structure constants with respect to the canonical bases are not only in $\mathbf{Z}[v, v^{-1}]$, but in $\mathbf{N}[v, v^{-1}]$. These specialize, for $v = 1$, to canonical bases for the objects in the classical theory, in which the structure constants are not only in $\mathbf{Z}$, but in $\mathbf{N}$. (As we will see, the non-simply laced case can be regarded as merely a twisted version of the simply laced case; thus, the simply laced case is the really fundamental one.)

The theory of semisimple Lie algebras also includes nowadays an extension to the case of affine Lie algebras. These Lie algebras, which appeared in the works of physicists, can be treated simultaneously with the usual semisimple algebras in the framework of Kac-Moody Lie algebras. The algebras of Drinfeld and Jimbo are also defined in this more general context.

This book contains an exposition of the topics above with emphasis on canonical bases. We will develop the theory without assuming any knowledge of semisimple Lie algebras or Kac-Moody Lie algebras (except towards the end). On the other hand, to construct canonical bases, we will make use of the theory of perverse sheaves, which will be reviewed but not explained.

The readers who are not comfortable with the theory of perverse sheaves are advised to skip Chapters 8–13, and accept the theorems in Chapter 14 without proof (the statement of those theorems do not involve perverse sheaves, only their proofs do).

The book is divided into six parts. Part I contains an elementary treatment of the algebras of Drinfeld and Jimbo. Part II contains the construction of canonical bases using perverse sheaves. Part III deals with results of Kashiwara and their applications. Part IV is concerned with the canonical basis of (a modified form of) the quantized enveloping algebra. Part V is concerned mainly with phenomena at roots of 1. Part VI is concerned with the action of the braid group.

References to the literature are given at the end of each Part.

Newton, Massachusetts, February 26, 1993

Acknowledgments. The author's work has been supported in part by the National Science Foundation. A part of the writing was done while the author enjoyed the hospitality of the Institut des Hautes Etudes Scientifiques in Bures sur Yvette.

I wish to thank Ann Kostant for her dedicated work in preparing this book for printing. I also wish to thank Henning H. Andersen and Toshiyuki Tanisaki for their very helpful comments on the manuscript.

Part I

THE DRINFELD-JIMBO ALGEBRA U

Chapters 1 and 3 contain a treatment of the Drinfeld-Jimbo algebras $\mathbf{U}$ over $\mathbf{Q}(v)$. We use a definition somewhat different from the original ones, which were in terms of quantum Serre relations. (Drinfeld worked in a topological setting, over power series, while Jimbo worked over $\mathbf{Q}(v)$.) Drinfeld introduced a non-degenerate pairing between the upper triangular part and the lower triangular part of $\mathbf{U}$ (in his setting). This can be also viewed as a non-degenerate pairing on the lower triangular part. We use such a pairing to give a definition of the lower triangular part in which the quantum Serre relations are not imposed but are automatically satisfied. The identification with the definition in terms of quantum Serre relations will be given in Part V.

The universal $\mathcal{R}$-matrix was introduced originally by Drinfeld as an intertwiner between the comultiplication of $\mathbf{U}$ and its transpose. We found it useful to look for an intertwiner between the comultiplication of $\mathbf{U}$ and the new comultiplication obtained by conjugating with a certain antilinear involution $^{-} : \mathbf{U} \to \mathbf{U}$.

This leads to what we call the "quasi-$\mathcal{R}$-matrix". It has a simpler characterization than Drinfeld's $\mathcal{R}$-matrix (see Chapter 4) and is equal to it except for the diagonal part.

In Chapter 5, we introduce some symmetries of an integrable module of $\mathbf{U}$; in Part VI, it will be seen that these symmetries define braid group actions.

Chapter 6 contains a proof of the quantum analogue of the Weyl-Kac complete reducibility theorem.

The higher order quantum Serre relations are introduced in Chapter 7; they will be used in the discussion of the quantum Frobenius homomorphism in Part V and of the braid group actions in Part VI.

CHAPTER 1

The Algebra f

1.1. Cartan Datum

1.1.1. A *Cartan datum* is a pair $(I, \cdot)$ consisting of a finite set I and a symmetric bilinear form $\nu, \nu' \mapsto \nu \cdot \nu'$ on the free abelian group $\mathbf{Z}[I]$, with values in $\mathbf{Z}$. It is assumed that:

(a) $i \cdot i \in \{2, 4, 6, \dots\}$ for any $i \in I$;

(b) $2\frac{i \cdot j}{i \cdot i} \in \{0, -1, -2, \dots\}$ for any $i \neq j$ in I.

Two Cartan data $(I, \cdot)$ and $(I, \circ)$ are said to be *proportional* if there exist integers $a, b \geq 1$ such that $ai \circ j = bi \cdot j$ for all $i, j \in I$.

1.1.2. We assume that a Cartan datum $(I, \cdot)$ is given. Let v be an indeterminate.

For any $i \in I$, we set $v_i = v^{i \cdot i/2}$. This notation will be extended in two different ways:

(a) for any rational function $P \in \mathbf{Q}(v)$ we shall write P_i for the rational function obtained from P by substituting v by v_i;

(b) for any $\nu = \sum_i \nu_i i \in \mathbf{Z}[I]$, we shall write $v_\nu = \prod_i v_i^{\nu_i}$ (if $\nu = i$, then v_ν is v_i in the earlier sense).

We shall use the notation

(c) $\operatorname{tr} \nu = \sum_i \nu_i \in \mathbf{Z}$ for $\nu = \sum_i \nu_i i \in \mathbf{Z}[I]$.

1.2. The Algebras $'\mathbf{f}$ and $\mathbf{f}$

1.2.1. We denote by $'\mathbf{f}$ the free associative $\mathbf{Q}(v)$-algebra with 1 with generators $\theta_i \quad (i \in I)$.

Let $\mathbf{N}[I]$ be the submonoid of $\mathbf{Z}[I]$ consisting of all linear combinations of elements of I with coefficients in $\mathbf{N}$. For any $\nu = \sum_i \nu_i i \in \mathbf{N}[I]$, we denote by $'\mathbf{f}_\nu$ the $\mathbf{Q}(v)$-subspace of $'\mathbf{f}$ spanned by the monomials $\theta_{i_1}\theta_{i_2}\cdots\theta_{i_r}$ such that for any $i \in I$, the number of occurrences of i in the sequence $i_1, i_2, \dots, i_r$ is equal to ν_i. Then each $'\mathbf{f}_\nu$ is a finite dimensional $\mathbf{Q}(v)$-vector space and we have a direct sum decomposition $'\mathbf{f} = \oplus_\nu {'\mathbf{f}_\nu}$ where ν runs over $\mathbf{N}[I]$. We have $'\mathbf{f}_\nu {'\mathbf{f}_{\nu'}} \subset {'\mathbf{f}_{\nu+\nu'}}$, $1 \in {'\mathbf{f}_0}$ and $\theta_i \in {'\mathbf{f}_i}$; these properties provide an alternative definition of $'\mathbf{f}_\nu$.

G. Lusztig, *Introduction to Quantum Groups*, Modern Birkhäuser Classics,
DOI 10.1007/978-0-8176-4717-9_1, © Springer Science+Business Media, LLC 2010

An element x of $'\mathbf{f}$ is said to be *homogeneous* if it belongs to $'\mathbf{f}_\nu$ for some ν. We then set $|x| = \nu$. In particular, we have $|0| = \nu$ for any ν.

1.2.2. The tensor product $'\mathbf{f} \otimes {'\mathbf{f}}$ can be regarded as a $\mathbf{Q}(v)$-algebra with multiplication

$$(x_1 \otimes x_2)(x'_1 \otimes x'_2) = v^{|x_2|\cdot|x'_1|} x_1 x'_1 \otimes x_2 x'_2$$

where $x_1, x_2, x'_1, x'_2 \in {'\mathbf{f}}$ are homogeneous; this algebra is associative since $\nu \cdot \nu'$ is bilinear. Similarly, $'\mathbf{f} \otimes {'\mathbf{f}} \otimes {'\mathbf{f}}$ is an associative $\mathbf{Q}(v)$-algebra with multiplication

$$(x_1 \otimes x_2 \otimes x_3)(x'_1 \otimes x'_2 \otimes x'_3) = v^{|x_2|\cdot|x'_1| + |x_3|\cdot|x'_2| + |x_3|\cdot|x'_1|} x_1 x'_1 \otimes x_2 x'_2 \otimes x_3 x'_3,$$

for homogeneous $x_1, x_2, x_3, x'_1, x'_2, x'_3$.

The following statement is easily verified: if $r : {'\mathbf{f}} \to {'\mathbf{f}} \otimes {'\mathbf{f}}$ is an algebra homomorphism, then $(r \otimes 1)r$ and $(1 \otimes r)r$ are algebra homomorphisms $'\mathbf{f} \to {'\mathbf{f}} \otimes {'\mathbf{f}} \otimes {'\mathbf{f}}$.

We apply this statement to the unique algebra homomorphism $r : {'\mathbf{f}} \to {'\mathbf{f}} \otimes {'\mathbf{f}}$ such that $r(\theta_i) = \theta_i \otimes 1 + 1 \otimes \theta_i$ for all i. For this r, the algebra homomorphisms $(r \otimes 1)r$ and $(1 \otimes r)r$ take the same value on any algebra generator θ_i, namely $\theta_i \otimes 1 \otimes 1 + 1 \otimes \theta_i \otimes 1 + 1 \otimes 1 \otimes \theta_i$; hence these two algebra homomorphisms coincide. Thus, we have the co-associativity property

$$(r \otimes 1)r = (1 \otimes r)r : {'\mathbf{f}} \to {'\mathbf{f}} \otimes {'\mathbf{f}} \otimes {'\mathbf{f}}.$$

Proposition 1.2.3. *There is a unique bilinear inner product* $(,)$ *on* $'\mathbf{f}$ *with values in* $\mathbf{Q}(v)$ *such that* $(1, 1) = 1$ *and*

(a) $(\theta_i, \theta_j) = \delta_{i,j}(1 - v_i^{-2})^{-1}$ *for all* $i, j \in I$;

(b) $(x, y'y'') = (r(x), y' \otimes y'')$ *for all* $x, y', y'' \in {'\mathbf{f}}$;

(c) $(xx', y'') = (x \otimes x', r(y''))$ *for all* $x, x', y'' \in {'\mathbf{f}}$.

(The bilinear form $({'\mathbf{f}} \otimes {'\mathbf{f}}) \times ({'\mathbf{f}} \otimes {'\mathbf{f}}) \to \mathbf{Q}(v)$ *given by* $x_1 \otimes x_2, x'_1 \otimes x'_2 \mapsto (x_1, x'_1)(x_2, x'_2)$ *is denoted again by* $(,)$.*)*

The bilinear form $(,)$ *on* $'\mathbf{f}$ *is symmetric.*

The linear maps $'\mathbf{f}_{\nu+\nu'} \to {'\mathbf{f}}_\nu \otimes {'\mathbf{f}}_{\nu'}$ defined by r give, by passage to dual spaces, linear maps $'\mathbf{f}^*_\nu \otimes {'\mathbf{f}}^*_{\nu'} \to {'\mathbf{f}}^*_{\nu+\nu'}$. These define the structure of an associative algebra with 1 on $\oplus_\nu {'\mathbf{f}}^*_\nu$. For any i, let $\xi_i \in {'\mathbf{f}}^*_i$ be the linear form given by $\xi_i(\theta_i) = (1 - v_i^{-2})^{-1}$.

Let $\phi : {}'\mathbf{f} \to \oplus_\nu {}'\mathbf{f}_\nu^*$ be the unique algebra homomorphism preserving 1, such that $\phi(\theta_i) = \xi_i$ for all i. For $x, y \in {}'\mathbf{f}$, we set $(x, y) = \phi(y)(x)$. Then (b) holds, since ϕ is an algebra homomorphism. Clearly,

(d) $(x, y) = 0$ if x, y are homogeneous, with $|x| \neq |y|$.

We now show that (c) holds. Assume that (c) is known for y'' replaced by y or by y' (both homogeneous) and for any x, x'. We prove then that (c) holds for $y'' = yy'$ and any x, x'. We can assume that x, x' are homogeneous. We write

$$r(x) = \sum x_1 \otimes x_2, \quad r(x') = \sum x'_1 \otimes x'_2,$$
$$r(y) = \sum y_1 \otimes y_2, \quad r(y') = \sum y'_1 \otimes y'_2,$$

all factors being homogeneous. Then

$$r(xx') = \sum v^{|x_2|\cdot|x'_1|} x_1 x'_1 \otimes x_2 x'_2, \quad r(yy') = \sum v^{|y_2|\cdot|y'_1|} y_1 y'_1 \otimes y_2 y'_2.$$

We have

$$(xx', yy') = \phi(yy')(xx') = (\phi(y)\phi(y'))(xx') = (\phi(y) \otimes \phi(y'))(r(xx'))$$
$$= \sum v^{|x_2|\cdot|x'_1|}(x_1 x'_1, y)(x_2 x'_2, y')$$
$$= \sum v^{|x_2|\cdot|x'_1|}(x_1 \otimes x'_1, r(y))(x_2 \otimes x'_2, r(y'))$$

(e)

$$= \sum v^{|x_2|\cdot|x'_1|}(x_1, y_1)(x'_1, y_2)(x_2, y'_1)(x'_2, y'_2).$$

On the other hand,

$$(x \otimes x', r(yy')) = \sum v^{|y_2|\cdot|y'_1|}(x \otimes x', y_1 y'_1 \otimes y_2 y'_2)$$
$$= \sum v^{|y_2|\cdot|y'_1|}(x, y_1 y'_1)(x', y_2 y'_2) = \sum v^{|y_2|\cdot|y'_1|}(r(x), y_1 \otimes y'_1)(r(x'), y_2 \otimes y'_2)$$

(f)

$$= \sum v^{|y_2|\cdot|y'_1|}(x_1, y_1)(x'_1, y_2)(x_2, y'_1)(x'_2, y'_2).$$

By (d), we may assume that $|x'_1| = |y_2|, |y'_1| = |x_2|$ in the last sum in (e) and (f); hence in (e) we have $|x_2| \cdot |x'_1| = |x_2| \cdot |y_2|$ and in (f) we have $|y_2| \cdot |y'_1| = |y_2| \cdot |x_2| = |x_2| \cdot |y_2|$ (by the symmetry of the form $\cdot$). Hence the two sums are equal and our assertion follows.

We now see that it suffices to verify (c) in the special case where y'' is one of the generators θ_i of ${}'\mathbf{f}$. In that case, we may assume that either $x = \theta_i$ and $x' = 1$, or $x = 1$ and $x' = \theta_i$; in either case, (c) follows from our definition of ξ_i. Property (a) is clear from the definition. The existence of $(,)$ is thus proved. The uniqueness of $(,)$ is immediate. The fact that $(,)$ is symmetric follows from uniqueness.

1.2.4. The ideal $\mathcal{I}$. Let $\mathcal{I}$ be the radical of the form $(,)$. We show that $\mathcal{I}$ is a two-sided ideal of $'\mathbf{f}$.

Let $x \in \mathcal{I}$ and $y \in {}'\mathbf{f}$. Let $z \in {}'\mathbf{f}$; we write $r(z) = \sum z' \otimes z''$. We have $(xy, z) = (x \otimes y, r(z)) = \sum(x, z')(y, z'') = 0$ and $(yx, z) = (y \otimes x, r(z)) = \sum(y, z')(x, z'') = 0$. This shows that $xy \in \mathcal{I}$ and $yx \in \mathcal{I}$; our assertion follows.

1.2.5. The algebra f. Let $\mathbf{f} = {}'\mathbf{f}/\mathcal{I}$ be the quotient algebra of $'\mathbf{f}$ by the ideal $\mathcal{I}$. Since $\mathcal{I}$ is compatible with the decomposition $\oplus_\nu {}'\mathbf{f}_\nu$, we have a direct sum decomposition $\mathbf{f} = \oplus_\nu \mathbf{f}_\nu$ where $\mathbf{f}_\nu$ is the image of $'\mathbf{f}_\nu$ under the natural map $'\mathbf{f} \to \mathbf{f}$. Each subspace $\mathbf{f}_\nu$ is finite dimensional since $'\mathbf{f}_\nu$ is finite dimensional. The form $(,)$ on $'\mathbf{f}$ defines a symmetric bilinear form on $\mathbf{f}$ denoted again by $(,)$; this form is non-degenerate on each summand $\mathbf{f}_\nu$.

We denote again by θ_i the image of θ_i in $\mathbf{f}$. If $x \in \mathbf{f}_\nu$, we say that x is homogeneous and we write $|x| = \nu$.

1.2.6. The homomorphism $r : \mathbf{f} \to \mathbf{f} \otimes \mathbf{f}$. If $x \in \mathcal{I}$ and $y, z \in {}'\mathbf{f}$, then $(r(x), y \otimes z) = (x, yz) = 0$. Thus, $r(x)$ is contained in the radical of the pairing on $'\mathbf{f} \otimes {}'\mathbf{f}$ defined by $(,)$. This radical is clearly equal to $\mathcal{I} \otimes {}'\mathbf{f} + {}'\mathbf{f} \otimes \mathcal{I}$. Thus $r(\mathcal{I}) \subset \mathcal{I} \otimes {}'\mathbf{f} + {}'\mathbf{f} \otimes \mathcal{I}$. It follows that r induces an algebra homomorphism $\mathbf{f} \to \mathbf{f} \otimes \mathbf{f}$, which is denoted again by r. We regard $\mathbf{f} \otimes \mathbf{f}$ as an algebra by the same rule (1.2.2) as the one defining $'\mathbf{f} \otimes {}'\mathbf{f}$.

The coassociativity property of 1.2.2 continues of course to hold for this r.

1.2.7. Let $\sigma : {}'\mathbf{f} \to {}'\mathbf{f}^{opp}$ be the homomorphism of algebras with 1 which takes each θ_i to θ_i. Let $^t r : {}'\mathbf{f} \to {}'\mathbf{f} \otimes {}'\mathbf{f}$ be the composition of r with the linear map $x \otimes y \to y \otimes x$ of $'\mathbf{f} \otimes {}'\mathbf{f}$ into itself.

Lemma 1.2.8. (a) *We have $r(\sigma(x)) = (\sigma \otimes \sigma)^t r(x)$ for all $x \in {}'\mathbf{f}$.*

(b) *We have $(\sigma(x), \sigma(x')) = (x, x')$ for all $x, x' \in {}'\mathbf{f}$.*

Assume that the equality in (a) holds for $x = x'$ and $x = x''$ (both homogeneous); we show that it holds for $x = x'x''$. Write $r(x') = \sum x'_1 \otimes x'_2$ and $r(x'') = \sum x''_1 \otimes x''_2$, all factors being homogeneous. By our hypothesis, we have $r(\sigma(x')) = \sum \sigma(x'_2) \otimes \sigma(x'_1)$ and $r(\sigma(x'')) = \sum \sigma(x''_2) \otimes \sigma(x''_1)$. We have

$$r(\sigma(x'x'')) = r(\sigma(x''))r(\sigma(x')) = \sum v^{|x''_1| \cdot |x'_2|} \sigma(x''_2)\sigma(x'_2) \otimes \sigma(x''_1)\sigma(x'_1)$$

and

$$(\sigma \otimes \sigma)^t r(x'x'') = \sum v^{|x''_1| \cdot |x'_2|} (\sigma \otimes \sigma)(x'_2 x''_2 \otimes x'_1 x''_1);$$

our assertion follows. It is now enough to verify (a) in the case where x is one of the algebra generators θ_i; this is obvious. Now (b) follows easily from (a) and the definition of $(,)$.

1.2.9. From 1.2.8(b) we see that $\sigma : {}'\mathbf{f} \to {}'\mathbf{f}$ maps $\mathcal{I}$ into itself; hence it induces an isomorphism $\mathbf{f} \cong \mathbf{f}^{opp}$ with square 1, denoted again by σ. The identities in 1.2.8(a),(b) continue to hold in $\mathbf{f}$.

1.2.10. Let $^- : \mathbf{Q}(v) \to \mathbf{Q}(v)$ be the unique $\mathbf{Q}$-algebra involution such that $\overline{v^n} = v^{-n}$ for all n. Let $^- : {}'\mathbf{f} \to {}'\mathbf{f}$ be the unique $\mathbf{Q}$-algebra involution such that $\overline{p\theta_i} = \bar{p}\theta_i$ for all $p \in \mathbf{Q}(v)$ and $i \in I$. Let ${}'\mathbf{f}\bar{\otimes}{}'\mathbf{f}$ be the $\mathbf{Q}(v)$-vector space ${}'\mathbf{f} \otimes {}'\mathbf{f}$ with the associative $\mathbf{Q}(v)$-algebra structure given by

$$(x_1 \otimes x_2)(x_1' \otimes x_2') = v^{-|x_2|\cdot|x_1'|}x_1x_1' \otimes x_2x_2',$$

where $x_1, x_2, x_1', x_2' \in {}'\mathbf{f}$ are homogeneous. Let $^- : {}'\mathbf{f} \otimes {}'\mathbf{f} \to {}'\mathbf{f}\bar{\otimes}{}'\mathbf{f}$ be the $\mathbf{Q}$-algebra isomorphism $^- \otimes ^-$. Let $\bar{r} : {}'\mathbf{f} \to {}'\mathbf{f}\bar{\otimes}{}'\mathbf{f}$ be the $\mathbf{Q}(v)$-algebra homomorphism defined as the composition

$${}'\mathbf{f} \xrightarrow{-} {}'\mathbf{f} \xrightarrow{r} {}'\mathbf{f} \otimes {}'\mathbf{f} \xrightarrow{-} {}'\mathbf{f}\bar{\otimes}{}'\mathbf{f}.$$

Thus $\bar{r}(x) = \overline{r(\bar{x})}$. The coassociativity property of r implies the coassociativity property $(\bar{r} \otimes 1)\bar{r} = (1 \otimes \bar{r})\bar{r}$ for $\bar{r}$.

Let $\{,\} : {}'\mathbf{f} \times {}'\mathbf{f} \to \mathbf{Q}(v)$ be the symmetric bilinear form defined by

$$\{x, y\} = \overline{(\bar{x}, \bar{y})}.$$

From the definitions we deduce $\{1, 1\} = 1$,

(a) $\{\theta_i, \theta_j\} = \delta_{i,j}(1 - v_i^2)^{-1}$ and

(b) $\{x, y'y''\} = \{\bar{r}(x), y' \otimes y''\}$ for all $x, y', y'' \in {}'\mathbf{f}$.

Lemma 1.2.11. (a) *Let $x \in {}'\mathbf{f}$ be homogeneous; write $r(x) = \sum x_1 \otimes x_2$ with x_1, x_2 homogeneous. We have $\bar{r}(x) = \sum v^{-|x_1|\cdot|x_2|}x_2 \otimes x_1$.*

(b) *Let $x, y \in {}'\mathbf{f}$ be homogeneous. We have*

$$\{x, y\} = (-1)^{\operatorname{tr}|x|}v^{-|x|\cdot|y|/2}v_{-|x|}(x, \sigma(y)).$$

Assume that (a) holds for $x = x'$ and for $x = x''$ (both homogeneous). We show that it also holds for $x = x'x''$. Write $r(x') = \sum x_1' \otimes x_2'$ and $r(x'') = \sum x_1'' \otimes x_2''$, all factors being homogeneous. By our hypothesis, we have $r(\bar{x}') = \sum v^{|x_1'|\cdot|x_2'|}\bar{x}_2' \otimes \bar{x}_1'$ and $r(\bar{x}'') = \sum v^{|x_1''|\cdot|x_2''|}\bar{x}_2'' \otimes \bar{x}_1''$.

We have $r(x) = \sum v^{|x'_2|\cdot|x''_1|} x'_1 x''_1 \otimes x'_2 x''_2$ and

$$r(\bar{x}) = r(\bar{x}')r(\bar{x}'') = \sum v^{|x'_1|\cdot|x'_2|+|x''_1|\cdot|x''_2|+|x''_2|\cdot|x'_1|} \bar{x}'_2\bar{x}''_2 \otimes \bar{x}'_1\bar{x}''_1.$$

The exponent of v is equal to $|x'_1x''_1|\cdot|x'_2x''_2| - |x''_1|\cdot|x'_2|$. Hence

$$\overline{r(\bar{x})} = \sum v^{-|x'_1x''_1|\cdot|x'_2x''_2|} v^{|x'_2|\cdot|x''_1|} x'_2x''_2 \otimes x'_1x''_1$$

and our assertion follows. It remains to show that (a) holds when x is one of the algebra generators θ_i of $'\mathbf{f}$. This follows immediately from the definitions.

We now prove (b). Assume that (b) holds for $y = y'$ and any $x = x'$ and also for $y = y''$ and any $x = x''$ (all homogeneous). We show that it holds for $y = y'y''$ and any homogeneous x. Write $r(x) = \sum x' \otimes x''$ with x', x'' homogeneous. We have

$$\begin{aligned}
(\bar{x}, \bar{y}) &= (r(\bar{x}), \bar{y}' \otimes \bar{y}'') = \sum v^{|x''|\cdot|x'|}(\bar{x}'' \otimes \bar{x}', \bar{y}' \otimes \bar{y}'') \\
&= \sum v^{|x''|\cdot|x'|}(\bar{x}'', \bar{y}')(\bar{x}', \bar{y}'') \\
&= (-1)^{\operatorname{tr}|x'| + \operatorname{tr}|x''|} \\
&\quad \times \sum v^{|x''|\cdot|x'|+|x''|\cdot|y'|/2+|x'|\cdot|y''|/2} v_{|x'|} v_{|x''|} \overline{(x'', \sigma(y'))(x', \sigma(y''))}
\end{aligned}$$

and

$$\begin{aligned}
&(-1)^{\operatorname{tr}|x|} v^{|x|\cdot|y|/2} v_{|x|} \overline{(x, \sigma(y))} = (-1)^{\operatorname{tr}|x|} v^{|x|\cdot|y|/2} v_{|x|} \overline{(r(x), \sigma(y'') \otimes \sigma(y'))} \\
&= \sum (-1)^{\operatorname{tr}|x|} v^{|x|\cdot|y|/2} v_{|x|} \overline{(x', \sigma(y''))(x'', \sigma(y'))}.
\end{aligned}$$

We may assume that $|x'| = |y''|$ and $|x''| = |y'|$. Hence the exponents $|x|\cdot|y|/2$ and $|x''|\cdot|x'| + |x''|\cdot|y'|/2 + |x'|\cdot|y''|/2$ are equal and our assertion follows. Thus, to verify (b), we may assume that y is a generator of our algebra. Similarly, we may assume that x is also a generator of our algebra. In this case, (b) is obvious.

1.2.12. By 1.2.11(b), the involution $^- : {'\mathbf{f}} \to {'\mathbf{f}}$ carries $\mathcal{I}$ onto itself; hence it induces an involution of $\mathbf{f}$ denoted again by $^-$. Again by 1.2.11(b), the form $\{,\}$ has a radical equal to $\mathcal{I}$; hence it induces a symmetric bilinear form on $\mathbf{f}$, denoted again by $\{,\}$, which is non-degenerate on each homogeneous component.

Let $\mathbf{f}\bar{\otimes}\mathbf{f}$ be the $\mathbf{Q}(v)$-vector space $\mathbf{f}\otimes\mathbf{f}$ with the associative $\mathbf{Q}(v)$-algebra structure given by the same rule as for $'\mathbf{f}$ in 1.2.10.

Then $\bar{r} : {'\mathbf{f}} \to {'\mathbf{f}}\bar{\otimes}{'\mathbf{f}}$ induces an algebra homomorphism $\bar{r} : \mathbf{f} \to \mathbf{f}\bar{\otimes}\mathbf{f}$. The identities in 1.2.11(a),(b) continue to hold for $\mathbf{f}$. We have $\sigma(\bar{x}) = \overline{\sigma(x)}$ for all $x \in \mathbf{f}$. Indeed, it suffices to check this on the generators θ_i, where it is obvious.

1.2.13. The maps r_i and ${}_ir$. Let $i \in I$. Clearly, there is a unique $\mathbf{Q}(v)$-linear map ${}_ir : {}'\mathbf{f} \to {}'\mathbf{f}$ such that ${}_ir(1) = 0$, ${}_ir(\theta_j) = \delta_{i,j}$ for all j and ${}_ir(xy) = {}_ir(x)y + v^{|x|\cdot i}x{}_ir(y)$ for all homogeneous x, y. If $x \in {}'\mathbf{f}_\nu$, we have ${}_ir(x) \in {}'\mathbf{f}_{\nu-i}$ if $\nu_i \geq 1$ and ${}_ir(x) = 0$ if $\nu_i = 0$; moreover, $r(x) = \theta_i \otimes {}_ir(x)$ plus terms of other bi-homogeneities.

Similarly, there is a unique $\mathbf{Q}(v)$-linear map $r_i : {}'\mathbf{f} \to {}'\mathbf{f}$ such that $r_i(1) = 0$, $r_i(\theta_j) = \delta_{i,j}$ for all j and $r_i(xy) = v^{|y|\cdot i}r_i(x)y + xr_i(y)$ for all homogeneous x, y. If $x \in {}'\mathbf{f}_\nu$ we have $r_i(x) \in {}'\mathbf{f}_{\nu-i}$ if $\nu_i \geq 1$ and $r_i(x) = 0$ if $\nu_i = 0$; moreover, $r(x) = r_i(x) \otimes \theta_i$ plus terms of other bi-homogeneities.

From the definition we see that

(a) $$(\theta_i y, x) = (\theta_i, \theta_i)(y, {}_ir(x)), \quad (y\theta_i, x) = (\theta_i, \theta_i)(y, r_i(x))$$

for all x, y, and

(b) $$\sigma r_i = {}_ir\sigma.$$

For any $i \in I$, the linear maps ${}_ir, r_i : {}'\mathbf{f} \to {}'\mathbf{f}$ leave $\mathcal{I}$ stable (by (a)); hence they induce linear maps ${}_ir, r_i : \mathbf{f} \to \mathbf{f}$. The identities above continue to hold in $\mathbf{f}$.

Lemma 1.2.14. *For any homogeneous $x \in \mathbf{f}$, we have*
(a) $r_i(x) = v^{|x|\cdot i - i\cdot i}\overline{{}_ir(\bar{x})}$.

If $x = 1$, then both sides of (a) are 0; if $x = \theta_j$, then both sides of (a) are $\delta_{i,j}$. Clearly, if (a) holds for x, x' homogeneous of the same degree, then it holds for any linear combination of x, x'.

Assume that (a) holds for x and x' (homogeneous); we show that it also holds for xx'. We have

$$\begin{aligned} v^{|xx'|\cdot i - i\cdot i}\overline{{}_ir(\overline{xx'})} &= v^{|xx'|\cdot i - i\cdot i}(\overline{{}_ir(\bar{x})\bar{x}' + v^{-|x|\cdot i}\bar{x}{}_ir(\bar{x}')}) \\ &= v^{|x'|\cdot i}v^{|x|\cdot i - i\cdot i}\overline{{}_ir(\bar{x})}x' + xv^{|x'|\cdot i - i\cdot i}\overline{{}_ir(\bar{x}')} \\ &= v^{|x'|\cdot i}r_i(x)x' + xr_i(x') = r_i(xx'). \end{aligned}$$

The lemma follows.

Lemma 1.2.15. *Let $x \in \mathbf{f}_\nu$, where $\nu \in \mathbf{N}[I]$ is different from 0.*
(a) *If $r_i(x) = 0$ for all i, then $x = 0$.*
(b) *If ${}_ir(x) = 0$ for all i, then $x = 0$.*

If x is as in (a) then, by 1.2.13(a), we have $(y\theta_i, x) = 0$ for all $y \in \mathbf{f}$. By our assumption on ν, we have $\mathbf{f}_\nu \subset \sum_i \mathbf{f}\theta_i$. It follows that $(\mathbf{f}_\nu, x) = 0$. Hence, by the non-degeneracy of $(,)$, we have $x = 0$. This proves (a). The proof of (b) is entirely similar.

1.3. Preliminaries on Gaussian Binomial Coefficients

1.3.1. Let $\mathcal{A} = \mathbf{Z}[v, v^{-1}]$. For $a \in \mathbf{Z}$ and $t \in \mathbf{N}$, we set

$$\begin{bmatrix} a \\ t \end{bmatrix} = \frac{\prod_{s=0}^{t-1}(v^{a-s} - v^{-a+s})}{\prod_{s=1}^{t}(v^{s} - v^{-s})} \in \mathbf{Q}(v).$$

We have

(a) $$\begin{bmatrix} a \\ t \end{bmatrix} = (-1)^t \begin{bmatrix} -a+t-1 \\ t \end{bmatrix};$$

(b) $$\begin{bmatrix} a \\ t \end{bmatrix} = 0 \quad \text{if } 0 \le a < t;$$

(c) $$\prod_{j=0}^{a-1}(1 + v^{2j}z) = \sum_{t=0}^{a} v^{t(a-1)} \begin{bmatrix} a \\ t \end{bmatrix} z^t \quad \text{if } a \ge 0.$$

Here z is another indeterminate. From (c), (a), it follows that

(d) $$\begin{bmatrix} a \\ t \end{bmatrix} \in \mathcal{A}.$$

If a', a'' are integers and $t \in \mathbf{N}$, then

(e) $$\begin{bmatrix} a'+a'' \\ t \end{bmatrix} = \sum_{t'+t''=t} v^{a't''-a''t'} \begin{bmatrix} a' \\ t' \end{bmatrix} \begin{bmatrix} a'' \\ t'' \end{bmatrix}.$$

Assume first that a', a'' are ≥ 0. Then (e) follows from (c) by the computation:

$$\sum_{t=0}^{a'+a''} v^{t(a'+a''-1)} \begin{bmatrix} a'+a'' \\ t \end{bmatrix} z^t = \prod_{j=0}^{a'+a''-1} (1 + v^{2j}z)$$

$$= \prod_{j=0}^{a'-1}(1 + v^{2j}z) \prod_{h=0}^{a''-1} (1 + v^{2h}(v^{2a'}z))$$

$$= \sum_{t'=0}^{a'} v^{t'(a'-1)} \begin{bmatrix} a' \\ t' \end{bmatrix} z^{t'} \sum_{t''=0}^{a''} v^{t''(a''-1)} \begin{bmatrix} a' \\ t' \end{bmatrix} v^{2a't''} z^{t''}.$$

For fixed t, we may regard (e) as an identity involving rational functions in three variables: $v, v^{a'}, v^{a''}$. Since this identity is already known to hold for all $a' \ge 0, a'' \ge 0$, it must hold as a formal identity in the three variables; hence it holds as an identity in v, for arbitrary a', a''.

1.3.2. We have $\begin{bmatrix} -1 \\ t \end{bmatrix} = (-1)^t$ for any $t \geq 0$. This follows from 1.3.1(a).

1.3.3. We shall use the notation

$$[n] = \begin{bmatrix} n \\ 1 \end{bmatrix} = \frac{v^n - v^{-n}}{v - v^{-1}} \quad \text{for } n \in \mathbf{Z},$$

$$[n]^! = \prod_{s=1}^{n} [s] \quad \text{for } n \in \mathbf{N}.$$

With this notation we have

$$\begin{bmatrix} a \\ t \end{bmatrix} = \frac{[a]^!}{[t]^! [a-t]^!} \text{ for } 0 \leq t \leq a.$$

1.3.4. If $a \geq 1$, we have

(a) $$\sum_{t=0}^{a} (-1)^t v^{t(1-a)} \begin{bmatrix} a \\ t \end{bmatrix} = 0.$$

This follows from 1.3.1(c) by setting $z = -1$.

1.3.5. If x, y are two elements in a $\mathbf{Q}(v)$-algebra such that $xy = v^2 yx$, then, for any $a \geq 0$, we have the *quantum binomial formula*:

$$(x+y)^a = \sum_{t=0}^{a} v^{t(a-t)} \begin{bmatrix} a \\ t \end{bmatrix} y^t x^{a-t}.$$

The proof is by induction on a.

1.4. Quantum Serre Relations

1.4.1. For any $p \in \mathbf{Z}$, let $\theta_i^{(p)}$ (in $'\mathbf{f}$ or $\mathbf{f}$) be defined as $\theta_i^p / [p]_i^!$ if $p \geq 0$ and as 0 if $p < 0$. (The notation $[p]_i^!$ is in accordance with 1.1.2.)

Lemma 1.4.2. *For any $p \in \mathbf{Z}$ we have*

(a) $$r(\theta_i^{(p)}) = \sum_{t+t'=p} v_i^{tt'} \theta_i^{(t)} \otimes \theta_i^{(t')};$$

(b) $$\bar{r}(\theta_i^{(p)}) = \sum_{t+t'=p} v_i^{-tt'} \theta_i^{(t)} \otimes \theta_i^{(t')}.$$

(a) (resp. (b)) follows from the quantum binomial formula 1.3.5, applied to the elements $\theta_i \otimes 1$ and $1 \otimes \theta_i$ of $\mathbf{f} \otimes \mathbf{f}$ (resp. $\mathbf{f} \bar{\otimes} \mathbf{f}$).

Proposition 1.4.3. *The generators θ_i of* $\mathbf{f}$ *satisfy the identities* $\sum_{p+p'=1-2i\cdot j/(i\cdot i)}(-1)^{p'}\theta_i^{(p)}\theta_j\theta_i^{(p')}=0$ *for any* $i\neq j$ *in* I.

(The identities above are called the *quantum Serre relations.*) The proof will be given in 1.4.6.

Lemma 1.4.4. *For any* $p\geq 0$ *we have*

$$\text{(a)}\qquad (\theta_i^{(p)},\theta_i^{(p)})=\prod_{s=1}^{p}(1-v_i^{-2s})^{-1}=v_i^{p(p+1)/2}(v_i-v_i^{-1})^{-p}([p]_i^!)^{-1}.$$

Note that (a) holds for $p=0$ or 1, by definition. Assume that (a) holds for p and for p'. We show that it holds for $p+p'$, using 1.4.2:

$$(\theta_i^{(p+p')},\theta_i^{(p+p')})=\begin{bmatrix}p+p'\\p\end{bmatrix}_i^{-1}(r(\theta_i^{(p+p')}),\theta_i^{(p)}\otimes\theta_i^{(p')})$$
$$=v_i^{pp'}\begin{bmatrix}p+p'\\p\end{bmatrix}_i^{-1}(\theta_i^{(p)},\theta_i^{(p)})(\theta_i^{(p')},\theta_i^{(p')})$$
$$=v_i^{pp'}\begin{bmatrix}p+p'\\p\end{bmatrix}_i^{-1}\prod_{s=1}^{p}(1-v_i^{-2s})^{-1}\prod_{s=1}^{p'}(1-v_i^{-2s})^{-1}$$
$$=\prod_{s=1}^{p+p'}(1-v_i^{-2s})^{-1}.$$

The lemma is proved.

Lemma 1.4.5. *Let* $n\in\mathbf{N}$ *and let* $p,p',q,q'\in\mathbf{N}$ *be such that* $p+p'=q+q'=n$. *Let* $i\neq j$. *We have*

$$\text{(a)}\qquad (\theta_i^{(p)}\theta_j\theta_i^{(p')},\theta_i^{(q)}\theta_j\theta_i^{(q')})$$
$$=\frac{v_i^{q(q-1)/2+q'(q'-1)/2}}{(v_i-v_i^{-1})^n(v_j-v_j^{-1})}\sum\frac{v_i^{-s(q-1)-t'(q'-1)}v^{(t'+s)(i\cdot j+(n-1)i\cdot i/2)}}{[t]_i^![t']_i^![s]_i^![s']_i^!}$$

where the sum is taken over all t,t',s,s' *in* $\mathbf{N}$ *such that* $t+s=q,t'+s'=q',t+t'=p,s+s'=p'$. *The left hand side is an inner product of elements of* $\mathbf{f}$.

We have

$$r(\theta_i^{(q)}\theta_j\theta_i^{(q')})$$
$$=\left(\sum_{t+s=q} v_i^{ts}\theta_i^{(t)}\otimes\theta_i^{(s)}\right)(\theta_j\otimes 1+1\otimes\theta_j)\left(\sum_{t'+s'=q'} v_i^{t's'}\theta_i^{(t')}\otimes\theta_i^{(s')}\right)$$
$$=\sum v_i^{ts+t's'+2st'}v^{si\cdot j}\theta_i^{(t)}\theta_j\theta_i^{(t')}\otimes\theta_i^{(s)}\theta_i^{(s')}$$
$$+\sum v_i^{ts+t's'+2st'}v^{t'i\cdot j}\theta_i^{(t)}\theta_i^{(t')}\otimes\theta_i^{(s)}\theta_j\theta_i^{(s')}.$$

Hence the left hand side of (a) is equal to

$$(\theta_i^{(p)}\theta_j\otimes\theta_i^{(p')},r(\theta_i^{(q)}\theta_j\theta_i^{(q')}))$$
$$=\sum v_i^{ts+t's'+2st'}v^{si\cdot j}(\theta_i^{(p)}\theta_j,\theta_i^{(t)}\theta_j\theta_i^{(t')})(\theta_i^{(s)}\theta_i^{(s')},\theta_i^{(p')})$$

where the sum is as in the lemma. We have

$$\begin{aligned}(\theta_i^{(p)}\theta_j,\theta_i^{(t)}\theta_j\theta_i^{(t')})&=(r(\theta_i^{(p)})r(\theta_j),\theta_i^{(t)}\theta_j\otimes\theta_i^{(t')})\\&=v_i^{tt'}v^{t'i\cdot j}(\theta_i^{(t)}\theta_j,\theta_i^{(t)}\theta_j)(\theta_i^{(t')},\theta_i^{(t')})\\&=v_i^{tt'}v^{t'i\cdot j}(\theta_i^{(t)},\theta_i^{(t)})(\theta_j,\theta_j)(\theta_i^{(t')},\theta_i^{(t')}).\end{aligned}$$

Introducing this in our earlier computation we see that the left hand side of (a) is equal to

$$\sum v_i^{ts+t's'+2st'+tt'+ss'}v^{(t'+s)i\cdot j}(\theta_i^{(s)},\theta_i^{(s)})(\theta_i^{(s')},\theta_i^{(s')})(\theta_i^{(t)},\theta_i^{(t)})(\theta_j,\theta_j)(\theta_i^{(t')},\theta_i^{(t')})$$
$$=\sum\frac{v_i^{ts+t's'+2st'+tt'+ss'+s(s+1)/2+s'(s'+1)/2+t(t+1)/2+t'(t'+1)/2}v^{(t'+s)i\cdot j}}{(v_i-v_i^{-1})^n(v_j-v_j^{-1})[s]_i^![s']_i^![t]_i^![t']_i^!}.$$

The lemma follows.

1.4.6. Proof of Proposition 1.4.3. Let $i\neq j$ be elements of I. Set $\alpha=-2i\cdot j/(i\cdot i)\in\mathbf{N}$. It suffices to show that the element $\sum_{p+p'=1+\alpha}(-1)^{p'}\theta_i^{(p)}\theta_j\theta_i^{(p')}\in\mathbf{f}$ is orthogonal under $(,)$ to $\theta_i^{(q)}\theta_j\theta_i^{(q')}$ for any q,q' such that $q+q'=1+\alpha$. (These elements span $\mathbf{f}_{j+(1+\alpha)i}$.)

Using Lemma 1.4.5, we see that it is enough to verify the identity

$$\sum(-1)^{s+s'}v_i^{-s(q-1)-t'(q'-1)}([t]_i^![t']_i^![s]_i^![s']_i^!)^{-1}=0$$

where the sum is taken over all t, t', s, s' in $\mathbf{N}$ such that $t+s = q, t'+s' = q'$. The left hand side is a product

$$\left(\sum_{t+s=q} (-1)^s v_i^{-s(q-1)}([t]_i^![s]_i^!)^{-1}\right)\left(\sum_{t'+s'=q'} (-1)^{s'} v_i^{-t'(q'-1)}([t']_i^![s']_i^!)^{-1}\right).$$

If $q > 0$, the first sum is zero, while if $q' > 0$, the second sum is zero using 1.3.4(a). Note that q, q' cannot both be zero, since their sum is $\alpha + 1 > 0$. The proposition is proved.

1.4.7. The algebra $_{\mathcal{A}}\mathbf{f}$. Let $_{\mathcal{A}}\mathbf{f}$ be the $\mathcal{A}$-subalgebra of $\mathbf{f}$ generated by the elements $\theta_i^{(s)}$ for various $i \in I$ and $s \in \mathbf{Z}$. Since the generators $\theta_i^{(s)}$ are homogeneous, we have $_{\mathcal{A}}\mathbf{f} = \oplus_\nu ({_{\mathcal{A}}\mathbf{f}_\nu})$ where ν runs over $\mathbf{N}[I]$ and $_{\mathcal{A}}\mathbf{f}_\nu = {_{\mathcal{A}}\mathbf{f}} \cap \mathbf{f}_\nu$.

CHAPTER 2

Weyl Group, Root Datum

2.1. The Weyl Group

2.1.1. Assume that a Cartan datum $(I, \cdot)$ is given. For any $i \neq j$ in I such that $(i \cdot i)(j \cdot j) - (i \cdot j)^2 > 0$ we define an integer $h(i,j) \in \{2,3,4,\dots\}$ by $\cos^2 \frac{\pi}{h(i,j)} = \frac{i \cdot j}{i \cdot i} \frac{j \cdot i}{j \cdot j}$. We have $h(i,j) = h(j,i) = 2,3,4$ or 6 according to whether $\frac{2i \cdot j}{i \cdot i} \frac{2j \cdot i}{j \cdot j}$ is $0,1,2$ or 3.

For any $i \neq j$ in I such that $(i \cdot i)(j \cdot j) - (i \cdot j)^2 \leq 0$, we set $h(i,j) = \infty$.

The *braid group* is the group defined by the generators $s_i \quad (i \in I)$ and the following relations (one for each $i \neq j$ in I such that $h(i,j) < \infty$):

(a) $s_i s_j s_i \cdots = s_j s_i s_j \cdots$ (both products have $h(i,j)$ factors).

We define W to be the group defined by the generators $s_i \quad (i \in I)$ and the relations (a) together with the relations $s_i^2 = 1$ for all i. Thus W is a *Coxeter group* of a special type, called the *Weyl group*. It is naturally a quotient group of the braid group.

2.1.2. Let $w \in W$. The *length* of w is the smallest integer $p \geq 0$ such that there exist $i_1, i_2, \dots, i_p$ in I with $w = s_{i_1} s_{i_2} \cdots s_{i_p}$. We then set $l(w) = p$ and we say that $s_{i_1} s_{i_2} \cdots s_{i_p}$ is a *reduced expression* of w. Note that $l(1) = 0$, $l(s_i) = 1$ and $l(ww') \leq l(w) + l(w')$ for $w, w' \in W$.

The following theorem will be used many times.

There is a unique map $w \mapsto \tilde{w}$ from W into the braid group such that $\tilde{1} = 1, \tilde{s}_i = s_i$ and $\widetilde{ww'} = \tilde{w}\tilde{w}'$ whenever $w, w' \in W$ satisfy $l(ww') = l(w) + l(w')$.

The uniqueness is trivial; the theorem asserts that, if $s_{i_1} s_{i_2} \cdots s_{i_p}$ and $s_{i'_1} s_{i'_2} \cdots s_{i'_p}$ are two reduced expressions of $w \in W$, then the equality $s_{i_1} s_{i_2} \cdots s_{i_p} = s_{i'_1} s_{i'_2} \cdots s_{i'_p}$ holds in the braid group.

2.1.3. A Cartan datum $(I, \cdot)$ is said to be *symmetric* if $i \cdot i = 2$ for all $i \in I$.

A Cartan datum $(I, \cdot)$ is said to be *simply laced* if it is symmetric and $i \cdot j \in \{0, -1\}$ for all $i \neq j$.

A Cartan datum $(I, \cdot)$ is said to be *irreducible* if I is non-empty and for any $i \neq j$ in I there exists a sequence $i = i_1, i_2, \dots, i_n = j$ in I such that $i_p \cdot i_{p+1} < 0$ for $p = 1, 2, \dots, n-1$.

G. Lusztig, *Introduction to Quantum Groups*, Modern Birkhäuser Classics,
DOI 10.1007/978-0-8176-4717-9_2, © Springer Science+Business Media, LLC 2010

A Cartan datum $(I, \cdot)$ is said to be *without odd cycles* if we cannot find a sequence $i_1, i_2, \dots, i_p, i_{p+1} = i_1$ in I such that $p \geq 3$ is odd, and $i_s \cdot i_{s+1} < 0$ for $s = 1, 2, \dots, p$ or, equivalently, if there exists a function $i \mapsto a_i$ from I to $\{0, 1\}$ such that $a_i + a_j = 1$ whenever $i \cdot j < 0$.

A Cartan datum $(I, \cdot)$ is said to be of *finite type* if the symmetric matrix $(i \cdot j)$ indexed by $I \times I$ is positive definite. This is equivalent to the requirement that W is a finite group. In this case, W has a unique element of maximal length; we denote it by w_0. We have $w_0^2 = 1$. A Cartan datum that is not of finite type is said to be of *infinite type.*

A Cartan datum $(I, \cdot)$ is said to be of *affine type* if it is irreducible and the symmetric matrix $(i \cdot j)$ indexed by $I \times I$ is positive semi-definite, but not positive definite.

2.2. Root Datum

2.2.1. A *root datum* of type $(I, \cdot)$ consists, by definition, of

(a) two finitely generated free abelian groups Y, X and a perfect bilinear pairing $\langle , \rangle : Y \times X \to \mathbf{Z}$;

(b) an imbedding $I \subset X \quad (i \mapsto i')$ and an imbedding $I \subset Y \quad (i \mapsto i)$ such that

(c) $\langle i, j' \rangle = 2\frac{i \cdot j}{i \cdot i}$ for all $i, j \in I$.

In particular, we have

(d) $\langle i, i' \rangle = 2$ for all i;

(e) $\langle i, j' \rangle \in \{0, -1, -2, \dots\}$ for all $i \neq j$.

Thus $(\langle i, j' \rangle)$ is a symmetrizable generalized Cartan matrix.

The imbeddings (b) induce homomorphisms $\mathbf{Z}[I] \to Y, \mathbf{Z}[I] \to X$; we shall often denote, again by ν, the image of $\nu \in \mathbf{Z}[I]$ by either of these homomorphisms.

2.2.2. A root datum as above is said to be X*-regular* (resp. Y*-regular*) if the image of the imbedding $I \subset X$ is linearly independent in X (resp. the image of the imbedding $I \subset Y$ is linearly independent in Y).

For example, we can take $Y = \mathbf{Z}[I]$ with the obvious imbedding $I \to Y$; $X = \mathrm{Hom}(Y, \mathbf{Z})$ with the obvious bilinear pairing $\langle , \rangle : Y \times X \to \mathbf{Z}$ (evaluation) and with the imbedding $I \to X$ defined by the condition (c) above. This root datum is Y-regular. We say that it is the *simply connected* root datum.

As another example, we can take $X = \mathbf{Z}[I]$ with the obvious imbedding $I \to X$; $Y = \mathrm{Hom}(X, \mathbf{Z})$ with the obvious bilinear pairing $\langle , \rangle : Y \times X \to \mathbf{Z}$

(evaluation) and with the imbedding $I \to Y$ defined by the condition (c) above. This root datum is X-regular. We say that it is the *adjoint* root datum.

A morphism from $(Y, X, \langle,\rangle, \dots)$ to $(Y', X', \langle,\rangle', \dots)$ (two root data of type $(I, \cdot)$) is, by definition, a pair of group homomorphisms $f : Y \to Y', g : X' \to X$ such that $\langle f(\mu), \lambda' \rangle' = \langle \mu, g(\lambda') \rangle$ for all $\mu \in Y, \lambda' \in X'$ and $f(i) = i, g(i') = i'$ for all $i \in I$. Thus the root data of type $(I, \cdot)$ form a category.

Given a root datum $(Y, X, \langle,\rangle, \dots)$ of type $(I, \cdot)$, there is a unique morphism from the simply connected root datum of type $(I, \cdot)$ to $(Y, X, \langle,\rangle, \dots)$ and a unique morphism from $(Y, X, \langle,\rangle, \dots)$ to the adjoint root datum of type $(I, \cdot)$.

If $(Y, X, \langle,\rangle, \dots)$ and $(Y', X', \langle,\rangle', \dots)$ are two root data of type $(I, \cdot)$, we can define a third root datum $(Y \oplus Y', X \oplus X', \langle,\rangle'', \dots)$ where $\langle,\rangle'' = \langle,\rangle \oplus \langle,\rangle'$, the imbedding $I \to Y \oplus Y'$ has as components the given imbeddings $I \to Y, I \to Y'$ and the imbedding $I \to X \oplus X'$ has as first component the given imbedding $I \to X$ and as second component zero.

Clearly,

(a) if $(Y', X', \langle,\rangle', \dots)$ is Y-regular, then $(Y \oplus Y', X \oplus X', \langle,\rangle'', \dots)$ is Y-regular;

(b) if $(Y, X, \langle,\rangle, \dots)$ is X-regular, then $(Y \oplus Y', X \oplus X', \langle,\rangle'', \dots)$ is X-regular.

Taking $(Y, X, \langle,\rangle, \dots)$ to be adjoint and $(Y', X', \langle,\rangle', \dots)$ to be simply connected, we see that $(Y \oplus Y', X \oplus X', \langle,\rangle'', \dots)$ is both X-regular and Y-regular. Thus there exist root data of type $(I, \cdot)$ which are both Y-regular and X-regular.

2.2.3. In the case where the root datum is X-regular, we define a partial order on X as follows: $\lambda \le \lambda'$ if and only if $\lambda' - \lambda \in \sum_i \mathbf{N} i'$. Without our assumption on the root datum, this would be only a preorder.

2.2.4. Given a Cartan datum $(I, \cdot)$ and an integer $l \ge 1$, we define a new Cartan datum $(I, \circ)$ with the same I and with

$$i \circ j = (i \cdot j) l_i l_j$$

where, for any $i \in I$, l_i denotes the smallest integer ≥ 1 such that $l_i(i \cdot i/2) \in l\mathbf{Z}$.

We show that this is indeed a Cartan datum. It is obvious that $i \circ i \in \{2,4,6,\dots\}$ for any $i \in I$. Now let $i \neq j$ in I. It is clear that

$$a = 2\frac{i \circ j}{i \circ i} = 2\frac{i \cdot j}{i \cdot i}\frac{l_j}{l_i}$$

is a rational number ≤ 0 and that $l_i a \in \mathbf{Z}$. To prove that a is an integer, it is enough to show, by the definition of l_i, that $l_i a(i \cdot i/2) \in l\mathbf{Z}$, or equivalently, that $l_j(i \cdot j) \in l\mathbf{Z}$. But $l_j(i \cdot j)$ is the product of the integer $2i \cdot j/(j \cdot j)$ with the integer $l_j(j \cdot j/2)$ which is in $l\mathbf{Z}$ by the definition of l_j. We have thus proved that $(I, \circ)$ is a Cartan datum. Note that l_i divides l.

2.2.5. Given a root datum $(Y, X, \dots)$ of type $(I, \cdot)$ and an integer $l \geq 1$, we define a new root datum $(Y^*, X^*, \dots)$ of type $(I, \circ)$ (see 2.2.4) as follows.

By definition, $X^* = \{\zeta \in X | \langle i, \zeta \rangle \in l_i\mathbf{Z} \text{ for all } i \in I\}$ and $Y^* = \text{Hom}(X^*, \mathbf{Z})$. The pairing $Y^* \times X^* \to \mathbf{Z}$ is the obvious one. The map $I \to X^*$ is given by $i \mapsto i'^* = l_i i' \in X$. The map $I \to Y^*$ associates to $i \in I$ the element $i^* \in Y^*$ whose value at any $\zeta \in X^*$ is given by $\langle i, \zeta \rangle / l_i$. The value of i^* at j'^* is, from the definition, the integer $\langle i, j' \rangle l_j / l_i = 2 i \circ j/(i \circ i)$.

We have an obvious imbedding $g : X^* \to X$; this induces by duality a homomorphism $f : Y \to Y^*$. Note that (f, g) is not a morphism of root data since the two root data, in general, correspond to different Cartan data. Clearly, if $(Y, X, \dots)$ is Y-regular (resp. X-regular), then $(Y^*, X^*, \dots)$ is Y-regular (resp. X-regular).

2.2.6. Let $s_i : Y \to Y$ be the homomorphism given by $s_i(\mu) = \mu - \langle \mu, i' \rangle i$. Note that $s_i^2 = 1$. Similarly, the homomorphism $s_i : X \to X$ given by $s_i(\lambda) = \lambda - \langle i, \lambda \rangle i'$ satisfies $s_i^2 = 1$. We have $\langle s_i(\mu), \lambda \rangle = \langle \mu, s_i(\lambda) \rangle$ for all μ, λ. It is easily checked that these formulas define homomorphisms

$$W \to \text{Aut }(Y), W \to \text{Aut }(X).$$

Hence W acts naturally on Y and X and we have $\langle w(\mu), \lambda \rangle = \langle \mu, w^{-1}(\lambda) \rangle$ for all μ, λ, w.

2.2.7. Let $s_{i_1} s_{i_2} \cdots s_{i_n}$ be a reduced expression in W where $n \geq 1$. Then $s_{i_n} s_{i_{n-1}} \cdots s_{i_2}(i_1) \in Y$ (resp. $s_{i_n} s_{i_{n-1}} \cdots s_{i_2}(i'_1) \in X$) is a linear combination of elements $i \in Y$ (resp. $i' \in X$) with coefficients in $\mathbf{N}$.

2.2.8. In the case where $(I, \cdot)$ is of finite type, there is a unique permutation $i \to \tilde{i}$ of I such that $w_0(i') = -\tilde{i}'$ for all i. Its square is 1. This implies the following property: if $\lambda', \lambda'' \in X$, then we have $\lambda' \leq \lambda''$ if and only if $w_0(\lambda'') \leq w_0(\lambda')$.

2.3. Coroots

2.3.1. Assume that $(I, \cdot)$ is of finite type. Let $\mathcal{R}$ be the set of all $\mu \in Y$ such that $\mu = w(i)$ for some $w \in W$, or equivalently, the set of *coroots*. $\mathcal{R}$ is a finite set. Let $\mathcal{R}^+$ be the set of all $\alpha \in \mathcal{R}$ such that $\alpha \in \sum_i \mathbf{N}i \subset Y$. We have a partition $\mathcal{R} = \mathcal{R}^+ \cup (-\mathcal{R}^+)$. Let $2\rho \in Y$ be the sum of all coroots in $\mathcal{R}^+$. This is not necessarily two times an element of Y, but it has the following well-known evenness property:

(a) $\langle 2\rho, i' \rangle = 2$ for all $i \in I$. In particular, $\langle 2\rho, \lambda \rangle \in 2\mathbf{Z}$ if $\lambda \in \mathbf{Z}[I] \subset X$.

If $i, j \in I$ are such that $j = w(i)$ for some $w \in I$, then it is known that $i \cdot i = j \cdot j$; hence for any $\alpha \in \mathcal{R}$ we can define $\alpha \cdot \alpha \in 2\mathbf{N}$ to be $i \cdot i$ where $i \in I$ is such that $\alpha = w(i)$ for some $i \in I$ and $w \in W$. This is independent of the choices made.

Let $\mathbf{n} : X \to \mathbf{Z}$ be the homomorphism given by

$\mathbf{n}(\lambda) = \sum (\alpha \cdot \alpha/2) \langle \alpha, \lambda \rangle$ (sum over all $\alpha \in \mathcal{R}^+$).

We have

(b) $\mathbf{n}(-w_0(\lambda)) = \mathbf{n}(\lambda)$ for all $\lambda \in X$;

(c) $\mathbf{n}(i') = i \cdot i$ for all $i \in I$.

2.3.2. For any $i \in I$, we define an element $\mu(i) \in Y$ as follows: we choose a sequence $(i_1, i_2, \dots, i_N)$ in I such that $s_{i_1} s_{i_2} \cdots s_{i_N}$ is a reduced expression of $w_0 \in W$ and we set

$$\mu(i) = \sum s_{i_N} s_{i_{N-1}} \dots s_{i_{p+1}}(i_p)$$

where the sum is taken over all $p \in [1, N]$ such that $i_p = i$. One checks that $\mu(i)$ is independent of the choice of reduced expression.

The verification of the following identity is left to the reader:

(a) $\sum_{i,j \in I} \langle \mu(i), \lambda \rangle \langle \mu(j), \lambda \rangle i \cdot j = \sum_{i \in I} \langle \mu(i), \lambda \rangle \langle i, \lambda \rangle i \cdot i$ for any $\lambda \in X$.

CHAPTER 3

The Algebra U

3.1. The Algebras 'U and U

3.1.1. Assume that a root datum $(Y, X, \langle , \rangle, \dots)$ of type $(I, \cdot)$ is given. We consider the associative $\mathbf{Q}(v)$-algebra $'\mathbf{U}$ (with 1) defined by the generators

$$E_i \quad (i \in I), \quad F_i \quad (i \in I), \quad K_\mu \quad (\mu \in Y)$$

and the relations (a)–(d) below.

(a) $$K_0 = 1, \quad K_\mu K_{\mu'} = K_{\mu+\mu'} \text{ for all } \mu, \mu' \in Y.$$

(b) $$K_\mu E_i = v^{\langle \mu, i' \rangle} E_i K_\mu \text{ for all } i \in I, \mu \in Y.$$

(c) $$K_\mu F_i = v^{-\langle \mu, i' \rangle} F_i K_\mu \text{ for all } i \in I, \mu \in Y.$$

(d) $$E_i F_j - F_j E_i = \delta_{ij} \frac{\tilde{K}_i - \tilde{K}_{-i}}{v_i - v_i^{-1}}.$$

(For any element $\nu = \sum_i \nu_i i \in \mathbf{Z}[I]$, we set $\tilde{K}_\nu = \prod_i K_{(i \cdot i/2)\nu_i i'}$. In particular, $\tilde{K}_{\pm i} = K_{\pm (i \cdot i/2) i'}$.)

We also consider the associative $\mathbf{Q}(v)$-algebra $\mathbf{U}$ (with 1) defined by the generators

$$E_i \quad (i \in I), \quad F_i \quad (i \in I), \quad K_\mu \quad (\mu \in Y)$$

and the relations (a)–(d) above, together with the following relations:

(e)

for any $f(\theta_i) \in \mathcal{I} \subset {'\mathbf{f}}$ (see 1.2.4) we have $f(E_i) = 0$ and $f(F_i) = 0$ in $\mathbf{U}$.

From (e), we see that there are well-defined algebra homomorphisms $\mathbf{f} \to \mathbf{U}$ $(x \mapsto x^+)$ (with image denoted $\mathbf{U}^+$) and $\mathbf{f} \to \mathbf{U}$ $(x \mapsto x^-)$ (with image denoted $\mathbf{U}^-$) which respect 1 and are such that $E_i = \theta_i^+, F_i = \theta_i^-$ for all $i \in I$. Clearly, there are well defined algebra homomorphisms $'\mathbf{f} \to {'\mathbf{U}}$ $(x \mapsto x^+)$ with image denoted $'\mathbf{U}^+$ and $'\mathbf{f} \to {'\mathbf{U}}$ $(x \mapsto x^-)$ with image denoted $'\mathbf{U}^-$ which respect 1 and are such that $E_i = \theta_i^+, F_i = \theta_i^-$ for all $i \in I$.

For any $p \in \mathbf{Z}$ we set $E_i^{(p)} = (\theta_i^{(p)})^+$ (in $'\mathbf{U}^+$ or $\mathbf{U}^+$) and $F_i^{(p)} = (\theta_i^{(p)})^-$ (in $'\mathbf{U}^-$ or $\mathbf{U}^-$).

G. Lusztig, *Introduction to Quantum Groups*, Modern Birkhäuser Classics,
DOI 10.1007/978-0-8176-4717-9_3, © Springer Science+Business Media, LLC 2010

3.1.2. If we are given a morphism (f,g) of root data of type $(I,\cdot)$ from $(Y,X,\langle,\rangle,\dots)$ to $(Y',X',\langle,\rangle',\dots)$, then there is a natural algebra homomorphism from $\mathbf{U}$, defined in terms of the first root datum, to $\mathbf{U}$, defined in terms of the second one: it is given by $E_i\mapsto E_i, F_i\mapsto F_i, K_\mu\mapsto K_{f(\mu)}$.

3.1.3. Clearly, there is a unique algebra automorphism (with square 1) $\omega:\mathbf{U}\to\mathbf{U}$ such that $\omega(E_i)=F_i,\omega(F_i)=E_i,\omega(K_\mu)=K_{-\mu}$ for $i\in I$, $\mu\in Y$. We have $\omega(x^+)=x^-$ and $\omega(x^-)=x^+$ for all $x\in\mathbf{f}$. (The same formulas define an algebra automorphism $\omega:{}'\mathbf{U}\to{}'\mathbf{U}$ with square 1.)

There is a unique isomorphism of $\mathbf{Q}(v)$-vector spaces $\sigma:\mathbf{U}\to\mathbf{U}$ such that $\sigma(E_i)=E_i,\sigma(F_i)=F_i,\sigma(K_\mu)=K_{-\mu}$ for $i\in I$, $\mu\in Y$ and $\sigma(uu')=\sigma(u')\sigma(u)$ for $u,u'\in\mathbf{U}$. We have $\sigma(x^+)=\sigma(x)^+$ and $\sigma(x^-)=\sigma(x)^-$ for all $x\in\mathbf{f}$.

Lemma 3.1.4 (Comultiplication). *There is a unique algebra homomorphism $\Delta:{}'\mathbf{U}\to{}'\mathbf{U}\otimes{}'\mathbf{U}$ (resp. $\Delta:\mathbf{U}\to\mathbf{U}\otimes\mathbf{U}$) where ${}'\mathbf{U}\otimes{}'\mathbf{U}$ (resp. $\mathbf{U}\otimes\mathbf{U}$) is regarded as an algebra in the standard way, which takes the generators E_i,F_i,K_μ respectively to the elements $\Delta(E_i),\Delta(F_i),\Delta(K_\mu)$ given by*

$$\Delta(E_i)=E_i\otimes 1+\tilde K_i\otimes E_i\quad (i\in I),$$
$$\Delta(F_i)=F_i\otimes\tilde K_{-i}+1\otimes F_i\quad (i\in I),$$
$$\Delta(K_\mu)=K_\mu\otimes K_\mu\quad (\mu\in Y).$$

We must show that the elements $\Delta(E_i),\Delta(F_i),\Delta(K_\mu)$ of ${}'\mathbf{U}\otimes{}'\mathbf{U}$ satisfy the defining relations of ${}'\mathbf{U}$. The relations 3.1.1(a),(b),(c) are easily checked. The relation 3.1.1(d) follows from the equality $\tilde K_iF_j\otimes E_i\tilde K_{-j}=F_j\tilde K_i\otimes\tilde K_{-j}E_i$ which, in turn, follows from the relations 3.1.1(b),(c) of ${}'\mathbf{U}$. This proves the assertion of the lemma concerning ${}'\mathbf{U}$. It remains to verify the relations 3.1.1(e) in the case of $\mathbf{U}$.

Using the definition, we see that the $\mathbf{Q}(v)$-linear map

$$'\mathbf{f}\otimes{}'\mathbf{f}\xrightarrow{j^+}{}'\mathbf{U}\otimes{}'\mathbf{U},$$

given by $x\otimes y\mapsto x^+\tilde K_{|y|}\otimes y^+$ for x,y homogeneous, is an algebra homomorphism. Similarly, the $\mathbf{Q}(v)$-linear map $'\mathbf{f}\bar\otimes{}'\mathbf{f}\xrightarrow{j^-}{}'\mathbf{U}\otimes{}'\mathbf{U}$ given by $x\otimes y\mapsto x^-\otimes\tilde K_{-|x|}y^-$ for x,y homogeneous, is an algebra homomorphism. The compositions $j^+r:{}'\mathbf{f}\to{}'\mathbf{U}\otimes{}'\mathbf{U}$ and $j^-\bar r:{}'\mathbf{f}\to{}'\mathbf{U}\otimes{}'\mathbf{U}$ are then algebra homomorphisms, and we have from the definitions, $j^+r(\theta_i)=\Delta(E_i)$, $j^-\bar r(\theta_i)=\Delta(F_i)$.

Now j^+r factors through an algebra homomorphism $\mathbf{f} \to \mathbf{U} \otimes \mathbf{U}$ (see 1.2.6). It follows that the elements $\Delta(E_i) \in \mathbf{U} \otimes \mathbf{U}$ satisfy the relations 3.1.1(e). Similarly, $j^-\bar{r}$ factors through an algebra homomorphism $\mathbf{f} \to \mathbf{U} \otimes \mathbf{U}$, hence the elements $\Delta(F_i) \in \mathbf{U} \otimes \mathbf{U}$ satisfy the relations 3.1.1(e). The lemma is proved.

3.1.5. The previous proof shows that, for any $x \in {}'\mathbf{f}$, we have

$$j^+r(x) = \Delta(x^+) \quad \text{and} \quad j^-\bar{r}(x) = \Delta(x^-),$$

or equivalently,

$$\Delta(x^+) = \sum x_1^+ \tilde{K}_{|x_2|} \otimes x_2^+ \quad \text{and} \quad \Delta(x^-) = \sum x_3^- \otimes \tilde{K}_{-|x_3|} x_4^-$$

where $r(x) = \sum x_1 \otimes x_2$ and $\bar{r}(x) = \sum x_3 \otimes x_4$ with x_1, x_2, x_3, x_4 homogeneous. In particular, using 1.4.2, we have

$$\Delta(E_i^{(p)}) = \sum_{p'+p''=p} v_i^{p'p''} E_i^{(p')} \tilde{K}_i^{p''} \otimes E_i^{(p'')},$$

$$\Delta(F_i^{(p)}) = \sum_{p'+p''=p} v_i^{-p'p''} F_i^{(p')} \otimes \tilde{K}_{-i}^{p'} F_i^{(p'')}.$$

Proposition 3.1.6. *For $x \in {}'\mathbf{f}$ and $i \in I$, we have (in $'\mathbf{U}$)*

(a) $$x^+ F_i - F_i x^+ = \frac{r_i(x)^+ \tilde{K}_i - \tilde{K}_{-i}({}_ir(x)^+)}{v_i - v_i^{-1}};$$

(b) $$x^- E_i - E_i x^- = \frac{r_i(x)^- \tilde{K}_{-i} - \tilde{K}_i({}_ir(x)^-)}{v_i - v_i^{-1}}.$$

Assume that (a) is known for x' and for x''; we prove it for $x = x'x''$. We have

$$\begin{aligned} x^+F_i - F_ix^+ &= x'^+x''^+F_i - F_ix'^+x''^+ \\ &= x'^+F_ix''^+ + x'^+\frac{r_i(x'')^+\tilde{K}_i - \tilde{K}_{-i}({}_ir(x'')^+)}{v_i - v_i^{-1}} - F_ix'^+x''^+ \\ &= \frac{r_i(x')^+\tilde{K}_i - \tilde{K}_{-i}({}_ir(x')^+)}{v_i - v_i^{-1}}x''^+ + x'^+\frac{r_i(x'')^+\tilde{K}_i - \tilde{K}_{-i}({}_ir(x'')^+)}{v_i - v_i^{-1}} \\ &= \frac{r_i(x'x'')^+\tilde{K}_i - \tilde{K}_{-i}({}_ir(x'x'')^+)}{v_i - v_i^{-1}}. \end{aligned}$$

We are thus reduced to proving (a) in the case where x is either 1 or θ_j; in both cases, (a) follows from the definitions. Now (b) follows from (a) by applying the involution ω.

The following result gives a generalization of the identity (a) above.

Proposition 3.1.7. *Let $x, y \in {}'\mathbf{f}$ be homogeneous. Write*

$$r_{(2)}(x) = (r \otimes 1)r(x) = \sum x_1 \otimes x_2 \otimes x_3$$

with $x_k \in {}'\mathbf{f}$ homogeneous and

$$\bar{r}_{(2)}(y) = (\bar{r} \otimes 1)\bar{r}(y) = \sum y_1 \otimes y_2 \otimes y_3$$

with $y_k \in {}'\mathbf{f}$ homogeneous. The following equality holds in ${}'\mathbf{U}$:

(a)

$$x^+ y^- = \sum (-1)^{\operatorname{tr}|x_1| - \operatorname{tr}|x_3|} v_{-|x_1|+|x_3|}(x_1, y_1)\tilde{K}_{-|x_1|} y_2^- x_2^+ \{x_3, y_3\} \tilde{K}_{|x_3|}.$$

Assume that this is known for $y = y'$ and any x and also for $y = y''$ and any x; we prove it for $y = y'y''$ and any x. Write

$$\bar{r}_{(2)}(y') = \sum y_1' \otimes y_2' \otimes y_3',$$
$$\bar{r}_{(2)}(y'') = \sum y_1'' \otimes y_2'' \otimes y_3'',$$
$$r_{(2)}(x_2) = \sum x_{21} \otimes x_{22} \otimes x_{23}.$$

We have

$$\begin{aligned} & x^+ y'^- y''^- \\ & = \sum (-1)^{\operatorname{tr}|x_1| - \operatorname{tr}|x_3|} v_{-|x_1|+|x_3|}(x_1, y_1')\tilde{K}_{-|x_1|} y_2'^- x_2^+ \{x_3, y_3'\} \tilde{K}_{|x_3|} y''^- \\ & = \sum (-1)^{\operatorname{tr}|x_1| - \operatorname{tr}|x_3| + \operatorname{tr}|x_{21}| - \operatorname{tr}|x_{23}|} v^{-|x_3|\cdot|y''|} v_{-|x_1|+|x_3|-|x_{21}|+|x_{23}|} \\ & \times (x_1, y_1')(x_{21}, y_1'')\tilde{K}_{-|x_1|} y_2'^- \tilde{K}_{-|x_{21}|} y_2''^- x_{22}^+ \{x_{23}, y_3''\}\{x_3, y_3'\} \tilde{K}_{|x_{23}|} \tilde{K}_{|x_3|}. \end{aligned}$$

Hence

$$\begin{aligned} & x^+ y'^- y''^- \\ & = \sum (-1)^{\operatorname{tr}|x_1| - \operatorname{tr}|x_3| + \operatorname{tr}|x_{21}| - \operatorname{tr}|x_{23}|} v^{-|x_3|\cdot(|y_1''|+|y_2''|+|y_3''|) - |x_{21}|\cdot|y_2'|} \\ & \times v_{-|x_1|+|x_3|-|x_{21}|+|x_{23}|}(x_1, y_1')(x_{21}, y_1'') \end{aligned}$$

(b)

$$\times \tilde{K}_{-|x_1|-|x_{21}|} y_2'^- y_2''^- x_{22}^+ \{x_{23}, y_3''\}\{x_3, y_3'\} \tilde{K}_{|x_{23}|+|x_3|}.$$

Note that $\bar{r}_{(2)}(y) = \sum v^{-|y_2'|\cdot|y_1''| - |y_3'|\cdot|y_1''| - |y_3'|\cdot|y_2''|} y_1' y_1'' \otimes y_2' y_2'' \otimes y_3' y_3''$; hence the right hand side of (a) for $y = y'y''$ is

$$\begin{aligned} P = \sum (-1)^{\operatorname{tr}|x_1| - \operatorname{tr}|x_3|} & v^{-|y_2'|\cdot|y_1''| - |y_3'|\cdot|y_1''| - |y_3'|\cdot|y_2''|} v_{-|x_1|+|x_3|}(x_1, y_1' y_1'') \\ & \times \tilde{K}_{-|x_1|} y_2'^- y_2''^- x_2^+ \{x_3, y_3' y_3''\} \tilde{K}_{|x_3|}. \end{aligned}$$

We write $r(x_1) = \sum x_{11} \otimes x_{12}$ and $r(x_3) = \sum x_{31} \otimes x_{32}$; then

$$\bar{r}(x_3) = \sum v^{-|x_{31}|\cdot|x_{32}|} x_{32} \otimes x_{31}.$$

We have

$$\begin{aligned}(x_1, y_1'y_1'') &= (r(x_1), y_1' \otimes y_1'') \\ &= \sum (x_{11}, y_1')(x_{12}, y_1'')\end{aligned}$$

and

$$\begin{aligned}\{x_3, y_3'y_3''\} &= \{\bar{r}(x_3), y_3' \otimes y_3''\} \\ &= \sum v^{-|x_{31}|\cdot|x_{32}|}\{x_{31}, y_3''\}\{x_{32}, y_3'\}.\end{aligned}$$

Introducing this in the previous expression for P, we see that

$$\begin{aligned}&P = \\ &\sum (-1)^{\operatorname{tr}|x_{11}| + \operatorname{tr}|x_{12}| - \operatorname{tr}|x_{31}| - \operatorname{tr}|x_{32}|} v^{-|y_2'|\cdot|y_1''| - |y_3'|\cdot|y_1''| - |y_3'|\cdot|y_2''| - |x_{31}|\cdot|x_{32}|} \\ &\qquad \times v_{-|x_{11}|-|x_{12}|+|x_{31}|+|x_{32}|}(x_{11}, y_1')(x_{12}, y_1'') \\ &\text{(c)} \\ &\qquad \times \tilde{K}_{-|x_{11}|-|x_{12}|} y_2'^{-} y_2''^{-} x_2^{+} \{x_{31}, y_3''\}\{x_{32}, y_3'\}\tilde{K}_{|x_{31}|+|x_{32}|}.\end{aligned}$$

By the coassociativity of r, the sum $\sum x_1 \otimes x_{21} \otimes x_{22} \otimes x_{23} \otimes x_3$ in (b) is equal to the sum $\sum x_{11} \otimes x_{12} \otimes x_2 \otimes x_{31} \otimes x_{32}$ in (c). Hence in (b) we may replace $x_1, x_{21}, x_{22}, x_{23}, x_3$ by $x_{11}, x_{12}, x_2, x_{31}, x_{32}$, respectively. Moreover, we may assume both in the sum in (b) and in (c) that $|x_{11}| = |y_1'|, |x_{12}| = |y_1''|, |x_{31}| = |y_3''|, |x_{32}| = |y_3'|$. We see that (b) is equal to (c).

We are thus reduced to proving (a) in the case where y is either 1 (when (a) is trivial) or θ_i (when (a) follows from 3.1.6). The proposition is proved.

Applying ω to the identity 3.1.7(a), we obtain the following result.

Corollary 3.1.8. *With notations as in Proposition 3.1.7, we have*

$$\begin{aligned}&x^-y^+ \\ &\quad = \sum (-1)^{\operatorname{tr}|x_1| - \operatorname{tr}|x_3|} v_{-|x_1|+|x_3|}(x_1, y_1)\tilde{K}_{|x_1|} y_2^+ x_2^- \{x_3, y_3\}\tilde{K}_{-|x_3|}.\end{aligned}$$

Corollary 3.1.9. *For any $N, M \geq 0$ we have in* **U** *or* $'\mathbf{U}$,

$$E_i^{(N)} F_i^{(M)}$$
$$= \sum_{t \geq 0} F_i^{(M-t)} \prod_{s=1}^{t} \frac{v_i^{2t-N-M-s+1}\tilde{K}_i - v_i^{-2t+N+M+s-1}\tilde{K}_{-i}}{v_i^s - v_i^{-s}} E_i^{(N-t)};$$

$$F_i^{(N)} E_i^{(M)}$$
$$= \sum_{t \geq 0} E_i^{(M-t)} \prod_{s=1}^{t} \frac{v_i^{2t-N-M-s+1}\tilde{K}_{-i} - v_i^{-2t+N+M+s-1}\tilde{K}_i}{v_i^s - v_i^{-s}} F_i^{(N-t)};$$

$$E_i^{(N)} F_j^{(M)} = F_j^{(M)} E_i^{(N)} \quad \textit{if } i \neq j.$$

This can be deduced from 3.1.7, 3.1.8, or alternatively, it can be proved directly by induction.

3.1.10. Coassociativity. The two compositions

$$\mathbf{U} \xrightarrow{\Delta} \mathbf{U} \otimes \mathbf{U} \xrightarrow{1 \otimes \Delta} \mathbf{U} \otimes \mathbf{U} \otimes \mathbf{U}, \quad \mathbf{U} \xrightarrow{\Delta} \mathbf{U} \otimes \mathbf{U} \xrightarrow{\Delta \otimes 1} \mathbf{U} \otimes \mathbf{U} \otimes \mathbf{U}$$

coincide. Indeed, both are algebra homomorphisms; hence it suffices to check that they coincide on the algebra generators of **U**. But they both take E_i, F_i, K_μ respectively to

$$E_i \otimes 1 \otimes 1 + \tilde{K}_i \otimes E_i \otimes 1 + \tilde{K}_i \otimes \tilde{K}_i \otimes E_i,$$

$$F_i \otimes \tilde{K}_{-i} \otimes \tilde{K}_{-i} + 1 \otimes F_i \otimes \tilde{K}_{-i} + 1 \otimes 1 \otimes F_i,$$

$$K_\mu \otimes K_\mu \otimes K_\mu.$$

3.1.11. Counit. There is a unique algebra homomorphism, called counit, $\mathbf{e} : \mathbf{U} \to \mathbf{Q}(v)$ such that $\mathbf{e}(E_i) = 0, \mathbf{e}(F_i) = 0, \mathbf{e}(K_\mu) = 1$.

(The verification of the relations 3.1.1(a)–(d) is immediate; to verify the relations 3.1.1(e) we use the fact that $\mathcal{I} \cap {}'\mathbf{f}_0 = 0$.)

Both compositions

$$\mathbf{U} \xrightarrow{\Delta} \mathbf{U} \otimes \mathbf{U} \xrightarrow{1 \otimes \mathbf{e}} \mathbf{U},$$

$$\mathbf{U} \xrightarrow{\Delta} \mathbf{U} \otimes \mathbf{U} \xrightarrow{\mathbf{e} \otimes 1} \mathbf{U}$$

are equal to the identity (by checking on generators).

3.1.12. The involution $^-$. There is a unique homomorphism of $\mathbf{Q}$-algebras $^- : \mathbf{U} \to \mathbf{U}$ such that

$$\bar{E}_i = E_i, \bar{F}_i = F_i, \bar{K}_\mu = K_{-\mu} \text{ and } \overline{fx} = \bar{f}\bar{x} \text{ for all } f \in \mathbf{Q}(v), x \in \mathbf{U}.$$

Indeed, it is easy to verify that $^-$ respects the relations 3.1.1(a)–(d) of $\mathbf{U}$. To check that it respects the relations 3.1.1(e), it suffices to use the fact that the maps $\mathbf{f} \to \mathbf{U}$ given by $x \mapsto (\bar{x})^+$ (resp. $x \mapsto (\bar{x})^-$) are homomorphisms of $\mathbf{Q}$-algebras. By a verification on generators, we see that $^- : \mathbf{U} \to \mathbf{U}$ has square 1.

3.1.13. The algebra ${}_\mathcal{A}\mathbf{U}$. Let ${}_\mathcal{A}\mathbf{U}^\pm$ be the $\mathcal{A}$-subalgebra of $\mathbf{U}^\pm$ corresponding to ${}_\mathcal{A}\mathbf{f} \subset \mathbf{f}$ under the isomorphism $\mathbf{f} \to \mathbf{U}^\pm$ given by $x \mapsto x^\pm$. In the case where the root datum is simply connected, we define ${}_\mathcal{A}\mathbf{U}$ to be the $\mathcal{A}$-subalgebra of $\mathbf{U}$ generated by the elements $E_i^{(t)}, F_i^{(t)}$ for various $i \in I$ and $t \in \mathbf{Z}$ and by the elements K_μ for $\mu \in Y$. This algebra will not be used in the sequel.

3.2. Triangular Decomposition for $'\mathbf{U}$ and $\mathbf{U}$

3.2.1. If M', M are two $'\mathbf{U}$-modules, then $M' \otimes M$ is naturally a $'\mathbf{U} \otimes {}'\mathbf{U}$-module; hence by restriction to $'\mathbf{U}$ under Δ, it is a $'\mathbf{U}$-module.

Lemma 3.2.2. *Let $\lambda \in X$. There is a unique $'\mathbf{U}$-module structure on the $\mathbf{Q}(v)$-vector space $'\mathbf{f}$ such that for any homogeneous $z \in {}'\mathbf{f}$, any $\mu \in Y$ and any $i \in I$ we have $K_\mu(z) = v^{\langle \mu, \lambda - |z| \rangle} z$, $F_i(z) = \theta_i z$ and $E_i(1) = 0$.*

The uniqueness is immediate. To prove existence, we define $E_i : {}'\mathbf{f} \to {}'\mathbf{f}$ by $E_i(z) = \frac{-v_i^{-\langle i, \lambda \rangle} r_i(z) + v_i^{\langle i, \lambda - |z| + i' \rangle} {}_i r(z)}{v_i - v_i^{-1}}$. A straightforward verification, using the definition of $r_i, {}_i r$, shows that we have a $'\mathbf{U}$-module structure on $'\mathbf{f}$.

3.2.3. We denote this $'\mathbf{U}$-module by M. Similarly, to an element $\lambda' \in X$, we associate a unique $'\mathbf{U}$-module structure on $'\mathbf{f}$ such that for any homogeneous $z \in {}'\mathbf{f}$, any $\mu \in Y$ and any $i \in I$, we have $K_\mu(z) = v^{\langle \mu, -\lambda' + |z| \rangle} z$, $E_i(z) = \theta_i z$ and $F_i(1) = 0$. We denote this $'\mathbf{U}$-module by M'. We form the $'\mathbf{U}$-module $M' \otimes M$; we denote the unit element of $'\mathbf{f} = M$ by 1 and that of $'\mathbf{f} = M'$ by $1'$. Thus, we have a canonical element $1' \otimes 1 \in M' \otimes M$.

Proposition 3.2.4. *Let $\mathbf{U}^0$ be the associative $\mathbf{Q}(v)$-algebra with 1 defined by the generators K_μ $(\mu \in Y)$ and the relations 3.1.1(a). (This is the group algebra of Y over $\mathbf{Q}(v)$.)*

(a) *The* $\mathbf{Q}(v)$*-linear map* $'\mathbf{f}\otimes\mathbf{U}^0\otimes{}'\mathbf{f}\to{}'\mathbf{U}$ *given by* $u\otimes K_\mu\otimes w\mapsto u^-K_\mu w^+$ *is an isomorphism.*

(b) *The* $\mathbf{Q}(v)$*-linear map* $'\mathbf{f}\otimes\mathbf{U}^0\otimes{}'\mathbf{f}\to{}'\mathbf{U}$ *given by* $u\otimes K_\mu\otimes w\mapsto u^+K_\mu w^-$ *is an isomorphism.*

Note that (b) follows from (a) using the involution ω. We prove (a). As a $\mathbf{Q}(v)$-vector space, $'\mathbf{U}$ is spanned by words in E_i, F_i, K_μ. By using repeatedly the relations 3.1.1(b),(c),(d) we see that any word in E_i, F_i, K_μ is a linear combination of words in which all F_i's precede all K_μ's and all K_μ's precede all E_i's. This shows that the map in (a) is surjective.

We now prove injectivity. Let $\lambda,\lambda'\in X$; we attach to them a $'\mathbf{U}$-module $M'\otimes M={}'\mathbf{f}\otimes{}'\mathbf{f}$ with a distinguished vector $1'\otimes 1$ as in 3.2.3. We define a $\mathbf{Q}(v)$-linear map $\phi:{}'\mathbf{U}\to M'\otimes M$ by $\phi(u)=u(1'\otimes 1)$.

Let B be a $\mathbf{Q}(v)$-basis of $'\mathbf{f}$ consisting of homogeneous elements and containing 1. Assume that in $'\mathbf{U}$ we have a relation $\sum_{b',\mu,b}c_{b',\mu,b}b'^-K_\mu b^+=0$ where b',μ,b run over B,Y,B respectively, and $c_{b',\mu,b}\in\mathbf{Q}(v)$ are zero except for finitely many indices. We must prove that the coefficients $c_{b',\mu,b}$ are all zero.

Assume that this is not so. Then we may consider the largest integer N such that there exist b',μ,b with $c_{b',\mu,b}\neq 0$ and $\operatorname{tr}|b'|=N$.

We have $\phi(\sum_{b',\mu,b}c_{b',\mu,b}b'^-K_\mu b^+)=0$. In other words, we have

$$\sum_{b',\mu,b}c_{b',\mu,b}\Delta(b'^-K_\mu b^+)(1'\otimes 1)=0. \tag{c}$$

Now

$$\Delta(b^+)=\sum_{b_1,b_2}g(b,b_1,b_2)b_1^+\tilde K_{|b_2|}\otimes b_2^+,$$
$$\Delta(b'^-)=\sum_{b'_1,b'_2}g'(b',b'_1,b'_2)b_1'^-\otimes\tilde K_{-|b'_1|}b_2'^-$$

(equalities in $'\mathbf{U}\otimes{}'\mathbf{U}$) where b_1,b_2,b'_1,b'_2 are in B and $g(b,b_1,b_2), g'(b',b'_1,b'_2)$ are certain elements of $\mathbf{Q}(v)$. Hence (c) can be rewritten as follows:

(d)
$$\sum c_{b',\mu,b}g(b,b_1,b_2)g'(b',b'_1,b'_2)b_1'^-K_\mu b_1^+\tilde K_{|b_2|}(1')\otimes\tilde K_{-|b'_1|}b_2'^-K_\mu b_2^+(1)=0.$$

We have $b_2^+(1)=0$ unless $b_2=1$; the corresponding $g(b,b_1,1)$ is zero unless $b_1=b$ in which case it is 1. Hence (d) simplifies to

$$\sum c_{b',\mu,b}g'(b',b'_1,b'_2)b_1'^-K_\mu b^+(1')\otimes\tilde K_{-|b'_1|}b_2'^-K_\mu(1)=0. \tag{e}$$

We may assume that the sum is restricted to those indices for which $c_{b',\mu,b} \neq 0, g'(b', b'_1, b'_2) \neq 0$.

We identify $M = {}'\mathbf{f}$ in such a way that $b'^-(1)$ corresponds to b' for any $b' \in B$. The equation (e) becomes

$$\text{(f)} \qquad \sum c_{b',\mu,b} g'(b', b'_1, b'_2) v^{\langle \mu, \lambda - \lambda' + |b| \rangle} b_1'^- b^+(1') \otimes \tilde{K}_{-|b'_1|} b'_2 = 0$$

(equality in $M' \otimes {}'\mathbf{f}$) where the sum is restricted as above. We project the equation (f) onto the summand $M' \otimes {}'\mathbf{f}_{\nu'}$ in $M' \otimes {}'\mathbf{f}$ where $\operatorname{tr} \nu' = N$. Thus we may further restrict the sum in (f) to those b', μ, b, b'_1, b'_2 such that $|b'_2| = \nu'$ and we still get zero. For such an index we have $|b'_2| \leq |b'|$ (coefficient by coefficient); hence $N = \operatorname{tr} |b'_2| \leq \operatorname{tr} |b'|$. By the definition of N, we have $\operatorname{tr} |b'| \leq N$. It follows that $|b'_2| = |b'|$. This implies that $b'_1 = 1, b'_2 = b'$ and $g'(b', 1, b') = 1$. Hence we obtain the equation

$$\sum c_{b',\mu,b} v^{\langle \mu, \lambda - \lambda' + |b| \rangle} b \otimes b' = 0$$

in ${}'\mathbf{f} \otimes {}'\mathbf{f}$. Now the elements $b \otimes b'$ form a basis of ${}'\mathbf{f} \otimes {}'\mathbf{f}$; hence the last equation implies: $\sum_\mu c_{b',\mu,b} v^{\langle \mu, \lambda - \lambda' + |b| \rangle} = 0$ for all b, b' such that $\operatorname{tr} |b'| = N$.) Since in the last equation $\lambda - \lambda'$ is an arbitrary element of X, we deduce that $c_{b',\mu,b} = 0$ for all b', μ, b such that $\operatorname{tr} |b'| = N$. This contradicts the definition of N. The proposition is proved.

Corollary 3.2.5. (a) *The $\mathbf{Q}(v)$-linear map $\mathbf{f} \otimes \mathbf{U}^0 \otimes \mathbf{f} \to \mathbf{U}$ given by $u \otimes K_\mu \otimes w \mapsto u^- K_\mu w^+$ is an isomorphism.*

(b) *The $\mathbf{Q}(v)$-linear map $\mathbf{f} \otimes \mathbf{U}^0 \otimes \mathbf{f} \to \mathbf{U}$ given by $u \otimes K_\mu \otimes w \mapsto u^+ K_\mu w^-$ is an isomorphism.*

Again, (b) follows from (a) using the involution ω. We prove (a). Let J_+ (resp. J_-) be the two-sided ideal of ${}'\mathbf{U}$ generated by the subset $\mathcal{I}^+ = \{x^+ | x \in \mathcal{I}\}$ (resp. by the subset $\mathcal{I}^- = \{x^- | x \in \mathcal{I}\}$). From the definitions, we see that $\mathbf{U}$ is the quotient algebra of ${}'\mathbf{U}$ by the two-sided ideal $J_+ + J_-$. From 3.1.7, we have

(c) $({}'\mathbf{U}^+)\mathcal{I}^- \subset \mathcal{I}^- \mathbf{U}^0 ({}'\mathbf{U}^+)$ and

(d) $\mathcal{I}^+ ({}'\mathbf{U}^-) \subset ({}'\mathbf{U}^-) \mathbf{U}^0 \mathcal{I}^+$.

Using ${}'\mathbf{U} = ({}'\mathbf{U}^+)\mathbf{U}^0({}'\mathbf{U}^-) = ({}'\mathbf{U}^-)\mathbf{U}^0({}'\mathbf{U}^+)$ and the fact that $\mathcal{I}^-$ is a two-sided ideal in ${}'\mathbf{U}^-$, we see that (c) implies

$$\begin{aligned} {}'\mathbf{U}\mathcal{I}^-({}'\mathbf{U}) &= ({}'\mathbf{U}^+)\mathbf{U}^0({}'\mathbf{U}^-)\mathcal{I}^-({}'\mathbf{U}^-)\mathbf{U}^0({}'\mathbf{U}^+) \subset ({}'\mathbf{U}^+)\mathcal{I}^-\mathbf{U}^0({}'\mathbf{U}^+) \\ &\subset \mathcal{I}^-\mathbf{U}^0({}'\mathbf{U}^+). \end{aligned}$$

Thus, $J_- = \mathcal{I}^-\mathbf{U}^0('\mathbf{U}^+)$. Similarly, we see that (d) implies

$$'\mathbf{U}\mathcal{I}^+('\mathbf{U}) = ('\mathbf{U}^-)\mathbf{U}^0('\mathbf{U}^+)\mathcal{I}^+('\mathbf{U}^+)\mathbf{U}^0('\mathbf{U}^-) \subset ('\mathbf{U}^-)\mathbf{U}^0\mathcal{I}^+('\mathbf{U}^-)$$
$$\subset ('\mathbf{U}^-)\mathbf{U}^0\mathcal{I}^+.$$

Thus, $J_+ = ('\mathbf{U}^-)\mathbf{U}^0\mathcal{I}^+$. Using 3.2.4, we may therefore identify

$$\mathbf{U} = \frac{'\mathbf{U}^- \otimes \mathbf{U}^0 \otimes '\mathbf{U}^+}{\mathcal{I}^- \otimes \mathbf{U}^0 \otimes '\mathbf{U}^+ + '\mathbf{U}^- \otimes \mathbf{U}^0 \otimes \mathcal{I}^+} = ('\mathbf{U}^-/\mathcal{I}^-) \otimes \mathbf{U}^0 \otimes ('\mathbf{U}^+/\mathcal{I}^+)$$

and (a) follows. The corollary is proved.

Corollary 3.2.6. *The map* $\mathbf{f} \to \mathbf{U}^+$ $(x \mapsto x^+)$ *is an isomorphism; the map* $\mathbf{f} \to \mathbf{U}^-$ $(x \mapsto x^-)$ *is an isomorphism; the algebra homomorphism* $\mathbf{U}^0 \to \mathbf{U}$ *given by* $K_\mu \mapsto K_\mu$ *for all* μ *is an imbedding.*

For any $\nu \in \mathbf{N}[I]$ we shall denote by $\mathbf{U}_\nu^+$ (resp. $\mathbf{U}_\nu^-$) the image of $\mathbf{f}_\nu$ under the isomorphism $\mathbf{f} \to \mathbf{U}^+$ (resp. $\mathbf{f} \to \mathbf{U}^-$) considered above.

Proposition 3.2.7. *Let* $x \in \mathbf{f}_\nu$ *where* $\nu \in \mathbf{N}[I]$ *is different from* 0.
(a) *If* $x^+F_i = F_ix^+$ *for all* $i \in I$, *then* $x = 0$.
(b) *If* $x^-E_i = E_ix^-$ *for all* $i \in I$, *then* $x = 0$.

If x is as in (a) then, using 3.1.6, we see that, for some integer n, we have $r_i(x)^+\tilde{K}_i - v^n({}_ir(x)^+)\tilde{K}_{-i} = 0$ in $\mathbf{U}$. Since $\tilde{K}_i$ and $\tilde{K}_{-i}$ are linearly independent in $\mathbf{U}^0$ (see 3.2.6), we deduce, using triangular decomposition (3.2.5) that $r_i(x) = {}_ir(x) = 0$. This holds for any i; hence by 1.2.15, we have $x = 0$. This proves (a). The proof of (b) is entirely similar.

3.3. Antipode

Lemma 3.3.1. *For* $\nu \in \mathbf{N}[I]$ *we set* $c(\nu) = \nu\cdot\nu/2 - \sum_i \nu_i i\cdot i/2 \in \mathbf{Z}$.
(a) *There is a unique homomorphism of* $\mathbf{Q}(v)$*-algebras* $S : \mathbf{U} \to \mathbf{U}^{opp}$ *such that*

$$S(E_i) = -\tilde{K}_{-i}E_i, \quad S(F_i) = -F_i\tilde{K}_i, \quad S(K_\mu) = K_{-\mu}$$

for all $i \in I, \mu \in Y$.
(b) *For any* $x \in \mathbf{f}_\nu$, *we have* $S(x^+) = (-1)^{\mathrm{tr}\,\nu}v^{c(\nu)}\tilde{K}_{-\nu}\sigma(x)^+$ *and* $S(x^-) = (-1)^{\mathrm{tr}\,\nu}v^{-c(\nu)}\sigma(x)^-\tilde{K}_\nu$.
(c) *There is a unique homomorphism of* $\mathbf{Q}(v)$*-algebras* $S' : \mathbf{U} \to \mathbf{U}^{opp}$ *such that*

$$S'(E_i) = -E_i\tilde{K}_{-i}, \quad S'(F_i) = -\tilde{K}_iF_i, \quad S'(K_\mu) = K_{-\mu}$$

for all $i \in I, \mu \in Y$.

(d) *For any* $x \in \mathbf{f}_\nu$, *we have* $S'(x^-) = (-1)^{\operatorname{tr} \nu} v^{c(\nu)} \tilde{K}_\nu \sigma(x)^-$ *and* $S'(x^+) = (-1)^{\operatorname{tr} \nu} v^{-c(\nu)} \sigma(x)^+ \tilde{K}_{-\nu}$.

(e) *We have* $SS' = S'S = 1$.

(f) *If* $x \in \mathbf{f}_\nu$, *then* $S(x^+) = v^{-f(\nu)} S'(x^+)$ *and* $S(x^-) = v^{f(\nu)} S'(x^-)$ *where* $f(\nu) = \sum_i \nu_i i \cdot i$.

It is easy to verify that S, S' respect the relations 3.1.1(a)–(d) of $\mathbf{U}$. To check that they respect the relations 3.1.1(e), it suffices to check that the maps $\mathbf{f} \to \mathbf{U}^{opp}$ given by $x \mapsto S(x^+)$ or $x \mapsto S(x^-)$ and those given by $x \mapsto S'(x^+)$ or $x \mapsto S'(x^-)$ are algebra homomorphisms; this is checked using the fact that $\sigma : \mathbf{f} \to \mathbf{f}^{opp}$ is an algebra homomorphism. This proves (a)–(d). The assertion (e) is proved by verification on generators. Finally, (f) follows from (b),(d).

3.3.2. The map S (resp. S') is called the *antipode* (resp. the *skew-antipode*) of $\mathbf{U}$.

3.3.3. Note that

$S(E_i^{(n)}) = (-1)^n v_i^{n^2-n} \tilde{K}_{-ni} E_i^{(n)}$ and $S(F_i^{(n)}) = (-1)^n v_i^{-n^2+n} F_i^{(n)} \tilde{K}_{ni}$;

$S'(E_i^{(n)}) = (-1)^n v_i^{-n^2+n} E_i^{(n)} \tilde{K}_{-ni}$ and $S'(F_i^{(n)}) = (-1)^n v_i^{n^2-n} \tilde{K}_{ni} F_i^{(n)}$.

3.3.4. S, S' are related to Δ as follows. Let ${}^t\Delta : \mathbf{U} \to \mathbf{U} \otimes \mathbf{U}$ be the composition $\mathbf{U} \xrightarrow{\Delta} \mathbf{U} \otimes \mathbf{U} \to \mathbf{U} \otimes \mathbf{U}$, where the last map is the linear map given by $x \otimes y \mapsto y \otimes x$. Then

$$(S \otimes S)(\Delta(x)) = {}^t\Delta(S(x)), \quad (S' \otimes S')(\Delta(x)) = {}^t\Delta(S'(x)).$$

Each of the compositions

$$\mathbf{U} \xrightarrow{\Delta} \mathbf{U} \otimes \mathbf{U} \xrightarrow{1 \otimes S} \mathbf{U} \otimes \mathbf{U} \xrightarrow{m} \mathbf{U}, \quad \mathbf{U} \xrightarrow{\Delta} \mathbf{U} \otimes \mathbf{U} \xrightarrow{S \otimes 1} \mathbf{U} \otimes \mathbf{U} \xrightarrow{m} \mathbf{U}$$

$$\mathbf{U} \xrightarrow{{}^t\Delta} \mathbf{U} \otimes \mathbf{U} \xrightarrow{1 \otimes S'} \mathbf{U} \otimes \mathbf{U} \xrightarrow{m} \mathbf{U}, \quad \mathbf{U} \xrightarrow{{}^t\Delta} \mathbf{U} \otimes \mathbf{U} \xrightarrow{S' \otimes 1} \mathbf{U} \otimes \mathbf{U} \xrightarrow{m} \mathbf{U}$$

(with m being the multiplication) is equal to the map $x \mapsto \mathbf{e}(x)1$.

The identity $m(1 \otimes S)\Delta(x) = \mathbf{e}(x)1$ is checked as follows. First we note that if this holds for x' and x'', then it also holds for $x'x''$. Hence it suffices to check it in the case where x is one of the algebra generators and that is immediate. The other identities are checked in the same way.

Finally, we have

$$\mathbf{e}S = \mathbf{e}S' = \mathbf{e}.$$

We see that the algebra $\mathbf{U}$ with the additional structure given by the comultiplication Δ, the co-unit $\mathbf{e}$, the antipode S and the skew-antipode S', is a *Hopf algebra.*

3.4. The category $\mathcal{C}$

3.4.1. We define a category $\mathcal{C}$ as follows. An object of $\mathcal{C}$ is a $\mathbf{U}$-module M with a given direct sum decomposition $M = \oplus_{\lambda\in X} M^\lambda$ (as a $\mathbf{Q}(v)$-vector space) such that, for any $\mu \in Y, \lambda \in X$ and $m \in M^\lambda$, we have $K_\mu m = v^{\langle\mu,\lambda\rangle} m$. (The subspaces M^λ are called the *weight spaces* of M; they are uniquely determined by the $\mathbf{U}$-module structure.) A morphism in $\mathcal{C}$ is a $\mathbf{U}$-linear map; it automatically respects the decompositions into weight spaces.

For example, $M = \mathbf{Q}(v)$ may be regarded as an object of $\mathcal{C}$ with $um = \mathbf{e}(u)m$ for $u \in \mathbf{U}$ and $m \in M$ ($\mathbf{e}$ is the co-unit); we have $M = M^0$.

3.4.2. Let $M \in \mathcal{C}$ and let $m \in M^\lambda$. The following formulas follow from 3.1.9.

(a) $E_i^{(a)} F_j^{(b)} m = F_j^{(b)} E_i^{(a)} m$ if $i \neq j$;

(b) $E_i^{(a)} F_i^{(b)} m = \sum_{t\geq 0} \begin{bmatrix} a-b+\langle i,\lambda\rangle \\ t \end{bmatrix}_i F_i^{(b-t)} E_i^{(a-t)} m$;

(c) $F_i^{(b)} E_i^{(a)} m = \sum_{t\geq 0} \begin{bmatrix} -a+b-\langle i,\lambda\rangle \\ t \end{bmatrix}_i E_i^{(a-t)} F_i^{(b-t)} m$.

3.4.3. Tensor product of U-modules. If $M', M'' \in \mathcal{C}$, the tensor product $M' \otimes M''$ (over $\mathbf{Q}(v)$) is naturally a $\mathbf{U}\otimes\mathbf{U}$-module with $(u'\otimes u'')(m'\otimes m'') = u'm' \otimes u''m''$. We restrict it to a $\mathbf{U}$-module via the algebra homomorphism $\Delta : \mathbf{U} \to \mathbf{U}\otimes\mathbf{U}$. The resulting $\mathbf{U}$-module with the weight space decomposition $(M'\otimes M'')^\lambda = \oplus_{\lambda'+\lambda''=\lambda} M'^{\lambda'} \otimes M''^{\lambda''}$ is naturally an object of $\mathcal{C}$. If $m' \in M'^{\lambda'}, m'' \in M''^{\lambda''}, i \in I$, we have

$$E_i(m'\otimes m'') = E_i m' \otimes m'' + v_i^{\langle i,\lambda'\rangle} m' \otimes E_i m'', \quad \text{and}$$

$$F_i(m'\otimes m'') = m' \otimes F_i m'' + v_i^{-\langle i,\lambda''\rangle} F_i m' \otimes m''.$$

More generally,

$$E_i^{(a)}(m'\otimes m'') = \sum_{a'+a''=a} v_i^{a'a''+a''\langle i,\lambda'\rangle} E_i^{(a')} m' \otimes E_i^{(a'')} m'',$$

$$F_i^{(a)}(m'\otimes m'') = \sum_{a'+a''=a} v_i^{a'a''-a'\langle i,\lambda''\rangle} F_i^{(a')} m' \otimes F_i^{(a'')} m''.$$

3.4.4. To any object M of $\mathcal{C}$ we associate a new object ${}^\omega M$ of $\mathcal{C}$ as follows. ${}^\omega M$ has the same underlying $\mathbf{Q}(v)$-vector space as M. By definition, $({}^\omega M)^\lambda = M^{-\lambda}$ for any $\lambda \in X$. For any $u \in \mathbf{U}$, the operator u on ${}^\omega M$ coincides with the operator $\omega(u)$ on M. (See 3.1.3.)

3.4.5. Verma modules. Let $\lambda \in X$. We show that there is a unique $\mathbf{U}$-module structure on the $\mathbf{Q}(v)$-vector space $\mathbf{f}$ such that, for any homogeneous $y \in \mathbf{f}$, any $\mu \in Y$ and any $i \in I$, we have $K_\mu(y) = v^{\langle \mu, \lambda - |y| \rangle} y$, $F_i(y) = \theta_i y$, and $E_i(1) = 0$.

The uniqueness is immediate. We now prove existence. We consider the left ideal $J = \sum_i \mathbf{U}E_i + \sum_\mu \mathbf{U}(K_\mu - v^{\langle \mu, \lambda \rangle})$ of $\mathbf{U}$. Then $\mathbf{U}/J$ is naturally a $\mathbf{U}$-module. Using triangular decomposition, we see easily that the $\mathbf{Q}(v)$-linear map $\mathbf{f} \to \mathbf{U}/J$ given by $x \mapsto x^- + J$ is an isomorphism. Via this isomorphism, $\mathbf{f}$ becomes a $\mathbf{U}$-module; it is easy to see that this has the required properties.

The module constructed above is called a *Verma module* and is denoted M_λ. It is an object of $\mathcal{C}$: we have $M_\lambda^{\lambda'} = \oplus_{\nu \in \mathbf{N}[I]; \lambda' = \lambda - \nu} \mathbf{f}_\nu$; note that in the last direct sum there may be more than one summand since the natural map $\mathbf{N}[I] \to X$ is not necessarily injective unless the root datum is X-regular.

3.4.6. Let M be an object of $\mathcal{C}$ and let $m \in M^\lambda$ be a vector such that $E_i m = 0$ for all i. We show that there is a unique morphism $t : M_\lambda \to M$ (in $\mathcal{C}$) such that $t(1) = m$.

Let $t : M_\lambda \to M$ be the map defined by $t(x^- 1) = x^- m$ for all $x \in \mathbf{f}$. Then t is automatically compatible with the decomposition into weight spaces. For any $x \in \mathbf{f}_\nu$ we have $t(E_i^{(a)} x^- 1) = E_i^{(a)} t(x^- 1)$. We argue by induction on $\operatorname{tr} \nu$. In the case where $\nu = 0$ we use our assumption on m. The induction step is obtained using the commutation formulas between $E_i^{(a)}, F_i^{(b)}$ in 3.4.2 which hold both on M_λ and M. Thus, t is a morphism as desired. The uniqueness of t is obvious.

3.4.7. The category $\mathcal{C}^{hi}$. Let $\mathcal{C}^{hi}$ be the full subcategory of $\mathcal{C}$ whose objects are the M with the following property: for any $m \in M$ there exists a number $N \geq 0$ such that $x^+ m = 0$ for all $x \in \mathbf{f}_\nu$ with $\operatorname{tr} \nu \geq N$. Note that the Verma module M_λ belongs to $\mathcal{C}^{hi}$. The same holds for any quotient module of M_λ.

3.5. Integrable objects of $\mathcal{C}$

3.5.1. An object $M \in \mathcal{C}$ is said to be *integrable* if for any $m \in M$ and any $i \in I$, there exists $n_0 \geq 1$ such that $E_i^{(n)} m = F_i^{(n)} m = 0$ for all $n \geq n_0$.

Let $\mathcal{C}'$ be the full subcategory of $\mathcal{C}$ whose objects are the integrable $\mathbf{U}$-modules.

3.5.2. From the formulas in 3.4.3, we see that

(a) if M', M'' are integrable, then $M' \otimes M''$ is integrable.

Clearly,

(b) if M is integrable, then ${}^{\omega}M$ is integrable.

Lemma 3.5.3. *Given $(a_i) \in \mathbf{N}^I, (b_i) \in \mathbf{N}^I$ and $\lambda \in X$, let M be the quotient of $\mathbf{U}$ by the left ideal generated by the elements $F_i^{a_i+1}, E_i^{b_i+1}$ with $i \in I$ and $(K_\mu - v^{\langle \mu, \lambda \rangle})$ with $\mu \in Y$. Then M is an integrable $\mathbf{U}$-module.*

Let $i \neq j$ in I and let $\alpha = -\langle i, j' \rangle$. We show that for any $N \geq \alpha + 1$ we have

(a) $F_i^N F_j \in \sum_{p+p'=N; N-\alpha \leq p' \leq N} \mathbf{Q}(v) F_i^p F_j F_i^{p'}$.

For $N = \alpha + 1$, this follows from the quantum Serre relation 1.4.3. Assume that (a) is known for some $N \geq \alpha + 1$; we prove it for $N + 1$.

By our hypothesis, we have

$$F_i^{N+1} F_j \in \sum_{p+p'=N; N-\alpha \leq p' \leq N} \mathbf{Q}(v) F_i^{p+1} F_j F_i^{p'}.$$

All terms in this sum are of the required type except possibly for the term corresponding to $p' = N - \alpha$. For that term we write (using again the quantum Serre relation):

$$F_i^{\alpha+1} F_j F_i^{N-\alpha} \in \sum_{r+r'=\alpha+1; 1 \leq r' \leq \alpha+1} \mathbf{Q}(v) F_i^r F_j F_i^{r'+N-\alpha}$$

which is of the required type. Thus (a) is proved.

Next we note that for any $N \geq 0$ we have

(b) $F_i^N E_j \in E_j F_i^N + \mathbf{Q}(v) F_i^{N-1}$ and $F_i^N K_\mu = v^{N \langle \mu, i' \rangle} K_\mu F_i^N$.

We now consider a fixed element $x \in \mathbf{U}$ of form $x_1 x_2 \cdots x_n$ where each factor x_p is either E_i or F_j or K_μ. We consider the product $F_i^N x = F_i^N x_1 x_2 \cdots x_n$. We can move F_i^N across $x_1, x_2, \ldots$ successively (from left to right) using (a),(b) and we see that we finally get a linear combination of terms of the form $y F_i^{N'}$ with $N' \geq N - c$ where $y \in \mathbf{U}$ and $c \geq 0$ is a constant depending on $x_1 x_2 \cdots x_n$, but not on N. Hence, if $N \geq a_i + 1 + c$, we have that F_i^N acts as zero on the image of x in M. Thus F_i acts locally nilpotently on M. In an entirely similar way, we see that E_i acts locally nilpotently on M. The lemma is proved.

Proposition 3.5.4. *Let $u \in \mathbf{U}$ be an element such that u acts as zero on any integrable $\mathbf{U}$-module. Then $u = 0$.*

By assumption, u belongs to the left ideal of the previous lemma for all choices of $(a_i), (b_i), \lambda$. But the intersection of all these left ideals is zero, as one sees using the triangular decomposition. Thus, $u = 0$.

3.5.5. In the remainder of this section, the root datum is assumed to be Y-regular. We define $X^+ = \{\lambda \in X | \langle i, \lambda \rangle \in \mathbf{N} \quad \forall i\}$. (This definition could be given for a not necessarily Y-regular root datum, but it would be useless.) We say that $\lambda \in X$ is *dominant* if $\lambda \in X^+$.

Proposition 3.5.6. *Let $\lambda \in X^+$.*

(a) *Let $\mathcal{T}$ be the left ideal of* $\mathbf{f}$ *generated by the elements $\theta_i^{\langle i,\lambda\rangle+1}$ for various $i \in I$. Then $\mathcal{T}$ is a subobject of the Verma module $M_\lambda \in \mathcal{C}$.*

(b) *Let $\Lambda_\lambda = M_\lambda/\mathcal{T}$ be the quotient object. Then Λ_λ is integrable.*

Clearly, $\mathcal{T}$ is the sum of its intersections with the weight spaces of M_λ; moreover, it is clearly stable under the operators $F_i : M_\lambda \to M_\lambda$. We now show that it is stable under the operators $E_i : M_\lambda \to M_\lambda$. Using the commutation relation between E_i and F_j, we see that it suffices to show that $E_i F_j^{(\langle j,\lambda\rangle+1)} 1 = 0$ in M_λ, for any $j \in I$. If $i \neq j$, this is clear. If $i = j$, we have, by 3.4.2:

$$E_i F_i^{(\langle i,\lambda\rangle+1)} 1 = \sum_{t \geq 0} \begin{bmatrix} 0 \\ t \end{bmatrix}_i F_i^{(\langle i,\lambda\rangle+1-t)} E_i^{(1-t)} 1 = F_i^{(\langle i,\lambda\rangle+1)} E_i 1 = 0.$$

This proves (a). We now prove (b). From the definition, we see that Λ_λ is naturally a quotient of the $\mathbf{U}$-module M defined in 3.5.3, with $b_i = 0, a_i = \langle i, \lambda \rangle$, hence it is integrable by 3.5.3. The proposition is proved.

3.5.7. We will denote the image of $1 \in \mathbf{f}$ in Λ_λ by η_λ or simply by η.

We now consider the $\mathbf{U}$-module ${}^\omega\Lambda_\lambda$. As a vector space, we have ${}^\omega\Lambda_\lambda = \Lambda_\lambda$. The vector $\eta_\lambda \in \Lambda_\lambda$, regarded as a vector of ${}^\omega\Lambda_\lambda$ will be denoted by ξ or $\xi_{-\lambda}$. By 3.5.2(b), ${}^\omega\Lambda_\lambda$ is integrable.

Proposition 3.5.8. *Let M be an object of $\mathcal{C}'$ and let $m \in M^\lambda$ be a non-zero vector such that $E_i m = 0$ for all i. Then $\lambda \in X^+$ and there is a unique morphism (in $\mathcal{C}'$) $t' : \Lambda_\lambda \to M$ which carries $\eta_\lambda \in \Lambda_\lambda$ to m.*

The uniqueness of t' is clear. To prove existence, we consider the morphism $t : M_\lambda \to M$ such that $t(1) = m$ (see 3.4.6). It remains to show that

(a) $\lambda \in X^+$ and $F_i^{\langle i,\lambda\rangle+1} m = 0$ for all i.

(This will imply that t factors through Λ_λ.) Let $i \in I$. We set $a = \langle i, \lambda \rangle$. We can find an integer $b \geq 1$ such that $F_i^{(b-1)} m \neq 0, F_i^{(b)} m = 0$. Using 3.4.2, we have $0 = E_i F_i^{(b)} m = [1 - b + a]_i F_i^{(b-1)} m$; hence $[1 - b + a]_i = 0$ and $a = b - 1$. Thus $a \geq 0$ and $F_i^{(a+1)} m = 0$. The proposition is proved.

CHAPTER 4

The Quasi-$\mathcal{R}$-Matrix

4.1. The Element Θ

4.1.1. Completions. Let $(\mathbf{U}\otimes\mathbf{U})^{\widehat{}}$ be the completion of the vector space $\mathbf{U}\otimes\mathbf{U}$ with respect to the descending sequence of vector spaces

$$\mathcal{H}_N = (\mathbf{U}^+\mathbf{U}^0(\sum_{\mathrm{tr}\,\nu\geq N}\mathbf{U}^-_\nu))\otimes\mathbf{U}+\mathbf{U}\otimes(\mathbf{U}^-\mathbf{U}^0(\sum_{\mathrm{tr}\,\nu\geq N}\mathbf{U}^+_\nu))$$

for $N=1,2,\ldots$. Note that each $\mathcal{H}_N$ is a left ideal in $\mathbf{U}\otimes\mathbf{U}$; moreover, for any $u\in\mathbf{U}\otimes\mathbf{U}$, we can find $r\geq 0$ such that $\mathcal{H}_{N+r}u\subset\mathcal{H}_N$ for all $N\geq 0$. It follows that the $\mathbf{Q}(v)$-algebra structure on $\mathbf{U}\otimes\mathbf{U}$ extends by continuity to a $\mathbf{Q}(v)$-algebra structure on $(\mathbf{U}\otimes\mathbf{U})^{\widehat{}}$. We have an obvious imbedding of algebras $\mathbf{U}\otimes\mathbf{U}\to(\mathbf{U}\otimes\mathbf{U})^{\widehat{}}$.

Let $^-:\mathbf{U}\otimes\mathbf{U}\to\mathbf{U}\otimes\mathbf{U}$ be the $\mathbf{Q}$-algebra automorphism given by $^-\otimes^-$. This extends by continuity to a $\mathbf{Q}$-algebra automorphism $^-:(\mathbf{U}\otimes\mathbf{U})^{\widehat{}}\to(\mathbf{U}\otimes\mathbf{U})^{\widehat{}}$ (with square 1).

Let $\bar{\Delta}:\mathbf{U}\to\mathbf{U}\otimes\mathbf{U}$ be the $\mathbf{Q}(v)$-algebra homomorphism given by $\overline{\Delta(x)}=\bar{\Delta}(\bar{x})$ for all $x\in\mathbf{U}$. We have in general $\bar{\Delta}\neq\Delta$ where Δ is as in Lemma 3.1.4.

Theorem 4.1.2. (a) *There is a unique family of elements $\Theta_\nu\in\mathbf{U}^-_\nu\otimes\mathbf{U}^+_\nu$ (with $\nu\in\mathbf{N}[I]$) such that $\Theta_0=1\otimes 1$ and $\Theta=\sum_\nu\Theta_\nu\in(\mathbf{U}\otimes\mathbf{U})^{\widehat{}}$ satisfies $\Delta(u)\Theta=\Theta\bar{\Delta}(u)$ for all $u\in\mathbf{U}$ (identity in $(\mathbf{U}\otimes\mathbf{U})^{\widehat{}}$).*

(b) *Let B be a $\mathbf{Q}(v)$-basis of $\mathbf{f}$ such that $B_\nu=B\cap\mathbf{f}_\nu$ is a basis of $\mathbf{f}_\nu$ for any ν. Let $\{b^*|b\in B_\nu\}$ be the basis of $\mathbf{f}_\nu$ dual to B_ν under $(,)$. We have*

$$\Theta_\nu=(-1)^{\mathrm{tr}\,\nu}v_\nu\sum_{b\in B_\nu}b^-\otimes b^{*+}\in\mathbf{U}^-_\nu\otimes\mathbf{U}^+_\nu.$$

We consider an element $\Theta\in(\mathbf{U}\otimes\mathbf{U})^{\widehat{}}$ of the form $\Theta=\sum_\nu\Theta_\nu$ where $\Theta_\nu=\sum_{b,b'\in B_\nu}c_{b',b}b'^-\otimes b^{*+}$ and $c_{b',b}\in\mathbf{Q}(v)$.

The set of $u\in\mathbf{U}$ such that $\Delta(u)\Theta=\Theta\bar{\Delta}(u)$ is clearly a subalgebra of $\mathbf{U}$ containing all K_μ. Hence, in order for this set to be equal to $\mathbf{U}$, it is

G. Lusztig, *Introduction to Quantum Groups*, Modern Birkhäuser Classics,
DOI 10.1007/978-0-8176-4717-9_4, © Springer Science+Business Media, LLC 2010

necessary and sufficient that it contains E_i, F_i for all i, or in other words, that

$$\sum_{b_1,b_2\in B_\nu} c_{b_1,b_2} E_i b_1{}^- \otimes b_2^{*+} + \sum_{b_3,b_4\in B_{\nu-i}} c_{b_3,b_4} \tilde{K}_i b_3^- \otimes E_i b_4^{*+}$$
$$= \sum_{b_1,b_2\in B_\nu} c_{b_1,b_2} b_1{}^- E_i \otimes b_2^{*+} + \sum_{b_3,b_4\in B_{\nu-i}} c_{b_3,b_4} b_3^- \tilde{K}_{-i} \otimes b_4^{*+} E_i,$$

and

$$\sum_{b_1,b_2\in B_\nu} c_{b_1,b_2} b_1{}^- \otimes F_i b_2^{*+} + \sum_{b_3,b_4\in B_{\nu-i}} c_{b_3,b_4} F_i b_3^- \otimes \tilde{K}_{-i} b_4^{*+}$$
$$= \sum_{b_1,b_2\in B_\nu} c_{b_1,b_2} b_1{}^- \otimes b_2^{*+} F_i + \sum_{b_3,b_4\in B_{\nu-i}} c_{b_3,b_4} b_3^- F_i \otimes b_4^{*+} \tilde{K}_i,$$

for any ν, i, with the convention that $B_{\nu-i}$ is empty, if $\nu_i = 0$. By the non-degeneracy of $(,)$, these identitites are equivalent to the identities

$$\sum_{b_1,b_2\in B_\nu} c_{b_1,b_2} (b_2^*, z)(E_i b_1{}^- - b_1{}^- E_i)$$
$$+ \sum_{b_3,b_4\in B_{\nu-i}} c_{b_3,b_4} ((\theta_i b_4^*, z) \tilde{K}_i b_3^- - (b_4^* \theta_i, z) b_3^- \tilde{K}_{-i}) = 0,$$

and

$$\sum_{b_1,b_2\in B_\nu} c_{b_1,b_2} (b_1, z)(F_i b_2^{*+} - b_2^{*+} F_i)$$
$$+ \sum_{b_3,b_4\in B_{\nu-i}} c_{b_3,b_4} ((\theta_i b_3, z) \tilde{K}_{-i} b_4^{*+} - (b_3 \theta_i, z) b_4^{*+} \tilde{K}_i) = 0$$

for any ν, i and $z \in \mathbf{f}_\nu$.

We substitute $(\theta_i b_4^*, z) = (\theta_i, \theta_i)(b_4^*, {}_i r(z))$ and $(b_4^* \theta_i, z) = (\theta_i, \theta_i)(b_4^*, r_i(z))$ in the first equation, and we make a similar substitution in the second equation; using 3.1.6, we see that our conditions turn into the two conditions

$$- \sum_{b_1,b_2\in B_\nu} c_{b_1,b_2} (b_2^*, z) v_i^{-1} (\theta_i, \theta_i)(r_i(b_1)^- \tilde{K}_{-i} - \tilde{K}_i({}_i r(b_1)^-))$$
$$+ \sum_{b_3,b_4\in B_{\nu-i}} c_{b_3,b_4} (\theta_i, \theta_i)((b_4^*, {}_i r(z)) \tilde{K}_i b_3^- - (b_4^*, r_i(z)) b_3^- \tilde{K}_{-i}) = 0,$$

and

$$-\sum_{b_1,b_2\in B_\nu} c_{b_1,b_2}(b_1,z)v_i^{-1}(\theta_i,\theta_i)(r_i(b_2)^+\tilde{K}_i-\tilde{K}_{-i}({}_ir(b_2)^+))$$
$$+\sum_{b_3,b_4\in B_{\nu-i}} c_{b_3,b_4}(\theta_i,\theta_i)((b_3,{}_ir(z))\tilde{K}_{-i}b_4^{*+}-(b_3,r_i(z))b_4^{*+}\tilde{K}_i)=0,$$

or equivalently, into the four conditions

(c) $$\sum_{b_1,b_2\in B_\nu} c_{b_1,b_2}(b_2^*,z)v_i^{-1}r_i(b_1)+\sum_{b_3,b_4\in B_{\nu-i}} c_{b_3,b_4}(b_4^*,r_i(z))b_3=0,$$

(d) $$\sum_{b_1,b_2\in B_\nu} c_{b_1,b_2}(b_2^*,z)v_i^{-1}{}_ir(b_1)+\sum_{b_3,b_4\in B_{\nu-i}} c_{b_3,b_4}(b_4^*,{}_ir(z))b_3=0,$$

(e) $$\sum_{b_1,b_2\in B_\nu} c_{b_1,b_2}(b_1,z)v_i^{-1}{}_ir(b_2)+\sum_{b_3,b_4\in B_{\nu-i}} c_{b_3,b_4}(b_3,{}_ir(z))b_4^*=0,$$

(f) $$\sum_{b_1,b_2\in B_\nu} c_{b_1,b_2}(b_1,z)v_i^{-1}r_i(b_2)+\sum_{b_3,b_4\in B_{\nu-i}} c_{b_3,b_4}(b_3,r_i(z))b_4^*=0$$

for any ν,i and $z\in\mathbf{f}_\nu$. These equations are clearly satisfied by taking for all ν, $c_{b',b}=(-1)^{\operatorname{tr}\nu}v_\nu\delta_{b',b}$ for $b,b'\in B_\nu$. This proves the existence part of (a) and (b). To prove the uniqueness in (a), it is enough to show that, given a solution $(c_{b',b})$ of the system of equations (c)–(f) such that $c_{b,b}=0$ for the unique element $b\in B_0$, we necessarily have $c_{b',b}=0$ for all ν and all $b',b\in B_\nu$. We argue by induction on $\operatorname{tr}\nu\geq 0$. In the case where $\operatorname{tr}\nu=0$, there is nothing to prove. Assume now that $\operatorname{tr}\nu>0$. Using the induction hypothesis, the second sum in equation (c) is zero. Hence this equation becomes $r_i(\sum_{b_1,b_2\in B_\nu} c_{b_1,b_2}(b_2^*,z)b_1)=0$. Since this holds for all $i\in I$, we see from 1.2.15 that $\sum_{b_1,b_2\in B_\nu} c_{b_1,b_2}(b_2^*,z)b_1=0$. (We have used that $\nu\neq 0$.) Since b_1 are linearly independent, we can deduce that $\sum_{b_2\in B_\nu} c_{b_1,b_2}(b_2^*,z)=0$ for any $b_1\in B_\nu$ and any $z\in\mathbf{f}_\nu$. Taking $z=b_2$, we see that $c_{b_1,b_2}=0$ for any $b_1,b_2\in B_\nu$. This completes the proof.

Corollary 4.1.3. *We have $\Theta\bar{\Theta}=\bar{\Theta}\Theta=1\otimes 1$ (equality in $(\mathbf{U}\otimes\mathbf{U})^\wedge$).*

From the definition of Θ, we see that Θ is an invertible element of the ring $(\mathbf{U}\otimes\mathbf{U})^\wedge$. Let $\Theta'=\Theta^{-1}$.

From the identity $\Delta(\bar{u})\Theta=\Theta\bar{\Delta}(\bar{u})$, we deduce that $\Theta'\Delta(\bar{u})=\bar{\Delta}(\bar{u})\Theta'$ for all $u\in\mathbf{U}$. Applying $^-$, it follows that $\bar{\Theta}'\bar{\Delta}(u)=\Delta(u)\bar{\Theta}'$ for all $u\in\mathbf{U}$. It is clear that $\bar{\Theta}'=\sum_\nu\bar{\Theta}'_\nu$, where $\bar{\Theta}'_\nu\in\mathbf{U}^-_\nu\otimes\mathbf{U}^+_\nu$ and $\bar{\Theta}'_0=1\otimes 1$. Thus, $\bar{\Theta}'$ satisfies the defining properties of Θ. By the uniqueness of Θ, we have $\Theta=\bar{\Theta}'$. The corollary follows.

4.1.4. The element Θ defined in 4.1.2(a) is called the *quasi-$\mathcal{R}$-matrix.* For example, in the case where $I = \{i\}$ and $X = Y = \mathbf{Z}$ with $i = 1 \in Y, i' = 2 \in X$, we have

$$\Theta = \sum_n (-1)^n v_i^{-n(n-1)/2} \{n\}_i F_i^{(n)} \otimes E_i^{(n)},$$

with the notation $\{n\}_i = \prod_{a=1}^n (v_i^a - v_i^{-a})$ for $n \geq 0$.

4.2. Some Identities for Θ

4.2.1. We now study the image of the elements Θ_ν under the homomorphisms $\Delta \otimes 1, 1 \otimes \Delta : \mathbf{U} \otimes \mathbf{U} \to \mathbf{U} \otimes \mathbf{U} \otimes \mathbf{U}$. For an element $P = \sum x \otimes y \in \mathbf{U} \otimes \mathbf{U}$ we shall denote the elements

$\sum x \otimes y \otimes 1, \sum 1 \otimes x \otimes y, \sum x \otimes 1 \otimes y$ of $\mathbf{U} \otimes \mathbf{U} \otimes \mathbf{U}$ by P^{12}, P^{23}, P^{13}.

Proposition 4.2.2. (a) $(\Delta \otimes 1)(\Theta_\nu) = \sum_{\nu' + \nu'' = \nu} \Theta_{\nu'}^{23} (1 \otimes \tilde{K}_{-\nu''} \otimes 1) \Theta_{\nu''}^{13}$.

(b) $(1 \otimes \Delta)(\Theta_\nu) = \sum_{\nu' + \nu'' = \nu} \Theta_{\nu'}^{12} (1 \otimes \tilde{K}_{\nu''} \otimes 1) \Theta_{\nu''}^{13}$.

Let B be as in 4.1.2(b). For any $x \in \mathbf{f}$, we have

$$r(x) = \sum f(x, b_1, b_2) b_1 \otimes b_2$$

and

$$\bar{r}(x) = \sum f'(x, b_3, b_4) b_3 \otimes b_4$$

with $b_1, b_2, b_3, b_4 \in B$ and $f(x, b_1, b_2), f'(x, b_3, b_4) \in \mathbf{Q}(v)$. Then (a), (b) are equivalent to

$$\sum_{b, b_3, b_4; |b_3| + |b_4| = |b| = \nu} f'(b, b_3, b_4) b_3^- \otimes \tilde{K}_{-|b_3|} b_4^- \otimes b^{*+}$$
$$= \sum_{b_3, b_4; |b_3| + |b_4| = \nu} b_3{}^- \otimes b_4^- \tilde{K}_{-|b_3|} \otimes b_4^{*+} b_3^{*+}$$

and

$$\sum_{b, b_1, b_2; |b_1| + |b_2| = |b| = \nu} f(b^*, b_1, b_2) b^- \otimes b_1^+ \tilde{K}_{|b_2|} \otimes b_2^+$$
$$= \sum_{b_1, b_2; |b_1| + |b_2| = \nu} b_1^{*-} b_2^{*-} \otimes b_1^+ \tilde{K}_{|b_2|} \otimes b_2^+.$$

It is therefore enough to prove the following identities:

(c) $\sum_{b;|b|=\nu} f'(b, b_3, b_4)b^* = v^{-|b_3|\cdot|b_4|}b_4^*b_3^*$ for all b_3, b_4 such that $|b_3| + |b_4| = \nu$ and

(d) $\sum_{b;|b|=\nu} f(b^*, b_1, b_2)b = b_1^*b_2^*$ for all b_2, b_1 such that $|b_1| + |b_2| = \nu$.

By 1.2.11(a), we have $f'(b, b_3, b_4) = v^{-|b_3|\cdot|b_4|}f(b, b_4, b_3)$; hence (c) is equivalent to

(c′) $\sum_{b;|b|=\nu} f(b, b_4, b_3)b^* = b_4^*b_3^*$ for all b_3, b_4 such that $|b_3| + |b_4| = \nu$.

Now (c′) is equivalent to $(r(b), b_4^* \otimes b_3^*) = (b, b_4^*b_3^*)$ and (d) is equivalent to $(r(b^*), b_1^* \otimes b_2^*) = (b_1^*b_2^*, b^*)$; these equalities follow from the definition of $(,)$. The proposition is proved.

Proposition 4.2.3. $(\bar{\Delta} \otimes 1)(\Theta_\nu) = \sum_{\nu'+\nu''=\nu} \Theta^{13}_{\nu'}(1 \otimes \tilde{K}_{\nu'} \otimes 1)\Theta^{23}_{\nu''}$.

From $\Theta\bar{\Theta} = 1$ we deduce that

(a) $\sum_{\nu_1+\nu_2=\nu} \Theta_{\nu_1}\bar{\Theta}_{\nu_2} = \delta_{\nu,0}1 \otimes 1$.

Applying $(\Delta \otimes 1)$ to both sides of this identity gives

$$\sum_{\nu_1+\nu_2=\nu} (\Delta \otimes 1)\Theta_{\nu_1}(\Delta \otimes 1)\bar{\Theta}_{\nu_2} = \delta_{\nu,0}1 \otimes 1 \otimes 1.$$

We multiply both sides on the left by $\sum_{\nu'+\nu''=\nu_3} \bar{\Theta}^{13}_{\nu'}(1 \otimes \tilde{K}_{-\nu'} \otimes 1)\bar{\Theta}^{23}_{\nu''}$, we substitute $(\Delta \otimes 1)(\Theta_{\nu_1}) = \sum_{\nu_1'+\nu_1''=\nu_1} \Theta^{23}_{\nu_1'}(1 \otimes \tilde{K}_{-\nu_1''} \otimes 1)\Theta^{13}_{\nu_1''}$, (see 4.2.2) and sum over ν, ν_3 subject to $\nu + \nu_3 = \nu_4$. We obtain for any ν_4:

$$\sum_{\nu_1'+\nu_1''+\nu_2+\nu'+\nu''=\nu_4} \bar{\Theta}^{13}_{\nu'}(1 \otimes \tilde{K}_{-\nu'} \otimes 1)\bar{\Theta}^{23}_{\nu''}\Theta^{23}_{\nu_1'}(1 \otimes \tilde{K}_{-\nu_1''} \otimes 1)\Theta^{13}_{\nu_1''}(\Delta \otimes 1)\bar{\Theta}_{\nu_2}$$
$$= \sum_{\nu'+\nu''=\nu_4} \bar{\Theta}^{13}_{\nu'}(1 \otimes \tilde{K}_{-\nu'} \otimes 1)\bar{\Theta}^{23}_{\nu''}.$$

Using again (a) (twice), this becomes

$$(\Delta \otimes 1)\bar{\Theta}_{\nu_4} = \sum_{\nu'+\nu''=\nu_4} \bar{\Theta}^{13}_{\nu'}(1 \otimes \tilde{K}_{-\nu'} \otimes 1)\bar{\Theta}^{23}_{\nu''}.$$

We apply ${}^- \otimes {}^- \otimes {}^-$ to both sides and obtain

$$(\bar{\Delta} \otimes 1)\Theta_{\nu_4} = \sum_{\nu'+\nu''=\nu_4} \Theta^{13}_{\nu'}(1 \otimes \tilde{K}_{\nu'} \otimes 1)\Theta^{23}_{\nu''}.$$

The proposition is proved.

Proposition 4.2.4. *For any ν we have*

(a) $\sum_{\nu'+\nu''=\nu}(\Delta\otimes 1)(\Theta_{\nu'})\Theta^{12}_{\nu''} = \sum_{\nu'+\nu''=\nu}(1\otimes\Delta)(\Theta_{\nu'})\Theta^{23}_{\nu''}$.

Using the defining property of Θ (see 4.1.2(a)), we can write the left hand side of (a) as $\sum_{\nu'+\nu''=\nu}\Theta^{12}_{\nu''}(\bar{\Delta}\otimes 1)(\Theta_{\nu'})$. By 4.2.3, this equals $\sum_{\nu'+\nu''+\nu'''=\nu}\Theta^{12}_{\nu'''}\Theta^{13}_{\nu'}(1\otimes\tilde{K}_{\nu'}\otimes 1)\Theta^{23}_{\nu''}$. This is equal to the right hand side of (a), by 4.2.2(b).

4.2.5. Specializing the identities $\Delta(u)\Theta = \Theta\bar{\Delta}(u)$, we deduce

(a) $$(E_i\otimes 1)\Theta_\nu + (\tilde{K}_i\otimes E_i)\Theta_{\nu-i} = \Theta_\nu(E_i\otimes 1) + \Theta_{\nu-i}(\tilde{K}_{-i}\otimes E_i),$$

(b) $$(1\otimes F_i)\Theta_\nu + (F_i\otimes\tilde{K}_{-i})\Theta_{\nu-i} = \Theta_\nu(1\otimes F_i) + \Theta_{\nu-i}(F_i\otimes\tilde{K}_i)$$

(identities in $\mathbf{U}\otimes\mathbf{U}$, with the convention that $\Theta_{\nu-i}=0$ if $\nu_i=0$).

From this we deduce that, with the notation $\Theta_{\le p} = \sum_{\nu:\ \mathrm{tr}\ \nu\le p}\Theta_\nu$, we have for all $p\ge 0$

(c) $$\begin{aligned}&(E_i\otimes 1+\tilde{K}_i\otimes E_i)\Theta_{\le p} - \Theta_{\le p}(E_i\otimes 1+\tilde{K}_{-i}\otimes E_i)\\ &= \sum_{\nu:\ \mathrm{tr}\ \nu=p}(\tilde{K}_i\otimes E_i)\Theta_\nu - \sum_{\nu:\ \mathrm{tr}\ \nu=p}\Theta_\nu(\tilde{K}_{-i}\otimes E_i),\end{aligned}$$

(d) $$\begin{aligned}&(1\otimes F_i+F_i\otimes\tilde{K}_{-i})\Theta_{\le p} - \Theta_{\le p}(1\otimes F_i+F_i\otimes\tilde{K}_i)\\ &= \sum_{\nu:\ \mathrm{tr}\ \nu=p}(F_i\otimes K_{-i})\Theta_\nu - \sum_{\nu:\ \mathrm{tr}\ \nu=p}\Theta_\nu(F_i\otimes\tilde{K}_i).\end{aligned}$$

CHAPTER 5

The Symmetries $T'_{i,e}$, $T''_{i,e}$ of an Integrable U-Module

5.1. THE CATEGORY $\mathcal{C}'_i$

5.1.1. In this chapter we fix $i \in I$.

Let $\mathcal{C}'_i$ be the category whose objects are $\mathbf{Z}$-graded $\mathbf{Q}(v)$-vector spaces $M = \oplus_{n \in \mathbf{Z}} M^n$ provided with two locally nilpotent $\mathbf{Q}(v)$-linear maps $E_i, F_i : M \to M$ such that

(a) $E_i(M^n) \subset M^{n+2}$ and $F_i(M^n) \subset M^{n-2}$ for all n;

(b) $E_iF_i - F_iE_i : M^n \to M^n$ is multiplication by $[n]_i$ for all n;

the morphisms in the category are $\mathbf{Q}(v)$-linear maps preserving the $\mathbf{Z}$-grading and commuting with E_i, F_i.

This is the same as the category $\mathcal{C}'$ of integrable $\mathbf{U}$-modules in the case where $I = \{i\}$ and $X = Y = \mathbf{Z}$ with $i = 1 \in Y, i' = 2 \in X$. On the other hand, in the general case, any object M in $\mathcal{C}'$ may be regarded, for any i, as an object of $\mathcal{C}'_i$ with the $\mathbf{Z}$-grading $M^n = \oplus_{\lambda \in X; \langle i, \lambda \rangle = n} M^\lambda$. (We forget the action of E_j, F_j for $j \neq i$.)

For $M \in \mathcal{C}'_i$ and $p \in \mathbf{Z}$, the operators $E_i^{(p)}, F_i^{(p)} : M \to M$ are given by $E_i^p/[p]_i^!, F_i^p/[p]_i^!$ if $p \geq 0$, and by 0 if $p < 0$.

5.1.2. For $M \in \mathcal{C}'_i$, let $c : M \to M$ be the $\mathbf{Q}(v)$-linear map given by

$$c(x) = E_iF_i(x) + \frac{v_i^{n-1} + v_i^{-n+1} - 2}{(v_i - v_i^{-1})^2} x = F_iE_i(x) + \frac{v_i^{n+1} + v_i^{-n-1} - 2}{(v_i - v_i^{-1})^2} x$$

for $x \in M^n$. It is easy to check that c is a morphism in $\mathcal{C}'_i$.

For $n \in \mathbf{Z}$, we set $s_n = \frac{v_i^{n+1} + v_i^{-n-1} - 2}{(v_i - v_i^{-1})^2}$ and $s'_n = \frac{v_i^{n-1} + v_i^{-n+1} - 2}{(v_i - v_i^{-1})^2}$.

Proposition 5.1.3. *Let $n \in \mathbf{Z}$ and $m \in \mathbf{N}$.*

(a) *The subspace $M^n(m) = \{x \in M^n | E_i^{(m)} x = 0\}$ is c-stable and c satisfies the identity $(c - s_n)(c - s_{n+2}) \cdots (c - s_{n+2m}) = 0$ on $M^n(m)$.*

(b) *The subspace $M^n[m] = \{x \in M^n | F_i^{(m)} x = 0\}$ is c-stable and c satisfies the identity $(c - s'_n)(c - s'_{n-2}) \cdots (c - s'_{n-2m}) = 0$ on $M^n[m]$.*

G. Lusztig, *Introduction to Quantum Groups*, Modern Birkhäuser Classics,
DOI 10.1007/978-0-8176-4717-9_5, © Springer Science+Business Media, LLC 2010

(c) *$c: M^n \to M^n$ is a locally finite, semisimple linear map.*

(d) *If $n \geq 0$, the eigenvalues of $c: M^n \to M^n$ are contained in $\{s_n, s_{n+2}, s_{n+4}, \cdots\}$; if $n \leq 0$, the eigenvalues of $c: M^n \to M^n$ are contained in $\{s_{-n}, s_{-n+2}, s_{-n+4}, \ldots\}$.*

Since c commutes with E_i, it leaves $M^n(m)$ stable. We prove (a) by induction on m. Assume first that $m = 0$. If $x \in M^n(0)$, then by definition, $(c - s_n)x = 0$. Assume now that $m > 0$ and that the result is already known for $m - 1$. If $x \in M^n(m)$, then $E_i x \in M^{n+2}(m-1)$. By the induction hypothesis, we have $(c - s_{n+2})(c - s_{n+4}) \cdots (c - s_{n+2m})E_i x = 0$. Since $cE_i = E_i c$, it follows that $E_i(c - s_{n+2})(c - s_{n+4}) \cdots (c - s_{n+2m})x = 0$. Applying F_i, we get $F_i E_i (c - s_{n+2})(c - s_{n+4}) \cdots (c - s_{n+2m})x = 0$. Hence, by the definition of c, we have $(c - s_n)(c - s_{n+2})(c - s_{n+4}) \cdots (c - s_{n+2m})x = 0$. This proves (a). The proof of (b) is entirely similar.

We prove (c). Assume first that $n \geq 0$. Since $s_0, s_1, s_2, \ldots$ are distinct elements of $\mathbf{Q}(v)$, we see from (a) that $c: M^n(m) \to M^n(m)$ is locally finite, semisimple (note that $M^n = \cup_{m \geq 0} M^n(m)$).

Assume next that $n \leq 0$. Since $s'_0, s'_{-1}, s'_{-2}, \ldots$ are distinct elements of $\mathbf{Q}(v)$, we see from (b) that $c: M^n[m] \to M^n[m]$ is locally finite, semisimple. It follows that $c: M^n \to M^n$ is locally finite, semisimple (note that $M^n = \cup_{m \geq 0} M^n[m]$). This proves (c). Now (d) follows from the earlier points and the identity $s'_n = s_{-n}$. The proposition is proved.

5.1.4. By taking the eigenspaces of $c: M \to M$ we obtain a canonical direct sum decomposition of M (as an object in $\mathcal{C}'_i$) into subobjects with the property that c acts on each subobject as scalar multiplication by s_m for some $m \in \mathbf{N}$.

Proposition 5.1.5. *Let $M \in \mathcal{C}'_i$ be such that $c = s_m$ on M for some $m \in \mathbf{N}$.*

(a) *If $M^n \neq 0$, then $n \in \{-m, -m+2, -m+4, \ldots, m\}$.*

(b) *If n and $n+2$ belong to $\{-m, -m+2, -m+4, \ldots, m\}$, then both $F_i E_i : M^n \to M^n$ and $E_i F_i : M^{n+2} \to M^{n+2}$ are given by multiplication by $[1 + m/2 + n/2]_i [m/2 - n/2]_i \neq 0$. Hence $E_i : M^n \to M^{n+2}$ and $F_i : M^{n+2} \to M^n$ are isomorphisms.*

(c) *Let $x \in M^n$ where $n \in \{-m, -m+2, -m+4, \ldots, m\}$. There are unique elements $z \in M^{-m}$ and $z' \in M^m$ such that $x = F_i^{(m/2-n/2)} z' = E_i^{(m/2+n/2)} z$.*

(d) *In the setup of (c), we have $F_i^{(1+m/2-n/2)} z' = E_i^{(-1+m/2+n/2)} z$ and $F_i^{(-1+m/2-n/2)} z' = E_i^{(1+m/2+n/2)} z$.*

(d) *In the setup of* (c), *we have* $F_i^{(1+m/2-n/2)}z' = E_i^{(-1+m/2+n/2)}z$ *and* $F_i^{(-1+m/2-n/2)}z' = E_i^{(1+m/2+n/2)}z$.

(a) follows immediately from 5.1.3(d). If $x \in M^n$, we have $F_iE_i(x) = (c - s_n)x = (s_m - s_n)x$; if $y \in M^{n+2}$, we have $E_iF_i(y) = (c - s'_{n+2})(y) = (s_m - s'_{n+2})y = (s_m - s_{-n-2})y = (s_m - s_n)y$. We have $s_m - s_n = [1+m/2+n/2]_i[m/2 - n/2]_i$ and (b) follows.

Now (c),(d) clearly follow from (b).

Lemma 5.1.6. *Let* $M \in \mathcal{C}'_i$ *and let* $y \in M^n - \{0\}$ *be such that* $E_iy = 0$. *Then*

(a) $n \geq 0$ *and* $F_i^{n+1}y = 0$ *and*

(b) $y \notin E_i^{n+1}M$.

(a) is already contained in the proof of 3.5.8. We prove (b). By 5.1.4, we may assume that M is as in 5.1.5. Assume that $y = E_i^{n+1}y'$ for some $y' \in M$. We may assume that $y' \in M^{-n-2} - \{0\}$. By 5.1.5(b), we then have that $F_i^{n+1}E_i^{n+1}y'$ is a non-zero multiple of y'; in particular, $F_i^{n+1}y \neq 0$, a contradiction.

5.2. First Properties of $T'_{i,e}, T''_{i,e}$

5.2.1. We fix $e = \pm 1$. Let $M \in \mathcal{C}'_i$. We define two $\mathbf{Q}(v)$-linear maps $T'_{i,e}, T''_{i,e} : M \to M$ (called *symmetries*) by

$$T'_{i,e}(z) = \sum_{a,b,c;a-b+c=n} (-1)^b v_i^{e(-ac+b)} F_i^{(a)} E_i^{(b)} F_i^{(c)} z,$$

$$T''_{i,e}(z) = \sum_{a,b,c;-a+b-c=n} (-1)^b v_i^{e(-ac+b)} E_i^{(a)} F_i^{(b)} E_i^{(c)} z$$

for $z \in M^n$; the integers a, b, c are restricted as shown; although the sums are infinite, for any given z, all but finitely many terms of either sum are zero.

Proposition 5.2.2. *Let* $m \geq 0$ *and let* $j, h \in [0, m]$ *be such that* $j + h = m$.
(a) *If* $\eta \in M^m$ *is such that* $E_i\eta = 0$, *then* $T'_{i,e}(F_i^{(j)}\eta) = (-1)^j v_i^{e(jh+j)} F_i^{(h)}\eta$.
(b) *If* $\xi \in M^{-m}$ *is such that* $F_i\xi = 0$, *then* $T''_{i,e}(E_i^{(j)}\xi) = (-1)^j v_i^{e(jh+j)} E_i^{(h)}\xi$.

We prove (a). Assume that $a-b+c=m-2j$. Using 3.4.2, we have

$$F_i^{(a)}E_i^{(b)}F_i^{(c)}F_i^{(j)}\eta = \begin{bmatrix} c+j \\ c \end{bmatrix}_i F_i^{(a)}E_i^{(b)}F_i^{(c+j)}\eta$$
$$= \sum_{t\geq 0} \begin{bmatrix} c+j \\ j \end{bmatrix}_i \begin{bmatrix} b-c+h \\ t \end{bmatrix}_i F_i^{(a)}F_i^{(c+j-t)}E_i^{(b-t)}\eta.$$

Now $E_i^{(b-t)}\eta = 0$ if $b \neq t$; thus

$$F_i^{(a)}E_i^{(b)}F_i^{(c)}F_i^{(j)}\eta = \begin{bmatrix} c+j \\ c \end{bmatrix}_i \begin{bmatrix} b-c+h \\ b \end{bmatrix}_i F_i^{(a)}F_i^{(c+j-b)}\eta$$
$$= \begin{bmatrix} c+j \\ j \end{bmatrix}_i \begin{bmatrix} b-c+h \\ b \end{bmatrix}_i \begin{bmatrix} h \\ a \end{bmatrix}_i F_i^{(h)}\eta.$$

It remains to show that

$$\sum_{a,b,c\geq 0; a-b+c=h-j} (-1)^b v_i^{e(-ac+b)} \begin{bmatrix} c+j \\ c \end{bmatrix}_i \begin{bmatrix} a+j \\ b \end{bmatrix}_i \begin{bmatrix} h \\ a \end{bmatrix}_i = (-1)^j v_i^{e(j+1)h}$$

for any $j,h \geq 0$.

The term corresponding to a,b,c is zero unless $a \leq h$, $b \leq a+j$ and $c \leq h$, hence the sum is finite. We replace $\left[\begin{smallmatrix} a+j \\ b \end{smallmatrix}\right]_i = \left[\begin{smallmatrix} a+j \\ h-c \end{smallmatrix}\right]_i$ and $\left[\begin{smallmatrix} c+j \\ c \end{smallmatrix}\right]_i$ by $(-1)^c\left[\begin{smallmatrix} -j-1 \\ c \end{smallmatrix}\right]_i$. The sum becomes

$$\sum_{a\geq 0}(-1)^{h-j-a}v_i^{e(a+hj+j)}\begin{bmatrix} h \\ a \end{bmatrix}_i \sum_{c=0}^{h} v_i^{e(-ac+c-hj-h)}\begin{bmatrix} -j-1 \\ c \end{bmatrix}_i \begin{bmatrix} a+j \\ h-c \end{bmatrix}_i.$$

We replace the sum over c by $\left[\begin{smallmatrix} a-1 \\ h \end{smallmatrix}\right]_i$ and we obtain

$$\sum_{a\geq 0}(-1)^{h-j-a}v_i^{e(a+hj+j)}\begin{bmatrix} h \\ a \end{bmatrix}_i \begin{bmatrix} a-1 \\ h \end{bmatrix}_i.$$

If $a>h$, we have $\left[\begin{smallmatrix} h \\ a \end{smallmatrix}\right]_i = 0$; if $1 \leq a \leq h$, we have $\left[\begin{smallmatrix} a-1 \\ h \end{smallmatrix}\right]_i = 0$; thus, if $a>0$, we have $\left[\begin{smallmatrix} h \\ a \end{smallmatrix}\right]_i \left[\begin{smallmatrix} a-1 \\ h \end{smallmatrix}\right]_i = 0$. Hence only $a=0$ contributes to the sum, which becomes

$$(-1)^{h-j}v_i^{e(hj+j)}\begin{bmatrix} -1 \\ h \end{bmatrix}_i = (-1)^j v_i^{e(hj+j)},$$

as required. This proves (a). The proof of (b) is entirely similar (it can also be reduced to (a)).

Proposition 5.2.3. *We have*

(a) $T'_{i,e}T''_{i,-e} = T''_{i,-e}T'_{i,e} = 1 : M \to M$.

(b) $T''_{i,e}x = (-1)^t v_i^{et} T'_{i,e}x$ for all $x \in M^t$.

Let m, j, h, η be as in 5.2.2(a). Let $\xi = F^{(m)}\eta$. We have, using 5.2.2(a),(b)

$$\begin{aligned} T''_{i,-e}T'_{i,e}(F_i^{(j)}\eta) &= T''_{i,-e}((-1)^j v_i^{e(jh+j)} F_i^{(h)}\eta) \\ &= (-1)^j v_i^{e(jh+j)} T''_{i,-e}(E_i^{(j)}\xi) \\ &= (-1)^j v_i^{e(jh+j)} (-1)^j v_i^{-e(jh+j)} E_i^{(h)}\xi = F_i^{(j)}\eta. \end{aligned}$$

Since the vector space M is generated by vectors of the form $F_i^{(j)}\eta$ as above, we have $T''_{i,-e}T'_{i,e} = 1$. The identity $T'_{i,e}T''_{i,-e} = 1$ is proved in a similar way (or it can be deduced from the previous identity). This proves (a).

To prove (b), we may assume that $x = F_i^{(j)}\eta = E_i^{(h)}\xi$. We have $T''_{i,e}x = (-1)^h v_i^{e(hj+h)} E_i^{(j)}\xi$ and $T'_{i,e}x = (-1)^j v_i^{e(jh+j)} F_i^{(h)}\eta = (-1)^j v_i^{e(jh+j)} E_i^{(j)}\xi$. It remains to note that $h = j + t$.

Proposition 5.2.4. *For any $z \in M^t$ we have*

(a) $-v_i^{e(t-2)} T''_{i,-e}(F_i z) = E_i T''_{i,-e}(z)$;

(b) $-v_i^{-et} T''_{i,-e}(E_i z) = F_i T''_{i,-e}(z)$;

(c) $-v_i^{-et} T'_{i,e}(F_i z) = E_i T'_{i,e}(z)$;

(d) $-v_i^{e(t+2)} T'_{i,e}(E_i z) = F_i T'_{i,e}(z)$;

(e) $T''_{i,-e} z \in M^{-t}$;

(f) $T'_{i,e} z \in M^{-t}$.

(e),(f) are clear from the definition. To prove (a),(b), we may assume that $z = F_i^{(j)}\eta = E_i^{(h)}\xi$ where m, j, h, η are as in 5.2.2(a) and $\xi = F^{(m)}\eta$; then $h = j + t$. We have

$$T''_{i,-e}(z) = (-1)^h v_i^{-e(jh+h)} E_i^{(j)}\xi = (-1)^h v_i^{-e(jh+h)} F_i^{(h)}\eta.$$

If $j = m$, then both sides of (a) are zero. So it suffices to prove (a) assuming $j < m$ and $h > 0$. We have

$$\begin{aligned} T''_{i,-e}(F_i z) &= [j+1]_i T''_{i,-e}(F_i^{(j+1)}\eta) = [j+1]_i T''_{i,-e}(E_i^{(h-1)}\xi) \\ &= [j+1]_i (-1)^{h-1} v_i^{-e(h-1)(j+2)} E_i^{(j+1)}\xi, \end{aligned}$$

and $E_iT''_{i,-e}(z) = (-1)^h v_i^{-e(jh+h)}[j+1]_i E_i^{(j+1)}\xi$; (a) follows. If $h = m$, then both sides of (b) are zero. So it suffices to prove (b) assuming $h < m$ and $j > 0$. We have

$$T''_{i,-e}(E_iz) = [h+1]_iT''_{i,-e}(E_i^{(h+1)}\xi) = (-1)^{h+1}[h+1]_iv_i^{-e(h+1)j}E_i^{(j-1)}\xi,$$

and $F_i(T''_{i,-e}z) = (-1)^h v_i^{-e(jh+h)}[h+1]_iF_i^{(h+1)}\eta$; (b) follows. The proof of (c),(d) is entirely similar; alternatively, we can deduce (c),(d) from (a),(b) using 5.2.3.

5.2.5. If M is an object of $\mathcal{C}'$ (an integrable **U**-module), then we can regard it as an object of $\mathcal{C}'_i$ as in 5.1.1; hence the symmetry operators $T'_{i,e}, T''_{i,e} : M \to M$ are well-defined. All the previous results will hold for these operators.

Lemma 5.2.6. *Let M be an integrable* **U**-*module and let $z \in M$. Let $\mu \in Y$ and let $\mu' = \mu - \langle \mu, i' \rangle i \in Y$. Then*

(a) $T''_{i,-e}(K_{\mu'}z) = K_\mu T''_{i,-e}(z)$;

(b) $T'_{i,e}(K_{\mu'}z) = K_\mu T'_{i,e}(z)$.

We may assume that $z \in M^\lambda$. Then, from the definition, $T''_{i,-e}z$ and $T'_{i,e}z$ belong to $M^{\lambda - \langle i,\lambda\rangle i'}$. Hence K_μ acts on either of these two vectors as multiplication by $\langle \mu, \lambda\rangle - \langle i, \lambda\rangle\langle \mu, i'\rangle = \langle \mu - \langle \mu, i'\rangle i, \lambda\rangle = \langle \mu', \lambda\rangle$. The lemma follows.

Proposition 5.2.7. *Let M be as in the previous lemma. For any $\lambda \in X$, the operator $T''_{i,-e} : M \to M$ defines an isomorphism of the λ-weight space of M onto the $s_i(\lambda)$-weight space of M. The inverse of this isomorphism is the restriction of $T'_{i,e}$.*

This follows immediately from the previous lemma.

5.3. The Operators L'_i, L''_i

5.3.1. Let M, N be two objects of $\mathcal{C}'_i$. By 3.4.3 and 3.5.2(a), the tensor product $M \otimes N$ is again an object of $\mathcal{C}'_i$; if $x \in M^t, y \in N^s$, the degree of $x \otimes y$ is $t + s$ and

$$E_i(x \otimes y) = E_ix \otimes y + v_i^t x \otimes E_iy, \quad F_i(x \otimes y) = x \otimes F_iy + v_i^{-s}F_ix \otimes y.$$

We define linear maps $L''_i, L'_i : M \otimes N \to M \otimes N$ by

$L''_i(x \otimes y) = \sum_n (-1)^n v_i^{-n(n-1)/2}\{n\}_iF_i^{(n)}x \otimes E_i^{(n)}y$,

$L'_i(x \otimes y) = \sum_n v_i^{n(n-1)/2}\{n\}_i F_i^{(n)} x \otimes E_i^{(n)} y,$

where $\{n\}_i$ is as in 4.1.4. Although the sums are infinite, any vector in $M \otimes N$ is annihilated by all but finitely many terms in the sum; hence the operators are well defined. These maps are a special case of the operators defined by $\Theta, \bar{\Theta}$ of 4.1, in the case $I = \{i\}$. (See 4.1.4.)

Lemma 5.3.2. *We have $L'_i L''_i = L''_i L'_i = 1$.*

This follows immediately from the identity 1.3.4(a); alternatively, it can be deduced from 4.1.3.

Lemma 5.3.3. *For $x \in M^t, y \in N^s$, we have*

$$F_i L''_i(x \otimes y) = L''_i(x \otimes F_i y + v_i^s F_i x \otimes y).$$

This can be deduced from the defining identity for Θ (see 4.1.2) or it can be checked directly.

Proposition 5.3.4. *Let M, N be objects of $\mathcal{C}'_i$. For any $z \in M \otimes N$, we have*

(a) $T''_{i,1}(L''_i(z)) = (T''_{i,1} \otimes T''_{i,1})(z)$.

(In the left hand side of (a) we have the action of $T''_{i,1}$ on $M \otimes N$; in the right hand side we have the action of $T''_{i,1}$ on M and on N.)

Assume that (a) holds for $z = x \otimes y$ where $x \in M^m, y \in N^n$; we claim that (a) must also hold for $z' = x \otimes F_i y + v_i^s F_i x \otimes y$. Using 5.3.3, 5.2.4(a), and our hypothesis, we have

$$\begin{aligned} T''_{i,1}(L''_i(z')) &= T''_{i,1}(F_i L''_i(z)) \\ &= -v_i^{s+t-2} E_i T''_{i,1}(L''_i(z)) \\ &= -v_i^{s+t-2} E_i (T''_{i,1} \otimes T''_{i,1})(z). \end{aligned}$$

On the other hand, again using 5.2.4(a), we have

$$\begin{aligned} (T''_{i,1} \otimes T''_{i,1})(z') &= T''_{i,1} x \otimes T''_{i,1}(F_i y) + v_i^s T''_{i,1}(F_i x) \otimes T''_{i,1} y \\ &= -v_i^{s-2} T''_{i,1} x \otimes E_i T''_{i,1} y - v_i^s v_i^{t-2} E_i T''_{i,1} x \otimes T''_{i,1} y \\ &= -v_i^{s+t-2} E_i (T''_{i,1} \otimes T''_{i,1})(z). \end{aligned}$$

Our claim is proved.

For $q \geq 0$, let Z_q be the subspace of $M \otimes N$ spanned by vectors of the form $x \otimes F^{(q)} y$ where x, y are homogeneous and $E_i y = 0$. It is clear that

$M \otimes N = \sum_q Z_q$. Hence to prove (a) it suffices to prove the statements (b),(c) below.

(b) The identity (a) holds for any $z \in Z_0$.

(c) If (a) holds for all $z \in Z_q$, then it holds for all $z \in Z_{q+1}$.

We prove (c). Let $x \in M^t, y \in M^s$ be such that $E_i y = 0$. By the assumption of (c), we have that (a) holds for $z = x \otimes F^{(q)} y$. By the earlier part of the proof, it follows that (a) holds for $z_1 = [q+1]_i x \otimes F_i^{(q+1)} y + v_i^{s-2q} F_i x \otimes F_i^{(q)} y$. Again by the assumption of (c), we have that (a) holds for $z_2 = v_i^{s-2q} F_i x \otimes F_i^{(q)} y$. It follows that (a) holds for $z_3 = [q+1]_i x \otimes F_i^{(q+1)} y$; since $[q+1]_i \neq 0$, it also holds for $x \otimes F_i^{(q+1)} y$. Thus (c) is proved.

It remains to prove (b). Let $x \in M^m, y \in N^n$ be such that $E_i y = 0$. From the definition, we have $L''_i(x \otimes y) = x \otimes y$. We must prove that $T''_{i,1}(x \otimes y) = T''_{i,1}(x) \otimes T''_{i,1}(y)$. We may assume that $n \geq 0$. We have

$$T''_{i,1}(x \otimes y) = \sum_{a,b,c;-a+b-c=m+n} (-1)^b v_i^{-ac+b} E_i^{(a)} F_i^{(b)} (E_i^{(c)} x \otimes y)$$

$$= \sum_{a',a'',b',b'',c;-a'-a''+b'+b''-c=m+n} (-1)^{b'+b''} v_i^{-a'c-a''c+b'+b''+b'b''-b'n+a'a''+a''(m+2c-2b')}$$
$$\times E_i^{(a')} F_i^{(b')} E_i^{(c)} x \otimes E_i^{(a'')} F_i^{(b'')} y.$$

Here we substitute $E_i^{(a'')} F_i^{(b'')} y = \left[{a''-b''+n \atop a''}\right]_i F_i^{(b''-a'')} y$ and we set $b'' = a'' + g$. We have $F_i^{(g)} y = 0$ unless $g \leq n$. We obtain

$$\sum_{a',a'',b',g,c;-a'+b'-c=m+n-g;g\leq n} (-1)^{b'+a''+g} v_i^{-a'c-a''c+b'+a''+g-b'a''+b'g-b'n+a'a''+a''(m+2c)}$$
$$\times \begin{bmatrix} n-g \\ a'' \end{bmatrix}_i E_i^{(a')} F_i^{(b')} E_i^{(c)} x \otimes F_i^{(g)} y$$
$$= \sum_{a',b',g,c;-a'+b'-c=m+n-g;g\leq n} (-1)^{b'+g} v_i^{-a'c+b'+g+b'g-b'n} \sum_{a''} (-1)^{a''} v_i^{a''(1+g-n)} \begin{bmatrix} n-g \\ a'' \end{bmatrix}_i$$
$$\times E_i^{(a')} F_i^{(b')} E_i^{(c)} x \otimes F_i^{(g)} y$$
$$= \sum_{a',b',c;-a'+b'-c=m} (-1)^{b'} v_i^{-a'c+b'} E_i^{(a')} F_i^{(b')} E_i^{(c)} x \otimes (-1)^n v_i^n F_i^{(n)} y$$
$$= T''_{i,1}(x) \otimes (-1)^n v_i^n F_i^{(n)} y = T''_{i,1}(x) \otimes T''_{i,1}(y).$$

The proposition is proved.

CHAPTER 6

Complete Reducibility Theorems

6.1. The Quantum Casimir Operator

6.1.1. In this chapter we assume that the root datum is both Y-regular and X-regular.

Let B, B_ν be as in 4.1.2. Applying $m(S \otimes 1)$ to the identities at the end of 4.2.5, where $m : \mathbf{U} \otimes \mathbf{U} \to \mathbf{U}$ is multiplication, we obtain for any $p \geq 0$:

$$\sum_{\nu:\ \mathrm{tr}\ \nu \leq p} \sum_{b \in B_\nu} (-1)^{\mathrm{tr}\ \nu} v_\nu (S(E_i b^-) b^{*+} + S(\tilde{K}_i b^-) E_i b^{*+} - S(b^- E_i) b^{*+} - S(b^- \tilde{K}_{-i}) b^{*+} E_i) = \sum_{\nu:\ \mathrm{tr}\ \nu = p} \sum_{b \in B_\nu} (-1)^p v_\nu (S(\tilde{K}_i b^-) E_i b^{*+} - S(b^- \tilde{K}_{-i}) b^{*+} E_i),$$

and

$$\sum_{\nu:\ \mathrm{tr}\ \nu \leq p} \sum_{b \in B_\nu} (-1)^{\mathrm{tr}\ \nu} v_\nu (S(b^-) F_i b^{*+} + S(F_i b^-) \tilde{K}_{-i} b^{*+} - S(b^-) b^{*+} F_i - S(b^- F_i) b^{*+} \tilde{K}_i) = \sum_{\nu:\ \mathrm{tr}\ \nu = p} \sum_{b \in B_\nu} (-1)^p v_\nu (S(F_i b^-) \tilde{K}_{-i} b^{*+} - S(b^- F_i) b^{*+} \tilde{K}_i);$$

equivalently, setting $\Omega_{\leq p} = \sum_{\nu:\ \mathrm{tr}\ \nu \leq p} \sum_{b \in B_\nu} (-1)^{\mathrm{tr}\ \nu} v_\nu S(b^-) b^{*+}$, we have

$$\tilde{K}_{-i} E_i \Omega_{\leq p} - \tilde{K}_i \Omega_{\leq p} E_i = \sum_{\nu:\ \mathrm{tr}\ \nu = p} \sum_{b \in B_\nu} (-1)^p v_\nu (S(\tilde{K}_{-i} b^-) E_i b^{*+} - S(b^- \tilde{K}_{-i}) b^{*+} E_i),$$

and

$$-\Omega_{\leq p} F_i + F_i \tilde{K}_i \Omega_{\leq p} \tilde{K}_i = \sum_{\nu:\ \mathrm{tr}\ \nu = p} \sum_{b \in B_\nu} (-1)^p v_\nu (S(F_i b^-) \tilde{K}_{-i} b^{*+} - S(b^- F_i) b^{*+} \tilde{K}_i).$$

G. Lusztig, *Introduction to Quantum Groups*, Modern Birkhäuser Classics,
DOI 10.1007/978-0-8176-4717-9_6, © Springer Science+Business Media, LLC 2010

6.1.2. If $M \in \mathcal{C}^{hi}$, then for any $m \in M$, we have that $\Omega(m) = \Omega_{\leq p}(m) \in M$ is independent of p for large enough p. We can write $\Omega(m) = \sum_b (-1)^{\text{tr }|b|} v_{|b|} S(b^-) b^{*+} m$ and we have

$$\text{(a)} \qquad \tilde{K}_{-i} E_i \Omega = \tilde{K}_i \Omega E_i, \quad \Omega F_i = F_i \tilde{K}_i \Omega \tilde{K}_i, \quad \Omega K_\mu = K_\mu \Omega$$

as operators on M. It follows that for $m \in M^\lambda$, we have $\Omega E_i(m) = v_i^{-2\langle i, \lambda + i' \rangle} E_i \Omega(m)$ and $\Omega F_i(m) = v_i^{2\langle i, \lambda \rangle} F_i \Omega(m)$.

6.1.3. Remark. Let us define an isomorphism of $\mathbf{Q}(v)$-algebras $\bar{S} : \mathbf{U} \to \mathbf{U}^{opp}$ by $\bar{S}(\bar{u}) = \overline{S(u)}$ (S is the antipode.) For any $u \in \mathbf{U}$, we have $S(u)\Omega = \Omega \bar{S}(u) : M \to M$. Indeed, it suffices to check this for the generators E_i, F_i, K_μ where it follows from the formulas above.

6.1.4. Let C be a fixed coset of X with respect to the subgroup $\mathbf{Z}[I] \subset X$. Let $G : C \to \mathbf{Z}$ be a function such that

(a) $G(\lambda) - G(\lambda - i') = i \cdot i \langle i, \lambda \rangle$

for all $\lambda \in C$ and all $i \in I$. Clearly, such a function exists and is unique up to addition of an arbitrary constant function $C \to \mathbf{Z}$.

Lemma 6.1.5. *Let $\lambda, \lambda' \in C \cap X^+$. Assume that $\lambda \geq \lambda'$ and $G(\lambda) = G(\lambda')$. Then $\lambda = \lambda'$.*

We can write $\lambda' = \lambda - i_1' - i_2' - \cdots - i_n'$ for some sequence $i_1, i_2, \ldots, i_n$ in I. Using 6.1.4(a) repeatedly, we see that

$$G(\lambda) - G(\lambda - i_1' - i_2' - \cdots - i_n') = \sum_{p=1}^{n} i_p \cdot i_p \langle i_p, \lambda \rangle - \sum_{1 \leq p < q \leq n} i_p \cdot i_q.$$

Using our assumption, we have therefore that

$$\text{(a)} \qquad \sum_{p=1}^{n} i_p \cdot i_p \langle i_p, \lambda \rangle = \sum_{1 \leq p < q \leq n} i_p \cdot i_q.$$

Since $\lambda \in X^+$, we have $\langle i, \lambda \rangle \in \mathbf{N}$ for all i, hence

$$\text{(b)} \qquad \sum_{p=1}^{n} i_p \cdot i_p \langle i_p, \lambda \rangle \geq 0.$$

Similarly, since $\lambda' \in X^+$, we have

$$\sum_{p=1}^{n} i_p \cdot i_p \langle i_p, \lambda' \rangle \geq 0,$$

or equivalently,

(c) $$\sum_{p=1}^{n} i_p \cdot i_p \langle i_p, \lambda - i'_1 - i'_2 - \cdots - i'_n \rangle \geq 0.$$

Adding (b),(c) term by term gives

$$2\sum_{p=1}^{n} i_p \cdot i_p \langle i_p, \lambda \rangle - 2 \sum_{1 \leq p < q \leq n} i_p \cdot i_q - 2\sum_{p=1}^{n} i_p \cdot i_p \geq 0.$$

Introducing here the equality (a), we obtain $-2\sum_{p=1}^{n} i_p \cdot i_p \geq 0$. Since $i_p \cdot i_p > 0$ for all p, it follows that $n = 0$; hence $\lambda = \lambda'$ as required.

6.1.6. Let $M \in \mathcal{C}$. For each $\mathbf{Z}[I]$-coset C in X we define $M_C = \oplus_{\lambda \in C} M^\lambda$. It is clear that $M = \oplus_C M_C$ as a vector space and that each M_C is a $\mathbf{U}$-submodule. Hence $M = \oplus_C M_C$ as an object in $\mathcal{C}$.

Proposition 6.1.7. *Let $M \in \mathcal{C}^{hi}$.*

(a) *Assume that there exists C as above such that $M = M_C$. Let $G : C \to \mathbf{Z}$ be as in 6.1.4. We define a linear map $\Xi : M \to M$ by $\Xi(m) = v^{G(\lambda)} m$ for all $\lambda \in C$ and all $m \in M^\lambda$. Then the operator $\Omega\Xi : M \to M$ is in the commutant of the $\mathbf{U}$-module M. Moreover, the $\mathbf{Q}(v)$-linear map $\Omega\Xi : M \to M$ is locally finite.*

(b) *Assume that M is a quotient of the Verma module $M_{\lambda'}$. Then $\Omega\Xi : M \to M$ is equal to $v^{G(\lambda')}$ times identity.*

(c) *Let M be as in* (a). *Then the eigenvalues of $\Omega\Xi : M \to M$ are of the form v^c for various integers c.*

We prove (a). We have for λ, m as above:

$$\begin{aligned}\Omega\Xi E_i(m) &= v^{G(\lambda+i')}\Omega E_i(m) = v^{G(\lambda+i')-i\cdot i\langle i,\lambda+i'\rangle} E_i\Omega(m) \\ &= v^{G(\lambda+i')-G(\lambda)-i\cdot i\langle i,\lambda+i'\rangle} E_i\Omega\Xi(m) = E_i\Omega\Xi(m)\end{aligned}$$

and

$$\begin{aligned}\Omega\Xi F_i(m) &= v^{G(\lambda-i')}\Omega F_i(m) = v^{G(\lambda-i')+i\cdot i\langle i,\lambda\rangle} F_i\Omega(m) \\ &= v^{G(\lambda-i')-G(\lambda)+i\cdot i\langle i,\lambda\rangle} F_i\Omega\Xi(m) = F_i\Omega\Xi(m).\end{aligned}$$

Moreover, $\Omega\Xi$ maps each weight space of M into itself. This proves the first assertion of (a). To prove the second assertion, it suffices to show that the restriction of $\Omega\Xi$ to any weight space is locally finite. Let $m \in M^\lambda$. Let M' be the $\mathbf{U}^+$-submodule of M generated by m. Let M'' be the $\mathbf{U}$-submodule of M generated by m. We have $M'' = \mathbf{U}^- M'$. We have $\dim M' < \infty$ since $M \in \mathcal{C}^{hi}$. It follows easily that all weight spaces of M'' are finite dimensional. In particular, the λ-weight space of M'' is finite dimensional. This weight space is stable under $\Omega\Xi$ and it contains m. Thus, $\Omega\Xi : M \to M$ is locally finite.

We prove (b). From the definition, $\Omega\Xi$ acts on the λ'-weight space of M as multiplication by $v^{G(\lambda')}$ times identity. Since this weight space generates M as a $\mathbf{U}$-module, we see that (b) follows from (a). (Note that (a) is applicable to M.)

We prove (c). Let $\tilde{M}$ be the sum of the generalized eigenspaces of $\Omega\Xi : M \to M$ corresponding to eigenvalues of form v^c for various integers c. We must show that $\tilde{M} = M$. By the argument in the proof of (a), we may assume that, for any $\lambda \in \mathcal{C}$, we have $d_M(\lambda) = \sum_{\lambda' \geq \lambda} \dim M^{\lambda'} < \infty$. We will prove that, for any $\lambda \in \mathcal{C}$, we have $M^\lambda \subset \tilde{M}$, by induction on $d = d_M(\lambda)$. If $d = 0$, there is nothing to prove. Assume now that $d \geq 1$. Let $\lambda_1 \in \mathcal{C}$ be maximal such that $\lambda_1 \geq \lambda$ and $M^{\lambda_1} \neq 0$. Let m_1 be a non-zero vector in M^{λ_1}. Let M_1 be the $\mathbf{U}$-submodule of M generated by m_1. Clearly, $d_{M/M_1}(\lambda) < d_M(\lambda)$. Hence, by the induction hypothesis, we have $(M/M_1)^\lambda \subset (M/M_1)\tilde{}$. On the other hand, by (b), we have $M_1 \subset \tilde{M}$. It follows that $M^\lambda \subset \tilde{M}$. The proposition is proved.

The operator $\Omega\Xi : M \to M$ is called the *quantum Casimir operator.*

6.2. Complete Reducibility in $\mathcal{C}^{hi} \cap \mathcal{C}'$

Lemma 6.2.1. *Let $M \in \mathcal{C}$. Assume that M is a non-zero quotient of the Verma module M_λ and that M is integrable. Then*
(a) $\lambda \in X^+$ *and*
(b) M *is simple.*

(a) follows from 3.5.8 applied to a non-zero vector $m \in M^\lambda$.

We prove (b). Assume that M' is a subobject of M distinct from M and 0. Then clearly, $M'^\lambda = 0$. We can find $\lambda' \in X$ maximal with the property that $M'^{\lambda'} \neq 0$. Then $\lambda' < \lambda$. Let m' be a non-zero vector in $M'^{\lambda'}$. By the maximality of λ', we have $E_i m' = 0$ for all i. By 3.4.6, there exists a morphism of $\mathbf{U}$-modules from the Verma module $M_{\lambda'}$ into M' whose image contains m'. Let M'' be the image of this homomorphism. Clearly M'' is integrable (since M is integrable). Applying (a) to M'' we see that $\lambda' \in X^+$.

Applying 6.1.7(b) to M and to M'' and to the $\mathbf{Z}[I]$-coset of λ (or λ') in X, we see that $\Omega\Xi(m) = v^{G(\lambda)}m$ for all $m \in M$ and $\Omega\Xi(m) = v^{G(\lambda')}m$ for all $m \in M''$. (G as in 6.1.4.) It follows that $G(\lambda) = G(\lambda')$. This contradicts 6.1.5 since $\lambda' < \lambda$. The lemma is proved.

Theorem 6.2.2. *Let M be an integrable* **U***-module in $\mathcal{C}^{hi}$. Then M is a sum of simple* **U***-submodules.*

We may assume that $M \neq 0$. By 6.1.6, we may also assume that $M = M_C$ for some $\mathbf{Z}[I]$-coset C in X. We choose a function $G : C \to \mathbf{Z}$ as in 6.1.4 and we define $\Omega\Xi : M \to M$ (in the commutant of M) as in 6.1.7.

By writing M as a direct sum of the generalized eigenspaces of $\Omega\Xi : M \to M$ (see 6.1.7), we may further assume that there exists $c \in \mathbf{Z}$ such that $(\Omega\Xi - v^c) : M \to M$ is locally nilpotent.

Let $P = \{m \in M | E_i m = 0 \quad \forall i\}$. We have $P = \oplus_{\lambda \in C} P^\lambda$ where $P^\lambda = P \cap M^\lambda$. For any non-zero element $m \in P^\lambda$, the **U**-submodule of M generated by m is of the type considered in 6.2.1; hence it is a simple subobject of M. Thus the **U**-submodule M' of M generated by P is a sum of simple **U**-submodules. Let $M'' = M/M'$.

Assume that $M'' \neq 0$. Then we can find $\lambda_1 \in C$ maximal such that $M''^{\lambda_1} \neq 0$. Let m_1 be a non-zero vector of M''^{λ_1}. We have $E_i m_1 = 0$ for all i. Applying 6.2.1 and 6.1.7 to the **U**-submodule of M'' generated by m_1, we see that $\lambda_1 \in X^+$ and $\Omega\Xi(m_1) = v^{G(\lambda_1)}m_1$. Since $(\Omega\Xi - v^c) : M \to M$ is locally nilpotent we see that $(\Omega\Xi - v^c) : M'' \to M''$ is locally nilpotent. Hence we must have $c = G(\lambda_1)$.

Let $\tilde{m}_1 \in M^{\lambda_1}$ be a representative for m_1. As in the proof of 6.1.7, the $\mathbf{U}^+$-submodule M_1 of M generated by $\tilde{m}_1$ is a finite dimensional $\mathbf{Q}(v)$-vector space which is the sum of its intersections with the weight spaces of M. Hence we can find $\lambda_2 \in C$ maximal such that $M_1 \cap M^{\lambda_2} \neq 0$. Let m_2 be a non-zero vector in $M_1 \cap M^{\lambda_2}$. We have $E_i m_2 = 0$ for all i. Applying 6.2.1 and 6.1.7 to the **U**-submodule of M generated by m_2, we see that $\lambda_2 \in X^+$ and $\Omega\Xi(m_2) = v^{G(\lambda_2)}m_2$. From the definition of c, we have that $G(\lambda_2) = c$. Hence $G(\lambda_1) = G(\lambda_2)$. Note that $\lambda_1 \in X^+, \lambda_2 \in X^+$; from the definitions, we see that $\lambda_2 \geq \lambda_1$. Using 6.1.5, we deduce that $\lambda_1 = \lambda_2$. It follows that M_1 is the one dimensional subspace spanned by $\tilde{m}_1$; hence we must have $E_i(\tilde{m}_1) = 0$ for all i, or equivalently, $\tilde{m}_1 \in P$. This implies that $m_1 = 0$, a contradiction.

We have proved that $M'' = 0$. Hence $M = M'$ and therefore M is a sum of simple **U**-submodules. The theorem is proved.

Corollary 6.2.3. (a) *For any $\lambda \in X^+$, the* **U***-module Λ_λ is a simple object* *of $\mathcal{C}'$.*

(b) *If $\lambda, \lambda' \in X^+$, the* **U**-*modules $\Lambda_\lambda, \Lambda_{\lambda'}$ are isomorphic if and only if $\lambda = \lambda'$.*

(c) *Any integrable module in $\mathcal{C}^{hi}$ is a direct sum of simple modules of the form Λ_λ for various $\lambda \in X^+$.*

(a) follows from 6.2.1 since Λ_λ is integrable. To prove (b), it suffices to note the following property which follows from the definitions: given the **U**-module $M = \Lambda_\lambda$ where $\lambda \in X^+$, there is a unique element $\lambda_1 \in X$ such that $M^{\lambda_1} \neq 0$ and λ_1 is maximal with this property; we have $\lambda_1 = \lambda$.

We prove (c). From Theorem 6.2.2, it follows that any integrable module in $\mathcal{C}^{hi}$ is a direct sum of simple objects of $\mathcal{C}^{hi}$ which are necessarily integrable. Let M' be one of these simple summands. Let $\lambda \in X$ be maximal such that $M'^\lambda \neq 0$. Let m be a non-zero vector in M'^λ. Then $E_i m = 0$ for all i. Using 3.5.8, we can find a non-zero morphism $\Lambda_\lambda \to M'$ (in $\mathcal{C}'$). Since both Λ_λ and M' are simple, this must be an isomorphism.

6.3. Affine or Finite Cartan Data

6.3.1. In this section we assume that $(I, \cdot)$ has the following positivity property: $\sum_{i,j} i \cdot j x_i x_j \geq 0$ for all $(x_i) \in \mathbf{N}^I$. This certainly holds if $(I, \cdot)$ is of affine or finite type. We first prove an irreducibility result for certain Verma modules.

Proposition 6.3.2. *Let $\lambda \in X$ be such that $\langle i, \lambda \rangle \leq -1$ for all i. Then M_λ is simple.*

Assume that M has a non-zero **U**-submodule M' distinct from M. Let $\lambda' \in X$ be maximal with the property that $M'^{\lambda'} \neq 0$. Let m' be a non-zero vector in $M'^{\lambda'}$. Then $E_i m' = 0$ for all i. Hence there is a homomorphism of **U**-modules $M_{\lambda'} \to M'$ whose image contains m'. Using 6.1.7 for M_λ and $M_{\lambda'}$, we see that $\Omega\Xi(m') = v^{G(\lambda)} m'$ and $\Omega\Xi(m') = v^{G(\lambda')} m'$. It follows that $G(\lambda) = G(\lambda')$. We have $\lambda' < \lambda$ hence we can write $\lambda' = \lambda - i_1' - i_2' - \cdots - i_n'$ for some sequence $i_1, i_2, \ldots, i_n$ in I with $n \geq 1$. As in 6.1.5, from $G(\lambda) = G(\lambda')$, we deduce

(a) $\sum_{p=1}^n i_p \cdot i_p \langle i_p, \lambda \rangle = \sum_{p<q \in [1,n]} i_p \cdot i_q$.

Hence $(\sum_{p=1}^n i_p) \cdot (\sum_{q=1}^n i_q) = \sum_{p=1}^n i_p \cdot i_p (2\langle i_p, \lambda \rangle + 1)$. By our assumption, the left hand side is ≥ 0 and the right hand side is < 0. This contradiction proves the proposition.

6.3.3. *In the remainder of this section, we assume that $(I, \cdot)$ is of finite type.* In this case the root datum is automatically Y-regular and X-regular.

Proposition 6.3.4. (a) *For any* $\lambda \in X^+$, *we have* $\dim \Lambda_\lambda < \infty$.

(b) *Let* $M \in \mathcal{C}$ *be such that* $\dim M < \infty$. *Then* M *is integrable and* $M \in \mathcal{C}^{hi}$, *hence (by 6.2.2), it is a direct sum of simple* **U**-*modules isomorphic to* Λ_λ *for various* $\lambda \in X^+$.

Let $\lambda' \in X$ be such that $\Lambda_\lambda^{\lambda'} \neq 0$. Using 5.2.7 several times, we see that we also have $\Lambda_\lambda^{w(\lambda')} \neq 0$ for any $w \in W$. In particular, we have $\Lambda_\lambda^{w_0(\lambda')} \neq 0$. It follows that $\lambda' \le \lambda$ and $w_0(\lambda') \le \lambda$. The last inequality implies, in view of 2.2.8, that $w_0(\lambda) \le \lambda'$. Thus, we have $w_0(\lambda) \le \lambda' \le \lambda$. These inequalities are satisfied by only finitely many λ'. Since each weight space of Λ_λ is finite dimensional, we see that (a) holds. Now (b) is immediate since the root datum is X-regular. The proposition follows.

6.3.5. The following result is a variant of the complete reducibility theorem 6.2.2: we assume (see 6.3.3) that the Cartan matrix is of finite type but we do not need the condition that our module is in $\mathcal{C}^{hi}$.

Proposition 6.3.6. *Let* M *be an integrable* **U**-*module. Then* M *is a sum of simple* **U**-*modules of form* Λ_λ *for various* $\lambda \in X^+$.

Let $m \in M^\zeta$ and let M' be the $\mathbf{U}^+$-submodule of M generated by m. Since M is integrable, there exist $a_i \in \mathbf{N}$ such that $E_i^{(a_i+1)}m = 0$ for $i \in I$. Hence there exists $\lambda' \in X^+$ such that $E_i^{(\langle i,\lambda'\rangle+1)}m = 0$ for all i. It follows that $u \to um$ gives a surjective linear map $\mathbf{U}^+/(\sum_i \mathbf{U}^+E_i^{(\langle i,\lambda'\rangle+1)}) \to M'$. Using 3.5.6, we see that $\mathbf{U}^+/(\sum_i \mathbf{U}^+E_i^{(\langle i,\lambda'\rangle+1)})$ is isomorphic as a vector space to $\Lambda_{\lambda'}$, hence it is finite dimensional, by 6.3.4. Thus, $\dim M' < \infty$. Let M'' be the **U**-submodule generated by M'. Since M' is stable under $\mathbf{U}^+$ and $\mathbf{U}^0$, M'' is equal to the $\mathbf{U}^-$-submodule generated by M'. By the argument above, the $\mathbf{U}^-$-submodule generated by a vector in M is finite dimensional. Since M' is finite dimensional, the $\mathbf{U}^-$-submodule generated by M' is finite dimensional. Thus, $\dim M'' < \infty$. We have shown that m is contained in a finite dimensional **U**-submodule of M. Thus, M is a sum of finite dimensional **U**-submodules. By 6.3.4(b), each of these is a sum of simple **U**-modules of form Λ_λ for various $\lambda \in X^+$. The proposition follows.

CHAPTER 7

Higher Order Quantum Serre Relations

7.1.1. In this chapter we assume that we are given $i \neq j$ in I and $e = \pm 1$. Given $n, m \in \mathbf{Z}$, we set

$$f_{i,j;n,m;e} = \sum_{r+s=m} (-1)^r v_i^{er(-\langle i,j'\rangle n - m+1)} \theta_i^{(r)} \theta_j^{(n)} \theta_i^{(s)} \in \mathbf{f}.$$

To simplify notation we shall write $f_{n,m;e}$ instead of $f_{i,j;n,m;e}$ when convenient, and we shall set $\alpha = -\langle i, j'\rangle \in \mathbf{N}$, $\alpha' = -\langle j, i'\rangle \in \mathbf{N}$.

Lemma 7.1.2. *We have (in* $\mathbf{U}$*)*

(a) $-v_i^{e(\alpha n-2m)} E_i f^+_{n,m;e} + f^+_{n,m;e} E_i = [m+1]_i f^+_{n,m+1;e}$;

(b) $-F_i f^+_{n,m;e} + f^+_{n,m;e} F_i = [\alpha n - m+1]_i \tilde{K}_{-ei} f^+_{n,m-1;e}$.

We prove (a). The left hand side of (a) is

$$\sum_{r+s=m} (-1)^{r+1} (v_i^{er(\alpha n-m+1)+e(\alpha n-2m)} [r+1]_i E_i^{(r+1)} E_j^{(n)} E_i^{(s)}$$
$$- v_i^{er(\alpha n-m+1)} [s+1]_i E_i^{(r)} E_j^{(n)} E_i^{(s+1)})$$
$$= \sum_{r+s=m+1} (-1)^r (v_i^{er(\alpha n-m+1)-m-1)} [r]_i + v_i^{er(\alpha n-m+1)} [s]_i) E_i^{(r)} E_j^{(n)} E_i^{(s)}.$$

It remains to observe that

$$v_i^{er(\alpha n-m+1)-e(m+1))} [r]_i + v_i^{er(\alpha n-m+1)} [s]_i = v_i^{er(\alpha n-m)} [m+1]_i$$

if $r+s=m+1$.

We prove (b). Using the identity

$$F_i E_i^{(N)} = E_i^{(N)} F_i - \frac{v_i^{-N+1} \tilde{K}_i - v_i^{N-1} \tilde{K}_{-i}}{v_i - v_i^{-1}} E_i^{(N-1)},$$

G. Lusztig, *Introduction to Quantum Groups*, Modern Birkhäuser Classics,
DOI 10.1007/978-0-8176-4717-9_7,

we see that the left hand side of (b) is

$$\sum_{r+s=m}(-1)^{r+1}v_i^{er(\alpha n-m+1)}(E_i^{(r)}F_iE_j^{(n)}E_i^{(s)}$$
$$-\frac{v_i^{-r+1}\tilde K_i-v_i^{r-1}\tilde K_{-i}}{v_i-v_i^{-1}}E_i^{(r-1)}E_j^{(n)}E_i^{(s)})$$
$$+\sum_{r+s=m}(-1)^{r}v_i^{er(\alpha n-m+1)}E_i^{(r)}E_j^{(n)}E_i^{(s)}F_i$$
$$=\sum_{r+s=m}(-1)^{r+1}v_i^{er(\alpha n-m+1)}(-E_i^{(r)}E_j^{(n)}\frac{v_i^{-s+1}\tilde K_i-v_i^{s-1}\tilde K_{-i}}{v_i-v_i^{-1}}E_i^{(s-1)}$$
$$-\frac{v_i^{-r+1}\tilde K_i-v_i^{r-1}\tilde K_{-i}}{v_i-v_i^{-1}}E_i^{(r-1)}E_j^{(n)}E_i^{(s)})$$
$$=\sum_{r+s=m}(-1)^{r}v_i^{er(\alpha n-m+1)}\frac{v_i^{-s+1-2r+\alpha n}\tilde K_i-v_i^{s-1+2r-\alpha n}\tilde K_{-i}}{v_i-v_i^{-1}}E_i^{(r)}E_j^{(n)}E_i^{(s-1)}$$
$$+\sum_{r+s=m}(-1)^{r}v_i^{er(\alpha n-m+1)}\frac{v_i^{-r+1}\tilde K_i-v_i^{r-1}\tilde K_{-i}}{v_i-v_i^{-1}}E_i^{(r-1)}E_j^{(n)}E_i^{(s)}$$
$$=\sum_{r+s=m-1}(-1)^{r}v_i^{er(\alpha n-m+1)}\frac{v_i^{-s-2r+\alpha n}\tilde K_i-v_i^{s+2r-\alpha n}\tilde K_{-i}}{v_i-v_i^{-1}}E_i^{(r)}E_j^{(n)}E_i^{(s)}$$
$$+\sum_{r+s=m-1}(-1)^{r-1}v_i^{e(r+1)(\alpha n-m+1)}\frac{v_i^{-r}\tilde K_i-v_i^{r}\tilde K_{-i}}{v_i-v_i^{-1}}E_i^{(r)}E_j^{(n)}E_i^{(s)}$$

and (b) follows.

From Lemma 7.1.2 we deduce by induction on $p\geq 0$ the following result.

Lemma 7.1.3. *We have*
(a) $E_i^{(p)}f^+_{n,m;e}=\sum_{p'=0}^{p}(-1)^{p'}v_i^{e(2pm-\alpha pn+pp'-p')}\begin{bmatrix}m+p'\\p'\end{bmatrix}_i f^+_{n,m+p';e}E_i^{(p-p')}$;
(b) $F_i^{(p)}f^+_{n,m;e}=\sum_{p'=0}^{p}(-1)^{p'}v_i^{-e(pp'-p')}\begin{bmatrix}\alpha n-m+p'\\p'\end{bmatrix}_i\tilde K_{-ep'i}f^+_{n,m-p';e}F_i^{(p-p')}$.

Lemma 7.1.4. *We have*

$$F_jf^+_{n,m;e}-f^+_{n,m;e}F_j=\tilde K_{-ej}\frac{v_j^{e(n-1)}}{v_j^{e}-v_j^{-e}}f^+_{n-1,m;-e}-\tilde K_{ej}\frac{v_j^{-e(n-1)}}{v_j^{-e}-v_j^{e}}f^+_{n-1,m;e}.$$

We have

$$F_j f^+_{n,m;e} - f^+_{n,m;e} F_j$$
$$= - \sum_{r+s=m} (-1)^r v_i^{er(\alpha n-m+1)} E_i^{(r)} \frac{v_j^{-n+1}\tilde{K}_j - v_j^{n-1}\tilde{K}_{-j}}{v_j - v_j^{-1}} E_j^{(n-1)} E_i^{(s)}$$
$$= - \sum_{r+s=m} (-1)^r v_i^{er(\alpha n-m+1)} \frac{v_j^{-n+1+\alpha' r}\tilde{K}_j - v_j^{n-1-\alpha' r}\tilde{K}_{-j}}{v_j - v_j^{-1}} E_i^{(r)} E_j^{(n-1)} E_i^{(s)}.$$

We now use the identity $v_j^{\alpha'} = v_i^{\alpha}$; the lemma follows.

Proposition 7.1.5. (a) *If $n < 0$ or $m < 0$, then $f_{n,m;e} = 0$.*
(b) *If $m > \alpha n$, then $f_{n,m;e} = 0$.*

(a) is obvious. In particular (b) holds for $n < 0$. Hence, to prove (b), we may assume that $n \geq 0$ and that (b) holds with n replaced by $n-1$. For such fixed n, we see from 7.1.2(b) that $f^+_{n,\alpha n+1;e}$ commutes with F_i and from 7.1.4 and the induction hypothesis, that $f^+_{n,\alpha n+1;e}$ commutes with F_j. (We have $\alpha n + 1 > \alpha(n-1)$ hence the induction hypothesis is applicable to $f_{n-1,\alpha n+1;\pm 1}$.) It is trivial that $f^+_{n,\alpha n+1;e}$ commutes with F_h for any $h \neq i,j$. Thus, $f^+_{n,\alpha n+1;e}$ commutes with F_h for any $h \in I$. Using 3.2.7(a), it follows that $f_{n,\alpha n+1;e}$ is a scalar multiple of 1. On the other hand, it belongs to $\mathbf{f}_{(\alpha n+1)i+nj}$ and $(\alpha n+1)i + nj \neq 0$. It follows that $f_{n,\alpha n+1;e} = 0$.

We now show, for our fixed n, that $f_{n,m;e} = 0$ whenever $m > \alpha n$. We argue by induction on m. If $m = \alpha n + 1$, this has been just proved. Hence we may assume that $m > \alpha n+1$. Using the induction hypothesis we see that the left hand side of the identity $-v_i^{e(\alpha n-2m+2)} E_i f^+_{n,m-1;e} + f^+_{n,m-1;e} E_i = [m]_i f^+_{n,m;e}$ (see 7.1.2) is zero. Hence we have $[m]_i f^+_{n,m;e} = 0$. We have $m \neq 0$, hence $f^+_{n,m;e} = 0$. It follows that $f_{n,m;e} = 0$ and the induction is completed. The proposition is proved.

7.1.6. The identities $f_{n,m;e} = 0 \quad (m > \alpha n; n \geq 1)$ in $\mathbf{f}$ are called the *higher order quantum Serre relations.* For $n = 1$ and $m = \alpha + 1$, they reduce to the usual quantum Serre relations.

Corollary 7.1.7. *For any $n, m \geq 0$ such that $m \geq \alpha n + 1$, we have*

$$\theta_i^{(m)} \theta_j^{(n)} = \sum_{r+s'=m; m-\alpha n \leq s' \leq m} \gamma_{s'} \theta_i^{(r)} \theta_j^{(n)} \theta_i^{(s')}$$

where $\gamma_{s'} = \sum_{q=0}^{m-\alpha n-1} (-1)^{s'+1+q} v_i^{-s'(\alpha n-m+1+q)+q} \begin{bmatrix} s' \\ q \end{bmatrix}_i$ *(identity in $\mathbf{f}$).*

From 7.1.5 we see that $f_{n,m-q;1} = 0$ for $0 \leq q \leq m - \alpha n - 1$; hence

$$g = \sum_{q=0}^{m-\alpha n-1} (-1)^q v_i^{-mq+q} f_{n,m-q;1} \theta_i^{(q)}$$

is zero. On the other hand, using the definitions, we have

$$\begin{aligned} g &= \sum_{q=0}^{m-\alpha n-1} \sum_{r+s=m-q} (-1)^r (-1)^q v_i^{r(\alpha n-m+q+1)} v_i^{-mq+q} \theta_i^{(r)} \theta_j^{(n)} \theta_i^{(s)} \theta_i^{(q)} \\ &= \sum_{r+s'=m} c_{r,s'} \theta_i^{(r)} \theta_j^{(n)} \theta_i^{(s')} \end{aligned}$$

where $c_{r,s'} = \sum_{q=0}^{m-\alpha n-1} (-1)^{r+q} v_i^{r(\alpha n-m+q+1)-mq+q} \begin{bmatrix} s' \\ q \end{bmatrix}_i$.

If $0 \leq s' \leq m - \alpha n - 1$, we may replace the range of summation above to $0 \leq q \leq s'$ and the sum will not change, since for $0 \leq s' < q$, the binomial coefficient $\begin{bmatrix} s' \\ q \end{bmatrix}_i$ is zero. Hence for such s' we have $c_{r,s'} = (-1)^r v_i^{r(\alpha n-m+1)} \sum_{q=0}^{s'} (-1)^q v_i^{q(1-s')} \begin{bmatrix} s' \\ q \end{bmatrix}_i$. By 1.3.4, the last sum is zero unless $s' = 0$. Thus, for $0 \leq s' \leq m - \alpha n - 1$, we have $c_{r,s'} = \delta_{0,s'} (-1)^m v_i^{m(\alpha n-m+1)}$. The corollary follows.

Notes on Part I

1. The Hopf algebra **U** has been defined in the simplest case (quantum analogue of SL_2) by Kulish and Reshetikhin [10] and Sklyanin [14] and, in the general case, by Drinfeld [2] and Jimbo [5], [6]. The definition given here is different from the original one; the two definitions will be reconciled in Part V.
2. The bilinear form (,) in 1.2.3 turns out eventually to be the same as that of Drinfeld [3].
3. The idea of defining the $\mathcal{A}$-form ${}_{\mathcal{A}}\mathbf{f}$ and ${}_{\mathcal{A}}\mathbf{U}$ of $\mathbf{f}$ and $\mathbf{U}$ (see 1.4.7, 3.1.13) in terms of v-analogues of divided powers appeared in [12]. (In the classical case, the **Z**-forms of enveloping algebras were defined in terms of divided powers with ordinary factorials by Chevalley and Kostant [9], for finite types, and by Tits, for infinite types.)
4. The theorem in 2.1.2 is due to Iwahori, for finite types, and to Matsumoto and Tits [1], in the general case. The statement in 2.2.7 can be deduced from a theorem of Tits on Coxeter groups, see [1], ch. 4, p.93, statement P_n.
5. The notion of Cartan datum (resp. root datum), see 1.1.1 (resp. 2.2.1), is closely related to (but not the same as) that of a generalized Cartan matrix (resp. a realization of it) in [7]. In fact, an irreducible generalized Cartan matrix is the same as an irreducible Cartan datum up to proportionality (see 1.1.1).
6. The commutation formulas in 3.1.7, 3.1.8, are closely connected with Drinfeld's description [3] of **U** (in the formal setting) as a quantum double. Their consequence, Corollary 3.1.9, is the quantum analogue of an identity of Kostant [9] (it was shown to me by V. Kac).
7. The definition 3.5.1 of integrable **U**-modules is the quantum analogue of Kac's definition [7] of integrable modules of a Kac-Moody Lie algebra.
8. The definition of universal $\mathcal{R}$-matrices is due to Drinfeld [3]. The characterization of a modified form of the $\mathcal{R}$-matrix given in 4.1.2, as well as in 4.1.3, appeared in [13]. Propositions 4.2.2 and 4.2.4 are due to Drinfeld [3].
9. The formulas for the operators $T'_{i,e}, T''_{i,e}$ (in 5.2.1) are new (they are classical for $v=1$). An identity like 5.3.4(a) (with a different definition of $T''_{i,1}$) is stated in [8] and [11].
10. The definition of the quantum Casimir operator (see 6.1) is due to Drinfeld [4]. The proof of the complete reducibility theorem 6.2.2 is inspired by the proof of the analogous result in the non-quantum case (Kac [7]).
11. A number of statements of Drinfeld in [3] were given without proof; some of the proofs were supplied by Tanisaki [15].

References

1. N. Bourbaki, *Groupes et algèbres de Lie, Ch. 4-6*, Hermann, 1968.
2. V. G. Drinfeld, *Hopf algebras and the quantum Yang-Baxter equation*, Soviet Math. Dokl. **32** (1985), 254–258.
3. ———, *Quantum groups*, Proc. Int. Congr. Math. Berkeley 1986, vol. 1, Amer. Math. Soc., 1988, pp. 798–820.
4. ———, *On almost cocommutative Hopf algebras*, Algebra and analysis **1** (1989), 30–46.
5. M. Jimbo, *A q-difference analogue of $U(\mathfrak{g})$ and the Yang-Baxter equation*, Lett. Math. Phys. **10** (1985), 63–69.
6. ———, *A q-analog of $U(\mathfrak{gl}(N+1))$, Hecke algebras and the Yang-Baxter equation*, Lett. Math. Phys. **11** (1986), 247–252.
7. V. G. Kac, *Infinite dimensional Lie algebras*, Birkhäuser, Boston, 1983.
8. A. N. Kirillov and N. Yu. Reshetikhin, *q-Weyl group and a multiplicative formula for universal R-matrices*, Commun. Math. Phys. **134** (1990), 421–431.
9. B. Kostant, *Groups over* **Z**, Proc. Symp. Pure Math. **9** (1966), 90–98, Amer. Math. Soc., Providence, R. I..
10. P. P. Kulish and N. Yu. Reshetikhin, *The quantum linear problem for the sine-Gordon equation and higher representations*, (Russian), Zap. Nauchn. Sem. LOMI **101** (1981), 101–110.
11. S. Z. Levendorskii and I. S. Soibelman, *Some applications of quantum Weyl groups*, J. Geom. and Phys. **7** (1990), 241–254.
12. G. Lusztig, *Quantum deformations of certain simple modules over enveloping algebras*, Adv. Math. **70** (1988), 237–249.
13. ———, *Canonical bases in tensor products*, Proc. Nat. Acad. Sci. **89** (1992), 8177–8179.
14. E. K. Sklyanin, *On an algebra generated by quadratic relations*, Uspekhi Mat. Nauk **40** (1985), 214.
15. T. Tanisaki, *Killing forms, Harish-Chandra isomorphisms and universal R-matrices for quantum algebras*, Infinite Analysis, World Scientific, 1992, pp. 941-961.

Part II

GEOMETRIC REALIZATION OF f

The algebra $\mathbf{f}$ has a canonical basis $\mathbf{B}$ with very remarkable properties. This gives an extremely rigid structure for $\mathbf{f}$ and also (in the Y-regular case) for each Λ_λ. Part II will introduce the canonical basis of $\mathbf{f}$. At the same time, $\mathbf{f}$ will be constructed in a purely geometric way, in terms of perverse sheaves on the moduli space of representations of a quiver.

Chapter 8 contains a review of the theory of perverse sheaves over an algebraic variety in positive characteristic. As far as definitions are concerned, it would have been possible to stay in characteristic zero and use $\mathcal{D}$-modules instead of perverse sheaves. This would certainly have been more elementary, but would have deprived us of the possibility of using the Weil conjecture and its consequences which are available in the framework of perverse sheaves on varieties in positive characteristic.

In Chapter 9 we introduce a class of perverse sheaves attached to a quiver and the operations of induction and restriction for perverse sheaves in this class. In Chapter 10, we study the Fourier-Deligne transform of perverse sheaves in our class. This is necessary for understanding the effect of changing the orientation of the quiver. In Chapter 11 we study linear categories with a given periodic functor (a functor which has some power equal to identity). These are needed to handle the case where the Cartan datum is not symmetric. (The geometry associated to a non-symmetric Cartan datum is very closely related to that associated to a symmetric Cartan datum, together with an action of a finite cyclic group.)

In Chapter 12 we study quivers with a cyclic group action. The geometric construction of $\mathbf{f}$ and of its canonical basis (up to signs) is given in Chapter 13.

In Chapter 14, we discuss various properties of the canonical basis. For example, the property expressed in Theorem 14.3.2 is responsible for the existence of a canonical basis in the simple integrable $\mathbf{U}$-modules (see Theorem 14.4.11). Perhaps the deepest property of the canonical basis is expressed by the positivity theorem 14.4.13, which states (for symmetric Cartan data) that the structure constants of **f** are given by polynomials with positive integer coefficients.

Theorem 14.4.9 gives a natural bijection between the canonical basis of **f** for a non-symmetric Cartan datum and the fixed point set of a cyclic group action on the canonical basis of the analogous algebra corresponding to a symmetric Cartan datum.

CHAPTER 8

Review of the Theory of Perverse Sheaves

8.1.1. Let p be a fixed prime number. All algebraic varieties will be over an algebraic closure k of the finite field F_p with p elements.

Let X be an algebraic variety. We denote by $\mathcal{D}(X) = \mathcal{D}^b_c(X)$ the bounded derived category of $\bar{\mathbf{Q}}_l$-(constructible) sheaves on X (see [1, 2.2.18]); here, l denotes a fixed prime number distinct from p and $\bar{\mathbf{Q}}_l$ is an algebraic closure of the field of l-adic numbers. Objects of $\mathcal{D}(X)$ are referred to as *complexes*. For a complex $K \in \mathcal{D}(X)$, we denote by $\mathcal{H}^n K$ the n-th cohomology sheaf of K (a $\bar{\mathbf{Q}}_l$-sheaf on X). We denote by $D(K) \in \mathcal{D}(X)$ the Verdier dual of K. The constant sheaf $\bar{\mathbf{Q}}_l$ on X will be denoted by $\mathbf{1}$.

For any integer j, let $K \mapsto K[j]$ be the shift functor $\mathcal{D}(X) \to \mathcal{D}(X)$; it satisfies $\mathcal{H}^n(K[j]) = \mathcal{H}^{n+j}K$. Let $f : X \to Y$ be a morphism of algebraic varieties. There are induced functors $f^* : \mathcal{D}(Y) \to \mathcal{D}(X)$, $f_* : \mathcal{D}(X) \to \mathcal{D}(Y)$, $f_! : \mathcal{D}(X) \to \mathcal{D}(Y)$. If f is proper, we have $f_* = f_!$ and $f_!(DK) = D(f_!K)$ for $K \in \mathcal{D}(X)$.

8.1.2. Let $\mathcal{M}(X)$ be the full subcategory of $\mathcal{D}(X)$ whose objects are those K in $\mathcal{D}(X)$ such that, for any integer n, the supports of both $\mathcal{H}^n K$ and $\mathcal{H}^n D(K)$ have dimension $\leq -n$. In particular, $\mathcal{H}^n K$ and $\mathcal{H}^n D(K)$ are zero for $n > 0$. The objects of $\mathcal{M}(X)$ are called *perverse sheaves* on X.

$\mathcal{M}(X)$ is an abelian category [1, 2.14, 1.3.6] in which all objects have finite length. The simple objects of $\mathcal{M}(X)$ are given by the Deligne-Goresky-MacPherson intersection cohomology complexes corresponding to various smooth irreducible subvarieties of X and to irreducible local systems on them.

For any $n \in \mathbf{Z}$, let $\tau_{\leq n} : \mathcal{D}(X) \to \mathcal{D}(X)$ and $H^n : \mathcal{D}(X) \to \mathcal{M}(X)$ be the functors of *truncation* and *perverse cohomology*, which in [1] are denoted by ${}^p\tau_{\leq n}$ and ${}^pH^n$.

There are natural morphisms

$$\tau_{\leq n-1}K \xrightarrow{\alpha_n} \tau_{\leq n}K \xrightarrow{\beta_n} (H^n K)[-n]$$

for any $K \in \mathcal{D}(X)$ and any n. For fixed K, we have $\tau_{\leq n}K = K$ for $n \gg 0$, $\tau_{\leq n}K = 0$ for $n \ll 0$ and $H^n K = 0$ for all but finitely many values of n.

G. Lusztig, *Introduction to Quantum Groups*, Modern Birkhäuser Classics,
DOI 10.1007/978-0-8176-4717-9_8, © Springer Science+Business Media, LLC 2010

For any $n \in \mathbf{Z}$, let $\mathcal{M}(X)[n]$ be the full subcategory of $\mathcal{D}(X)$ whose objects are of the form $K[n]$ for some $K \in \mathcal{M}(X)$.

8.1.3. A complex $K \in \mathcal{D}(X)$ is said to be *semisimple* if for each n,

(a) there exists $\gamma_n : (H^n K)[-n] \to \tau_{\leq n} K$ such that (α_n, γ_n) define an isomorphism $\tau_{\leq n-1} K \oplus (H^n K)[-n] \cong \tau_{\leq n} K$ and

(b) $H^n K$ is a semisimple object of $\mathcal{M}(X)$.

It follows that K is isomorphic to $\oplus_n (H^n K)[-n]$ in $\mathcal{D}(X)$.

8.1.4. Let $f : X \to Y$ be a smooth morphism with connected fibres of dimension d. We have $D(f^*K) = f^*(D(K))[2d]$ for $K \in \mathcal{D}(Y)$. (We will ignore the Tate twist.) If $K \in \mathcal{M}(Y)$, then $f^*K \in \mathcal{M}(X)[-d]$ (see [1, 4.2.4]) and $K \mapsto f^*K$ defines a fully faithful functor from $\mathcal{M}(Y)$ to $\mathcal{M}(X)[-d]$ (see [1, 4.2.5]).

8.1.5. Let $f : X \to Y$ be a proper morphism with Y smooth. Then $f_! \mathbf{1} \in \mathcal{D}(Y)$ is a semisimple complex. (See [1, 5.4.5, 5.3.8].)

8.1.6. More generally, let $f : X \to Y$ be a morphism. Assume that we are given a partition $X = X_0 \cup X_1 \cup \cdots \cup X_m$ such that $X_{\leq j} = X_0 \cup X_1 \cup \cdots \cup X_j$ is closed for $j = 0, 1, \ldots, m$. (We define $X_{\leq j} = \emptyset$ for $j < 0$.) Assume that, for each j, we are given morphisms $X_j \xrightarrow{f_j''} Z_j \xrightarrow{f_j'} Y$ such that Z_j is smooth, f_j'' is a vector bundle, f_j' is proper and $f_j' f_j'' = f_j$ where $f_j : X_j \to Y$ is the restriction of f''. Then $f_! \mathbf{1} \in \mathcal{D}(Y)$ is a semisimple complex. Moreover, for any n and j, there is a canonical exact sequence (in $\mathcal{M}(Y)$):

(a) $$0 \to H^n(f_j)_! \mathbf{1} \to H^n(f_{\leq j})_! \mathbf{1} \to H^n(f_{\leq j-1})_! \mathbf{1} \to 0$$

where $f_{\leq j} : X_{\leq j} \to Y$ is the restriction of f. The proof is essentially the same as that in [5, 3.7]; it is based on the theory of weights in [1].

8.1.7. G-equivariant complexes. Let $m : G \times X \to X$ be the action of a connected algebraic group G on X; let $\pi : G \times X \to X$ be the second projection. A perverse sheaf K on X is said to be *G-equivariant* if the perverse sheaves $\pi^* K[\dim G]$ and $m^* K[\dim G]$ are isomorphic. More generally, a complex $K \in \mathcal{M}(X)[n]$ is said to be G-equivariant if the perverse sheaf $K[-n]$ is G-equivariant.

We denote by $\mathcal{M}_G(X)$ the full subcategory of $\mathcal{M}(X)$ whose objects are the G-equivariant perverse sheaves on X. More generally, we denote by

$\mathcal{M}_G(X)[n]$ the full subcategory of $\mathcal{M}(X)[n]$ whose objects are of the form $K[n]$ where $K \in \mathcal{M}_G(X)$.

Here are some properties of G-equivariant complexes.

(a) If $A \in \mathcal{M}_G(X)$, and $B \in \mathcal{M}(X)$ is a subquotient of A, then $B \in \mathcal{M}_G(X)$.

(b) Assume that G acts on two varieties X', X and that $f : X' \to X$ is a morphism compatible with the G-actions. If $K \in \mathcal{M}_G(X)$, then $H^n(f^*K) \in \mathcal{M}_G(X')$ for all n. If $K' \in \mathcal{M}_G(X')$, then $H^n(f_!K') \in \mathcal{M}_G(X)$ for all n.

(c) Assume that $f : X \to Y$ is a locally trivial principal G-bundle (in particular G acts freely on X and trivially on Y). Let $d = \dim G$. If $K \in \mathcal{M}(X)[n]$, then we have $K \in \mathcal{M}_G(X)[n]$ if and only if K is isomorphic to f^*K' for some $K' \in \mathcal{M}(Y)[n+d]$. The functor $\mathcal{M}(Y)[n+d] \to \mathcal{M}_G(X)[n]$ $(K' \to f^*K')$ and the functor $\mathcal{M}_G(X)[n] \to \mathcal{M}(Y)[n+d]$ $(K \to f_\flat K := (H^{-n-d}f_*K)[n+d])$ define an equivalence of the categories $\mathcal{M}_G(X)[n], \mathcal{M}(Y)[n+d]$.

8.1.8. A semisimple complex K on a variety X with a G-action is said to be G-equivariant if for any $n \in \mathbf{Z}$, H^nK is a G-equivariant perverse sheaf on X.

Let $f : X \to Y$ be as in 8.1.7(c). If K' is a semisimple complex on Y, then f^*K' is a G-equivariant semisimple complex on X. Conversely, if K is a semisimple G-equivariant complex on X, then K is isomorphic to f^*K' for some semisimple complex K' on Y, which is unique up to isomorphism. In fact, we have $K' \cong f_\flat K$ where, by definition, $f_\flat K = \oplus_n f_\flat((H^nK)[-n])$ and $f_\flat((H^nK)[-n]) \in \mathcal{M}(Y)[-n+d]$ is as in 8.1.7(c).

8.1.9. Let A, B be two G-equivariant semisimple complexes on a variety X with G-action; let j be an integer. We choose a smooth irreducible algebraic variety Γ with a free action of G such that the $\bar{\mathbf{Q}}_l$-cohomology of Γ is zero in degrees $1, 2, \dots, m$ where m is a large integer (compared to $|j|$).

Let us consider the diagram

$$X \xleftarrow{s} \Gamma \times X \xrightarrow{t} G\backslash(\Gamma \times X)$$

where the maps s, t are the obvious ones. Then s^*A, s^*B are semisimple G-equivariant complexes; since t is a principal G-bundle, the semisimple complexes $t_\flat s^*A, t_\flat s^*B$ on $G\backslash(\Gamma \times X)$ are well-defined. Let $u : G\backslash(\Gamma \times X) \to$ {point} be the obvious map. Consider the $\bar{\mathbf{Q}}_l$-vector space

$$\mathcal{H}^{j+2\dim(G\backslash\Gamma)}(u_!(t_\flat s^*A \otimes t_\flat s^*B)).$$

By a standard argument (see [6, 1.1, 1.2]), we can show that this vector space is canonically attached to A, B, j: it is independent of the choice of m and Γ provided that m is sufficiently large. We denote this vector space by $\mathbf{D}_j(X, G; A, B)$.

8.1.10. We give some properties of $\mathbf{D}_j(X, G; A, B)$.

(a) $\mathbf{D}_j(X, G; A, B) = \mathbf{D}_j(X, G; B, A)$.

(b) $\mathbf{D}_j(X, G; A[n], B[m]) = \mathbf{D}_{j+n+m}(X, G; A, B)$ for all $n, m \in \mathbf{Z}$.

(c) $\mathbf{D}_j(X, G; A \oplus A_1, B) = \mathbf{D}_j(X, G; A, B) \oplus \mathbf{D}_j(X, G; A_1, B)$.

(d) If A, B are perverse sheaves, then $\mathbf{D}_j(X, G; A, B) = 0$ for $j > 0$; if, in addition, A, B are simple, then $\mathbf{D}_0(X, G; A, B) = \bar{\mathbf{Q}}_l$, if $B \cong DA$ and $\mathbf{D}_0(X, G; A, B) = 0$, otherwise.

(e) There exists $j_0 \in \mathbf{Z}$ such that $\mathbf{D}_j(X, G; A, B) = 0$ for $j \geq j_0$.

(f) If A', B' (resp. A'', B'') are G'-equivariant (resp. G''-equivariant) semisimple complexes on a variety X' (resp. X'') with a G'-action (resp. G''-action) where G', G'' are connected algebraic groups, then $A' \otimes A''$ and $B' \otimes B''$ are $G' \times G''$-equivariant semisimple complexes on $X' \times X''$ and we have a canonical isomorphism

$$\mathbf{D}_j(X' \times X'', G' \times G''; A' \otimes A'', B' \otimes B'')$$
$$= \sum_{j'+j''=j} \mathbf{D}_{j'}(X', G'; A', B') \otimes \mathbf{D}_{j''}(X'', G''; A'', B'').$$

The sum is finite by (e).

Properties (a),(b),(c) are obvious; (d) follows from [5, 7.4]; (e) follows from (d); (f) follows from the Künneth formula.

8.1.11. Fourier-Deligne transform. We fix a non-trivial character $F_p \to \bar{\mathbf{Q}}_l^*$. The Artin-Schreier covering $k \to k$ given by $x \to x^p - x$ has F_p as a group of covering transformations. Hence our character $F_p \to \bar{\mathbf{Q}}_l^*$ gives rise to a $\bar{\mathbf{Q}}_l$-local system of rank 1 on k; its inverse image under any morphism $T : X' \to k$ of algebraic varieties is a local system $\mathcal{L}_T$ of rank 1 on X'.

Let $E \to X$ and $E' \to X$ be two vector bundles of constant fibre dimension d over the variety X. Assume that we are given a bilinear map $T : E \times_X E' \to k$ which defines a duality between the two vector bundles. We have a diagram $E \xleftarrow{s} E \times_X E' \xrightarrow{t} E'$ where s, t are the obvious projections.

The Fourier-Deligne transform is the functor $\Phi : \mathcal{D}(E) \to \mathcal{D}(E')$ defined by $\Phi(K) = t_!(s^*(K) \otimes \mathcal{L}_T)[d]$. Interchanging the roles of E, E' (and keeping the same T) we have a Fourier-Deligne transform $\Phi : \mathcal{D}(E') \to \mathcal{D}(E)$; it is known that the Fourier inversion formula $\Phi(\Phi(K)) = j^*K$ holds for $K \in \mathcal{D}(E)$, where $j : E \to E$ is multiplication by -1 on each fibre of E.

Φ restricts to an equivalence of categories $\mathcal{M}(E) \to \mathcal{M}(E')$; hence it defines a bijection between the set of isomorphism classes of simple objects in $\mathcal{M}(E)$ and the analogous set for $\mathcal{M}(E')$. It also commutes with the functors $K \mapsto H^n K$.

8.1.12. Let A (resp. A') be an object of $\mathcal{D}(E)$ (resp. $\mathcal{D}(E')$). Let $u, u', \dot{u}$ be the obvious maps of $E, E', E \times_X E'$ to the point. We have

$$u_!(A \otimes \Phi(A')) = u'_!(\Phi(A) \otimes A').$$

Indeed, from the definitions, we see that both sides may be identified with $\dot{u}_!(s^*A \otimes t^*A' \otimes \mathcal{L}_T[d])$.

8.1.13. Let $T : k^n \to k$ be a non-constant affine-linear function. Let $u : k^n \to \{\text{point}\}$ be the obvious map. We have $u_!(\mathcal{L}_T) = 0$. The proof is left to the reader.

CHAPTER 9

Quivers and Perverse Sheaves

9.1. The Complexes L_ν

9.1.1. By definition, a (finite) *graph* is a pair consisting of two finite sets $\mathbf{I}$ (*vertices*) and H (*edges*) and a map which to each $h \in H$ associates a two-element subset $[h]$ of $\mathbf{I}$.

We say that h is an edge joining the two vertices in $[h]$. We assume given a finite graph $(\mathbf{I}, H, h \mapsto [h])$. An *orientation* of our graph consists of two maps $H \to \mathbf{I}$ denoted $h \mapsto h'$ and $h \mapsto h''$ such that for any $h \in H$, the two elements of $[h]$ are precisely h', h''. We assume given an orientation of our graph. Thus we have an oriented graph (=quiver). Note that

(a) for any $h \in H$, we have $h' \neq h''$.

9.1.2. Let $\mathcal{V}$ be the category of finite dimensional $\mathbf{I}$-graded k-vector spaces $\mathbf{V} = \oplus_{\mathbf{i}\in\mathbf{I}}\mathbf{V}_\mathbf{i}$; the morphisms in $\mathcal{V}$ are isomorphisms of vector spaces compatible with the grading.

For each $\nu = \sum_\mathbf{i} \nu_\mathbf{i}\mathbf{i} \in \mathbf{N}[\mathbf{I}]$ we denote by $\mathcal{V}_\nu$ the full subcategory of $\mathcal{V}$ whose objects are those $\mathbf{V}$ such that $\dim \mathbf{V}_\mathbf{i} = \nu_\mathbf{i}$ for all $\mathbf{i} \in \mathbf{I}$. Then each object of $\mathcal{V}$ belongs to $\mathcal{V}_\nu$ for a unique $\nu \in \mathbf{N}[\mathbf{I}]$ and any two objects of $\mathcal{V}_\nu$ are isomorphic to each other. Moreover, $\mathcal{V}_\nu$ is non-empty for any $\nu \in \mathbf{N}[\mathbf{I}]$.

Given $\mathbf{V} \in \mathcal{V}$, we define $G_\mathbf{V} = \{g \in GL(\mathbf{V}) | g(\mathbf{V}_\mathbf{i}) = \mathbf{V}_\mathbf{i} \text{ for all } \mathbf{i} \in \mathbf{I}\}$ and

$$\mathbf{E}_\mathbf{V} = \oplus_{h\in H}\mathrm{Hom}(\mathbf{V}_{h'}, \mathbf{V}_{h''}).$$

Then $G_\mathbf{V}$ is an algebraic group (isomorphic to $\prod_\mathbf{i} GL(\mathbf{V}_\mathbf{i})$) acting naturally on the vector space $\mathbf{E}_\mathbf{V}$ by

$$(g, x) \mapsto gx = x' \text{ where } x'_h = g_{h''}x_h g_{h'}^{-1} \text{ for all } h \in H.$$

9.1.3. Flags. A subset $\mathbf{I}'$ of $\mathbf{I}$ is said to be *discrete* if there is no $h \in H$ such that $[h] \subset \mathbf{I}'$.

If $\nu \in \mathbf{N}[\mathbf{I}]$, we define the support of ν as $\{\mathbf{i} \in \mathbf{I} | \nu_\mathbf{i} \neq 0\}$. We say that ν is discrete if its support is a discrete subset of $\mathbf{I}$.

G. Lusztig, *Introduction to Quantum Groups*, Modern Birkhäuser Classics,
DOI 10.1007/978-0-8176-4717-9_9, © Springer Science+Business Media, LLC 2010

Let $\mathcal{X}$ be the set of all sequences $\boldsymbol{\nu} = (\nu^1, \nu^2, \dots, \nu^m)$ in $\mathbf{N}[\mathbf{I}]$ such that ν^l is discrete for all l. Now let $\mathbf{V} \in \mathcal{V}$ and let $\boldsymbol{\nu} \in \mathcal{X}$ be such that $\dim \mathbf{V}_\mathbf{i} = \sum_l \nu^l_\mathbf{i}$ for all $\mathbf{i} \in \mathbf{I}$. A *flag* of type $\boldsymbol{\nu}$ in $\mathbf{V}$ is by definition a sequence

(a) $$f = (\mathbf{V} = \mathbf{V}^0 \supset \mathbf{V}^1 \supset \cdots \supset \mathbf{V}^m = 0)$$

of $\mathbf{I}$-graded subspaces of $\mathbf{V}$ such that, for $l = 1, 2, \dots, m$, the graded vector space $\mathbf{V}^{l-1}/\mathbf{V}^l$ belongs to $\mathcal{V}_{\nu^l}$. If $x \in \mathbf{E_V}$, we say that f is x-stable if $x_h(\mathbf{V}^l_{h'}) \subset \mathbf{V}^l_{h''}$ for all $l = 0, 1, \dots, m$ and all h.

Let $\mathcal{F}_\nu$ be the variety of all flags of type ν in $\mathbf{V}$. Let $\tilde{\mathcal{F}}_\nu$ be the variety of all pairs (x, f) such that $x \in \mathbf{E_V}$ and $f \in \mathcal{F}_\nu$ is x-stable. Note that $G_\mathbf{V}$ acts (transitively) on $\mathcal{F}_\nu$ by $g : f \to gf$ where f is as in (a) and $gf = (\mathbf{V} = g\mathbf{V}^0 \supset g\mathbf{V}^1 \supset \cdots \supset g\mathbf{V}^m = 0)$. Hence $G_\mathbf{V}$ acts on $\tilde{\mathcal{F}}_\nu$ by $g : (x, f) \to (gx, gf)$.

Let $\pi_\nu : \tilde{\mathcal{F}}_\nu \to \mathbf{E_V}$ be the first projection. We note the following properties which are easily checked.

(b) $\mathcal{F}_\nu$ is a smooth, irreducible, projective variety of dimension

$$\sum_{\mathbf{i}; l < l'} \nu^{l'}_\mathbf{i} \nu^l_\mathbf{i};$$

the second projection $\tilde{\mathcal{F}}_\nu \to \mathcal{F}_\nu$ is a vector bundle of dimension

$$\sum_{h; l' < l} \nu^{l'}_{h'} \nu^l_{h''}.$$

(c) $\tilde{\mathcal{F}}_\nu$ is a smooth, irreducible variety of dimension

$$f(\nu) = \sum_{h; l' < l} \nu^{l'}_{h'} \nu^l_{h''} + \sum_{\mathbf{i}; l < l'} \nu^{l'}_\mathbf{i} \nu^l_\mathbf{i}.$$

(d) π_ν is a proper $G_\mathbf{V}$-equivariant morphism.

Let $\tilde{L}_\nu = (\pi_\nu)_! \mathbf{1} \in \mathcal{D}(\mathbf{E_V})$. By (c),(d) and by 8.1.5, $\tilde{L}_\nu$ is a semisimple complex on $\mathbf{E_V}$. Let $L_\nu = \tilde{L}_\nu[f(\nu)]$. Since $D(\mathbf{1}[f(\nu)]) = \mathbf{1}[f(\nu)]$ on $\tilde{\mathcal{F}}_\nu$ (see (c)) we have $D(L_\nu) = L_\nu$.

We denote by $\mathcal{P}_\mathbf{V}$ the full subcategory of $\mathcal{M}(\mathbf{E_V})$ consisting of perverse sheaves which are direct sums of simple perverse sheaves L that have the following property: $L[d]$ appears as a direct summand of L_ν for some $d \in \mathbf{Z}$ and some $\nu \in \mathcal{X}$ such that $\dim \mathbf{V}_\mathbf{i} = \sum_l \nu^l_\mathbf{i}$ for all $\mathbf{i} \in \mathbf{I}$.

We denote by $\mathcal{Q}_\mathbf{V}$ the full subcategory of $\mathcal{D}(\mathbf{E_V})$ whose objects are the complexes that are isomorphic to finite direct sums of complexes of the form $L[d']$ for various simple perverse sheaves $L \in \mathcal{P}_\mathbf{V}$ and various $d' \in \mathbf{Z}$. Any complex in $\mathcal{Q}_\mathbf{V}$ is semisimple and $G_\mathbf{V}$-equivariant. From 8.1.4, we see that $\mathcal{P}_\mathbf{V}$ and $\mathcal{Q}_\mathbf{V}$ are stable under Verdier duality.

9.1.4. Let $\boldsymbol{\nu} = (\nu^1, \nu^2, \dots, \nu^m) \in \mathcal{X}$. Assume that for some j we write $\nu^j = \nu_1^j + \nu_2^j$ where $\nu_1^j, \nu_2^j \in \mathbf{N}[\mathbf{I}]$ have disjoint support. Let $\boldsymbol{\nu}' = (\nu^1, \nu^2, \dots, \nu^{j-1}, \nu_1^j, \nu_2^j, \nu^{j+1}, \dots, \nu^m) \in \mathcal{X}$. It is clear that $\tilde{L}_{\boldsymbol{\nu}} = \tilde{L}_{\boldsymbol{\nu}'}$ and $f(\boldsymbol{\nu}) = f(\boldsymbol{\nu}')$. Hence $L_{\boldsymbol{\nu}} = L_{\boldsymbol{\nu}'}$. Thus, in the definition of $\mathcal{P}_{\mathbf{V}}$, we may restrict ourselves to sequences $\boldsymbol{\nu} = (\nu^1, \nu^2, \dots, \nu^m) \in \mathcal{X}$ such that each ν^j is of the form $n\mathbf{i}$ for some $\mathbf{i} \in \mathbf{I}$ and some $n > 0$. Since there are only finitely many such $\boldsymbol{\nu}$ (subject to $\dim \mathbf{V}_{\mathbf{i}} = \sum_l \nu_{\mathbf{i}}^l$ for all $\mathbf{i} \in \mathbf{I}$) we see that $\mathcal{P}_{\mathbf{V}}$ has only finitely many simple objects, up to isomorphism.

9.1.5. In the special case where $\mathbf{V}$ is such that $\sum_{\mathbf{i}} \dim \mathbf{V}_{\mathbf{i}} \mathbf{i}$ is discrete, we have $\mathbf{E}_{\mathbf{V}} = 0$ and $\mathcal{P}_{\mathbf{V}}$ has exactly one simple object up to isomorphism, namely $\mathbf{1}$.

9.1.6. Let $K, K' \in \mathcal{Q}_{\mathbf{V}}$. The following two conditions are equivalent:

(a) $K \cong K'$;

(b) $\dim \mathbf{D}_j(\mathbf{E}_{\mathbf{V}}, G_{\mathbf{V}}; K, DB) = \dim \mathbf{D}_j(\mathbf{E}_{\mathbf{V}}, G_{\mathbf{V}}; K', DB)$ for all simple objects $B \in \mathcal{P}_{\mathbf{V}}$ and all $j \in \mathbf{Z}$.

It is clear that (a) implies (b). Assume now that K, K' are not isomorphic. Now K is a direct sum of complexes $L[n]$ where L runs over the isomorphism classes of simple objects L of $\mathcal{P}_{\mathbf{V}}$ and $n \in \mathbf{Z}$; let $m(L,n) \in \mathbf{N}$ be the number of times that $L[n]$ appears in this direct sum. We define similarly $m'(L,n)$ by replacing K by K'. Since K, K' are not isomorphic, we can find L_0, n_0 such that $m(L_0, n_0) \neq m'(L_0, n_0)$ and such that $m(L,n) = m'(L,n)$ for all L and all $n < n_0$. By (b), we have

$$\sum_{L,n} m(L,n) \dim \mathbf{D}_{j+n}(\mathbf{E}_{\mathbf{V}}, G_{\mathbf{V}}; L, DB) = \sum_{L,n} m'(L,n) \dim \mathbf{D}_{j+n}(\mathbf{E}_{\mathbf{V}}, G_{\mathbf{V}}; L, DB)$$

for all simple objects $B \in \mathcal{P}_{\mathbf{V}}$ and all $j \in \mathbf{Z}$.

Using 8.1.10(d), we rewrite this as follows:

$$m(B,-j) + \sum_L \sum_{n; n < -j} m(L,n) \dim \mathbf{D}_{j+n}(\mathbf{E}_{\mathbf{V}}, G_{\mathbf{V}}; L, DB)$$
$$= m'(B,-j) + \sum_L \sum_{n; n < -j} m'(L,n) \dim \mathbf{D}_{j+n}(\mathbf{E}_{\mathbf{V}}, G_{\mathbf{V}}; L, DB). \quad \text{(c)}$$

We apply this to $B = L_0$ and $j = -n_0$. Since by our assumption, $m(L,n) = m'(L,n)$ for $n < n_0$, we see that (c) implies $m(L_0, n_0) = m'(L_0, n_0)$. This is a contradiction. Thus the equivalence of (a),(b) is proved.

9.2. The functors Ind and Res

9.2.1. Let $\mathbf{T}, \mathbf{W}$ be two objects of $\mathcal{V}$. We can form $\mathbf{E_T}, \mathbf{E_W}$ and their product $\mathbf{E_T} \times \mathbf{E_W}$. This has an action of $G_\mathbf{T} \times G_\mathbf{W}$ (product of actions as in 9.1.2).

We define a full subcategory $\mathcal{P}_{\mathbf{T},\mathbf{W}}$ of $\mathcal{M}(\mathbf{E_T} \times \mathbf{E_W})$ and a full subcategory $\mathcal{Q}_{\mathbf{T},\mathbf{W}}$ of $\mathcal{D}(\mathbf{E_T} \times \mathbf{E_W})$, as a special case of the definitions of $\mathcal{P}_\mathbf{V}, \mathcal{Q}_\mathbf{V}$ in 9.1.3; indeed, $\mathbf{T} \times \mathbf{W}$ and $\mathbf{E_T} \times \mathbf{E_W}$ are special cases of $\mathbf{V}$ and $\mathbf{E_V}$ where the oriented graph in 9.1.1 has been replaced by the disjoint union of two copies of that oriented graph.

From the definitions it is clear that any simple object $B \in \mathcal{P}_{\mathbf{T},\mathbf{W}}$ is the external tensor product $B' \otimes B''$ of two simple objects $B' \in \mathcal{P}_\mathbf{T}$ and $B'' \in \mathcal{P}_\mathbf{W}$ (and conversely). Note that any complex in $\mathcal{Q}_{\mathbf{T},\mathbf{W}}$ is semisimple and $G_\mathbf{T} \times G_\mathbf{W}$-equivariant.

9.2.2. We assume that we are given $\mathbf{V}, \mathbf{T}, \mathbf{W}$ in $\mathcal{V}$, that $\mathbf{W}$ is a subspace of $\mathbf{V}$ and that $\mathbf{T} = \mathbf{V}/\mathbf{W}$. We also assume that the obvious maps $\mathbf{W} \to \mathbf{V}$ and $\mathbf{V} \to \mathbf{T}$ preserve the $\mathbf{I}$-grading. Let Q be the stabilizer of $\mathbf{W}$ in $G_\mathbf{V}$ (a parabolic subgroup of $G_\mathbf{V}$). We denote by U the unipotent radical of Q. We have canonically $Q/U = G_\mathbf{T} \times G_\mathbf{W}$.

Let F be the closed subvariety of $\mathbf{E_V}$ consisting of all $x \in \mathbf{E_V}$ such that $x_h(\mathbf{W}_{h'}) \subset \mathbf{W}_{h''}$ for all $h \in H$. We denote by $\iota : F \to \mathbf{E_V}$ the inclusion. Note that Q acts on F (restriction of the $G_\mathbf{V}$-action on $\mathbf{E_V}$).

If $x \in F$, then x induces elements $x' \in \mathbf{E_T}$ and $x'' \in \mathbf{E_W}$; the map $x \mapsto (x', x'')$ is a vector bundle $\kappa : F \to \mathbf{E_T} \times \mathbf{E_W}$. Now Q acts on $\mathbf{E_T} \times \mathbf{E_W}$ through its quotient $Q/U = G_\mathbf{T} \times G_\mathbf{W}$. The map κ is compatible with the Q-actions.

We set $G_\mathbf{V} = G, Q/U = \bar{G}, \mathbf{E_V} = E, \mathbf{E_T} \times \mathbf{E_W} = \bar{E}$. We have a diagram

$$\bar{E} \xleftarrow{\kappa} F \xrightarrow{\iota} E.$$

Let $E'' = G \times_P F, E' = G \times_U F$. We have a diagram

$$\bar{E} \xleftarrow{p_1} E' \xrightarrow{p_2} E'' \xrightarrow{p_3} E$$

where $p_1(g, f) = \kappa(f)$; $p_2(g, f) = (g, f)$; $p_3(g, f) = g(\iota(f))$. Note that p_1 is smooth with connected fibres, p_2 is a $\bar{G}$-principal bundle and p_3 is proper.

Let A be a complex in $\mathcal{Q}_{\mathbf{T},\mathbf{W}}$ and let B be a complex in $\mathcal{Q}_\mathbf{V}$. We can form $\kappa_!(\iota^* B) \in \mathcal{D}(\bar{E})$. Now $p_1^* A$ is a $\bar{G}$-equivariant semisimple complex on E'; hence $(p_2)_\flat p_1^* A$ is a well-defined semisimple complex on E'' (see 8.1.7(c)). We can form $(p_3)_!(p_2)_\flat p_1^* A \in \mathcal{D}(\mathbf{V})$.

Lemma 9.2.3. $(p_3)_!(p_2)_\flat p_1^*A \in \mathcal{Q}_{\mathbf{V}}$.

The general case can be immediately reduced to the case where A is a simple perverse sheaf in $\mathcal{P}_{\mathbf{T},\mathbf{W}}$ and this is immediately reduced to the case where $A = L_{\boldsymbol{\nu}'} \otimes L_{\boldsymbol{\nu}''}$. (Note that a direct summand of a complex in $\mathcal{Q}_{\mathbf{V}}$ belongs to $\mathcal{Q}_{\mathbf{V}}$.) Thus, it suffices to prove that $(p_3)_!(p_2)_\flat p_1^*(\tilde{L}_{\boldsymbol{\nu}'} \otimes \tilde{L}_{\boldsymbol{\nu}''}) \in \mathcal{Q}_{\mathbf{V}}$, where $\boldsymbol{\nu}' = (\nu_1', \nu_2', \ldots, \nu_{m'}') \in \mathcal{X}$ and $\boldsymbol{\nu}'' = (\nu_1'', \nu_2'', \ldots, \nu_{m''}'') \in \mathcal{X}$ satisfy $\dim \mathbf{T}_\mathbf{i}' = \sum_l \nu_\mathbf{i}'^l$, $\dim \mathbf{W}_\mathbf{i} = \sum_l \nu_\mathbf{i}''^l$ for all $\mathbf{i} \in \mathbf{I}$.

Let $\boldsymbol{\nu}'\boldsymbol{\nu}''$ be the sequence of elements in $\mathbf{N}[\mathbf{I}]$ formed by the elements of the sequence $\boldsymbol{\nu}'$ followed by the elements of the sequence $\boldsymbol{\nu}''$. Recall that $\tilde{\mathcal{F}}_{\boldsymbol{\nu}'\boldsymbol{\nu}''}$ consists of pairs (x, f) where $x \in \mathbf{E}_{\mathbf{V}}$ and f is a flag of type $\boldsymbol{\nu}'\boldsymbol{\nu}''$ in $\mathbf{V}$ which is x-stable. Now the subspace with index m' in $f = (\mathbf{V} = \mathbf{V}^0 \supset \mathbf{V}^1 \supset \ldots)$ is in the G-orbit of $\mathbf{W}$. The pairs (x, f) for which this subspace is equal to $\mathbf{W}$ form a closed subvariety $\tilde{\mathcal{F}}_{\boldsymbol{\nu}'\boldsymbol{\nu}'',0}$ of $\tilde{\mathcal{F}}_{\boldsymbol{\nu}'\boldsymbol{\nu}''}$; for such (x, f) we have $x \in F$, hence $(x, f) \to x$ defines a (proper) morphism $\tilde{\mathcal{F}}_{\boldsymbol{\nu}'\boldsymbol{\nu}'',0} \to F$. This morphism is Q-equivariant (for the natural actions of Q). Hence it induces a proper morphism $u : G \times_Q \tilde{\mathcal{F}}_{\boldsymbol{\nu}'\boldsymbol{\nu}'',0} \to G \times_Q F = E''$. Since $G \times_Q \tilde{\mathcal{F}}_{\boldsymbol{\nu}'\boldsymbol{\nu}'',0}$ is smooth, the complex $\tilde{L} = u_! \mathbf{1} \in \mathcal{D}(E'')$ is semisimple. (See 8.1.5.) It is clear from the definitions that $p_2^*\tilde{L} = p_1^*(\tilde{L}_{\boldsymbol{\nu}'} \otimes \tilde{L}_{\boldsymbol{\nu}''})$ and $(p_2)_\flat p_1^*(\tilde{L}_{\boldsymbol{\nu}'} \otimes \tilde{L}_{\boldsymbol{\nu}''}) \cong \tilde{L}$.

It remains to show that $(p_3)_!\tilde{L} \in \mathcal{Q}_{\mathbf{V}}$, or equivalently, that $(p_3 u)_! \mathbf{1} \in \mathcal{Q}_{\mathbf{V}}$. We may identify in a natural way $G \times_Q \tilde{\mathcal{F}}_{\boldsymbol{\nu}'\boldsymbol{\nu}'',0} = \tilde{\mathcal{F}}_{\boldsymbol{\nu}'\boldsymbol{\nu}''}$; then $p_3 u = \pi_{\boldsymbol{\nu}'\boldsymbol{\nu}''}$. It follows that $(p_3 u)_! \mathbf{1} = \tilde{L}_{\boldsymbol{\nu}'\boldsymbol{\nu}''}$ which is in $\mathcal{Q}_{\mathbf{V}}$ by definition. The lemma is proved.

Lemma 9.2.4. $\kappa_!(\iota^* B) \in \mathcal{Q}_{\mathbf{T},\mathbf{W}}$.

We may assume that B is a simple perverse sheaf in $\mathcal{P}_{\mathbf{V}}$. Since a direct summand of a complex in $\mathcal{Q}_{\mathbf{T},\mathbf{W}}$ belongs to $\mathcal{Q}_{\mathbf{T},\mathbf{W}}$ we see that it suffices to prove that $\kappa_!(\iota^* \tilde{L}_{\boldsymbol{\nu}}) \in \mathcal{Q}_{\mathbf{T},\mathbf{W}}$, where $\boldsymbol{\nu} \in \mathcal{X}$ satisfies $\dim \mathbf{V}_\mathbf{i} = \sum_l \nu_\mathbf{i}^l$ for all $\mathbf{i} \in \mathbf{I}$.

Let $\tilde{F} \subset \tilde{\mathcal{F}}_{\boldsymbol{\nu}}$ be the inverse image of $F \subset E$ under $\pi_{\boldsymbol{\nu}}$. Let $\tilde{\pi} : \tilde{F} \to F$ be the restriction of $\pi_{\boldsymbol{\nu}}$. We have $\iota^* \tilde{L}_{\boldsymbol{\nu}} = \tilde{\pi}_! \mathbf{1}$; hence

$$\kappa_!(\iota^* \tilde{L}_{\boldsymbol{\nu}}) = \kappa_! \tilde{\pi}_! \mathbf{1} = (\kappa \tilde{\pi})_! \mathbf{1}.$$

Let $\boldsymbol{\nu} = (\nu^1, \nu^2, \ldots, \nu^m)$. For any $\boldsymbol{\tau}, \boldsymbol{\omega} \in \mathcal{X}$ of the form

$$\boldsymbol{\tau} = (\tau^1, \tau^2, \ldots, \tau^m), \boldsymbol{\omega} = (\omega^1, \omega^2, \ldots, \omega^m)$$

such that $\tau^l + \omega^l = \nu^l$ for all l, we define a subvariety $\tilde{F}(\boldsymbol{\tau}, \boldsymbol{\omega})$ of $\tilde{F}$ as the set of all pairs (x, f) where $x \in F$ and $f = (\mathbf{V} = \mathbf{V}^0 \supset \mathbf{V}^1 \supset \cdots \supset \mathbf{V}^m = 0) \in$

$\tilde{\mathcal{F}}_{\nu}$ is x-stable and is such that the graded vector space $(\mathbf{V}^{l-1} \cap \mathbf{W})/(\mathbf{V}^l \cap \mathbf{W})$ belongs to $\mathcal{V}_{\omega^l}$ for $l = 1, 2, \ldots, m$.

If (x, f) is as above, then there are induced elements $(x', f') \in \tilde{\mathcal{F}}_{\boldsymbol{\tau}}$ and $(x'', f'') \in \tilde{\mathcal{F}}_{\boldsymbol{\omega}}$; here x'' is deduced from x by restriction to $\mathbf{W}$ and x' is deduced from x by passage to quotient; f' is given by the images of the subspaces in f under the projection $\mathbf{V} \to \mathbf{T}$ and f'' is given by the intersections of the subspaces in f with $\mathbf{W}$. Thus we have a morphism $\alpha : \tilde{F}(\boldsymbol{\tau}, \boldsymbol{\omega}) \to \tilde{\mathcal{F}}_{\boldsymbol{\tau}} \times \tilde{\mathcal{F}}_{\boldsymbol{\omega}}$. We have a commutative diagram

$$\begin{array}{ccc} \tilde{F}(\boldsymbol{\tau}, \boldsymbol{\omega}) & \longrightarrow & \tilde{F} \\ {\scriptstyle\alpha}\big\downarrow & & \big\downarrow{\scriptstyle\kappa\tilde{\pi}} \\ \tilde{\mathcal{F}}_{\boldsymbol{\tau}} \times \tilde{\mathcal{F}}_{\boldsymbol{\omega}} & \longrightarrow & \bar{E}\,. \end{array}$$

where the upper horizontal arrow is the obvious inclusion and the lower horizontal arrow is $\pi_{\tau} \times \pi_{\omega}$.

It is not difficult to verify (as in [9, 4.4]) that α is a (locally trivial) vector bundle of dimension $M(\boldsymbol{\tau}, \boldsymbol{\omega}) = \sum_{h; l' < l} \tau_{h'}^{l'} \omega_{h''}^{l} + \sum_{\mathbf{i}; l < l'} \tau_{\mathbf{i}}^{l'} \omega_{\mathbf{i}}^{l}$.

It is clear that the locally closed subvarieties $\tilde{F}(\boldsymbol{\tau}, \boldsymbol{\omega})$ form a partition of $\tilde{F}$. Let $\tilde{F}_j$ be the union of all subvarieties $\tilde{F}(\boldsymbol{\tau}, \boldsymbol{\omega})$ of fixed dimension j. Let Z_j be the disjoint union of the varieties $\tilde{\mathcal{F}}_{\boldsymbol{\tau}} \times \tilde{\mathcal{F}}_{\boldsymbol{\omega}}$ (union over those $(\boldsymbol{\tau}, \boldsymbol{\omega})$ such that $\tilde{F}(\boldsymbol{\tau}, \boldsymbol{\omega}) \subset \tilde{F}_j$). The maps α above can be assembled together to form a vector bundle $\tilde{F}_j \to Z_j$. The maps $\pi_{\boldsymbol{\tau}} \times \pi_{\boldsymbol{\omega}}$ can be assembled together to form a (proper) morphism $Z_j \to \bar{E}$. We have a commutative diagram

$$\begin{array}{ccc} \tilde{F}_j & \longrightarrow & \tilde{F} \\ \big\downarrow & & \big\downarrow{\scriptstyle\kappa\tilde{\pi}} \\ Z_j & \longrightarrow & \bar{E} \end{array}$$

We may therefore use 8.1.6 to conclude that $(\kappa\tilde{\pi})_!\mathbf{1}$ is a semisimple complex and that, for any i and j, there is a canonical exact sequence (in $\mathcal{M}(\bar{E})$):

(a) $$0 \to H^n(f_j)_!\mathbf{1} \to H^n(f_{\le j})_!\mathbf{1} \to H^n(f_{\le j-1})_!\mathbf{1} \to 0$$

where $f_j : \tilde{F}_j \to \bar{E}$ and $f_{\le j} : \cup_{j'; j' \le j} \tilde{F}_{j'} \to \bar{E}$ are the restrictions of $\kappa\tilde{\pi}$.

The earlier arguments show that

(b) $$(f_j)_!\mathbf{1} = \oplus(\tilde{L}_{\boldsymbol{\tau}} \otimes \tilde{L}_{\boldsymbol{\omega}})[-2M(\boldsymbol{\tau}, \boldsymbol{\omega})]$$

where the direct sum is taken over all $(\boldsymbol{\tau}, \boldsymbol{\omega})$ such that $\tilde{F}(\boldsymbol{\tau}, \boldsymbol{\omega}) \subset \tilde{F}_j$.

From (a),(b) we see by induction on j that all composition factors of $H^n(f_{\le j})_!\mathbf{1}$ are in $\mathcal{P}_{\mathbf{V}}$. Taking j large enough we see that all composition factors of $H^n(\kappa\tilde{\pi})_!\mathbf{1}$ are in $\mathcal{P}_{\mathbf{V}}$. Since $(\kappa\tilde{\pi})_!\mathbf{1}$ is semisimple, it follows that $(\kappa\tilde{\pi})_!\mathbf{1} \in \mathcal{Q}_{\mathbf{T},\mathbf{W}}$. The lemma is proved.

9.2.5. By Lemmas 9.2.3, 9.2.4, we have well-defined functors

$$\tilde{\mathrm{Ind}}^{\mathbf{V}}_{\mathbf{T},\mathbf{W}} : \mathcal{Q}_{\mathbf{T},\mathbf{W}} \to \mathcal{Q}_{\mathbf{V}} \quad (A \mapsto (p_3)_!(p_2)_\flat p_1^* A)$$

and

$$\tilde{\mathrm{Res}}^{\mathbf{V}}_{\mathbf{T},\mathbf{W}} : \mathcal{Q}_{\mathbf{V}} \to \mathcal{Q}_{\mathbf{T},\mathbf{W}} \quad (B \mapsto \kappa_!(\iota^* B)).$$

Since $\tilde{\mathrm{Ind}}^{\mathbf{V}}_{\mathbf{T},\mathbf{W}}$ is defined using a direct image under a proper map and inverse images under smooth morphisms with connected fibres, it commutes with Verdier duality up to shift (see 8.1.1, 8.1.4); more precisely,

$$D(\tilde{\mathrm{Ind}}^{\mathbf{V}}_{\mathbf{T},\mathbf{W}}(A)) = \tilde{\mathrm{Ind}}^{\mathbf{V}}_{\mathbf{T},\mathbf{W}}(D(A))[2d_1 - 2d_2]$$

where d_1 is the dimension of the fibres of p_1 and d_2 is the dimension of the fibres of p_2. We have

$$d_1 - d_2 = \sum_h \dim \mathbf{T}_{h'} \dim \mathbf{W}_{h''} + \sum_{\mathbf{i}} \dim \mathbf{T}_{\mathbf{i}} \dim \mathbf{W}_{\mathbf{i}}.$$

We set

$$\mathrm{Ind}^{\mathbf{V}}_{\mathbf{T},\mathbf{W}} = \tilde{\mathrm{Ind}}^{\mathbf{V}}_{\mathbf{T},\mathbf{W}}[d_1 - d_2].$$

Then

$$D(\mathrm{Ind}^{\mathbf{V}}_{\mathbf{T},\mathbf{W}}(A)) = \mathrm{Ind}^{\mathbf{V}}_{\mathbf{T},\mathbf{W}}(D(A)).$$

The functor $\mathrm{Ind}^{\mathbf{V}}_{\mathbf{T},\mathbf{W}}$ is called *induction*.

9.2.6. From the proof of 9.2.3 and of 9.2.4, we see that

(a) $$\tilde{\mathrm{Ind}}^{\mathbf{V}}_{\mathbf{T},\mathbf{W}}(\tilde{L}_{\nu'} \otimes \tilde{L}_{\nu''}) = \tilde{L}_{\nu'\nu''}$$

(b) $$\tilde{\mathrm{Res}}^{\mathbf{V}}_{\mathbf{T},\mathbf{W}} \tilde{L}_{\nu} \cong \oplus(\tilde{L}_{\tau} \otimes \tilde{L}_{\omega})[-2M(\tau,\omega)]$$

where the sum is taken over all $\tau = (\tau^1, \tau^2, \dots, \tau^m), \omega = (\omega^1, \omega^2, \dots, \omega^m)$ such that $\dim \mathbf{T}_{\mathbf{i}} = \sum_l \tau^l_{\mathbf{i}}, \dim \mathbf{W}_{\mathbf{i}} = \sum_l \omega^l_{\mathbf{i}}$ for all $\mathbf{i}$ and $\tau^l + \omega^l = \nu^l$ for all l.

9.2.7. We have $f(\nu'\nu'') = f(\nu) + f(\nu') + d_1 - d_2$; hence from 9.2.6(a) we deduce that

$$\operatorname{Ind}_{\mathbf{T},\mathbf{W}}^{\mathbf{V}}(L_{\nu'} \otimes L_{\nu''}) = L_{\nu'\nu''}.$$

9.2.8. Let Γ be a smooth irreducible variety with a free action of G. Let $\bar{\Gamma} = U\backslash\Gamma$. Then $\bar{\Gamma}$ is a smooth irreducible variety with a free action of $\bar{G}$ induced by that of G. Consider the diagram

$$E \xleftarrow{s} \Gamma \times E \xrightarrow{t} G\backslash(\Gamma \times E)$$

with the obvious maps s, t. As in 8.1.9, s^*A is a semisimple G-equivariant complex on $\Gamma \times E$ and, since t is a principal G-bundle, the semisimple complex $t_\flat s^*B \in \mathcal{D}(G\backslash(\Gamma \times E))$ is well-defined. In particular, we can replace B by $\tilde{\operatorname{Ind}}_{\mathbf{T},\mathbf{W}}^{\mathbf{V}} A$ and we obtain the semisimple complex

$$t_\flat s^*(\tilde{\operatorname{Ind}}_{\mathbf{T},\mathbf{W}}^{\mathbf{V}} A) \in \mathcal{D}(G\backslash(\Gamma \times E)).$$

Replacing E, Γ, G by $\bar{E}, \bar{\Gamma}, \bar{G}$, we obtain a similar diagram

$$\bar{E} \xleftarrow{\bar{s}} \bar{\Gamma} \times \bar{E} \xrightarrow{\bar{t}} \bar{G}\backslash(\bar{\Gamma} \times \bar{E})$$

and we can consider the semisimple complex $\bar{t}_\flat \bar{s}^* A \in \mathcal{D}(\bar{G}\backslash(\bar{\Gamma} \times \bar{E}))$. In particular, we can replace A by $\tilde{\operatorname{Res}}_{\mathbf{T},\mathbf{W}}^{\mathbf{V}} B$ and we obtain the semisimple complex

$$\bar{t}_\flat \bar{s}^*(\tilde{\operatorname{Res}}_{\mathbf{T},\mathbf{W}}^{\mathbf{V}} B) \in \mathcal{D}(\bar{G}\backslash(\bar{\Gamma} \times \bar{E})).$$

Let $u : G\backslash(\Gamma \times E) \to \{\text{point}\}$ and $\bar{u} : \bar{G}\backslash(\bar{\Gamma} \times \bar{E}) \to \{\text{point}\}$ be the obvious maps.

Lemma 9.2.9 (Adjunction). *We have a natural isomorphism*

$$\mathcal{H}^n u_!(t_\flat s^*(\tilde{Ind}_{\mathbf{T},\mathbf{W}}^{\mathbf{V}} A) \otimes t_\flat s^*(B)) \cong \mathcal{H}^n \bar{u}_!(\bar{t}_\flat \bar{s}^*(A) \otimes \bar{t}_\flat \bar{s}^*(\tilde{Res}_{\mathbf{T},\mathbf{W}}^{\mathbf{V}} B))$$

for $n \in \mathbf{Z}$. Hence, for any $j \in \mathbf{Z}$, we have

(a) $$\mathbf{D}_{j'}(E, G; \tilde{Ind}_{\mathbf{T},\mathbf{W}}^{\mathbf{V}} A, B) = \mathbf{D}_j(\bar{E}, \bar{G}; A, \tilde{Res}_{\mathbf{T},\mathbf{W}}^{\mathbf{V}} B)$$

where $j' = j + 2\dim G/Q$.

The proof (which uses 8.1.6) is given in [2]; we will not repeat it here. The shift from j to j' comes from the formula $\dim(G\backslash\Gamma) = \dim(\bar{G}\backslash\bar{\Gamma}) - \dim G/Q$.

9.2.10. In order to eliminate the shift from j to j' in the previous lemma, we define

$$\mathrm{Res}^{\mathbf{V}}_{\mathbf{T},\mathbf{W}}(B) = \tilde{\mathrm{Res}}^{\mathbf{V}}_{\mathbf{T},\mathbf{W}}[d_1 - d_2 - 2\dim G/Q],$$

where d_1, d_2 are as in 9.2.5. We can now rewrite the conclusion of the previous lemma as follows:

(a) $$\mathbf{D}_j(E, G; \mathrm{Ind}^{\mathbf{V}}_{\mathbf{T},\mathbf{W}} A, B) = \mathbf{D}_j(\bar{E}, \bar{G}; A, \mathrm{Res}^{\mathbf{V}}_{\mathbf{T},\mathbf{W}} B).$$

Note that $\dim G/Q = \sum_{\mathbf{i}} \dim \mathbf{T}_{\mathbf{i}} \dim \mathbf{W}_{\mathbf{i}}$; hence

$$d_1 - d_2 - 2\dim G/Q = \sum_h \dim \mathbf{T}_{h'} \dim \mathbf{W}_{h''} - \sum_{\mathbf{i}} \dim \mathbf{T}_{\mathbf{i}} \dim \mathbf{W}_{\mathbf{i}}.$$

The functor $\mathrm{Res}^{\mathbf{V}}_{\mathbf{T},\mathbf{W}}$ is called *restriction.*

9.2.11. We can rewrite 9.2.6(b) as follows:

$$\mathrm{Res}^{\mathbf{V}}_{\mathbf{T},\mathbf{W}} L_{\boldsymbol{\nu}} \cong \oplus (L_{\boldsymbol{\tau}} \otimes L_{\boldsymbol{\omega}})[M'(\boldsymbol{\tau}, \boldsymbol{\omega})]$$

where the sum is taken over all $\boldsymbol{\tau} = (\tau^1, \tau^2, \ldots, \tau^m), \boldsymbol{\omega} = (\omega^1, \omega^2, \ldots, \omega^m)$ such that $\dim \mathbf{T}_{\mathbf{i}} = \sum_l \tau^l_{\mathbf{i}}, \dim \mathbf{W}_{\mathbf{i}} = \sum_l \omega^l_{\mathbf{i}}$ for all $\mathbf{i}$ and $\tau^l + \omega^l = \nu^l$ for all l; we have

$$M'(\boldsymbol{\tau}, \boldsymbol{\omega}) = d_1 - d_2 - 2\dim G/Q + f(\boldsymbol{\nu}) - f(\boldsymbol{\tau}) - f(\boldsymbol{\omega}) - 2M(\boldsymbol{\tau}, \boldsymbol{\omega}).$$

We show that the last expression is independent of the orientation of our graph. From the definitions we have

$$M'(\boldsymbol{\tau}, \boldsymbol{\omega}) = \sum_h \dim \mathbf{T}_{h'} \dim \mathbf{W}_{h''} - \sum_{\mathbf{i}} \dim \mathbf{T}_{\mathbf{i}} \dim \mathbf{W}_{\mathbf{i}} + \sum_{h; l' < l} \tau^{l'}_{h'} \omega^l_{h''}$$
$$+ \sum_{h; l' < l} \omega^{l'}_{h'} \tau^l_{h''} + \sum_{\mathbf{i}; l < l'} \tau^{l'}_{\mathbf{i}} \omega^l_{\mathbf{i}} + \sum_{\mathbf{i}; l < l'} \omega^{l'}_{\mathbf{i}} \tau^l_{\mathbf{i}} - 2 \sum_{h; l' < l} \tau^{l'}_{h'} \omega^l_{h''} - 2 \sum_{\mathbf{i}; l < l'} \tau^{l'}_{\mathbf{i}} \omega^l_{\mathbf{i}}.$$

It follows that

$$M'(\boldsymbol{\tau}, \boldsymbol{\omega}) = - \sum_{h; l' < l} (\tau^{l'}_{h'} \omega^l_{h''} + \tau^{l'}_{h''} \omega^l_{h'})$$
$$+ \sum_h (\dim \mathbf{T}_{h'} \dim \mathbf{W}_{h''} + \dim \mathbf{T}_{h''} \dim \mathbf{W}_{h'})$$
$$- \sum_{\mathbf{i}; l < l'} \tau^{l'}_{\mathbf{i}} \omega^l_{\mathbf{i}} + \sum_{\mathbf{i}; l > l'} \tau^{l'}_{\mathbf{i}} \omega^l_{\mathbf{i}} - \sum_{\mathbf{i}} \dim \mathbf{T}_{\mathbf{i}} \dim \mathbf{W}_{\mathbf{i}},$$

which is clearly independent of orientation.

9.3. The Categories $\mathcal{P}_{\mathbf{V};\mathbf{I}';\geq\gamma}$ and $\mathcal{P}_{\mathbf{V};\mathbf{I}';\gamma}$

9.3.1. Let $\mathbf{I}'$ be a discrete subset of $\mathbf{I}$ (see 9.1.3). Let $\gamma = \sum_{\mathbf{i}} \gamma_{\mathbf{i}}\mathbf{i} \in \mathbf{N}[\mathbf{I}]$ be such that $\gamma_{\mathbf{i}} = 0$ for all $\mathbf{i} \in \mathbf{I} - \mathbf{I}'$. Let $\mathcal{P}_{\mathbf{V};\mathbf{I}';\geq\gamma}$ be the full subcategory of $\mathcal{P}_{\mathbf{V}}$ consisting of perverse sheaves which are direct sums of simple perverse sheaves L that have the following property: there exists a graded subspace $\mathbf{W} \subset \mathbf{V}$ and an object $A \in \mathcal{Q}_{\mathbf{T},\mathbf{W}}$ such that $\mathbf{T} = \mathbf{V}/\mathbf{W}$ satisfies $\dim \mathbf{T}_{\mathbf{i}} \geq \gamma_{\mathbf{i}}$ if $\mathbf{i} \in \mathbf{I}'$, and $\mathbf{T}_{\mathbf{i}'} = 0$ for $\mathbf{i}' \notin \mathbf{I}'$; moreover, L is a direct summand of $\widetilde{\mathrm{Ind}}_{\mathbf{T},\mathbf{W}}^{\mathbf{V}} A$.

Clearly,

(a) $\mathcal{P}_{\mathbf{V};\mathbf{I}';\geq\gamma} \supset \mathcal{P}_{\mathbf{V};\mathbf{I}';\geq\gamma'}$ if $\gamma' \in \mathbf{N}[\mathbf{I}]$ has support contained in $\mathbf{I}'$ and $\gamma_{\mathbf{i}} \leq \gamma'_{\mathbf{i}}$ for all $\mathbf{i} \in \mathbf{I}'$. Any object of $\mathcal{P}_{\mathbf{V}}$ is in $\mathcal{P}_{\mathbf{V};\mathbf{I}';\geq 0}$. Moreover, $\mathcal{P}_{\mathbf{V};\mathbf{I}',\geq\gamma}$ is empty if $\gamma_{\mathbf{i}} > \dim \mathbf{V}_{\mathbf{i}}$ for some $\mathbf{i} \in \mathbf{I}'$.

Let $\mathcal{P}_{\mathbf{V};\mathbf{I}';>\gamma}$ be the full subcategory of $\mathcal{P}_{\mathbf{V}}$ consisting of the objects which are in $\mathcal{P}_{\mathbf{V};\mathbf{i};\geq\gamma'}$ for some $\gamma' \in \mathbf{N}[\mathbf{I}]$ with support contained in $\mathbf{I}'$ such that $\gamma'(\mathbf{i}) \geq \gamma(\mathbf{i})$ for all $\mathbf{i} \in \mathbf{I}'$ and $\gamma'(\mathbf{i}) > \gamma(\mathbf{i})$ for some $\mathbf{i} \in \mathbf{I}'$.

Let $\mathcal{P}_{\mathbf{V};\mathbf{I}';\gamma}$ be the full subcategory of $\mathcal{P}_{\mathbf{V};\mathbf{i};\geq\gamma}$ consisting of those objects of $\mathcal{P}_{\mathbf{V};\mathbf{i};\geq\gamma}$ which are not in $\mathcal{P}_{\mathbf{V};\mathbf{i};>\gamma}$. If K is a simple object of $\mathcal{P}_{\mathbf{V}}$ and $\mathbf{V} \neq 0$, then K is a direct summand of some shift of $L_{\boldsymbol{\nu}}$ where $\boldsymbol{\nu}$ starts with $\nu^1 = \gamma$ which may be assumed to be of form $n\mathbf{i}$ for some $\mathbf{i} \in \mathbf{I}$ and some $n > 0$ (see 9.1.4); we see then that $K \in \mathcal{P}_{\mathbf{V};\{\mathbf{i}\};\geq n\mathbf{i}}$. Thus:

(b) if K is a simple object of $\mathcal{P}_{\mathbf{V}}$ and $\mathbf{V} \neq 0$, then there exists $\mathbf{i} \in \mathbf{I}$ such that $K \in \mathcal{P}_{\mathbf{V};\{\mathbf{i}\};\geq\mathbf{i}}$.

9.3.2. We now assume that $\mathbf{W} \subset \mathbf{V}$ and $\mathbf{T} = \mathbf{V}/\mathbf{W}$ are such that for any $h \in H$ we have $\mathbf{T}_{h'} = 0$ (hence $\mathbf{W}_{h'} = \mathbf{V}_{h'}$). It follows that $\mathbf{E}_{\mathbf{T}} = 0$. Moreover, we have a natural imbedding $\iota : \mathbf{E}_{\mathbf{W}} \to \mathbf{E}_{\mathbf{V}}$; if $x = (x_h) \in \mathbf{E}_{\mathbf{W}}$, then the h-component of $\iota(x) = x'$ is the composition $\mathbf{V}_{h'} = \mathbf{W}_{h'} \xrightarrow{x_h} \mathbf{W}_{h''} \subset \mathbf{V}_{h''}$. (In our case we have $\kappa : F \cong \mathbf{E}_{\mathbf{W}}$ and the imbedding ι above may be identified with the imbedding $F \to \mathbf{E}_{\mathbf{V}}$, see 9.2.2.) From our assumption it follows that the set $\{\mathbf{i} \in \mathbf{I} | \mathbf{T}_{\mathbf{i}} \neq 0\}$ is discrete; let $\mathbf{I}'$ be a discrete subset of $\mathbf{I}$ containing $\{\mathbf{i} \in \mathbf{I} | \mathbf{T}_{\mathbf{i}} \neq 0\}$.

We consider the locally closed subset Θ of $\mathbf{E}_{\mathbf{V}}$ consisting of all $x \in \mathbf{E}_{\mathbf{V}}$ such that $\dim \mathbf{V}_{\mathbf{i}}/(\sum_{h\in H:h''=\mathbf{i}} x_h(\mathbf{V}_{h'})) = \dim \mathbf{T}_{\mathbf{i}}$ for all $\mathbf{i} \in \mathbf{I}'$, and the open subset Ξ of $\mathbf{E}_{\mathbf{W}}$ consisting of all $x \in \mathbf{E}_{\mathbf{W}}$ such that $\sum_{h\in H:h''=\mathbf{i}} x_h(\mathbf{W}_{h'}) = \mathbf{W}_{\mathbf{i}}$ for all $\mathbf{i} \in \mathbf{I}'$.

Let $p : G \times_Q \mathbf{E}_{\mathbf{W}} \to \mathbf{E}_{\mathbf{V}}$ be the unique G-equivariant map such that $(1, x) \mapsto i(x)$ for all $x \in \mathbf{E}_{\mathbf{W}}$; let $p_0 : G \times_Q \Xi \to \Theta$ be the restriction of p. Note that p_0 is an isomorphism. The inverse map can be described as follows. Let $x \in \Theta$. The **I**-graded subspace

$$\oplus_{\mathbf{i}\in\mathbf{I}'}\Big(\sum_{h\in H:h''=\mathbf{i}} x_h(\mathbf{V}_{h'})\Big) \oplus (\oplus_{\mathbf{i}\in\mathbf{I}-\mathbf{I}'}\mathbf{V}_{\mathbf{i}})$$

of $\mathbf{V}$ has the same dimension in each degree as $\mathbf{W}$; hence it is equal to $g(\mathbf{W})$ for some $g \in G$. The element gx is equal to $\iota(x')$ for a well-defined $x' \in \mathbf{E_W}$. Then $p_0^{-1}(x) = (g, x')$. In particular, $\Theta, G \times_Q \Xi, G \times_Q \mathbf{E_W}$ are irreducible of the same dimension.

Since p is a proper map, its image $p(G \times_Q \mathbf{E_W})$ is a closed subset of $\mathbf{E_V}$ containing Θ. Hence $\dim G \times_Q \mathbf{E_W} \geq \dim p(G \times_Q \mathbf{E_W}) \geq \dim \Theta$. It follows that these inequalities are equalities; hence Θ is open dense in $p(G \times_Q \mathbf{E_W})$.

We have a commutative diagram

$$\begin{array}{ccccc} G \times_Q \Xi & \xrightarrow{p_0} & \Theta & \xleftarrow{\iota_0} & \Xi \\ {\scriptstyle j_0}\downarrow & & {\scriptstyle j}\downarrow & & {\scriptstyle m}\downarrow \\ G \times_Q \mathbf{E_W} & \xrightarrow{p} & \mathbf{E_V} & \xleftarrow{\iota} & \mathbf{E_W} \end{array}$$

where ι_0, j, j_0, m denote the inclusions. Both squares in the diagram are cartesian.

Let $\mathcal{P}^0_{\mathbf{W}}$ be the full subcategory of $\mathcal{P}_{\mathbf{W}}$ whose objects are those perverse sheaves A such that any simple constituent of A has support which meets Ξ. Let $\mathcal{P}^1_{\mathbf{W}}$ be the full subcategory of $\mathcal{P}_{\mathbf{W}}$ whose objects are those perverse sheaves A such that the support of A is contained in $\mathbf{E_W} - \Xi$.

Let $\mathcal{P}^0_{\mathbf{V}}$ be the full subcategory of $\mathcal{P}_{\mathbf{V}}$ whose objects are those perverse sheaves B such that any simple constituent of B has support which meets Θ and is contained in the closure of Θ. Let $\mathcal{P}^1_{\mathbf{V}}$ be the full subcategory of $\mathcal{P}_{\mathbf{V}}$ whose objects are those perverse sheaves B such that the support of B is disjoint from Θ and is contained in the closure of Θ. Clearly, any object $A \in \mathcal{P}_{\mathbf{W}}$ has a canonical decomposition $A = A^0 \oplus A^1$ where $A^0 \in \mathcal{P}^0_{\mathbf{W}}$ and $A^1 \in \mathcal{P}^1_{\mathbf{W}}$. Moreover, any object $B \in \mathcal{P}_{\mathbf{V}}$ with support contained in the closure of Θ has a canonical decomposition $B = B^0 \oplus B^1$ where $B^0 \in \mathcal{P}^0_{\mathbf{V}}$ and $B^1 \in \mathcal{P}^1_{\mathbf{V}}$.

Proposition 9.3.3. (a) *Let $A \in \mathcal{P}^0_{\mathbf{W}}$. If $n \neq 0$, we have $H^n \mathrm{Ind}^{\mathbf{V}}_{\mathbf{T},\mathbf{W}} A \in \mathcal{P}^1_{\mathbf{V}}$. If $n = 0$, then $H^n \mathrm{Ind}^{\mathbf{V}}_{\mathbf{T},\mathbf{W}} A \in \mathcal{P}_{\mathbf{V}}$ has support contained in the closure of Θ; hence $\xi(A) = (H^{\dim G/Q} \mathrm{Ind}^{\mathbf{V}}_{\mathbf{T},\mathbf{W}} A)^0 \in \mathcal{P}^0_{\mathbf{V}}$ is defined.*

(b) *Let $B \in \mathcal{P}^0_{\mathbf{V}}$. If $n \neq 0$, we have $H^n \mathrm{Res}^{\mathbf{V}}_{\mathbf{T},\mathbf{W}} B \in \mathcal{P}^1_{\mathbf{W}}$. If $n = 0$, then $H^n \mathrm{Res}^{\mathbf{V}}_{\mathbf{T},\mathbf{W}} B \in \mathcal{P}_{\mathbf{W}}$ hence $\rho(B) = (H^{-\dim G/Q} \mathrm{Res}^{\mathbf{V}}_{\mathbf{T},\mathbf{W}} B)^0 \in \mathcal{P}^0_{\mathbf{W}}$ is defined.*

(c) *The functors* $\xi : \mathcal{P}^0_{\mathbf{W}} \to \mathcal{P}^0_{\mathbf{V}}$ *and* $\rho : \mathcal{P}^0_{\mathbf{V}} \to \mathcal{P}^0_{\mathbf{W}}$ *establish equivalences of categories inverse to each other.*

Note that j^*B is a perverse sheaf on Θ, since the support of B is contained in the closure of Θ and Θ is open in its closure. Moreover, j^*B is a G-equivariant perverse sheaf. Since $\Theta = G \times_Q \Xi$, it follows that $\iota_0^*(j^*B)[-\dim G/Q]$ is a perverse sheaf on Ξ. But $m^*\iota^*B = \iota_0^* j^*B$, hence $m^*\iota^*B[-\dim G/Q]$ is a perverse sheaf on Ξ. Since m is an open imbedding, we have $m^*(H^n\iota^*B) = H^n(m^*\iota^*B)$ for any n, and this is zero if $n \neq -\dim G/Q$. Hence if $n \neq -\dim G/Q$, the support of $H^n\iota^*B$ is disjoint from Ξ. We have $\tilde{\mathrm{Res}}^{\mathbf{V}}_{\mathbf{T},\mathbf{W}}B = \mathrm{Res}^{\mathbf{V}}_{\mathbf{T},\mathbf{W}}B[\dim G/Q] = \iota^*B$ and (b) is proved.

Let $\tilde{A}$ be the perverse sheaf on $G \times_Q \mathbf{E}_{\mathbf{W}}$ such that $\tilde{A} = r'_\flat r^*A[\dim G/Q]$ in the diagram $\mathbf{E}_{\mathbf{W}} \xleftarrow{r} G/U \times \mathbf{E}_{\mathbf{W}} \xrightarrow{r'} G \times_Q \mathbf{E}_{\mathbf{W}}$. By definition,

$$\tilde{\mathrm{Ind}}^{\mathbf{V}}_{\mathbf{T},\mathbf{W}}A = p_! r'_\flat r^* A = p_! \tilde{A}[-\dim G/Q].$$

This shows that $\tilde{\mathrm{Ind}}^{\mathbf{V}}_{\mathbf{T},\mathbf{W}}A$ has support contained in the image of p, hence in the closure of Θ. We have

$$j^*(\tilde{\mathrm{Ind}}^{\mathbf{V}}_{\mathbf{T},\mathbf{W}}A) = j^* p_! \tilde{A}[-\dim G/Q] = (p_0)_! j_0^* \tilde{A}[-\dim G/Q].$$

Since j_0 is an open imbedding, we see that $j_0^*\tilde{A}$ is a perverse sheaf on $G \times_Q \Xi$. Since $p_0 : G \times_Q \Xi \to \Theta$ is an isomorphism, it follows that $(p_0)_! j_0^* \tilde{A}$ is a perverse sheaf on Θ. Thus $j^*(\tilde{\mathrm{Ind}}^{\mathbf{V}}_{\mathbf{T},\mathbf{W}}A)[\dim G/Q]$ is a perverse sheaf on Θ.

Since $\tilde{\mathrm{Ind}}^{\mathbf{V}}_{\mathbf{T},\mathbf{W}}A$ has support contained in the closure of Θ, it follows that $H^n\tilde{\mathrm{Ind}}^{\mathbf{V}}_{\mathbf{T},\mathbf{W}}A$ has support contained in the closure of Θ and $j^*(H^n\tilde{\mathrm{Ind}}^{\mathbf{V}}_{\mathbf{T},\mathbf{W}}A) = H^n(j^*\tilde{\mathrm{Ind}}^{\mathbf{V}}_{\mathbf{T},\mathbf{W}}A)$. This is zero if $n \neq \dim G/Q$. Hence for $n \neq \dim G/Q$, the support of $H^n\tilde{\mathrm{Ind}}^{\mathbf{V}}_{\mathbf{T},\mathbf{W}}A$ is disjoint from Θ. This proves (a), since $\mathrm{Ind}^{\mathbf{V}}_{\mathbf{T},\mathbf{W}}A = \tilde{\mathrm{Ind}}^{\mathbf{V}}_{\mathbf{T},\mathbf{W}}A[\dim G/Q]$.

From the proof of (b), we have

$$\begin{aligned} m^*(\rho(B)) &= m^*(H^{-\dim G/Q}\iota^*B) \\ &= m^*(\iota^*B[-\dim G/Q]) \\ &= \iota_0^* j^* B[-\dim G/Q]. \end{aligned}$$

This implies that $j_0^*\tilde{\rho}(B) = p_0^* j^* B$. From the proof of (a), we have

$$\begin{aligned} j^*(\xi(A)) &= j^*(H^{\dim G/Q}\tilde{\mathrm{Ind}}^{\mathbf{V}}_{\mathbf{T},\mathbf{W}}A) \\ &= j^*(\tilde{\mathrm{Ind}}^{\mathbf{V}}_{\mathbf{T},\mathbf{W}}A[\dim G/Q]) = (p_0)_! j_0^* \tilde{A}. \end{aligned}$$

Hence

$$m^*(\rho(\xi(A)) = \iota_0^* j^*(\xi(A))[-\dim G/Q] = \iota_0^*(p_0)_! j_0^* \tilde{A}[-\dim G/Q] = m^* A$$

and

$$j^*(\xi(\rho(B)) = (p_0)_! j_0^* \tilde{\rho}(B) = (p_0)_! p_0^* j^* B = j^* B.$$

Since $A \in \mathcal{P}^0_{\mathbf{W}}$ and $B \in \mathcal{P}^0_{\mathbf{V}}$, it follows that $\rho(\xi(A)) = A$ and $\xi(\rho(B)) = B$. The proposition is proved.

9.3.4. Assume that $\mathbf{I}'$ is a subset of $\mathbf{I}$ such that $h' \notin \mathbf{I}'$ for any $h \in H$, that is, $\mathbf{i}$ is a *sink* of our quiver, for any $\mathbf{i} \in \mathbf{I}'$. Let $\mathbf{V} \in \mathcal{V}$. For any $\gamma = \sum_{\mathbf{i}} \gamma_{\mathbf{i}} \mathbf{i} \in \mathbf{N}[\mathbf{I}]$ with support contained in $\mathbf{I}'$, let $\mathbf{E}_{\mathbf{V};\gamma}$ be the set of all $x \in \mathbf{E}_{\mathbf{V}}$ such that

$$\dim \mathbf{V}_{\mathbf{i}}/(\sum_{h \in H: h''=\mathbf{i}} x_h(\mathbf{V}_{h'})) = \gamma_{\mathbf{i}}$$

for any $\mathbf{i} \in \mathbf{I}'$. The sets $\mathbf{E}_{\mathbf{V};\gamma}$ form a partition of $\mathbf{E}_{\mathbf{V}}$ with the following property: for any γ as above, the union $\mathbf{E}_{\mathbf{V};\geq\gamma} = \cup_{\gamma'} \mathbf{E}_{\mathbf{V};\gamma'}$ (with γ' running over the elements of $\mathbf{N}[\mathbf{I}]$ with support contained in $\mathbf{I}'$ such that $\gamma'_{\mathbf{i}} \geq \gamma_{\mathbf{i}}$ for all $\mathbf{i} \in \mathbf{I}$) is a closed subset of $\mathbf{E}_{\mathbf{V}}$. Hence for any simple object B of $\mathcal{P}_{\mathbf{V}}$, there is a unique element $\gamma = \gamma^B \in \mathbf{N}[\mathbf{I}]$ with support contained in $\mathbf{I}'$ such that the support of B is contained in $\mathbf{E}_{\mathbf{V};\geq\gamma}$ and meets $\mathbf{E}_{\mathbf{V};\gamma}$. We have $\gamma_{\mathbf{i}} \leq \dim \mathbf{V}_{\mathbf{i}}$ for all $\mathbf{i} \in \mathbf{I}'$.

Lemma 9.3.5. *Assume that $B \in \mathcal{P}_{\mathbf{V};\mathbf{I}';\gamma'}$ where $\gamma' \in \mathbf{N}[\mathbf{I}]$ has support contained in $\mathbf{I}'$. Then $\gamma^B = \gamma'$.*

We write γ instead of γ^B. Let $\mathbf{W}$ be a graded subspace of $\mathbf{V}$ such that $\mathbf{T} = \mathbf{V}/\mathbf{W}$ satisfies $\dim \mathbf{T}_{\mathbf{i}} = \gamma_{\mathbf{i}}$ for all $\mathbf{i} \in \mathbf{I}'$ and $\mathbf{T}_{\mathbf{i}'} = 0$ for all $\mathbf{i}' \in \mathbf{I} - \mathbf{I}'$. We may apply Proposition 9.3.3 to $\mathbf{I}'$ and B. With notations there, let $A = \rho(B) \in \mathcal{P}_{\mathbf{W}}$; we have that some shift of B is a direct summand of $\widetilde{\mathrm{Ind}}_{\mathbf{T},\mathbf{W}}^{\mathbf{V}} A$. Hence $B \in \mathcal{P}_{\mathbf{V};\mathbf{I}';\geq\gamma}$.

From the definition of induction we see that any perverse sheaf in $\mathcal{P}_{\mathbf{V};\mathbf{I}';\geq\gamma'}$ has support contained in $\mathbf{E}_{\mathbf{V};\geq\gamma'}$. In particular, the support of B is contained in $\mathbf{E}_{\mathbf{V};\geq\gamma'}$. By definition, the support of B meets $\mathbf{E}_{\mathbf{V};\gamma}$; hence $\mathbf{E}_{\mathbf{V};\gamma}$ meets $\mathbf{E}_{\mathbf{V};\geq\gamma'}$, so that $\gamma_{\mathbf{i}} \geq \gamma'_{\mathbf{i}}$ for all $\mathbf{i} \in \mathbf{I}'$.

Assume that $\gamma_{\mathbf{i}} > \gamma'_{\mathbf{i}}$ for some $\mathbf{i} \in \mathbf{I}'$. Since $B \in \mathcal{P}_{\mathbf{V};\mathbf{I}';\geq\gamma}$, it follows that $B \in \mathcal{P}_{\mathbf{V};\mathbf{I}';>\gamma'}$, which contradicts our assumption that $B \in \mathcal{P}_{\mathbf{V};\mathbf{I}';\gamma'}$. Thus, we must have $\gamma_{\mathbf{i}} = \gamma'_{\mathbf{i}}$ for all $\mathbf{i} \in \mathbf{I}'$. The lemma is proved.

CHAPTER 10

Fourier-Deligne Transform

10.1. Fourier-Deligne Transform and Restriction

10.1.1. In addition to the orientation $h \to h'$, $h \to h''$ in 9.1.1, we shall consider a new orientation of our graph. Thus, we assume we are given two new maps $H \to I$ denoted $h \mapsto {}'h$ and $h \mapsto {}''h$, such that for any $h \in H$, the subset $[h]$ of $\mathbf{I}$ consists precisely of ${}'h, {}''h$. Let

$$H_1 = \{h \in H | {}'h = h' \text{ and } {}''h = h''\}; \quad H_2 = \{h \in H | {}'h = h'' \text{ and } {}''h = h'\}.$$

Then H_1, H_2 form a partition of H.

For $\mathbf{V} \in \mathcal{V}$, we define ${}'\mathbf{E_V}$ like $\mathbf{E_V}$ in 9.1.2, but using the new orientation: ${}'\mathbf{E_V} = \oplus_{h \in H} \mathrm{Hom}(\mathbf{V}_{'h}, \mathbf{V}_{''h})$. This has a natural $G_\mathbf{V}$-action just like $\mathbf{E_V}$.

We have

$$\mathbf{E_V} = \oplus_{h \in H_1} \mathrm{Hom}(\mathbf{V}_{h'}, \mathbf{V}_{h''}) \oplus (\oplus_{h \in H_2} \mathrm{Hom}(\mathbf{V}_{h'}, \mathbf{V}_{h''})),$$

$${}'\mathbf{E_V} = \oplus_{h \in H_1} \mathrm{Hom}(\mathbf{V}_{h'}, \mathbf{V}_{h''}) \oplus (\oplus_{h \in H_2} \mathrm{Hom}(\mathbf{V}_{h''}, \mathbf{V}_{h'})).$$

Let $\dot{\mathbf{E}}_\mathbf{V}$ be the vector space

$$\oplus_{h \in H_1} \mathrm{Hom}(\mathbf{V}_{h'}, \mathbf{V}_{h''}) \oplus (\oplus_{h \in H_2} \mathrm{Hom}(\mathbf{V}_{h'}, \mathbf{V}_{h''})) \oplus (\oplus_{h \in H_2} \mathrm{Hom}(\mathbf{V}_{h''}, \mathbf{V}_{h'})).$$

We have the diagram

$$\text{(a)} \qquad \mathbf{E_V} \xleftarrow{s} \dot{\mathbf{E}}_\mathbf{V} \xrightarrow{t} {}'\mathbf{E_V}$$

where s, t are the obvious projections.

Let $T : \dot{\mathbf{E}}_\mathbf{V} \to k$ be the map given by $T(e) = \sum_{h \in H_2} \mathrm{tr}\ (\mathbf{V}_{h'} \to \mathbf{V}_{h''} \to \mathbf{V}_{h'})$ where the two unnamed maps are components of e. Let us consider the Fourier-Deligne transform $\Phi : \mathcal{D}(\mathbf{E_V}) \to \mathcal{D}({}'\mathbf{E_V})$ defined by $\Phi(K) = t_!(s^*(K) \otimes \mathcal{L}_T)[d_\mathbf{V}]$ where $d_\mathbf{V} = \sum_{h \in H_2} \dim \mathbf{V}_{h'} \dim \mathbf{V}_{h''}$. (See 8.1.11.) Now let $\mathbf{T}, \mathbf{W}$ be as in 9.2.1. We may consider a diagram like (a) for $\mathbf{T}$ and for $\mathbf{W}$ instead of $\mathbf{V}$; taking direct products, we obtain the diagram

$$\mathbf{E_T} \times \mathbf{E_W} \xleftarrow{\bar{s}} \dot{\mathbf{E}}_\mathbf{T} \times \dot{\mathbf{E}}_\mathbf{W} \xrightarrow{\bar{t}} {}'\mathbf{E_T} \times {}'\mathbf{E_W}.$$

G. Lusztig, *Introduction to Quantum Groups*, Modern Birkhäuser Classics,
DOI 10.1007/978-0-8176-4717-9_10, © Springer Science+Business Media, LLC 2010

On each of $\dot{\mathbf{E}}_{\mathbf{T}}$ and $\dot{\mathbf{E}}_{\mathbf{W}}$ we have a linear form like T above; the sum of these gives a linear form $\bar{T} : \dot{\mathbf{E}}_{\mathbf{T}} \times \dot{\mathbf{E}}_{\mathbf{W}} \to k$. The Fourier-Deligne transform $\Phi : \mathcal{D}(\mathbf{E}_{\mathbf{T}} \times \mathbf{E}_{\mathbf{W}}) \to \mathcal{D}('\mathbf{E}_{\mathbf{T}} \times '\mathbf{E}_{\mathbf{W}})$ is given by

$$\Phi(K) = \bar{t}_!(\bar{s}^*(K) \otimes \mathcal{L}_{\bar{T}})[d_{\mathbf{T}} + d_{\mathbf{W}}].$$

The following result shows the relation between the Fourier-Deligne transform and the restriction functor.

Proposition 10.1.2. *For any $K \in \mathcal{Q}_{\mathbf{V}}$ we have*

$$\Phi(\tilde{Res}^{\mathbf{V}}_{\mathbf{T},\mathbf{W}} K) = \tilde{Res}^{\mathbf{V}}_{\mathbf{T},\mathbf{W}}(\Phi(K))[\pi]$$

where

$$\pi = \sum_{h \in H_2} (\dim \mathbf{T}_{h''} \dim \mathbf{W}_{h'} - \dim \mathbf{T}_{h'} \dim \mathbf{W}_{h''}).$$

We consider the commutative diagram of vector spaces and linear maps

$$\begin{array}{ccccccccc}
\mathbf{E}_{\mathbf{T}} \times \mathbf{E}_{\mathbf{W}} & \xleftarrow{p} & F & & & \xrightarrow{\iota} & & & \mathbf{E}_{\mathbf{V}} \\
\uparrow{\scriptstyle \bar{s}} & & \uparrow{\scriptstyle \dot{s}} & & & & & & \uparrow{\scriptstyle s} \\
\dot{\mathbf{E}}_{\mathbf{T}} \times \dot{\mathbf{E}}_{\mathbf{W}} & \xleftarrow{\dot{p}} & \Psi & \xleftarrow{\dot{q}} & \dot{F} & \xrightarrow{\zeta} & \Xi & \xrightarrow{i} & \dot{\mathbf{E}}_{\mathbf{V}} \\
\downarrow{\scriptstyle \bar{t}} & & & & & & \downarrow{\scriptstyle \dot{t}} & & \downarrow{\scriptstyle t} \\
'\mathbf{E}_{\mathbf{T}} \times '\mathbf{E}_{\mathbf{W}} & & & \xleftarrow{'p} & & & 'F & \xrightarrow{'\iota} & '\mathbf{E}_{\mathbf{V}}
\end{array}$$

where the following notation is used.

F is the set of all $x \in \mathbf{E}_{\mathbf{V}}$ such that $x_h(\mathbf{W}_{h'}) \subset \mathbf{W}_{h''}$ for all $h \in H$; p is the obvious surjective map and ι is the obvious imbedding.

$'F$ is the set of all $x \in '\mathbf{E}_{\mathbf{V}}$ such that $x_h(\mathbf{W}_{'h}) \subset \mathbf{W}_{''h}$ for all $h \in H$; $'p$ is the obvious surjective map and $'\iota$ is the obvious imbedding.

$\dot{F}$ is the set of all $x \in \dot{\mathbf{E}}_{\mathbf{V}}$ such that $sx \in F$ and $tx \in 'F$.

Ξ is defined by the condition that $(i, t, \dot{t}, '\iota)$ is a cartesian diagram.

Ψ is defined by the condition that $(\dot{s}, p, \dot{p}, \bar{s})$ is a cartesian diagram.

$\dot{q}$ is such that $\dot{s}\dot{q}$ and $\dot{p}\dot{q}$ are the obvious surjective maps.

ζ is such that $i\zeta$ and $\dot{t}\zeta$ are the obvious imbeddings.

We have $\Xi = 'F \oplus (\oplus_{h \in H_2} \mathrm{Hom}(\mathbf{V}_{h'}, \mathbf{V}_{h''}))$. Let Z be the subspace of Ξ consisting of the elements such that each component $\mathbf{V}_{h''} \to \mathbf{V}_{h'}$ $(h \in H_2)$ carries $\mathbf{W}_{h''}$ to 0 and all other components are zero. Let $c : \Xi \to \Xi/Z$ be the canonical map. Let $\tilde{T} : \Xi \to k$ be given by $\tilde{T}(x) = T(\iota(x))$. From definitions, it follows immediately that the restriction of $\tilde{T}$ to a fibre ($\cong Z$) of $c : \Xi \to \Xi/Z$ is an affine-linear function which is constant if and only if that fibre is contained in the subspace $\dot{\zeta}(\dot{F})$.

Let $\Xi' = \Xi - \dot{\zeta}(\dot{F})$, and let $(\Xi/Z)' = c(\Xi')$. We have $Z \subset \dot{\zeta}(\dot{F})$; hence all fibres of $c' : \Xi' \to (\Xi/Z)'$ (restriction of c) are isomorphic to Z.

Let $T' : \Xi' \to k$ be the restriction of $\tilde{T}$. As we have seen above, the restriction of T' to any fibre of $c' : \Xi' \to (\Xi/Z)'$ is a non-constant affine-linear function. Hence the local system $\mathcal{L}_{T'}$ on Ξ' satisfies $c'_!(\mathcal{L}_{T'}) = 0$ (see 8.1.13). Using the distinguished triangle associated to the partition $\Xi = \Xi' \cup \dot{\zeta}(\dot{F})$, we deduce that $c_!\dot{\zeta}_!(\dot{\zeta}^*\mathcal{L}_{\tilde{T}}) = c_!\mathcal{L}_{\tilde{T}}$. It is clear that the composition $si : \Xi \to \mathbf{E_V}$ factors through Ξ/Z; hence i^*s^*K is in the image of c^* so that the previous equality implies

$$c_!(\dot{\zeta}_!(\dot{\zeta}^*\mathcal{L}_{\tilde{T}}) \otimes i^*s^*K) = c_!(\mathcal{L}_{\tilde{T}} \otimes i^*s^*K).$$

It is also clear that the composition $'p\dot{t} : \Xi \to {}'\mathbf{E_T} \times {}'\mathbf{E_W}$ factors through Ξ/Z. Hence the previous equality implies

$$'p_!\dot{t}_!(\dot{\zeta}_!(\dot{\zeta}^*\mathcal{L}_{\tilde{T}}) \otimes i^*s^*K) = {}'p_!\dot{t}_!(\mathcal{L}_{\tilde{T}} \otimes i^*s^*K).$$

We have $Ti\dot{\zeta} = \bar{T}\dot{p}\dot{q}$; hence $\dot{p}^*\dot{q}^*\mathcal{L}_{\bar{T}} = \dot{\zeta}^*i^*\mathcal{L}_T = \dot{\zeta}^*\mathcal{L}_{\tilde{T}}$. Since $\dot{q}$ is a surjective linear map with kernel of dimension

$$m = \sum_{h \in H_2} \dim \mathbf{T}_{h''} \dim \mathbf{W}_{h'},$$

we obtain $\dot{q}_!\dot{q}^*L = L[-2m]$ for all $L \in \mathcal{D}(\Psi)$. We have

$$\begin{aligned}
\Phi(\mathrm{Res}^{\mathbf{V}}_{\mathbf{T},\mathbf{W}} K) &= \bar{t}_!(\mathcal{L}_T \otimes \bar{s}^* p_! \iota^* K)[d_{\mathbf{T}} + d_{\mathbf{W}}] \\
&= \bar{t}_!(\mathcal{L}_T \otimes \dot{p}_! \dot{s}^* \iota^* K)[d_{\mathbf{T}} + d_{\mathbf{W}}] \\
&= \bar{t}_!(\mathcal{L}_{\bar{T}} \otimes \dot{p}_! \dot{q}_! \dot{q}^* \dot{s}^* \iota^* K[2m])[d_{\mathbf{T}} + d_{\mathbf{W}}] \\
&= \bar{t}_! \dot{p}_! \dot{q}_! (\dot{p}^* \dot{q}^* (\mathcal{L}_{\bar{T}}) \otimes \dot{q}^* \dot{s}^* \iota^* K)[2m + d_{\mathbf{T}} + d_{\mathbf{W}}] \\
&= {}'p_! \dot{t}_! \dot{\zeta}_! (\dot{p}^* \dot{q}^* (\mathcal{L}_{\bar{T}}) \otimes \dot{\zeta}^* i^* s^* K)[2m + d_{\mathbf{T}} + d_{\mathbf{W}}] \\
&= {}'p_! \dot{t}_! (\dot{\zeta}_! (\dot{p}^* \dot{q}^* \mathcal{L}_{\bar{T}}) \otimes i^* s^* K)[2m + d_{\mathbf{T}} + d_{\mathbf{W}}] \\
&= {}'p_! \dot{t}_! (\dot{\zeta}_! (\dot{\zeta}^* \mathcal{L}_{\tilde{T}}) \otimes i^* s^* K)[2m + d_{\mathbf{T}} + d_{\mathbf{W}}] \\
&= {}'p_! \dot{t}_! (\mathcal{L}_{\tilde{T}} \otimes i^* s^* K)[2m + d_{\mathbf{T}} + d_{\mathbf{W}}]
\end{aligned}$$

and

$$\begin{aligned}
\mathrm{Res}^{\mathbf{V}}_{\mathbf{T},\mathbf{W}}(\Phi(K))[\pi] &= {}'p_! {}'\iota^* t_!(\mathcal{L}_T \otimes s^* K)[\pi + d_{\mathbf{V}}] \\
&= {}'p_! \dot{t}_! i^* (\mathcal{L}_T \otimes s^* K)[\pi + d_{\mathbf{V}}] \\
&= {}'p_! \dot{t}_! (\mathcal{L}_{\tilde{T}} \otimes i^* s^* K)[\pi + d_{\mathbf{V}}].
\end{aligned}$$

It remains for us to observe that $\pi + d_{\mathbf{V}} = 2m + d_{\mathbf{T}} + d_{\mathbf{W}}$. The proposition is proved.

10.1.3. We can reformulate the previous proposition using $\mathrm{Res}^{\mathbf{V}}_{\mathbf{T},\mathbf{W}}$ instead of $\widetilde{\mathrm{Res}}^{\mathbf{V}}_{\mathbf{T},\mathbf{W}}$; the shift by π will then disappear:

$$\Phi(\mathrm{Res}^{\mathbf{V}}_{\mathbf{T},\mathbf{W}} K) = \mathrm{Res}^{\mathbf{V}}_{\mathbf{T},\mathbf{W}}(\Phi(K)).$$

10.2. Fourier-Deligne Transform and Induction

10.2.1. Let $\nu = (\nu^1, \nu^2, \ldots, \nu^m) \in \mathcal{X}$ be such that $\dim \mathbf{V_i} = \sum_l \nu^l_{\mathbf{i}}$ for all $\mathbf{i}$. Recall that we have a natural proper morphism $\pi_\nu : \tilde{\mathcal{F}}_\nu \to \mathbf{E_V}$. The same definition with the new orientation for our graph gives a proper morphism $'\pi_\nu : '\tilde{\mathcal{F}}_\nu \to {'\mathbf{E_V}}$, where $'\tilde{\mathcal{F}}_\nu$ is the variety of all pairs (x, f) such that $x \in {'\mathbf{E_V}}$ and $f \in \mathcal{F}_\nu$ is x-stable; $'\pi_\nu$ is the first projection.

Recall the definition $\tilde{L}_\nu = (\pi_\nu)_! \mathbf{1} \in \mathcal{D}(\mathbf{E_V})$. Similarly, we set $'\tilde{L}_\nu = ('\pi_\nu)_! \mathbf{1} \in \mathcal{D}('\mathbf{E_V})$.

Proposition 10.2.2. $\Phi(\tilde{L}_\nu) = {'\tilde{L}_\nu}[M]$ *where*

$$M = \sum_{h \in H_2; l > l'} (\nu^l_{h'} \nu^{l'}_{h''} - \nu^{l'}_{h'} \nu^{l'}_{h''}).$$

Consider the commutative diagram

$$\begin{array}{ccccc}
\tilde{\mathcal{F}}_\nu & \longrightarrow & \Xi & \xrightarrow{c} & \bar{\Xi} \\
{\scriptstyle \pi_\nu}\downarrow & & {\scriptstyle \rho}\downarrow & & \downarrow \\
\mathbf{E_V} & \xleftarrow{s} & \dot{\mathbf{E}}_\mathbf{V} & \xrightarrow{t} & '\mathbf{E_V}
\end{array}$$

where the following notation is used.

Ξ is the set of all (x, y, f) where $x \in \mathbf{E_V}$, f is an x-stable flag in $\mathcal{F}_\nu$ and $y \in {'\mathbf{E_V}}$ is such that $y_h = x_h : \mathbf{V}_{h'} \to \mathbf{V}_{h''}$ for any $h \in H_1$.

$\bar{\Xi}$ is the set of all (y, f) where $y \in {'\mathbf{E_V}}$ and $f = (\mathbf{V} = \mathbf{V}^0 \supset \mathbf{V}^1 \supset \cdots \supset \mathbf{V}^m = 0)$ is a flag in $\mathcal{F}_\nu$ such that $y_h(\mathbf{V}^l_{h'}) \subset \mathbf{V}^l_{h''}$ for all l and all $h \in H_1$.

The lower horizontal maps are as in 10.1.1(a); the other maps are the obvious ones. The left square is cartesian. We have $s^*(\pi_\nu)_! \mathbf{1} = \rho_! \mathbf{1}$. Hence

$$\Phi(\tilde{L}_\nu) = t_!(\mathcal{L}_T \otimes \rho_! \mathbf{1})[d_\mathbf{V}] = \tilde{t}_!(\mathcal{L}_{\tilde{T}})[d_\mathbf{V}]$$

where $T : \dot{\mathbf{E}}_\mathbf{V} \to k$ is as in 10.1.1, $\tilde{T} : \Xi \to k$ is given by $\tilde{T} = T\rho$ and $\tilde{t} = t\rho : \Xi \to {'\mathbf{E_V}}$.

The fibres of c are affine spaces of dimension $N = \sum_{h \in H_2; l < l'} \nu^l_{h'} \nu^{l'}_{h''}$. (In the formula for N we have $\nu^l_{h'} \nu^{l'}_{h''} = 0$ for $l = l'$, since ν^l is discrete.)

We have a partition $\Xi = \Xi_0 \cup \Xi_1$ where Ξ_0 is the closed subset of Ξ consisting of those (x, y, f) such that f is y-stable. It can be verified that the restriction of $\tilde{T}$ to the fibre of c at $c(x, y, f)$ is an affine-linear function and that this function is constant if and only if $(x, y, f) \in \Xi_0$. Note that Ξ_0 is a union of fibres of c.

Using 8.1.13, it follows that $(c_1)_!(\mathcal{L}_{\tilde{T}}|_{\Xi_1}) = 0$, where $c' : \Xi_1 \to \bar{\Xi}$ is the restriction of c. Hence, if $j : \Xi_0 \to \Xi$ is the inclusion, we have $c_! \dot{j}_!(\dot{j}^* \mathcal{L}_{\tilde{T}}) = \dot{c}_! \mathcal{L}_{\tilde{T}}$. From the commutative diagram above, it then follows that

$$(t\rho)_!(\mathcal{L}_{\tilde{T}} = (t_0)_!(\mathcal{L}_{\tilde{T}}|_{\Xi_0})$$

where $t_0 : \Xi_0 \to {}'\mathbf{E_V}$ is the restriction of $t\rho$.

Let $(x, y, f) \in \Xi_0$ with f as above. We have

$$\tilde{T}(x, y, f) = T(x, y) = \sum_{h \in H_2} \text{tr } (y_h x_h : \mathbf{V}_{h'} \to \mathbf{V}_{h'}).$$

Since f is stable under both x and y, we have

$$\text{tr } (y_h x_h : \mathbf{V}_{h'} \to \mathbf{V}_{h'}) = \sum_l \text{tr } (y_h x_h : \mathbf{V}^{l-1}_{h'}/\mathbf{V}^l_{h'} \to \mathbf{V}^{l-1}_{h''}/\mathbf{V}^l_{h''}).$$

For any l, at least one of the vector spaces $\mathbf{V}^{l-1}_{h'}/\mathbf{V}^l_{h'}, \mathbf{V}^{l-1}_{h''}/\mathbf{V}^l_{h''}$ is zero, since ν^l is discrete. Thus, $\text{tr } (y_h x_h : \mathbf{V}_{h'} \to \mathbf{V}_{h'}) = 0$ for each $h \in H_2$, so that $\tilde{T}(x, y, f) = 0$. Since $\tilde{T}$ is identically zero on Ξ_0, we have $\mathcal{L}_{\tilde{T}}|_{\Xi_0} = \mathbf{1}$ and we see that

$$(t\rho)_!(\mathcal{L}_{\tilde{T}} = (t_0)_! \mathbf{1}.$$

Now t_0 can be factored as a composition $\Xi_0 \to {}'\tilde{\mathcal{F}}_\nu \xrightarrow{'\pi_\nu} {}'\mathbf{E_V}$, where the first map (restriction of c) is a vector bundle of dimension N. Hence

$$(t_0)_! \mathbf{1} = ({}'\pi_\nu)_! \mathbf{1}[-2N] = {}'L_\nu[-2N].$$

It follows that $(t\rho)_!(\mathcal{L}_{\tilde{T}} = (t_0)_! \mathbf{1} = {}'L_\nu[-2N]$. It remains for us to observe that $d_\mathbf{V} - 2N = M$. The proposition is proved.

10.2.3. Using the proposition and the general properties of the Fourier-Deligne transform (see 8.1.11) we see that $\Phi : \mathcal{D}(\mathbf{E_V}) \to \mathcal{D}({}'\mathbf{E_V})$ defines an equivalence of categories $\mathcal{Q}_\mathbf{V} \to {}'\mathcal{Q}_\mathbf{V}$ and $\mathcal{P}_\mathbf{V} \to {}'\mathcal{P}_\mathbf{V}$, where ${}'\mathcal{Q}_\mathbf{V}, {}'\mathcal{P}_\mathbf{V}$ are defined as $\mathcal{Q}_\mathbf{V}, \mathcal{P}_\mathbf{V}$ but using the new orientation of our graph. Hence Φ induces a bijection between the set of simple objects in $\mathcal{P}_\mathbf{V}$ and that in ${}'\mathcal{P}_\mathbf{V}$.

10.2.4. We have a natural action of $(k^*)^H$ on $\mathbf{E_V}$ (resp. on $\tilde{\mathcal{F}}_\nu$) given by $(\zeta_h) : (x_h) \mapsto (\zeta_h x_h)$ (resp. $(\zeta_h) : ((x_h), f) \mapsto ((\zeta_h x_h), f)$. The map π_ν is compatible with these actions. It follows that $H^n L_\nu$ is $(k^*)^H$-equivariant for any n. Hence any $K \in \mathcal{P}_\mathbf{V}$ is $(k^*)^H$-equivariant. In particular, we have $j^*K = K$, where $j : \mathbf{E_V} \to \mathbf{E_V}$ is the involution which acts as -1 on the summands $\mathrm{Hom}(\mathbf{V}_{h'}, \mathbf{V}_{h''})$ for $h \in H_2$ and as 1 on the other summands. Hence for $K \in \mathcal{P}_\mathbf{V}$, the Fourier inversion formula (see 8.1.11) simplifies to $\Phi(\Phi(K)) = K$.

10.2.5. Let $A \in \mathcal{Q}_\mathbf{V}$ and let $A' \in {}'\mathcal{Q}_\mathbf{V}$. For any $j \in \mathbf{Z}$, we have a canonical isomorphism

$$\mathbf{D}_j(\mathbf{E_V}, G_\mathbf{V}; A, \Phi(A')) = \mathbf{D}_j({}'\mathbf{E_V}, G_\mathbf{V}; \Phi(A), A').$$

This follows by applying 8.1.12 to the diagram

$$G_\mathbf{V}\backslash(\Gamma \times \mathbf{E_V}) \leftarrow G_\mathbf{V}\backslash(\Gamma \times \dot{\mathbf{E}}_\mathbf{V}) \to G_\mathbf{V}\backslash(\Gamma \times {}'\mathbf{E_V})$$

obtained from 10.1.1(a), where Γ is a suitable smooth variety with a free $G_\mathbf{V}$-action.

Proposition 10.2.6. *With the notations of Proposition 10.1.2, let* $L \in \mathcal{Q}_{\mathbf{T},\mathbf{W}}$. *There exists an isomorphism in* ${}'\mathcal{Q}_\mathbf{V}$*:*

$$\Phi(Ind^{\mathbf{V}}_{\mathbf{T},\mathbf{W}} L) \cong Ind^{\mathbf{V}}_{\mathbf{T},\mathbf{W}}(\Phi L).$$

Since ${}'\mathcal{P}_\mathbf{V}$ is stable under Verdier duality, we see from 9.1.6 that it suffices to check that

(a)

$$\dim \mathbf{D}_j({}'\mathbf{E_V}, G_\mathbf{V}; \Phi(\mathrm{Ind}^{\mathbf{V}}_{\mathbf{T},\mathbf{W}} L), \Phi K) = \dim \mathbf{D}_j({}'\mathbf{E_V}, G_\mathbf{V}; \mathrm{Ind}^{\mathbf{V}}_{\mathbf{T},\mathbf{W}}(\Phi L), \Phi K)$$

for any $K \in \mathcal{P}_\mathbf{V}$ and any $j \in \mathbf{Z}$.

By 10.2.5, the left hand side of (a) is equal to

$$\dim \mathbf{D}_j(\mathbf{E_V}, G_\mathbf{V}; \mathrm{Ind}^{\mathbf{V}}_{\mathbf{T},\mathbf{W}} L, K)$$

and by 9.2.9, this is equal to

$$\dim \mathbf{D}_j(\mathbf{E_T} \times \mathbf{E_W}, G_\mathbf{T} \times G_\mathbf{W}; L, \mathrm{Res}^{\mathbf{V}}_{\mathbf{T},\mathbf{W}} K).$$

By 9.2.9, the right hand side of (a) is equal to

$$\dim \mathbf{D}_j({}'\mathbf{E_T} \times {}'\mathbf{E_W}, G_\mathbf{T} \times G_\mathbf{W}; \Phi L, \mathrm{Res}^{\mathbf{V}}_{\mathbf{T},\mathbf{W}}(\Phi K))$$

and by 10.2.5, this is equal to

$$\dim \mathbf{D}_j(\mathbf{E_T} \times \mathbf{E_W}, G_\mathbf{T} \times G_\mathbf{W}; L, \Phi(\mathrm{Res}^{\mathbf{V}}_{\mathbf{T},\mathbf{W}}(\Phi K))).$$

Hence (a) is equivalent to

$$\begin{aligned}&\dim \mathbf{D}_j(\mathbf{E_T} \times \mathbf{E_W}, G_\mathbf{T} \times G_\mathbf{W}; L, \mathrm{Res}^{\mathbf{V}}_{\mathbf{T},\mathbf{W}} K)\\ &= \dim \mathbf{D}_j(\mathbf{E_T} \times \mathbf{E_W}, G_\mathbf{T} \times G_\mathbf{W}; L, \Phi(\mathrm{Res}^{\mathbf{V}}_{\mathbf{T},\mathbf{W}}(\Phi K))).\end{aligned}$$

But this follows from $\mathrm{Res}^{\mathbf{V}}_{\mathbf{T},\mathbf{W}} K = \Phi(\mathrm{Res}^{\mathbf{V}}_{\mathbf{T},\mathbf{W}}(\Phi K))$ (see 10.1.3). The proposition is proved.

10.3. A Key Inductive Step

Lemma 10.3.1. *Let $\mathbf{I}', \gamma$ be as in 9.3.1. The Fourier-Deligne transform $\Phi : \mathcal{P}_\mathbf{V} \to {}'\mathcal{P}_\mathbf{V}$ defines an equivalence of categories between $\mathcal{P}_{\mathbf{V};\mathbf{I}';\gamma}$ and the analogous category ${}'\mathcal{P}_{\mathbf{V};\mathbf{I}';\gamma}$ defined as $\mathcal{P}_{\mathbf{V};\mathbf{I}';\gamma}$ with respect to the new orientation.*

This follows immediately from the definitions since the Fourier-Deligne transform commutes with Ind.

Proposition 10.3.2. *Let $\mathbf{I}', \gamma$ be as in 9.3.1. Let $\mathbf{W}$ be a graded subspace of $\mathbf{V}$ such that $\mathbf{T} = \mathbf{V}/\mathbf{W}$ satisfies $\dim \mathbf{T_i} = \gamma_\mathbf{i}$ for all $\mathbf{i} \in \mathbf{I}'$ and $\mathbf{T_{i'}} = 0$ for all $\mathbf{i}' \in \mathbf{I} - \mathbf{I}'$.*

(a) *Let B be a simple object of $\mathcal{P}_{\mathbf{V};\mathbf{I}';\gamma}$. We have*

$$\mathit{Res}^{\mathbf{V}}_{\mathbf{T},\mathbf{W}} B \cong A \oplus (\oplus_j L_j[j])$$

where A is a simple object of $\mathcal{P}_{\mathbf{W};\mathbf{I}';0}$ and $L_j \in \mathcal{P}_{\mathbf{W};\mathbf{I}';>0}$ for all j.

(b) *Let A be a simple object of $\mathcal{P}_{\mathbf{W};\mathbf{I}';0}$. We have*

$$\mathit{Ind}^{\mathbf{V}}_{\mathbf{T},\mathbf{W}} A \cong B \oplus (\oplus_j C_j[j])$$

where B is a simple object of $\mathcal{P}_{\mathbf{V};\mathbf{I}';\gamma}$ and

$$C_j \in \mathcal{P}_{\mathbf{V};\mathbf{I}';>\gamma}$$

for all j.

(c) *The maps $B \mapsto A$ in (a) and $A \mapsto B$ in (b) are inverse bijections between the set of isomorphism classes of simple objects in $\mathcal{P}_{\mathbf{V};\mathbf{I}';\gamma}$ and the analogous set for $\mathcal{P}_{\mathbf{W};\mathbf{I}';0}$.*

This statement is independent of the orientation of our graph: we use the previous lemma and the fact that the Fourier-Deligne transform commutes with Ind and Res. Hence it is enough to prove the proposition under the additional assumption that $h' \notin \mathbf{I}'$ for any $h \in H$. We can achieve this by a change of orientation.

Let A be as in (b). By Lemma 9.3.5 , the support of A meets $\mathbf{E}_{\mathbf{W};0}$. Hence Proposition 9.3.3 is applicable to $A, \mathbf{I}'$; it shows that $\mathrm{Ind}^{\mathbf{V}}_{\mathbf{T},\mathbf{W}} A \cong B \oplus (\oplus_j C_j[j])$ where B is a simple object of $\mathcal{P}_{\mathbf{V}}$ such that the support of B is contained in $\mathbf{E}_{\mathbf{V};\geq\gamma}$ and meets $\mathbf{E}_{\mathbf{V};\gamma}$; $C_j \in \mathcal{P}_{\mathbf{V}}$ has support contained in $\mathbf{E}_{\mathbf{V};\geq\gamma}$ and is disjoint from $\mathbf{E}_{\mathbf{V};\gamma}$ for any j. By Lemma 9.3.5, we then have $B \in \mathcal{P}_{\mathbf{V};\mathbf{I}';\gamma}$ and $C_j \in \mathcal{P}_{\mathbf{V};\mathbf{I}';>\gamma}$.

Conversely, let B be as in (a). By Lemma 9.3.5, we have that the support of B is contained in $\mathbf{E}_{\mathbf{V};\geq\gamma}$ and meets $\mathbf{E}_{\mathbf{V};\gamma}$. Hence Proposition 9.3.3 is applicable to B and $\mathbf{I}'$. It shows that $\mathrm{Res}^{\mathbf{V}}_{\mathbf{T},\mathbf{W}} B \cong A \oplus (\oplus_j L_j[j])$ where A is a simple object of $\mathcal{P}_{\mathbf{W}}$ such that the support of A meets $\mathbf{E}_{\mathbf{W};0}$ and $L_j \in \mathcal{P}_{\mathbf{W}}$ has support disjoint from $\mathbf{E}_{\mathbf{W};0}$ for any j. By Lemma 9.3.5 we then have $A \in \mathcal{P}_{\mathbf{W};\mathbf{I}';0}$ and $L_j \in \mathcal{P}_{\mathbf{W};\mathbf{I}';>0}$. This proves (a), (b). Statement (c) follows from the last assertion of Proposition 9.3.3.

10.3.3. Remark. The previous proof shows that, given $\mathbf{I}'$ as above and a simple object B in $\mathcal{P}_{\mathbf{V}}$, there is a unique $\gamma \in \mathbf{N}[\mathbf{I}]$ with support contained in $\mathbf{I}'$ such that $B \in \mathcal{P}_{\mathbf{V};\mathbf{I}';\gamma}$.

The existence of γ is obvious. To prove uniqueness, we may assume that the orientation has been chosen as in the previous proof; but then γ is such that the support of B is contained in $\mathbf{E}_{\mathbf{V};\geq\gamma}$ and meets $\mathbf{E}_{\mathbf{V};\gamma}$ and these conditions determine γ uniquely since the support of B is irreducible.

10.3.4. Passage to the opposite orientation. Let $\mathbf{V} \in \mathcal{V}$. For each $\mathbf{i} \in \mathbf{I}$, let $\mathbf{V}^*_{\mathbf{i}}$ be the dual space of $\mathbf{V}_{\mathbf{i}}$ and let $\mathbf{V}^* = \oplus_{\mathbf{i}} \mathbf{V}^*_{\mathbf{i}} \in \mathcal{V}$. Assume now that the new orientation (see 10.1.1) of our graph is the opposite of the old one, that is, $'h = h''$ and $''h = h'$ for all $h \in H$. We have an isomorphism $\rho : \mathbf{E}_{\mathbf{V}} \cong {'\mathbf{E}_{\mathbf{V}^*}}$ given by $\rho(x) = x'$ where $x'_h : \mathbf{V}^*_{h''} \to \mathbf{V}^*_{h'}$ is the transpose of $x_h : \mathbf{V}_{h'} \to \mathbf{V}_{h''}$. This induces an equivalence of categories $\rho_! : \mathcal{D}(\mathbf{E}_{\mathbf{V}}) \cong \mathcal{D}(\mathbf{E}_{\mathbf{V}^*})$ with inverse ρ^*.

Let $\boldsymbol{\nu} = (\nu^1, \nu^2, \ldots, \nu^m) \in \mathcal{X}$ be such that $\dim \mathbf{V}_{\mathbf{i}} = \sum_l \nu^l_{\mathbf{i}}$ for all $\mathbf{i} \in \mathbf{I}$. Let $\boldsymbol{\nu}' = (\nu^m, \nu^{m-1}, \ldots, \nu^1) \in \mathcal{X}$. It follows immediately from definitions that $\rho_! L_{\boldsymbol{\nu}} = L_{\boldsymbol{\nu}'} \in \mathcal{D}(\mathbf{E}_{\mathbf{V}^*})$. From this we deduce that $\rho_!$ defines equivalences of categories $\mathcal{P}_{\mathbf{V}} \to {'\mathcal{P}_{\mathbf{V}^*}}$ and $\mathcal{Q}_{\mathbf{V}} \to {'\mathcal{Q}_{\mathbf{V}^*}}$.

CHAPTER 11

Periodic Functors

11.1.1. Let C be a category in which the space of morphisms between any two objects has a given $\bar{\mathbf{Q}}_l$-vector space structure such that composition of morphisms is bilinear and such that finite direct sums exist. We say that C is a *linear* category.

A functor from one linear category to another linear category is said to be linear if it respects the $\bar{\mathbf{Q}}_l$-vector space structures.

11.1.2. Assume that we are given an integer $\mathbf{n} \geq 1$ and a linear functor $a^* : C \to C$ such that $a^{*\mathbf{n}}$ is the identity functor from C to C. We say that a is a *periodic functor.*

We define a new category $\tilde{C}$ as follows. The objects of $\tilde{C}$ are pairs (A, ϕ) where A is an object of C and $\phi : a^*A \to A$ is an isomorphism in C such that the composition

$$a^{*\mathbf{n}}A \xrightarrow{a^{*(\mathbf{n}-1)}\phi} a^{*(\mathbf{n}-1)}A \xrightarrow{a^{*(\mathbf{n}-2)}\phi} a^{*(\mathbf{n}-2)}A \to \cdots \to a^*A \xrightarrow{\phi} A$$

is the identity map of A.

Let (A, ϕ) and (A', ϕ') be two objects of $\tilde{C}$. There is a natural automorphism $u : \mathrm{Hom}(A, A') \to \mathrm{Hom}(A, A')$ given by $u(f) = \phi' a^*(f) \phi^{-1}$. From the definitions it follows that $u^{\mathbf{n}} = 1$. By definition,

$$\mathrm{Hom}_{\tilde{C}}((A, \phi), (A', \phi')) = \{f \in \mathrm{Hom}_C(A, A') | u(f) = f\}.$$

The composition of morphisms in $\tilde{C}$ is induced by that in C. The direct sum of two objects (A, ϕ) and (A', ϕ') of $\tilde{C}$ is $(A \oplus A', \phi \oplus \phi')$. Thus, $\tilde{C}$ is in a natural way a linear category. Clearly, if (A, ϕ) is an object of $\tilde{C}$, then so is $(A, \zeta\phi)$ for any $\zeta \in \bar{\mathbf{Q}}_l$ such that $\zeta^{\mathbf{n}} = 1$.

11.1.3. Assume that we are given three objects $(A, \phi), (A', \phi'), (A'', \phi'')$ of $\tilde{C}$ and morphisms $i' : (A', \phi') \to (A, \phi)$, $p'' : (A, \phi) \to (A'', \phi'')$ in $\tilde{C}$ such that the following holds.

(a) There exist morphisms $i'' : A'' \to A$ and $p' : A \to A'$ in C such that

$$p'i' = 1_{A'},\ p'i'' = 0,\ p''i' = 0,\ p''i'' = 1_{A''},\ i'p' + i''p'' = 1_A.$$

G. Lusztig, *Introduction to Quantum Groups*, Modern Birkhäuser Classics,
DOI 10.1007/978-0-8176-4717-9_11, © Springer Science+Business Media, LLC 2010

We show that $(A',\phi')\oplus(A'',\phi'')\cong(A,\phi)$ in $\tilde{C}$. Recall that $u^{\mathbf{n}}(i'')=i''$ where $u:\mathrm{Hom}_C(A'',A)\to\mathrm{Hom}_C(A'',A)$ is as in 11.1.2. Let $\tilde{i}''=\sum_{j=1}^{\mathbf{n}}u^j(i'')/\mathbf{n}:A''\to A$. Then $\tilde{i}''\in\mathrm{Hom}_{\tilde{C}}(A'',A')$ and $p''\tilde{i}''=1_{A''}$.

We set the $\tilde{p}'=p'-p'\tilde{i}''p'':A\to A'$. Then

$$\tilde{p}'i'=1_{A'},\ \tilde{p}'\tilde{i}''=0,\ p''i'=0,\ i'\tilde{p}'+\tilde{i}''p''=1_A.$$

It follows that $(i',\tilde{i}'')$ define an isomorphism $(A',\phi')\oplus(A'',\phi'')\to(A,\phi)$ (in $\tilde{C}$). Our assertion is proved.

11.1.4. An object (A,ϕ) of $\tilde{C}$ is said to be *traceless* if there exists an object B of C, an integer $t\geq 2$ dividing $\mathbf{n}$, such that $a^{*t}B\cong B$, and an isomorphism $A\cong B\oplus a^*B\oplus\cdots\oplus a^{*(t-1)}B$ under which ϕ corresponds to an isomorphism $a^*B\oplus a^{*2}B\oplus\cdots\oplus a^{*t}B\cong B\oplus a^*B\oplus\cdots\oplus a^{*(t-1)}B$ carrying the summand $a^{*j}B$ onto the summand $a^{*j}B$ (for $1\leq j\leq t-1$) and the summand $a^{*t}B$ onto the summand B.

11.1.5. Let $\mathcal{O}$ be the subring of $\bar{\mathbf{Q}}_l$ consisting of all $\mathbf{Z}$-linear combinations of $\mathbf{n}$-th roots of 1. We associate to C and a^* an $\mathcal{O}$-module $\mathcal{K}(C)$. By definition, $\mathcal{K}(C)$ is the $\mathcal{O}$-module generated by symbols $[B,\phi]$, one for each isomorphism class of objects (B,ϕ) of $\tilde{C}$, subject to the following relations:

(a) $[B,\phi]+[B',\phi']=[B\oplus B',\phi\oplus\phi']$;

(b) $[B,\phi]=0$ if (B,ϕ) is traceless;

(c) $[B,\zeta\phi]=\zeta[B,\phi]$ if $\zeta\in\bar{\mathbf{Q}}_l$ satisfies $\zeta^{\mathbf{n}}=1$.

This definition is similar to that of a Grothendieck group.

11.1.6. Now let C' be another linear category with a given functor $a^*:C'\to C'$ such that $a^{*\mathbf{n}}$ is the identity functor from C' to C'. Let $b:C\to C'$ be a linear functor. Assume that we are given an isomorphism of functors $ba^*=a^*b:C\to C'$. Then b induces a linear functor $\tilde{b}:\tilde{C}\to\tilde{C}'$ by $b(A,\phi)=(bA,\phi')$ where $\phi':a^*bA\to bA$ is the composition $a^*bA=ba^*A\xrightarrow{b(\phi)}bA$. It is clear that $[A,\phi]\mapsto[bA,\phi']$ respects the relations of $\mathcal{K}(C),\mathcal{K}(C')$ and hence defines an $\mathcal{O}$-linear map $\mathcal{K}(C)\to\mathcal{K}(C')$.

11.1.7. Assume now that C is, in addition, an abelian category in which any object is a direct sum of finitely many simple objects. Let B be a simple object of C. Let t_B be the smallest integer ≥ 1 such that $(a^*)^{t_B}B$ is isomorphic to B; let $f_B:(a^*)^{t_B}B\to B$ be an isomorphism. We have $\mathbf{n}=n't_B$ where n' is an integer ≥ 1.

The composition

$$\text{(a)}\qquad B = (a^*)^{n' t_B} B \longrightarrow \cdots \xrightarrow{a^{*2t_B} f_B} a^{*2t_B} B \xrightarrow{a^{*t_B} f_B} a^{*t_B} B \xrightarrow{f_B} B$$

is a non-zero scalar times identity (since B is simple); hence by changing f_B by a non-zero multiple of f_B, we can assume that the composition (a) is the identity.

Consider the isomorphism

$$\phi_B : a^*(B \oplus a^*B \oplus \cdots (a^*)^{t_B-1}B) = a^*B \oplus a^{*2}B \oplus \cdots (a^*)^{t_B}B$$
$$\to B \oplus a^*B \oplus \cdots (a^*)^{t_B-1}B$$

which maps the summand $a^{*j}B$ onto the summand $a^{*j}B$ by the identity map (for $1 \le j \le t_B - 1$) and maps the summand $(a^*)^{t_B}B$ onto the summand B by f_B. From the definitions we see that

$$(B \oplus a^*B \oplus \cdots (a^*)^{t_B-1}B, \phi_B)$$

is an object of $\tilde{C}$.

Let $\mathcal{S}$ be a set of simple objects of C with the following property: any simple object in C is isomorphic to $a^{*j}B$ for a unique B in $\mathcal{S}$ and some $j \ge 0$. For each B in $\mathcal{S}$ we choose ϕ_B as above. It is easy to see that any object of $\tilde{C}$ is isomorphic to

$$\text{(b)}\qquad \oplus_{B\in\mathcal{S}}((B \oplus a^*B \oplus \cdots (a^*)^{t_B-1}B) \otimes E_B, \phi_B \otimes \psi_B)$$

where for each B, E_B is a finite dimensional $\bar{\mathbf{Q}}_l$-vector space with a given automorphism $\psi_B : E_B \to E_B$ such that $\psi_B^{\mathbf{n}} = 1$ and $E_B = 0$ for all but finitely many B. Note that the summands corresponding to B such that $t_B \ge 2$ are traceless.

Let B be a simple object of C such that $a^*B \cong B$. The isomorphisms $\phi : a^*B \cong B$ such that $(B, \phi) \in \tilde{C}$, generate a free $\mathcal{O}$-submodule of rank 1 of $\mathrm{Hom}_C(a^*B, B)$; we denote this $\mathcal{O}$-submodule by $\mathcal{O}_B$. It is easy to see that $\mathcal{O}_B$ depends only on the isomorphism class of B.

11.1.8. From 11.1.7 it follows easily that

$$\text{(a)}\qquad \mathcal{K}(C) = \oplus_B \mathcal{O}_B$$

as $\mathcal{O}$-modules (the sum is taken over the isomorphism classes of simple objects B such that $a^*B \cong B$); to the element $\phi \in \mathcal{O}_B$ such that $(B, \phi) \in \tilde{C}$ corresponds the element $[B, \phi] \in \mathcal{K}(C)$.

CHAPTER 12

Quivers with Automorphisms

12.1. The Group $\mathcal{K}(\mathcal{Q}_{\mathbf{V}})$

12.1.1. An *admissible automorphism* of the graph $(\mathbf{I}, H, h \mapsto [h])$ consists, by definition, of a permutation $a : \mathbf{I} \to \mathbf{I}$ and a permutation $a : H \to H$ such that for any $h \in H$, we have $[a(h)] = a[h]$ as subsets of $\mathbf{I}$ and such that there is no edge joining two vertices in the same a-orbit.

In this chapter we assume that we are given an integer $\mathbf{n} \geq 1$ and an admissible automorphism a of the graph $(\mathbf{I}, H, h \mapsto [h])$ in 9.1.1. We assume that $a^{\mathbf{n}} = 1$ both on $\mathbf{I}$ and on H. From the definition it follows that

(a) any a-orbit on $\mathbf{I}$ is a discrete subset of $\mathbf{I}$, in the sense of 9.1.3.

An orientation $h \to h', h \to h''$ (see 9.1.1) of our graph is said to be compatible with a if, for any $h \in H$, we have $(a(h))' = a(h')$ and $(a(h))'' = a(h'')$. From property (a) we can deduce that there is at least one orientation of our graph which is compatible with a. This is seen as follows. Choose a set of representatives H_0 for the a-orbits on H. For each $h \in H_0$, choose one element h' of $[h]$; let h'' be the other element of $[h]$. Now let $h \in H$. We can find $n \in \mathbf{Z}$ such that $h = a^n h_0$ where $h_0 \in H_0$ is uniquely determined. We set $h' = a^n(h'_0)$ and $h'' = a^n(h''_0)$. We must prove that h', h'' are independent of the choice of n. We are reduced to verifying the following statement: if $m \in \mathbf{Z}$ satisfies $a^m h_0 = h_0$, then we have $a^m(h'_0) = h'_0$ and $a^m(h''_0) = h''_0$. If this is not the case, we have $a^m(h'_0) = h''_0$, which contradicts property (a).

12.1.2. Let $\mathcal{V}^a$ be the category of finite dimensional $\mathbf{I}$-graded k-vector spaces $\mathbf{V} = \oplus_{\mathbf{i} \in \mathbf{I}} \mathbf{V}_\mathbf{i}$ with a given linear map $a : \mathbf{V} \to \mathbf{V}$ such that $a(\mathbf{V}_\mathbf{i}) = \mathbf{V}_{a(\mathbf{i})}$ for all $\mathbf{i} \in \mathbf{I}$ and $a^j|_{\mathbf{V}_\mathbf{i}} = 1_{\mathbf{V}_\mathbf{i}}$ for all $\mathbf{i} \in \mathbf{I}$ and all $j \geq 1$ such that $a^j(\mathbf{i}) = \mathbf{i}$. The morphisms in $\mathcal{V}^a$ are isomorphisms of vector spaces compatible with the grading and with a. Note that $a : \mathbf{V} \to \mathbf{V}$ automatically satisfies $a^{\mathbf{n}} = 1$. Let

$$\mathbf{N}[\mathbf{I}]^a = \{\nu \in \mathbf{N}[\mathbf{I}] | \nu_\mathbf{i} = \nu_{a(\mathbf{i})} \quad \forall \mathbf{i} \in \mathbf{I}\}.$$

For each $\nu \in \mathbf{N}[\mathbf{I}]^a$, we denote by $\mathcal{V}^a_\nu$ the full subcategory of $\mathcal{V}^a$ whose objects are those $\mathbf{V}$ such that $\dim \mathbf{V}_\mathbf{i} = \nu_\mathbf{i}$ for all $\mathbf{i} \in \mathbf{I}$. Then each object

G. Lusztig, *Introduction to Quantum Groups*, Modern Birkhäuser Classics,
DOI 10.1007/978-0-8176-4717-9_12, © Springer Science+Business Media, LLC 2010

of $\mathcal{V}^a$ belongs to $\mathcal{V}^a_\nu$ for a unique $\nu \in \mathbf{N}[\mathbf{I}]^a$ and any two objects of $\mathcal{V}^a_\nu$ are isomorphic to each other. Moreover $\mathcal{V}^a_\nu$ is non-empty for any $\nu \in \mathbf{N}[\mathbf{I}]^a$.

We choose an orientation of our graph, compatible with a. If $\mathbf{V} \in \mathcal{V}^a$, then we may regard $\mathbf{V}$ as an object of $\mathcal{V}$ (by forgetting a); in particular, $G_{\mathbf{V}}$ and $\mathbf{E}_{\mathbf{V}}$ are defined.

There is a natural automorphism $a : G_{\mathbf{V}} \to G_{\mathbf{V}}$ given by $a(gz) = a(g)(a(z))$ for all $g \in G_{\mathbf{V}}$ and $z \in \mathbf{V}$; moreover, there is a natural automorphism $a : \mathbf{E}_{\mathbf{V}} \to \mathbf{E}_{\mathbf{V}}$ $(x \mapsto a(x))$, such that for any $x = (x_h) \in \mathbf{E}_{\mathbf{V}}$ and any $h \in H$, the compositions

$$\mathbf{V}_{h'} \xrightarrow{x_h} \mathbf{V}_{h''} \xrightarrow{a} \mathbf{V}_{a(h'')} \text{ and } \mathbf{V}_{h'} \xrightarrow{a} \mathbf{V}_{a(h')} \xrightarrow{a(x)_{a(h)}} \mathbf{V}_{a(h'')}$$

coincide. Both of these automorphisms satisfy $a^{\mathbf{n}} = 1$. Note that $a(gx) = a(g)(a(x))$ for all $g \in G_{\mathbf{V}}$ and $x \in \mathbf{E}_{\mathbf{V}}$. Taking the inverse image under $a : \mathbf{E}_{\mathbf{V}} \to \mathbf{E}_{\mathbf{V}}$ gives us a functor $a^* : \mathcal{D}(\mathbf{E}_{\mathbf{V}}) \to \mathcal{D}(\mathbf{E}_{\mathbf{V}})$.

Let $\boldsymbol{\nu} = (\nu^1, \nu^2, \dots, \nu^m) \in \mathcal{X}$ be such that $\dim \mathbf{V}_{\mathbf{i}} = \sum_l \nu^l_{\mathbf{i}}$ for all $\mathbf{i} \in \mathbf{I}$; here, $\nu^l \in \mathbf{N}[\mathbf{I}]$ for each l. Let $\boldsymbol{\nu}' = (\nu'^1, \nu'^2, \dots, \nu'^m) \in \mathcal{X}$ be defined by $\nu'^l_{\mathbf{i}} = \nu^l_{a(\mathbf{i})}$. We have natural isomorphisms $a : \mathcal{F}_{\boldsymbol{\nu}'} \to \mathcal{F}_{\boldsymbol{\nu}}$ and $a : \tilde{\mathcal{F}}_{\boldsymbol{\nu}'} \to \tilde{\mathcal{F}}_{\boldsymbol{\nu}}$ given by $a(f) = (\mathbf{V} = a(\mathbf{V}^0) \supset a(\mathbf{V}^1) \supset \cdots \supset a(\mathbf{V}^m) = 0)$ for $f = (\mathbf{V} = \mathbf{V}^0 \supset \mathbf{V}^1 \supset \cdots \supset \mathbf{V}^m = 0)$ and $a(x, f) = (ax, af)$. It follows that $a^* \tilde{L}_{\boldsymbol{\nu}} = \tilde{L}_{\boldsymbol{\nu}'}$. From this we see that a^* takes $\mathcal{Q}_{\mathbf{V}}$ into itself and $\mathcal{P}_{\mathbf{V}}$ into itself. Applying the definitions in 11.1.2, 11.1.5 to the linear categories $\mathcal{Q}_{\mathbf{V}}, \mathcal{P}_{\mathbf{V}}$ and to the periodic functor a^* on them, we obtain the linear categories $\tilde{\mathcal{Q}}_{\mathbf{V}}, \tilde{\mathcal{P}}_{\mathbf{V}}$ and the $\mathcal{O}$-modules $\mathcal{K}(\mathcal{Q}_{\mathbf{V}}), \mathcal{K}(\mathcal{P}_{\mathbf{V}})$.

12.1.3. We shall use the notation $\mathcal{O}' = \mathcal{O}[v, v^{-1}]$ (v is an indeterminate). Now $\mathcal{K}(\mathcal{Q}_{\mathbf{V}})$ is naturally an $\mathcal{O}'$-module by $v^n[B, \phi] = [B', \phi']$, where (B', ϕ') is obtained from (B, ϕ) by applying the shift $[n]$.

We define $\mathcal{O}'$-linear maps

$$\mathcal{K}(\mathcal{Q}_{\mathbf{V}}) \xrightarrow{\rho} \mathcal{O}' \otimes_{\mathcal{O}} \mathcal{K}(\mathcal{P}_{\mathbf{V}}) \xrightarrow{\rho'} \mathcal{K}(\mathcal{Q}_{\mathbf{V}})$$

as follows: ρ' sends a symbol $[B, \phi]$ (where $B \in \mathcal{P}_{\mathbf{V}}$) to $[B, \phi]$; ρ sends a symbol $[B, \phi]$ (where $B \in \mathcal{Q}_{\mathbf{V}}$) to $\sum_n v^{-n}[H^n B, H^n(\phi)]$. It is easy to verify that ρ, ρ' respect the defining relations; hence they are well-defined. It is also clear that $\rho\rho' = 1$.

If $K \in \mathcal{Q}_{\mathbf{V}}$ and we are given an isomorphism $\phi : a^*K \to K$, then ϕ induces isomorphisms $\phi_n : a^*(H^n K[-n]) \cong H^n K[-n]$ for each n; moreover we have $(K, \phi) \cong \oplus_n (H^n K[-n], \phi_n)$ as objects of $\tilde{\mathcal{Q}}_{\mathbf{V}}$. This follows from

8.1.3(a) by repeated applications of 11.1.3. This shows that $\rho'\rho = 1$. Thus, ρ' is an isomorphism

$$\mathcal{O}' \otimes_{\mathcal{O}} \mathcal{K}(\mathcal{P}_{\mathbf{V}}) \cong \mathcal{K}(\mathcal{Q}_{\mathbf{V}}).$$

Using this and 11.1.8, we see that

(a) $$\oplus_B \mathcal{O}' \otimes_{\mathcal{O}} \mathcal{O}_B = \mathcal{K}(\mathcal{Q}_{\mathbf{V}})$$

where the sum is over a set of representatives B for the isomorphism classes of simple perverse sheaves in $\mathcal{P}_{\mathbf{V}}$ such that $a^*B \cong B$; $\mathcal{O}_B$ is a free $\mathcal{O}$-module of rank 1 defined as in 11.1.7 (to $\phi \in \mathcal{O}_B$ such that $(B,\phi) \in \tilde{\mathcal{P}}_{\mathbf{V}}$ corresponds $[B,\phi] \in \mathcal{K}(\mathcal{Q}_{\mathbf{V}})$).

12.1.4. Now let $\mathbf{T}, \mathbf{W}$ be two objects of $\mathcal{V}^a$. We can form $\mathbf{E}_{\mathbf{T}}, \mathbf{E}_{\mathbf{W}}$ and their product $\mathbf{E}_{\mathbf{T}} \times \mathbf{E}_{\mathbf{W}}$. This has an action of $G_{\mathbf{T}} \times G_{\mathbf{W}}$ (product of actions as in 9.1.2) and an automorphism a (product of the automorphisms a of the factors) such that $a^{\mathbf{n}} = 1$.

The functor $a^* : \mathcal{D}(\mathbf{E}_{\mathbf{T}} \times \mathbf{E}_{\mathbf{W}}) \to \mathcal{D}(\mathbf{E}_{\mathbf{T}} \times \mathbf{E}_{\mathbf{W}})$ takes the subcategories $\mathcal{P}_{\mathbf{T},\mathbf{W}}$ and $\mathcal{Q}_{\mathbf{T},\mathbf{W}}$ into themselves. Applying the definitions in 11.1.2, 11.1.5 to these linear categories and to a^*, we obtain linear categories $\tilde{\mathcal{P}}_{\mathbf{T},\mathbf{W}}, \tilde{\mathcal{Q}}_{\mathbf{T},\mathbf{W}}$ and $\mathcal{O}$-modules $\mathcal{K}(\mathcal{P}_{\mathbf{T},\mathbf{W}}), \mathcal{K}(\mathcal{Q}_{\mathbf{T},\mathbf{W}})$. These are special cases of the definitions of $\tilde{\mathcal{P}}_{\mathbf{V}}, \tilde{\mathcal{Q}}_{\mathbf{V}}$ and $\mathcal{K}(\mathcal{P}_{\mathbf{V}}), \mathcal{K}(\mathcal{Q}_{\mathbf{V}})$ in 12.1.2; indeed, $\mathbf{T} \times \mathbf{W}$ and $\mathbf{E}_{\mathbf{T}} \times \mathbf{E}_{\mathbf{W}}$ are special cases of $\mathbf{V}$ and $\mathbf{E}_{\mathbf{V}}$ where our graph has been replaced by the disjoint union of two copies of itself. Thus, $\mathcal{K}(\mathcal{Q}_{\mathbf{T},\mathbf{W}})$ is naturally a $\mathcal{O}'$-module and we have

$$\mathcal{O}' \otimes_{\mathcal{O}} \mathcal{K}(\mathcal{P}_{\mathbf{T},\mathbf{W}}) \cong \mathcal{K}(\mathcal{Q}_{\mathbf{T},\mathbf{W}})$$

and

$$\oplus_B \mathcal{O}' \otimes_{\mathcal{O}} \mathcal{O}_B = \mathcal{K}(\mathcal{Q}_{\mathbf{T},\mathbf{W}})$$

where the sum is over a set of representatives B for the isomorphism classes of simple perverse sheaves in $\mathcal{P}_{\mathbf{T},\mathbf{W}}$ such that $a^*B \cong B$; $\mathcal{O}_B$ is a free $\mathcal{O}$-module of rank 1 defined as in 11.1.7 (to $\phi \in \mathcal{O}_B$ such that $(B,\phi) \in \tilde{\mathcal{P}}_{\mathbf{T},\mathbf{W}}$ corresponds $[B,\phi] \in \mathcal{K}(\mathcal{Q}_{\mathbf{T},\mathbf{W}})$).

From the definitions it is clear that any simple object $B \in \mathcal{P}_{\mathbf{T},\mathbf{W}}$ is the external tensor product $B' \otimes B''$ of two simple objects $B' \in \mathcal{P}_{\mathbf{T}}$ and $B'' \in \mathcal{P}_{\mathbf{W}}$ (and conversely); we have $a^*B \cong B$ if and only if $a^*B' \cong B'$ and $a^*B'' \cong B''$ and then $\mathrm{Hom}(a^*B, B) = \mathrm{Hom}(a^*B', B') \otimes \mathrm{Hom}(a^*B'', B'')$ and $\mathcal{O}_B = \mathcal{O}_{B'} \otimes_{\mathcal{O}} \mathcal{O}_{B''}$. It follows that

$$\mathcal{K}(\mathcal{Q}_{\mathbf{T},\mathbf{W}}) = \mathcal{K}(\mathcal{Q}_{\mathbf{T}}) \otimes_{\mathcal{O}'} \mathcal{K}(\mathcal{Q}_{\mathbf{W}}).$$

12.1.5. Assume now that $\mathbf{W}$ is an $\mathbf{I}$-graded, a-stable subspace of $\mathbf{V}$ and that $\mathbf{T} = \mathbf{V}/\mathbf{W}$ with the induced grading and a-action. The functors $\mathrm{Ind}_{\mathbf{T},\mathbf{W}}^{\mathbf{V}} : \mathcal{Q}_{\mathbf{T},\mathbf{W}} \to \mathcal{Q}_{\mathbf{V}}$ and $\mathrm{Res}_{\mathbf{T},\mathbf{W}}^{\mathbf{V}} : \mathcal{Q}_{\mathbf{V}} \to \mathcal{Q}_{\mathbf{T},\mathbf{W}}$ are linear and compatible with a^*; hence by 11.1.6, we have induced linear functors $\tilde{\mathcal{Q}}_{\mathbf{T},\mathbf{W}} \to \tilde{\mathcal{Q}}_{\mathbf{V}}$, $\tilde{\mathcal{Q}}_{\mathbf{V}} \to \tilde{\mathcal{Q}}_{\mathbf{T},\mathbf{W}}$ and $\mathcal{O}$-linear maps

$$\mathrm{ind}_{\mathbf{T},\mathbf{W}}^{\mathbf{V}} : \mathcal{K}(\mathcal{Q}_{\mathbf{T},\mathbf{W}}) \to \mathcal{K}(\mathcal{Q}_{\mathbf{V}}) \quad \text{and} \quad \mathrm{res}_{\mathbf{T},\mathbf{W}}^{\mathbf{V}} : \mathcal{K}(\mathcal{Q}_{\mathbf{V}}) \to \mathcal{K}(\mathcal{Q}_{\mathbf{T},\mathbf{W}}).$$

These last two maps are in fact $\mathcal{O}'$-linear since $\mathrm{Ind}_{\mathbf{T},\mathbf{W}}^{\mathbf{V}}$ and $\mathrm{Res}_{\mathbf{T},\mathbf{W}}^{\mathbf{V}}$ commute with shifts.

12.2. Inner Product

12.2.1. Let $(B', \phi'), (B'', \phi'')$ be two objects of $\tilde{\mathcal{Q}}_{\mathbf{V}}$. The vector space $\mathbf{D}_j(\mathbf{E}_{\mathbf{V}}, G_{\mathbf{V}}; B', B'')$ (see 8.1.9) has a natural automorphism a such that $a^{\mathbf{n}} = 1$. This can be constructed as follows. Let V be an $\mathbf{I}$-graded k-vector space such that for each $\mathbf{i} \in \mathbf{I}$, $\dim V_{\mathbf{i}}$ is finite but large.

Let Γ be the smooth, irreducible variety consisting of all injective linear maps $\mathbf{V} \to V$ which respect the $\mathbf{I}$-grading. Then Γ has an obvious free $G_{\mathbf{V}}$-action and its $\bar{\mathbf{Q}}_l$-cohomology is zero in degrees $1, 2, \dots, m$ where m is a large integer.

By the definition of $\mathcal{V}^a$, we have an automorphism $a : \mathbf{V} \to \mathbf{V}$. This induces an automorphism $a : \Gamma \to \Gamma$. Taking the product with the automorphism $a : \mathbf{E}_{\mathbf{V}} \to \mathbf{E}_{\mathbf{V}}$, we obtain an automorphism $a : \Gamma \times \mathbf{E}_{\mathbf{V}} \to \Gamma \times \mathbf{E}_{\mathbf{V}}$. This induces an automorphism a of the orbit space $G_{\mathbf{V}} \backslash (\Gamma \times \mathbf{E}_{\mathbf{V}})$. The isomorphisms $\phi' : a^* B' \to B'$ and $\phi'' : a^* B'' \to B''$ induce isomorphisms $\tilde{\phi}' : a^* \tilde{B}' \to \tilde{B}'$ and $\tilde{\phi}'' : a^* \tilde{B}'' \to \tilde{B}''$ where $\tilde{B}' = t_\flat s^* B', \tilde{B}'' = t_\flat s^* B''$ are semisimple complexes in $G_{\mathbf{V}} \backslash (\Gamma \times \mathbf{E}_{\mathbf{V}})$ defined as in 8.1.9.

Now $\tilde{\phi}' \otimes \tilde{\phi}'' : a^*(\tilde{B}' \otimes \tilde{B}'') \to (\tilde{B}' \otimes \tilde{B}'')$ induces an isomorphism of

$$\mathcal{H}^{j+2\dim(G\backslash\Gamma)}(u_!(\tilde{B}' \otimes \tilde{B}''))$$

onto itself, or equivalently, an isomorphism of

$$\mathbf{D}_j(\mathbf{E}_{\mathbf{V}}, G_{\mathbf{V}}; B', B'')$$

onto itself, denoted again by a. Here u denotes the map of $G_{\mathbf{V}} \backslash (\Gamma \times \mathbf{E}_{\mathbf{V}})$ into the point. It is clear that $a : \mathbf{D}_j(\mathbf{E}_{\mathbf{V}}, G_{\mathbf{V}}; B', B'') \to \mathbf{D}_j(\mathbf{E}_{\mathbf{V}}, G_{\mathbf{V}}; B', B'')$ satisfies $a^{\mathbf{n}} = 1$.

We define

$$\{(B', \phi'), (B'', \phi'')\} = \sum_j \operatorname{tr}(a, \mathbf{D}_j(\mathbf{E}_{\mathbf{V}}, G_{\mathbf{V}}; B', B''))v^{-j} \in \mathcal{O}((v)).$$

The last inclusion follows from 8.1.10(e). (In general, given a ring R, the ring of power series $\sum_{n\in\mathbf{Z}} a_n v^n$ with $a_n \in R$ such that $a_n = 0$ for $n \ll 0$ (resp. $n < 0$; or $n \gg 0$; or $n > 0$) is denoted $R((v))$ (resp. $R[[v]]$; or $R((v^{-1}))$; or $R[[v^{-1}]]$).)

If (B', ϕ') is traceless, then we can write $B' = C \oplus a^*C \oplus \cdots \oplus a^{*(t-1)}C$ (where $t \geq 2$) and ϕ' acts like a matrix without diagonal terms. We have a corresponding decomposition

$$\begin{aligned}&\mathbf{D}_j(\mathbf{E_V}, G_\mathbf{V}; B', B'') = \\ &\mathbf{D}_j(\mathbf{E_V}, G_\mathbf{V}; C, B'') \oplus \mathbf{D}_j(\mathbf{E_V}, G_\mathbf{V}; a^*C, B'') \oplus \cdots \\ &\oplus \mathbf{D}_j(\mathbf{E_V}, G_\mathbf{V}; a^{*(t-1)}C, B'')\end{aligned}$$

Then $a : \mathbf{D}_j(\mathbf{E_V}, G_\mathbf{V}; B', B'') \to \mathbf{D}_j(\mathbf{E_V}, G_\mathbf{V}; B', B'')$ acts with respect to the above decomposition like a matrix without diagonal terms; hence it has trace zero. Thus, $\{(B', \phi'), (B'', \phi'')\} = 0$ if either (B', ϕ') or (B'', ϕ'') is traceless. It follows immediately that

$$\{[B', \phi'], [B'', \phi'']\} = \{(B', \phi'), (B'', \phi'')\}$$

is a well-defined symmetric $\mathcal{O}'$-bilinear pairing

$$\mathcal{K}(\mathcal{Q}_\mathbf{V}) \times \mathcal{K}(\mathcal{Q}_\mathbf{V}) \times \to \mathcal{O}((v)).$$

We can define in the same way a symmetric $\mathcal{O}'$-bilinear pairing

$$\{,\} : \mathcal{K}(\mathcal{Q}_{\mathbf{T},\mathbf{W}}) \times \mathcal{K}(\mathcal{Q}_{\mathbf{T},\mathbf{W}}) \times \to \mathcal{O}((v));$$

indeed, as we have seen, $\mathcal{K}(\mathcal{Q}_{\mathbf{T},\mathbf{W}})$ is a special case of $\mathcal{K}(\mathcal{Q}_\mathbf{V})$ in the case where our graph has been replaced by the disjoint union of two copies of itself.

Using the definitions and 8.1.10(f), we see that

$$\{A_1 \otimes A_3, A_2 \otimes A_4\} = \{A_1, A_2\}\{A_3, A_4\}$$

for any $A_1, A_2 \in \mathcal{K}(\mathcal{Q}_\mathbf{T})$ and $A_3, A_4 \in \mathcal{K}(\mathcal{Q}_\mathbf{W})$. We then have

$$A_1 \otimes A_3, A_2 \otimes A_4 \in \mathcal{K}(\mathcal{Q}_{\mathbf{T},\mathbf{W}}).$$

We have the following result.

Lemma 12.2.2. *Let $\alpha \in \mathcal{K}(\mathcal{Q}_{\mathbf{T},\mathbf{W}}), \beta \in \mathcal{K}(\mathcal{Q}_{\mathbf{V}})$. We have*

$$\{\alpha, res^{\mathbf{V}}_{\mathbf{T},\mathbf{W}}(\beta)\} = \{ind^{\mathbf{V}}_{\mathbf{T},\mathbf{W}}(\alpha), \beta\}.$$

Let (A, ϕ) be an object of $\tilde{\mathcal{Q}}_{\mathbf{T},\mathbf{W}}$ and let (B, ϕ') be an object of $\tilde{\mathcal{Q}}_{\mathbf{V}}$. We take the trace of the automorphism a in both sides of 9.2.10(a); the lemma follows.

12.3. Properties of $L_{\boldsymbol{\nu}}$

12.3.1. Let $\mathcal{X}^a$ be the set of all $\boldsymbol{\nu} = (\nu^1, \nu^2, \dots, \nu^m) \in \mathcal{X}$ such that $\nu^l \in \mathbf{N}[\mathbf{I}]^a$ for all l. Let $\boldsymbol{\nu} = (\nu^1, \nu^2, \dots, \nu^m) \in \mathcal{X}^a$ be such that $\dim \mathbf{V}_{\mathbf{i}} = \sum_l \nu^l_{\mathbf{i}}$ for all $\mathbf{i} \in \mathbf{I}$. We have natural automorphisms $a : \mathcal{F}_{\boldsymbol{\nu}} \to \mathcal{F}_{\boldsymbol{\nu}}$ and $a : \tilde{\mathcal{F}}_{\boldsymbol{\nu}} \to \tilde{\mathcal{F}}_{\boldsymbol{\nu}}$ defined as in 12.1.2. The obvious isomorphism $a^*\mathbf{1} \cong \mathbf{1}$ on $\tilde{\mathcal{F}}_{\boldsymbol{\nu}}$ induces an isomorphism

$$\phi_0 : a^*(\pi_{\boldsymbol{\nu}})_!\mathbf{1} = (\pi_{\boldsymbol{\nu}})_!(a^*\mathbf{1}) \cong (\pi_{\boldsymbol{\nu}})_!\mathbf{1}$$

in $\mathcal{D}(\mathbf{E}_{\mathbf{V}})$.

It is easy to see that $(\tilde{L}_{\boldsymbol{\nu}}, \phi_0)$ is an object of $\tilde{\mathcal{Q}}_{\mathbf{V}}$. Applying to ϕ_0 a shift by $f(\boldsymbol{\nu})$ (see 9.1.3(c)), we obtain an isomorphism $a^*L_{\boldsymbol{\nu}} \cong L_{\boldsymbol{\nu}}$ which is denoted again by ϕ_0. Note that $(L_{\boldsymbol{\nu}}, \phi_0)$ is an object of $\tilde{\mathcal{Q}}_{\mathbf{V}}$.

Let $\mathbf{W} \subset \mathbf{V}$ and $\mathbf{T} = \mathbf{V}/\mathbf{W}$ be as in 12.1.5. Let $\boldsymbol{\nu}' = (\nu'^1, \nu'^2, \dots, \nu'^m) \in \mathcal{X}^a$ and $\boldsymbol{\nu}'' = (\nu''^1, \nu''^2, \dots, \nu''^n) \in \mathcal{X}^a$ be such that $\dim \mathbf{T}_{\mathbf{i}} = \sum_l \nu'^l_{\mathbf{i}}$, $\dim \mathbf{W}_{\mathbf{i}} = \sum_l \nu''^l_{\mathbf{i}}$ for all $\mathbf{i} \in \mathbf{I}$.

Lemma 12.3.2. *The following equality holds in $\mathcal{K}(\mathcal{Q}_{\mathbf{V}})$:*

$$ind^{\mathbf{V}}_{\mathbf{T},\mathbf{W}}([L_{\boldsymbol{\nu}'}, \phi_0] \otimes [L_{\boldsymbol{\nu}''}, \phi_0]) = [L_{\boldsymbol{\nu}'\boldsymbol{\nu}''}, \phi_0].$$

This follows from 9.2.7.

Lemma 12.3.3. *Let $\boldsymbol{\nu} = (\nu^1, \nu^2, \dots, \nu^m) \in \mathcal{X}^a$ be such that $\dim \mathbf{V}_{\mathbf{i}} = \sum_l \nu^l_{\mathbf{i}}$ for all $\mathbf{i} \in \mathbf{I}$. The following equality holds in $\mathcal{K}(\mathcal{Q}_{\mathbf{T},\mathbf{W}})$:*

$$res^{\mathbf{V}}_{\mathbf{T},\mathbf{W}}[L_{\boldsymbol{\nu}}, \phi_0] = \sum v^{M'(\boldsymbol{\tau},\boldsymbol{\omega})}[L_{\boldsymbol{\tau}}, \phi_0] \otimes [L_{\boldsymbol{\omega}}, \phi_0],$$

where the sum is taken over all $\boldsymbol{\tau} = (\tau^1, \tau^2, \dots, \tau^m), \boldsymbol{\omega} = (\omega^1, \omega^2, \dots, \omega^m)$ in $\mathcal{X}^a$ such that $\dim \mathbf{T}_{\mathbf{i}} = \sum_l \tau^l_{\mathbf{i}}$, $\dim \mathbf{W}_{\mathbf{i}} = \sum_l \omega^l_{\mathbf{i}}$ for all $\mathbf{i}$ and $\tau^l + \omega^l = \nu^l$ for all l; $M'(\boldsymbol{\tau}, \boldsymbol{\omega})$ is as in 9.2.11.

This follows from the decomposition 9.2.11. The terms in that decomposition corresponding to $\boldsymbol{\tau}, \boldsymbol{\omega}$ with some coordinate not in $\mathbf{N}[\mathbf{I}]^a$ can be grouped according to non-trivial a-orbits and contribute traceless objects, hence they disappear from the final result.

Lemma 12.3.4. *Let i be an a-orbit on I and let $\gamma = \sum_{\mathbf{i}\in i} \mathbf{i} \in \mathbf{N}[\mathbf{I}]$. Let $\mathbf{V} \in \mathcal{V}^a$ be such that $\dim \mathbf{V}_\mathbf{i} = n$ for all $\mathbf{i} \in i$ and $\dim \mathbf{V}_\mathbf{i} = 0$ for all $\mathbf{i} \notin i$. Let d be the number of elements of i. The following equality holds in $\mathcal{K}(\mathcal{Q}_\mathbf{V})$:*

(a) $$[\tilde{L}_{\gamma,(n-1)\gamma}, \phi_0] = \sum_{s=0}^{n-1} v^{-2sd}[\tilde{L}_{n\gamma}, \phi_0].$$

Using the definitions and the known structure of the cohomology of a product of d copies of a projective $(n-1)$-space, we see that the left hand side of (a) is represented by $(\oplus \mathbf{1}[-2s_1 - 2s_2 - \cdots - 2s_d], \phi)$, a sum over all sequences $(s_1, s_2, \ldots, s_d)$ with $0 \le s_j \le n-1$; here ϕ maps the summand corresponding to $(s_1, s_2, \ldots, s_d)$ to the summand corresponding to $(s_2, s_3, \ldots, s_d, s_1)$. The summands corresponding to sequences whose terms are not all equal to each other, form traceless objects. The remaining terms give the right hand side of (a).

12.3.5. Remark. In the previous lemma we have $[L_{\gamma,(n-1)\gamma}, \phi_0] = v^{d(n-1)}[\tilde{L}_{\gamma,(n-1)\gamma}, \phi_0]$ and $[L_{n\gamma}, \phi_0] = [\tilde{L}_{n\gamma}, \phi_0]$; hence

(a) $$[L_{\gamma,(n-1)\gamma}, \phi_0] = v^{d(n-1)} \sum_{s=0}^{n-1} v^{-2sd}[L_{n\gamma}, \phi_0].$$

Lemma 12.3.6. *We preserve the assumptions of Lemma 12.3.4, and we take $n = 1$. We have*

(a) $$\{[L_\gamma, \phi_0], [L_\gamma, \phi_0]\} = \sum_{s=0}^{\infty} v^{2ds} = (1 - v^d)^{-1}.$$

Since i is discrete, we have $\mathbf{E}_\mathbf{V} = 0$ and $L_\gamma = \mathbf{1}$. Using the definitions we see that the left hand side of (a) is computed as follows: we consider the product of d copies of an infinite projective space with the automorphism given by cyclically permuting the factors; we must find the trace of the induced automorphism on each cohomology space. This is given by the same computation as in the previous lemma.

Lemma 12.3.7. *Let $\boldsymbol{\nu} = (\nu^1, \nu^2, \ldots, \nu^m)$ and $\boldsymbol{\nu}' = (\nu'^1, \nu'^2, \ldots, \nu'^n)$ be two elements of $\mathcal{X}^a$ such that $\dim \mathbf{V}_\mathbf{i} = \sum_l \nu^l_\mathbf{i} = \sum_l \nu'^l_\mathbf{i}$ for all $\mathbf{i} \in \mathbf{I}$. We have $\{[L_{\boldsymbol{\nu}}, \phi_0], [L_{\boldsymbol{\nu}'}, \phi_0]\} \in \mathbf{Z}((v)) \cap \mathbf{Q}(v)$.*

Using 9.1.4, we see that we may assume that each ν^l and each ν'^l is of the form $N \sum_{\mathbf{i}\in i} \mathbf{i}$ for some a-orbit i in $\mathbf{I}$. Using 12.3.5, we can further assume that we always have $N = 1$.

We argue by induction on $m+n$. The case where m or n is zero, is trivial. The case where $m=n=1$ follows from 12.3.6. Hence we may assume that m or n is ≥ 2. Assume first that $n \geq 2$. Let $\mathbf{W}$ be an $\mathbf{I}$-graded a-stable subspace of $\mathbf{V}$ such that $\mathbf{T}=\mathbf{V}/\mathbf{W}$ satisfies $\dim \mathbf{T}_{\mathbf{i}} = \nu^1_{\mathbf{i}}$ for all $\mathbf{i} \in \mathbf{I}$. Let $\boldsymbol{\nu}^1 = (\nu^1), \boldsymbol{\nu}^2 = (\nu^2, \nu^3, \dots, \nu^m)$. We have

$$\begin{aligned}\{[L_{\boldsymbol{\nu}}, \phi_0], [L_{\boldsymbol{\nu}'}, \phi_0]\} &= \{\mathrm{ind}^{\mathbf{V}}_{\mathbf{T},\mathbf{W}}([L_{\boldsymbol{\nu}^1}, \phi_0] \otimes [L_{\boldsymbol{\nu}^2}, \phi_0]), [L_{\boldsymbol{\nu}'}, \phi_0]\} \\ &= \{[L_{\boldsymbol{\nu}^1}, \phi_0] \otimes [L_{\boldsymbol{\nu}^2}, \phi_0], \mathrm{res}^{\mathbf{V}}_{\mathbf{T},\mathbf{W}}[L_{\boldsymbol{\nu}'}, \phi_0]\}\end{aligned}$$

and this is in $\mathbf{Z}((v)) \cap \mathbf{Q}(v)$, by 12.3.3 and the induction hypothesis. We can treat similarly the case where $m \geq 2$. The lemma is proved.

12.4. Verdier Duality

12.4.1. Let $\mathbf{V} \in \mathcal{V}^a$. Let (B, ϕ) be an object of $\tilde{\mathcal{Q}}_{\mathbf{V}}$. Recall that $\phi : a^*B \cong B$. Applying Verdier duality, we obtain $D(\phi) : D(B) \cong D(a^*B) = a^*(D(B))$. The inverse of this isomorphism is an isomorphism $D(\phi)^{-1} : a^*(D(B)) \cong D(B)$. It is clear that $(D(B), D(\phi)^{-1})$ is an object of $\tilde{\mathcal{Q}}_{\mathbf{V}}$. It is easy to see that we have a well-defined homomorphism of abelian groups $D : \mathcal{K}(\mathcal{Q}_{\mathbf{V}}) \to \mathcal{K}(\mathcal{Q}_{\mathbf{V}})$ given by $D[B, \phi] = [D(B), D(\phi)^{-1}]$. This homomorphism has square equal to 1. Moreover, from the definitions, we see that it is semi-linear with respect to the involution $^- : \mathcal{O}' \to \mathcal{O}'$ given by $v^n \mapsto v^{-n}$ and $\zeta \to \zeta^{-1}$ for ζ such that $\zeta^{\mathbf{n}} = 1$.

12.4.2. Let $\mathbf{W} \subset \mathbf{V}$ and $\mathbf{T} = \mathbf{V}/\mathbf{W}$ be as in 12.1.5. We have a homomorphism of abelian groups $D : \mathcal{K}(\mathcal{Q}_{\mathbf{T},\mathbf{W}}) \to \mathcal{K}(\mathcal{Q}_{\mathbf{T},\mathbf{W}})$ defined in the same way as $D : \mathcal{K}(\mathcal{Q}_{\mathbf{V}}) \to \mathcal{K}(\mathcal{Q}_{\mathbf{V}})$ (and in fact is a special case of it). From the definition we see that the following diagram is commutative:

$$\begin{array}{ccc}\mathcal{K}(\mathcal{Q}_{\mathbf{T}}) \otimes \mathcal{K}(\mathcal{Q}_{\mathbf{W}}) & \xrightarrow{D \otimes D} & \mathcal{K}(\mathcal{Q}_{\mathbf{T}}) \otimes \mathcal{K}(\mathcal{Q}_{\mathbf{W}}) \\ \downarrow & & \downarrow \\ \mathcal{K}(\mathcal{Q}_{\mathbf{T},\mathbf{W}}) & \xrightarrow{D} & \mathcal{K}(\mathcal{Q}_{\mathbf{T},\mathbf{W}})\end{array}$$

where the tensor products are over $\mathcal{O}'$ and the vertical maps are the isomorphisms in 12.1.4.

Lemma 12.4.3. *For any $\alpha \in \mathcal{K}(\mathcal{Q}_{\mathbf{T},\mathbf{W}})$, we have $D(ind^{\mathbf{V}}_{\mathbf{T},\mathbf{W}}(\alpha)) = ind^{\mathbf{V}}_{\mathbf{T},\mathbf{W}}(D\alpha)$ in $\mathcal{K}(\mathcal{Q}_{\mathbf{V}})$.*

This follows from 9.2.5.

12.5. Self-Dual Elements

Lemma 12.5.1. *Let $B \in \mathcal{P}_{\mathbf{V}}$ be a simple object such that $a^*B \cong B$.*

(a) *If $\mathbf{i} \in \mathbf{I}$ and $n \in \mathbf{N}$ are such that $B \in \mathcal{P}_{\mathbf{V};\{\mathbf{i}\};n\mathbf{i}}$, then $B \in \mathcal{P}_{\mathbf{V};i;n\gamma}$ where i is the a-orbit of $\mathbf{i}$ and $\gamma = \sum_{\mathbf{i}' \in i} \mathbf{i}' \in \mathbf{N}[\mathbf{I}]$.*

(b) *Assume that $B \in \mathcal{P}_{\mathbf{V};i;\gamma'}$ where i is as in (a) and $\gamma' \in \mathbf{N}[\mathbf{I}]$ has support contained in i. Then $\gamma' = n\gamma$ for some $n \geq 0$, where γ is as in (a).*

(c) *If $\mathbf{V} \neq 0$, then there exists an a-orbit i in $\mathbf{I}$ such that $B \in \mathcal{P}_{\mathbf{V};i;n\gamma}$ where γ is as in (a), for some $n \geq 1$.*

We prove (a). Our statement is independent of the orientation, as in the proof of 10.3.2. Hence we may assume that the orientation is such that $h' \notin i$ for any $h \in H$. Such an orientation exists since i is discrete. By our assumption and by Lemma 9.3.5 we have

$$\dim \mathbf{V}_{\mathbf{i}} / (\sum_{h \in H: h''=\mathbf{i}} x_h(\mathbf{V}_{h'})) \geq n$$

for all x in the support of B and

$$\dim \mathbf{V}_{\mathbf{i}} / (\sum_{h \in H: h''=\mathbf{i}} x_h(\mathbf{V}_{h'})) = n$$

for some x in the support of B.

Since the support of B is a-invariant, it follows that the same holds when $\mathbf{i}$ is replaced by any $\mathbf{i}'$ in i. Using Lemma 9.3.5 again, we see that the conclusion of (a) holds.

We prove (b). Let $\mathbf{i} \in i$. We can find $n \in \mathbf{N}$ such that $B \in \mathcal{P}_{\mathbf{V};\{\mathbf{i}\};n\mathbf{i}}$. By (a), we have $B \in \mathcal{P}_{\mathbf{V};i;n\gamma}$. By 9.3.5, the inclusions $B \in \mathcal{P}_{\mathbf{V};i;n\gamma}, B \in \mathcal{P}_{\mathbf{V};i;\gamma'}$ imply $\gamma' = n\gamma$ and (b) follows.

(c) follows immediately from (a) and 9.3.1(b).

Proposition 12.5.2. *Let B be a simple object of $\mathcal{P}_{\mathbf{V}}$ such that a^*B is isomorphic to B.*

(a) *There exists an isomorphism $\phi : a^*B \cong B$ such that $(B, \phi) \in \tilde{\mathcal{P}}_{\mathbf{V}}$ and such that $(D(B), D(\phi)^{-1})$ is isomorphic to (B, ϕ) as objects of $\tilde{\mathcal{P}}_{\mathbf{V}}$. Moreover, ϕ is unique, if $\mathbf{n}$ is odd, and unique up to multiplication by ± 1, if $\mathbf{n}$ is even.*

(b) *We have $D(B) \cong B$ as objects of $\mathcal{P}_{\mathbf{V}}$.*

(b) clearly follows from (a). We prove the existence of ϕ in (a). This is trivial when $\mathbf{V} = 0$. Hence we may assume that $\mathbf{V} \neq 0$. By 12.5.1, there exists an a-orbit i on $\mathbf{I}$ and an integer $n \geq 1$ such that $B \in \mathcal{P}_{\mathbf{V};i;n\gamma}$ where $\gamma = \sum_{\mathbf{i} \in i} \mathbf{i}$. Hence it is enough to prove the following statement for any fixed a-orbit i in $\mathbf{I}$.

For any $n \geq 1$, and any simple object $B \in \mathcal{P}_{\mathbf{V};i;n\gamma}$ such that $a^*B \cong B$, there exists $\phi : a^*B \cong B$ such that $(B, \phi) \in \tilde{\mathcal{P}}_{\mathbf{V}}$ and such that $(D(B), D(\phi)^{-1})$ is isomorphic to (B, ϕ).

We argue by descending induction on n. (We have $n \leq \dim \mathbf{V_i}$, for any $\mathbf{i} \in i$). Let $\mathcal{S}$ be the set of all simple objects $C' \in \mathcal{P}_{\mathbf{V};i;>n\gamma}$ (up to isomorphism) such that $a^*C' \cong C'$; for each such C', we denote by $f_{C'}$ some isomorphism $a^*C' \cong C'$ such that $(C', f_{C'}) \in \tilde{\mathcal{P}}_{\mathbf{V}}$. By 12.5.1(b), for each $C' \in \mathcal{S}$, we have $C' \in \mathcal{P}_{\mathbf{V};i;n'\gamma}$ for some $n' > n$, hence the induction hypothesis (on n) is applicable. Hence we may assume that $f_{C'}$ is such that $(D(C'), D(f_{C'})^{-1})$ is isomorphic to $(C', f_{C'})$ for each $C' \in \mathcal{S}$.

Let $\mathbf{W}$ be an $\mathbf{I}$-graded a-stable subspace of $\mathbf{V}$ such that $\mathbf{T} = \mathbf{V}/\mathbf{W}$ satisfies $\dim \mathbf{T_i} = n$ for all $\mathbf{i} \in i$ and $\mathbf{T_i} = 0$ for all $\mathbf{i} \in \mathbf{I} - i$. Such $\mathbf{W}$ exists since i is an a-orbit. By Proposition 10.3.2 (for $\mathbf{I}' = i$) we can find a simple object A of $\mathcal{P}_{\mathbf{W};i;0}$ such that $\mathrm{Ind}^{\mathbf{V}}_{\mathbf{T},\mathbf{W}} A \cong B \oplus (\oplus_j C_j[j])$ where $C_j \in \mathcal{P}_{\mathbf{V};i;>n\gamma}$ for all j. Since $\mathrm{Ind}^{\mathbf{V}}_{\mathbf{T},\mathbf{W}}$ commutes with a^* and $a^*B \cong B$, we have also $\mathrm{Ind}^{\mathbf{V}}_{\mathbf{T},\mathbf{W}} a^*A \cong B \oplus (\oplus_j a^*C_j[j])$. By the uniqueness of A in 10.3.2, we must then have $a^*A \cong A$. By the induction hypothesis (on $\dim \mathbf{V}$) we can find an isomorphism $\phi' : a^*A \cong A$ such that $(A, \phi') \in \tilde{\mathcal{P}}_{\mathbf{W}}$ and such that $(D(A), D(\phi')^{-1})$ is isomorphic to (A, ϕ').

Applying to $[A, \phi']$ the homomorphism

$$\mathrm{ind}^{\mathbf{V}}_{\mathbf{T},\mathbf{W}} : \mathcal{K}(\mathcal{Q}_{\mathbf{W}}) = \mathcal{K}(\mathcal{Q}_{\mathbf{T},\mathbf{W}}) \to \mathcal{K}(\mathcal{Q}_{\mathbf{V}}),$$

we necessarily obtain an element of the form $[B, \phi] + \sum_{C' \in \mathcal{S}} P_{C'}[C', f_{C'}]$ where $\phi : a^*B \cong B$ is some isomorphism such that $(B, \phi) \in \tilde{\mathcal{P}}_{\mathbf{V}}$. The coefficients $P_{C'}$ are in $\mathcal{O}'$.

Since $D[A, \phi'] = [A, \phi']$, we have $D(\mathrm{ind}^{\mathbf{V}}_{\mathbf{T},\mathbf{W}} A) = \mathrm{ind}^{\mathbf{V}}_{\mathbf{T},\mathbf{W}} A$ (see 12.4.3). Hence we have

$$D[B, \phi] + \sum \bar{P}_{C'} D[C', f_{C'}] = [B, \phi] + \sum P_{C'}[C', f_{C'}]$$

($\bar{P}_{C'}$ as in 12.4.1). Substituting here $D[C', f_{C'}] = [C', f_{C'}]$, we obtain

$$D[B, \phi] = [B, \phi] + \sum_{C'} (P_{C'} - \bar{P}_{C'})[C', f_{C'}].$$

Now the elements $[B, \phi]$ and $[C', f_{C'}]$ form a subset of a basis of the $\mathcal{O}'$-module $\mathcal{K}(\mathcal{Q}_{\mathbf{V}})$ and $D[B, \phi]$ is a scalar multiple of some element in that basis. Hence the previous equality forces $D[B,\phi]$ to be equal to $[B,\phi]$. This proves the existence of ϕ.

Assume that $\zeta\phi$ has the same property as ϕ, where ζ satisfies $\zeta^{\mathbf{n}} = 1$. We have

$$\zeta[B,\phi] = [B,\zeta\phi] = D[B,\zeta\phi] = D(\zeta[B,\phi]) = \zeta^{-1}D[B,\phi] = \zeta^{-1}[B,\phi];$$

hence $\zeta = \zeta^{-1}$ and $\zeta = \pm 1$. If $\mathbf{n}$ is odd, it follows that $\zeta = 1$. The proposition is proved.

Lemma 12.5.3. *Let $(B,\phi), (B',\phi')$ be objects of $\tilde{\mathcal{P}}_{\mathbf{V}}$ such that B, B' are simple objects of $\mathcal{P}_{\mathbf{V}}$.*

(a) *If B, B' are not isomorphic, then $\{[B,\phi],[B',\phi']\} \in v\mathcal{O}[[v]]$.*

(b) *If $(B',\phi') = (D(B), D(\phi)^{-1})$, then $\{[B,\phi],[B',\phi']\} \in 1 + v\mathcal{O}[[v]]$.*

(c) *If (B,ϕ) is as in 12.5.2, then $\{[B,\phi],[B,\phi]\} \in 1 + v\mathcal{O}[[v]]$.*

(a) and (b) follow from 8.1.10(d) and the definitions, using the fact that $D(B) \cong B$ (see 12.5.2(b)). (c) follows from (b).

12.6. L_ν AS ADDITIVE GENERATORS

12.6.1. Let $\mathbf{V} \in \mathcal{V}^a$. Let $M_{\mathbf{V}}$ be the $\mathcal{A}$-submodule of $\mathcal{K}(\mathcal{Q}_{\mathbf{V}})$ generated by the elements $[L_{\boldsymbol{\nu}}, \phi_0]$ for various $\boldsymbol{\nu} = (\nu^1, \nu^2, \dots, \nu^m) \in \mathcal{X}^a$ such that $\dim \mathbf{V}_{\mathbf{i}} = \sum_l \nu^l_{\mathbf{i}}$ for all $\mathbf{i} \in \mathbf{I}$.

Lemma 12.6.2. *For any $\alpha, \alpha' \in M_{\mathbf{V}}$ we have $\{\alpha,\alpha'\} \in \mathbf{Z}((v)) \cap \mathbf{Q}(v)$.*

This follows immediately from 12.3.7.

Proposition 12.6.3. *The following two $\mathcal{A}$-submodules of $\mathcal{K}(\mathcal{Q}_{\mathbf{V}})$ coincide:*

(a) *$M_{\mathbf{V}}$;*

(b) *the $\mathcal{A}$-submodule generated by the elements $[B,\phi]$ where (B,ϕ) is as in 12.5.2.*

We show by induction on $\dim \mathbf{V}$ that

(c) if (B,ϕ) is as in 12.5.2, then $[B,\phi] \in M_{\mathbf{V}}$.

This is trivial when $\mathbf{V} = 0$. Hence we may assume that $\mathbf{V} \neq 0$. By 12.5.1, there exists an a-orbit i on $\mathbf{I}$ and an integer $n \geq 1$ such that $B \in \mathcal{P}_{\mathbf{V};i;n\gamma}$ where $\gamma = \sum_{\mathbf{i}\in i} \mathbf{i}$. Hence it is enough to prove the following statement for any fixed a-orbit i in $\mathbf{I}$. For any $n \geq 1$, and any (B,ϕ) as in 12.5.2 such that $B \in \mathcal{P}_{\mathbf{V};i;n\gamma}$, we have $[B,\phi] \in M_{\mathbf{V}}$.

We argue by descending induction on n. We have $n \leq \dim \mathbf{V}_\mathbf{i}$, for any $\mathbf{i} \in i$. Let $\mathbf{W}$ be an $\mathbf{I}$-graded a-stable subspace of $\mathbf{V}$ such that $\mathbf{T} = \mathbf{V}/\mathbf{W}$ satisfies $\dim \mathbf{T}_\mathbf{i} = n$ for all $\mathbf{i} \in i$ and $\mathbf{T}_\mathbf{i} = 0$ for all $\mathbf{i} \in \mathbf{I} - i$. (Such $\mathbf{W}$ exists since i is an a-orbit.)

As in the proof of 12.5.2, we can find a self-dual element $(A, \phi') \in \tilde{\mathcal{P}}_\mathbf{W}$ such that $A \in \mathcal{P}_{\mathbf{W};i;0}$ and such that (with the notation there) $\text{ind}^\mathbf{V}_{\mathbf{T},\mathbf{W}}[A, \phi'] = [B, \phi_1] + \sum_{C' \in \mathcal{S}} P_{C'}[C', f_{C'}]$ where $\phi_1 = \pm\phi$. By 12.5.1(b), for each $C' \in \mathcal{S}$, we have $C' \in \mathcal{P}_{\mathbf{V};i;n'\gamma}$ for some $n' > n$; hence

(d) $$[C', f_{C'}] \in M_\mathbf{V},$$

by the induction hypothesis. Replacing, if necessary ϕ' by $-\phi'$, we can assume that $\phi_1 = \phi$. We now show that,

(e) for any integer $r \leq 0$, the coefficient of v^r in $P_{C'}$ is an integer, for any $C' \in \mathcal{S}$.

Assume that this is not so. Then we can find an integer $r \leq 0$ which is as small as possible with the following property: there exists $C' \in \mathcal{S}$ such that the coefficient $c(C')$ of v^r in $P_{C'}$ is in $\mathcal{O} - \mathbf{Z}$. For such C', we have

(f) $$\begin{aligned}\{\text{ind}^\mathbf{V}_{\mathbf{T},\mathbf{W}}[A, \phi'], [C', f_{C'}]\} &= \{[B, \phi], [C', f_{C'}]\} \\ &\quad + \sum_{C'' \in \mathcal{S}} P_{C''}\{[C'', f_{C''}], [C', f_{C'}]\}.\end{aligned}$$

By the induction hypothesis, we have $[A, \phi'] \in M_\mathbf{W}$ and by 12.3.2, $\text{ind}^\mathbf{V}_{\mathbf{T},\mathbf{W}}$ carries $M_\mathbf{W}$ into $M_\mathbf{V}$; hence, $\text{ind}^\mathbf{V}_{\mathbf{T},\mathbf{W}}[A, \phi'] \in M_\mathbf{V}$. Using now 12.6.2 and (d), we see that the left hand side of (f) is in $\mathbf{Z}((v))$. In particular, the coefficient of v^r in the left hand side of (f) is an integer.

By 12.5.3(a), we have $\{[B, \phi], [C', f_{C'}]\} \in v\mathcal{O}[[v]]$ since B, C' are not isomorphic. This implies that the coefficient of v^r in $\{[B, \phi], [C', f_{C'}]\}$ is zero since $r \leq 0$. If $C'' \in \mathcal{S}$ is not C' then, by 12.5.3(a) and 12.6.2, we have $\{[C'', f_{C''}], [C', f_{C'}]\} \in v\mathcal{O}[[v]] \cap \mathbf{Z}((v)) = v\mathbf{Z}[[v]]$, hence $P_{C''}\{[C'', f_{C''}], [C', f_{C'}]\} \in P_{C''}v\mathbf{Z}[[v]]$. Using the minimality of r, we see from this that the coefficient of v^r in $P_{C''}\{[C'', f_{C''}], [C', f_{C'}]\}$ is an integer.

Similarly, using 12.5.3(c) and 12.6.2, we have

$$\{[C', f_{C'}], [C', f_{C'}]\} \in (1 + v\mathcal{O}[[v]]) \cap \mathbf{Z}((v)) = 1 + v\mathbf{Z}[[v]];$$

hence $P_{C'}\{[C', f_{C'}], [C', f_{C'}]\} \in P_{C'}(1 + v\mathbf{Z}[[v]])$. Using the minimality of r, we see from this that the coefficient of v^r in $P_{C'}\{[C', f_{C'}], [C', f_{C'}]\}$ is $c(C')$ plus an integer.

Combining these results, we see that the coefficient of v^r in the right hand side of (f) is $c(C')$ plus an integer; on the other hand, we have seen that the coefficient of v^r in the left hand side of (f) is an integer. It follows that $c(C')$ is an integer; this is a contradiction. Thus, (e) holds.

Next we note that $D[A,\phi'] = [A,\phi']$. Using 12.4.3, we deduce that $D(\mathrm{ind}_{\mathbf{T},\mathbf{W}}^{\mathbf{V}}[A,\phi']) = \mathrm{ind}_{\mathbf{T},\mathbf{W}}^{\mathbf{V}}[A,\phi']$. Hence

$$D([B,\phi] + \sum_{C''\in\mathcal{S}} P_{C''}[C'',f_{C''}]) = [B,\phi] + \sum_{C''\in\mathcal{S}} P_{C''}[C'',f_{C''}]$$

or equivalently,

$$[B,\phi] + \sum_{C''\in\mathcal{S}} \bar{P}_{C''}[C'',f_{C''}] = [B,\phi] + \sum_{C''\in\mathcal{S}} P_{C''}[C'',f_{C''}]$$

where $\bar{P}_{C''}$ is as in 12.4.1. Comparing coefficients on the two sides, we obtain $\bar{P}_{C''} = P_{C''}$ for all C''.

Thus, the coefficients of v^r and of v^{-r} in $P_{C''}$ coincide. Since either r or $-r$ is ≤ 0, we see from (e) that the coefficient of v^r in $P_{C''}$ is an integer, for any r and any $C'' \in \mathcal{S}$. In other words, we have $P_{C''} \in \mathcal{A}$. In the identity

$$\mathrm{ind}_{\mathbf{T},\mathbf{W}}^{\mathbf{V}}[A,\phi'] = [B,\phi] + \sum_{C''\in\mathcal{S}} P_{C''}[C'',f_{C''}],$$

all terms except $[B,\phi]$ are in $M_{\mathbf{V}}$, as we have seen. It follows that so is $[B,\phi]$. Thus (c) is proved.

The remainder of the proof will be similar to that of (c). Consider the set $\mathcal{T}$ of all simple objects $C \in \mathcal{P}_{\mathbf{V}}$ (up to isomorphism) such that $a^*C \cong C$; for each such C we select $f_C : a^*C \cong C$ such that $(C, f_C) \in \tilde{\mathcal{P}}_{\mathbf{V}}$ and $D[C,f_C] = [C,f_C]$. Let $\alpha \in M_{\mathbf{V}}$. Using 12.1.3 and 12.5.2, we can write uniquely $\alpha = \sum_{C\in\mathcal{T}} p_C[C,f_C]$ with $p_C \in \mathcal{O}'$.

To complete the proof, it remains to show that $p_C \in \mathcal{A}$ for all C. Assume that this is not so. Then we can find an integer r_0 which is as small as possible with the following property: there exists $C_0 \in \mathcal{T}$ such that the coefficient $b(C_0)$ of v^{r_0} in p_{C_0} is in $\mathcal{O} - \mathbf{Z}$. For such C_0 we have

$$\text{(g)}\qquad \{\alpha, [C_0, f_{C_0}]\} = \sum_{C\in\mathcal{T}} p_C\{[C,f_C],[C_0,f_{C_0}]\}.$$

Using (c), 12.5.3, and the definition of r_0, we see that the coefficient of v^{r_0} in the right hand side of (g) is equal to $b(C_0)$ plus an integer. Using (c) and 12.6.2, we see that the coefficient of v^{r_0} in the left hand side of (g) is an integer. It follows that $b(C_0)$ is an integer. This is a contradiction. The proposition is proved.

12.6.4. Let M be a module over a ring R. A subset $\mathcal{B}$ of M is said to be a *signed basis* of M if there exists a basis $\mathcal{B}'$ of M such that $\mathcal{B} = \mathcal{B}' \cup (-\mathcal{B}')$. Let $\mathcal{B}_{\mathbf{V}}$ be the subset of $\mathcal{K}(\mathcal{Q}_{\mathbf{V}})$ consisting of all elements $[B, \phi]$ with (B, ϕ) as in 12.5.2 (if $\mathbf{n}$ is even) and of all elements $\pm[B, \phi]$ with (B, ϕ) as in 12.5.2 (if $\mathbf{n}$ is odd). From 12.1.3 and 12.5.2 we see that

(a) *$\mathcal{B}_{\mathbf{V}}$ is a signed basis of the $\mathcal{O}'$-module $\mathcal{K}(\mathcal{Q}_{\mathbf{V}})$,*

and from 12.6.3 we see that

(b) *$\mathcal{B}_{\mathbf{V}}$ is a signed basis of the $\mathcal{A}$-module $M_{\mathbf{V}}$.*

CHAPTER 13

The Algebras ${}_{\mathcal{O}'}\mathbf{k}$ and $\mathbf{k}$

13.1. The Algebra ${}_{\mathcal{O}'}\mathbf{k}$

13.1.1. We preserve the setup of the previous chapter. Given $\nu \in \mathbf{N}[\mathbf{I}]^a$, we may regard $\mathbf{V} \mapsto \mathcal{K}(\mathcal{Q}_\mathbf{V})$ as a functor on the category $\mathcal{V}^a_\nu$ with values in the category of $\mathcal{O}'$-modules. An isomorphism $\mathbf{V} \cong \mathbf{V}'$ in $\mathcal{V}^a_\nu$ induces an isomorphism $\mathbf{E}_\mathbf{V} \cong \mathbf{E}_{\mathbf{V}'}$ compatible with the a-actions; this induces an isomorphism $\mathcal{Q}_\mathbf{V} \cong \mathcal{Q}_{\mathbf{V}'}$ which induces an isomorphism $\mathcal{K}(\mathcal{Q}_\mathbf{V}) \cong \mathcal{K}(\mathcal{Q}_{\mathbf{V}'})$ that is actually independent of the choice of the isomorphism $\mathbf{V} \cong \mathbf{V}'$ by the equivariance properties of the complexes considered. Hence we may take the direct limit $\varinjlim_\mathbf{V} \mathcal{K}(\mathcal{Q}_\mathbf{V})$ over the category $\mathcal{V}^a_\nu$. This direct limit is denoted by ${}_{\mathcal{O}'}\mathbf{k}_\nu$. By the previous discussion, the natural homomorphism $\mathcal{K}(\mathcal{Q}_\mathbf{V}) \to {}_{\mathcal{O}'}\mathbf{k}_\nu$ is an isomorphism for any $\mathbf{V} \in \mathcal{V}^a_\nu$.

The signed basis $\mathcal{B}_\mathbf{V}$ of $\mathcal{K}(\mathcal{Q}_\mathbf{V})$ (see 12.6.4) (where $\mathbf{V} \in \mathcal{V}^a_\nu$) can be regarded as a signed basis of the $\mathcal{O}'$-module ${}_{\mathcal{O}'}\mathbf{k}_\nu$, independent of $\mathbf{V}$; we denote it by $\mathcal{B}_\nu$. It is a finite set.

13.1.2. Let ${}_{\mathcal{O}'}\mathbf{k} = \oplus_\nu({}_{\mathcal{O}'}\mathbf{k}_\nu)$ (ν runs over $\mathbf{N}[\mathbf{I}]^a$). Let $\mathcal{B} = \sqcup_\nu \mathcal{B}_\nu$, a signed basis of the $\mathcal{O}'$-module ${}_{\mathcal{O}'}\mathbf{k}$. An element $x \in {}_{\mathcal{O}'}\mathbf{k}$ is said to be *homogeneous* if it belongs to ${}_{\mathcal{O}'}\mathbf{k}_\nu$ for some ν; we then write $|x| = \nu$.

The homomorphisms $\mathrm{ind}^{\mathbf{V}}_{\mathbf{T},\mathbf{W}}$ can be regarded as $\mathcal{O}'$-linear maps ${}_{\mathcal{O}'}\mathbf{k}_\tau \otimes_{\mathcal{O}'} ({}_{\mathcal{O}'}\mathbf{k}_\omega) \to {}_{\mathcal{O}'}\mathbf{k}_\nu$, defined whenever $\tau, \omega, \nu \in \mathbf{N}[\mathbf{I}]^a$ satisfy $\tau + \omega = \nu$. They define a multiplication operation, hence they define a structure of $\mathcal{O}'$-algebra on ${}_{\mathcal{O}'}\mathbf{k}$. For any $\boldsymbol{\nu} = (\nu^1, \dots, \nu^m) \in \mathcal{X}^a$, we may regard $L_{\boldsymbol{\nu}}$ as an element in ${}_{\mathcal{O}'}\mathbf{k}_\nu$ where $\nu_\mathbf{i} = \sum_l \nu^l_\mathbf{i}$ for all $\mathbf{i}$.

Lemma 12.3.3 can be now restated as follows:

(a) $$L_{\boldsymbol{\nu}'} L_{\boldsymbol{\nu}''} = L_{\boldsymbol{\nu}'\boldsymbol{\nu}''}.$$

Since the elements $L_{\boldsymbol{\nu}}$ generate ${}_{\mathcal{O}'}\mathbf{k}$ as a $\mathcal{O}'$-module (see 12.6.3), it follows that the algebra structure on ${}_{\mathcal{O}'}\mathbf{k}$ is associative. One can also see this more directly.

13.1.3. The homomorphisms $\mathrm{res}^{\mathbf{V}}_{\mathbf{T},\mathbf{W}}$ can be regarded as $\mathcal{O}'$-linear maps ${}_{\mathcal{O}'}\mathbf{k}_\nu \to {}_{\mathcal{O}'}\mathbf{k}_\tau \otimes_{\mathcal{O}'} ({}_{\mathcal{O}'}\mathbf{k}_\omega)$, defined whenever $\tau, \omega, \nu \in \mathbf{N}[\mathbf{I}]^a$ satisfy $\tau + \omega = \nu$. By taking direct sums, we obtain an $\mathcal{O}'$-linear map $\bar{r} : {}_{\mathcal{O}'}\mathbf{k} \to {}_{\mathcal{O}'}\mathbf{k} \otimes_{\mathcal{O}'} ({}_{\mathcal{O}'}\mathbf{k})$.

G. Lusztig, *Introduction to Quantum Groups*, Modern Birkhäuser Classics,
DOI 10.1007/978-0-8176-4717-9_13,

13.1.4. We have a symmetric bilinear pairing $\mathbf{Z}[\mathbf{I}] \times \mathbf{Z}[\mathbf{I}] \to \mathbf{Z}$ given by

$$\nu \cdot \nu' = 2\sum_{\mathbf{i}} \nu_{\mathbf{i}}\nu'_{\mathbf{i}} - \sum_{h}(\nu_{h'}\nu'_{h''} + \nu_{h''}\nu'_{h'}).$$

This bilinear form is independent of orientation. Let ${}_{\mathcal{O}'}\mathbf{k}\bar{\otimes}_{\mathcal{O}'}({}_{\mathcal{O}'}\mathbf{k})$ be the $\mathcal{O}'$-module ${}_{\mathcal{O}'}\mathbf{k} \otimes_{\mathcal{O}'} ({}_{\mathcal{O}'}\mathbf{k})$ with the $\mathcal{O}'$-algebra structure given by

$$(x \otimes y)(x' \otimes y') = v^{-|x'|\cdot|y|} xx' \otimes yy'$$

for x, x', y, y' homogeneous.

Lemma 13.1.5. $\bar{r} : {}_{\mathcal{O}'}\mathbf{k} \to {}_{\mathcal{O}'}\mathbf{k}\bar{\otimes}_{\mathcal{O}'}({}_{\mathcal{O}'}\mathbf{k})$ *is a homomorphism of* $\mathcal{O}'$-*algebras.*

We must check that $\bar{r}(xy) = \bar{r}(x)\bar{r}(y)$ for any $x, y \in {}_{\mathcal{O}'}\mathbf{k}$. Since the elements $L_{\boldsymbol{\nu}}$ generate the $\mathcal{O}'$-module ${}_{\mathcal{O}'}\mathbf{k}$ (see 12.6.3), we may assume that $x = L_{\boldsymbol{\nu}'}, y = L_{\boldsymbol{\nu}''}$, where $\boldsymbol{\nu}' = (\nu'^1, \dots, \nu'^m)$ and $\boldsymbol{\nu}'' = (\nu''^1, \dots, \nu''^n)$ are elements of $\mathcal{X}^a$. We have

(a) $$\bar{r}(L_{\boldsymbol{\nu}'}) = \sum v^{M'(\boldsymbol{\tau}', \boldsymbol{\omega}')} L_{\boldsymbol{\tau}'} \otimes L_{\boldsymbol{\omega}'}$$

where the sum is taken over all $\boldsymbol{\tau}' = (\tau'^1, \dots, \tau'^m)$ and $\boldsymbol{\omega}' = (\omega'^1, \dots, \omega'^m)$ in $\mathcal{X}^a$ such that $\tau'^l + \omega'^l = \nu'^l$ for $1 \le l \le m$; $M'(\boldsymbol{\tau}', \boldsymbol{\omega}')$ is as in 9.2.11.

Similarly, we have

$$\bar{r}(L_{\boldsymbol{\nu}''}) = \sum v^{M'(\boldsymbol{\tau}'', \boldsymbol{\omega}'')} L_{\boldsymbol{\tau}''} \otimes L_{\boldsymbol{\omega}''}$$

where the sum is taken over all $\boldsymbol{\tau}'' = (\tau''^{(m+1)}, \dots, \tau''^{(m+n)})$ and $\boldsymbol{\omega}'' = (\omega''^{(m+1)}, \dots, \omega''^{(m+n)})$ in $\mathcal{X}^a$ such that $\tau''^l + \omega''^l = \nu''^{l-m}$ for $m+1 \le l \le m+n$. Hence

$$\bar{r}(L_{\boldsymbol{\nu}'})\bar{r}(L_{\boldsymbol{\nu}''}) = \sum v^{M'(\boldsymbol{\tau}', \boldsymbol{\omega}') + M'(\boldsymbol{\tau}'', \boldsymbol{\omega}'') + |L_{\boldsymbol{\omega}'}|\cdot|L_{\boldsymbol{\tau}''}|} L_{\boldsymbol{\tau}'\boldsymbol{\tau}''} \otimes L_{\boldsymbol{\omega}'\boldsymbol{\omega}''}$$

where the sum is taken over all $\boldsymbol{\tau}' = (\tau'^1, \dots, \tau'^m)$, $\boldsymbol{\omega}' = (\omega'^1, \dots, \omega'^m)$, $\boldsymbol{\tau}'' = (\tau''^{(m+1)}, \dots, \tau''^{(m+n)})$, $\boldsymbol{\omega}'' = (\omega''^{(m+1)}, \dots, \omega''^{(m+n)})$ in $\mathcal{X}^a$ such that $\tau'^l + \omega'^l = \nu'^l$ for $1 \le l \le m$ and $\tau''^l + \omega''^l = \nu''^{l-m}$ for $m+1 \le l \le m+n$.

We have

$$\bar{r}(L_{\boldsymbol{\nu}'}L_{\boldsymbol{\nu}''}) = \bar{r}(L_{\boldsymbol{\nu}'\boldsymbol{\nu}''}) = \sum v^{M'(\boldsymbol{\tau}, \boldsymbol{\omega})} L_{\boldsymbol{\tau}} \otimes L_{\boldsymbol{\omega}}$$

where the sum is taken over all $\boldsymbol{\tau} = (\tau^1, \dots, \tau^{m+n})$ and $\boldsymbol{\omega} = (\omega^1, \dots, \omega^{m+n})$ in $\mathcal{X}^a$ such that $\tau^l + \omega^l = \nu'^l$ for $1 \le l \le m$ and $\tau^l + \omega^l = \nu''^{l-m}$ for $m+1 \le l \le m+n$.

It remains to show that

$$|L_{\boldsymbol{\omega}'}| \cdot |L_{\boldsymbol{\tau}''}| = M'(\boldsymbol{\tau}'\boldsymbol{\tau}'', \boldsymbol{\omega}'\boldsymbol{\omega}'') - M'(\boldsymbol{\tau}', \boldsymbol{\omega}') - M'(\boldsymbol{\tau}'', \boldsymbol{\omega}'').$$

This follows by a straightforward computation. The lemma is proved.

13.1.6. The pairing $\{,\}$ on $\mathcal{K}(\mathcal{Q}_{\mathbf{V}})$ (see 12.1.2) (where $\mathbf{V} \in \mathcal{V}_\nu^a$) can be regarded as an $\mathcal{O}'$-bilinear pairing $\{,\} : {}_{\mathcal{O}'}\mathbf{k}_\nu \times {}_{\mathcal{O}'}\mathbf{k}_\nu \to \mathcal{O}((v))$, which is independent of $\mathbf{V}$. This extends to an $\mathcal{O}'$-bilinear pairing $\{,\} : {}_{\mathcal{O}'}\mathbf{k} \times {}_{\mathcal{O}'}\mathbf{k} \to \mathcal{O}((v))$ such that for homogeneous x, y, $\{x, y\}$ is given by the previous pairing if $|x| = |y|$, and is zero if $|x| \neq |y|$.

13.1.7. We define a $\mathcal{O}'$-bilinear pairing $\{,\}$ on ${}_{\mathcal{O}'}\mathbf{k} \otimes_{\mathcal{O}'} ({}_{\mathcal{O}'}\mathbf{k})$ by

$$\{x' \otimes x'', y' \otimes y''\} = \{x', y'\}\{x'', y''\}.$$

The identity

$$\{x, y'y''\} = \{\bar{r}(x), y' \otimes y''\}$$

for all $x, y', y'' \in {}_{\mathcal{O}'}\mathbf{k}$, follows immediately from 12.2.2.

13.1.8. The homomorphism $D : \mathcal{K}(\mathcal{Q}_{\mathbf{V}}) \to \mathcal{K}(\mathcal{Q}_{\mathbf{V}})$ (where $\mathbf{V} \in \mathcal{V}_\nu^a$) can be regarded as a group homomorphism $D : {}_{\mathcal{O}'}\mathbf{k}_\nu \to {}_{\mathcal{O}'}\mathbf{k}_\nu$ that has square 1 and is semi-linear with respect to the ring involution $^- : \mathcal{O}' \to \mathcal{O}'$ given by $v^n \mapsto v^{-n}$ and $\zeta \mapsto \zeta^{-1}$ for $\zeta \in \mathcal{O}, \zeta^{\mathbf{n}} = 1$. By taking direct sums we obtain $D : {}_{\mathcal{O}'}\mathbf{k} \to {}_{\mathcal{O}'}\mathbf{k}$ which, by 12.4.3, is a ring homomorphism.

13.1.9. We shall regard ${}_{\mathcal{O}'}\mathbf{k} \otimes_{\mathcal{O}'} ({}_{\mathcal{O}'}\mathbf{k})$ as an $\mathcal{O}'$-algebra with

$$(x \otimes y)(x' \otimes y') = v^{|x'|\cdot|y|} xx' \otimes yy'$$

for x, x', y, y' homogeneous. This should be distinguished from the algebra ${}_{\mathcal{O}'}\mathbf{k}\bar{\otimes}_{\mathcal{O}'}({}_{\mathcal{O}'}\mathbf{k})$. Let $D : {}_{\mathcal{O}'}\mathbf{k}\bar{\otimes}_{\mathcal{O}'}({}_{\mathcal{O}'}\mathbf{k}) \to {}_{\mathcal{O}'}\mathbf{k} \otimes_{\mathcal{O}'} ({}_{\mathcal{O}'}\mathbf{k})$ be the ring isomorphism given by $D(x \otimes y) = D(x) \otimes D(y)$ for all x, y.

Let $r : {}_{\mathcal{O}'}\mathbf{k} \to {}_{\mathcal{O}'}\mathbf{k} \otimes_{\mathcal{O}'} ({}_{\mathcal{O}'}\mathbf{k})$ be the $\mathcal{O}'$-algebra homomorphism defined as the composition

$${}_{\mathcal{O}'}\mathbf{k} \xrightarrow{D} {}_{\mathcal{O}'}\mathbf{k} \xrightarrow{\bar{r}} {}_{\mathcal{O}'}\mathbf{k}\bar{\otimes}_{\mathcal{O}'}({}_{\mathcal{O}'}\mathbf{k}) \xrightarrow{D} {}_{\mathcal{O}'}\mathbf{k} \otimes_{\mathcal{O}'} ({}_{\mathcal{O}'}\mathbf{k}).$$

13.1.10. Let $(,) : {}_{\mathcal{O}'}\mathbf{k} \times {}_{\mathcal{O}'}\mathbf{k} \to \mathcal{O}((v^{-1}))$ be the $\mathcal{O}'$-bilinear pairing given by $(x, y) = \overline{\{D(x), D(y)\}}$. Here, $^- : \mathcal{O}((v)) \to \mathcal{O}((v^{-1}))$ is given by $\sum_n a_n v^n \mapsto \sum_n \bar{a}_n v^{-n} \quad (a_n \in \mathcal{O})$.

From 13.1.7, we deduce the identity

(a) $$(x, y'y'') = (r(x), y' \otimes y'')$$

for all $x, y', y'' \in {}_{\mathcal{O}'}\mathbf{k}$.

13.1.11. From the definition we have

(a) $D(b) = b$ for all $b \in \mathcal{B}$.

We have

(b) $\{b, b'\} \in v\mathbf{Z}[[v]] \cap \mathbf{Q}(v)$ for any $b, b' \in \mathcal{B}$ such that $b' \neq \pm b$.

Indeed, $\{b, b'\}$ is in $v\mathcal{O}[[v]]$ by 12.5.3 and in $\mathbf{Z}((v))$ by 12.6.2 and 12.6.3, and hence in $v\mathbf{Z}[[v]]$.

We have

(c) $\{b, b\} \in 1 + v\mathbf{Z}[[v]] \cap \mathbf{Q}(v)$ for all $b \in \mathcal{B}$.

Indeed, $\{b, b\}$ is in $1 + v\mathcal{O}[[v]]$ by 12.5.3 and in $\mathbf{Z}((v))$ by 12.6.2 and 12.6.3, and hence in $1 + v\mathbf{Z}[[v]]$.

From (a),(b),(c) we deduce:

(b$'$) $(b, b') \in v^{-1}\mathbf{Z}[[v^{-1}]] \cap \mathbf{Q}(v)$ for any $b, b' \in \mathcal{B}$ such that $b' \neq \pm b$.

(c$'$) $(b, b) \in 1 + v^{-1}\mathbf{Z}[[v^{-1}]] \cap \mathbf{Q}(v)$ for all $b \in \mathcal{B}$.

13.1.12. Let i be an a-orbit on $\mathbf{I}$ and let $\gamma = \sum_{\mathbf{i} \in i} \mathbf{i}$. For any $n \geq 0$, ${}_{\mathcal{O}'}\mathbf{k}_{n\gamma}$ has a distinguished element denoted $\mathbf{1}_{ni}$; it corresponds to $[\mathbf{1}, 1] \in \mathcal{K}(\mathcal{Q}_{\mathbf{V}})$ where $\mathbf{V} \in \mathcal{V}^a_{n\gamma}$. This element forms a basis of the $\mathcal{O}'$-module ${}_{\mathcal{O}'}\mathbf{k}_{n\gamma}$. When $n = 0$, this is independent of i and is denoted simply by $\mathbf{1} \in {}_{\mathcal{O}'}\mathbf{k}_0$. Note that

(a) $\mathbf{1}$ is the unit element of the algebra ${}_{\mathcal{O}'}\mathbf{k}$.

(b) The elements $L_\nu \in {}_{\mathcal{O}'}\mathbf{k}$ are precisely the elements of ${}_{\mathcal{O}'}\mathbf{k}$ which are products of elements of form $\mathbf{1}_{ni}$ for various i, n. Hence the elements $\mathbf{1}_{ni}$ generate ${}_{\mathcal{O}'}\mathbf{k}$ as an $\mathcal{O}'$-algebra.

(c) We have $\mathbf{1}_i \mathbf{1}_{(n-1)i} = v^{d(n-1)}(\sum_{s=0}^{n-1} v^{-ds})\mathbf{1}_{ni}$ (for $n \geq 1$), where d is the number of elements in the orbit i. (See 12.3.4.)

From 12.3.6, we have

(d) $\{\mathbf{1}_i, \mathbf{1}_i\} = (1 - v^{2d})^{-1}$ (where d is as above)

or equivalently, since $D(\mathbf{1}_i) = \mathbf{1}_i$:

(d$'$) $(\mathbf{1}_i, \mathbf{1}_i) = (1 - v^{-2d})^{-1}$.

From (b) and (c) we see that

(e) $\bar{r}(\mathbf{1}_i) = r(\mathbf{1}_i) = \mathbf{1}_i \otimes \mathbf{1} + \mathbf{1} \otimes \mathbf{1}_i$.

This is obvious from the definitions. It is clear that

(f) $(\mathbf{1}, \mathbf{1}) = \{\mathbf{1}, \mathbf{1}\} = 1$.

13.1.13. From 10.3.4, it follows easily that there is a unique $\mathcal{O}'$-linear map $\sigma : {}_{\mathcal{O}'}\mathbf{k} \to {}_{\mathcal{O}'}\mathbf{k}$ such that $\sigma(L_{\boldsymbol{\nu}}) = L_{\boldsymbol{\nu}'}$ for any $\boldsymbol{\nu} = (\nu^1, \nu^2, \ldots, \nu^m) \in \mathcal{X}^a$, where $\boldsymbol{\nu}' = (\nu^m, \nu^{m-1}, \ldots, \nu^1) \in \mathcal{X}^a$; moreover, we have $\sigma(\mathcal{B}) = \mathcal{B}$. It follows that σ is the unique isomorphism of ${}_{\mathcal{O}'}\mathbf{k}$ onto the algebra opposed to ${}_{\mathcal{O}'}\mathbf{k}$ such that $\sigma(\mathbf{1}_{ni}) = \mathbf{1}_{ni}$ for all $i \in I$ and $n \geq 0$.

13.2. The Algebra k

13.2.1. Let ${}_{\mathcal{A}}\mathbf{k}$ be the $\mathcal{A}$-submodule of ${}_{\mathcal{O}'}\mathbf{k}$ spanned by $\mathcal{B}$, or equivalently (see 12.6.3), by the elements $L_{\boldsymbol{\nu}}$ for various $\boldsymbol{\nu} \in \mathcal{X}^a$. Thus, on the one hand, ${}_{\mathcal{A}}\mathbf{k}$ is the $\mathcal{A}$-subalgebra of ${}_{\mathcal{O}'}\mathbf{k}$ generated by the elements $\mathbf{1}_{ni}$ as in 13.1.12(b), and on the other hand, $\mathcal{B}$ is a signed basis for the $\mathcal{A}$-module ${}_{\mathcal{A}}\mathbf{k}$. We have ${}_{\mathcal{A}}\mathbf{k} = \oplus_{\nu}({}_{\mathcal{A}}\mathbf{k}_{\nu})$ where ${}_{\mathcal{A}}\mathbf{k}_{\nu}$ is the $\mathcal{A}$-submodule generated by $\mathcal{B}_{\nu}$.

13.2.2. From 13.1.5(a), we see that $\bar{r}$ restricts to an $\mathcal{A}$-linear map ${}_{\mathcal{A}}\mathbf{k} \to {}_{\mathcal{A}}\mathbf{k} \otimes_{\mathcal{A}} ({}_{\mathcal{A}}\mathbf{k})$, denoted again by $\bar{r}$; this is an $\mathcal{A}$-algebra homomorphism if ${}_{\mathcal{A}}\mathbf{k} \otimes_{\mathcal{A}} ({}_{\mathcal{A}}\mathbf{k})$ (which is naturally imbedded in ${}_{\mathcal{O}'}\mathbf{k} \bar{\otimes}_{\mathcal{O}'} ({}_{\mathcal{O}'}\mathbf{k})$) is given the induced $\mathcal{A}$-algebra structure (see 13.1.5).

13.2.3. By 13.1.11(a), the ring homomorphism $D : {}_{\mathcal{O}'}\mathbf{k} \to {}_{\mathcal{O}'}\mathbf{k}$ restricts to a ring homomorphism $D : {}_{\mathcal{A}}\mathbf{k} \to {}_{\mathcal{A}}\mathbf{k}$ which has square 1 and is semi-linear with respect to the ring involution $^{-} : \mathcal{A} \to \mathcal{A}$.

13.2.4. From 13.2.3 and 13.2.2, it follows that the $\mathcal{O}'$-algebra homomorphism $r : {}_{\mathcal{O}'}\mathbf{k} \to {}_{\mathcal{O}'}\mathbf{k} \otimes_{\mathcal{O}'} ({}_{\mathcal{O}'}\mathbf{k})$ (see 13.1.9) restricts to an $\mathcal{A}$-algebra homomorphism ${}_{\mathcal{A}}\mathbf{k} \to {}_{\mathcal{A}}\mathbf{k} \otimes_{\mathcal{A}} ({}_{\mathcal{A}}\mathbf{k})$, denoted again by r. This is an $\mathcal{A}$-algebra homomorphism if ${}_{\mathcal{A}}\mathbf{k} \otimes_{\mathcal{A}} ({}_{\mathcal{A}}\mathbf{k})$ (which is naturally imbedded in ${}_{\mathcal{O}'}\mathbf{k} \otimes_{\mathcal{O}'} ({}_{\mathcal{O}'}\mathbf{k})$) is given the induced $\mathcal{A}$-algebra structure (see 13.1.9).

13.2.5. The pairing $(,) : {}_{\mathcal{O}'}\mathbf{k} \times {}_{\mathcal{O}'}\mathbf{k} \to \mathcal{O}((v^{-1}))$ (see 13.1.10) restricts to an $\mathcal{A}$-bilinear pairing $(,) : {}_{\mathcal{A}}\mathbf{k} \times {}_{\mathcal{A}}\mathbf{k} \to \mathbf{Z}((v^{-1})) \cap \mathbf{Q}(v)$ (see 13.1.11(b′),(c′)). The equation analogous to 13.1.10(a) continues of course to hold over $\mathcal{A}$.

13.2.6. Let $\mathbf{k}$ be the $\mathbf{Q}(v)$-algebra $\mathbf{Q}(v) \otimes_{\mathcal{A}} ({}_{\mathcal{A}}\mathbf{k})$. Note that $\mathcal{B}$ is a signed basis of the $\mathbf{Q}(v)$-vector space $\mathbf{k}$. We have a direct sum decomposition $\mathbf{k} = \oplus_{\nu}\mathbf{k}_{\nu}$ where $\mathbf{k}_{\nu}$ is the subspace spanned by $\mathcal{B}_{\nu}$.

From 13.2.1 and 13.1.12(c), we see by induction on n that $\mathbf{k}$ is generated as a $\mathbf{Q}(v)$-algebra by the elements $\mathbf{1}_i$ for the various a-orbits on $\mathbf{I}$.

13.2.7. The homomorphism r in 13.2.4 extends to a $\mathbf{Q}(v)$-algebra homomorphism $\mathbf{k} \to \mathbf{k} \otimes_{\mathbf{Q}(v)} \mathbf{k}$ (denoted again by r) where $\mathbf{k} \otimes_{\mathbf{Q}(v)} \mathbf{k}$ is regarded as a $\mathbf{Q}(v)$-algebra by the same rule as in 13.1.9.

13.2.8. The pairing $(,)$ on ${}_{\mathcal{A}}\mathbf{k}$ extends to a $\mathbf{Q}(v)$-bilinear pairing $(,) : \mathbf{k} \times \mathbf{k} \to \mathbf{Q}(v)$. From 13.1.11(b′),(c′), we see that the restriction of this pairing to $\mathbf{k}_\nu$ is non-degenerate, for any ν. (Its determinant with respect to a basis contained in $\mathcal{B}_\nu$ belongs to $1 + v^{-1}\mathbf{Z}[[v^{-1}]] \cap \mathbf{Q}(v)$ and hence is non-zero.)

13.2.9. Let I be the set of a-orbits on $\mathbf{I}$. We identify $\mathbf{Z}[I]$ with the subgroup

$$\mathbf{Z}[\mathbf{I}]^a = \{\nu \in \mathbf{Z}[\mathbf{I}] | \nu_{\mathbf{i}} = \nu_{a(\mathbf{i})} \quad \forall \mathbf{i} \in \mathbf{I}\}$$

of $\mathbf{Z}[\mathbf{I}]$ by associating to each $\nu \in \mathbf{Z}[I]$ the element of $\mathbf{Z}[\mathbf{I}]$ (denoted again ν), in which the coefficient of $\mathbf{i}$ is ν_i where i is the a-orbit of $\mathbf{i}$.

For $\nu, \nu' \in \mathbf{Z}[I]$ we define $\nu \cdot \nu' \in \mathbf{Z}$ by regarding ν, ν' as elements of $\mathbf{Z}[\mathbf{I}]$ as above and then computing $\nu \cdot \nu'$ according to 13.1.4. (This is a symmetric bilinear form). According to this rule, we have, for $i, j \in I$: $i \cdot j =$ minus the number of $h \in H$ such that $[h]$ consists of a point in i and a point in j, if $i \neq j$ and $i \cdot i =$ twice the number of elements in the orbit i. Note that, if $i \neq j$ then $-2\frac{i \cdot j}{i \cdot i} \in \mathbf{N}$; indeed, this is the number of $h \in H$ such that $[h]$ consists of a given point in i and some point in j. Hence we have obtained a Cartan datum $(I, \cdot)$.

13.2.10. Let $\mathbf{f}$ be the $\mathbf{Q}(v)$-algebra, defined as in 1.2.5, in terms of the Cartan datum $(I, \cdot)$ just described. Recall that $\mathbf{f} = {}'\mathbf{f}/\mathcal{I}$ where ${}'\mathbf{f}$ is the free associative $\mathbf{Q}(v)$-algebra on the generators $\theta_i (i \in I)$ and $\mathcal{I}$ is a two-sided ideal defined as the radical of a certain symmetric bilinear form $(,)$ on ${}'\mathbf{f}$. Let $\chi : {}'\mathbf{f} \to \mathbf{k}$ be the unique homomorphism of $\mathbf{Q}(v)$-algebras with 1 such that $\chi(\theta_i) = \mathbf{1}_i$ for each $i \in I$.

Theorem 13.2.11. *χ induces an algebra isomorphism* $\mathbf{f} = {}'\mathbf{f}/\mathcal{I} \cong \mathbf{k}$.

The homomorphism χ is surjective, since $\mathbf{k}$ is generated by the $\mathbf{1}_i$ as a $\mathbf{Q}(v)$-algebra (see 13.2.6.)

The homomorphism $r : {}'\mathbf{f} \to {}'\mathbf{f} \otimes {}'\mathbf{f}$ (see 1.2.2) and the homomorphism $r : \mathbf{k} \to \mathbf{k} \otimes \mathbf{k}$ (see 13.2.7) make the following diagram commutative:

$$\begin{array}{ccc} {}'\mathbf{f} & \xrightarrow{r} & {}'\mathbf{f} \otimes {}'\mathbf{f} \\ \chi \downarrow & & \downarrow \chi\otimes\chi \\ \mathbf{k} & \xrightarrow{r} & \mathbf{k} \otimes \mathbf{k} \end{array}$$

Indeed, first we note that $\chi \otimes \chi$ is an algebra homomorphism, since $\nu \cdot \nu'$ on $\mathbf{Z}[I]$ has been defined in terms of the pairing on $\mathbf{Z}[\mathbf{I}]$. Hence the two possible compositions in the diagram are algebra homomorphisms; to check that they are equal, it suffices to do this on the generators θ_i. But they both take θ_i to $\mathbf{1}_i \otimes \mathbf{1} + \mathbf{1} \otimes \mathbf{1}_i$ (see 13.1.12(e)).

For $x, y \in {}'\mathbf{f}$, we set $((x,y)) = (\chi(x), \chi(y))$ (the right hand side is as in 13.2.8). We have

(a) $((\theta_i, \theta_i)) = (\mathbf{1}_i, \mathbf{1}_i) = (1 - v^{-i \cdot i/2})^{-1}$ (see 13.1.12(d′)); $((\theta_i, \theta_j)) = (\mathbf{1}_i, \mathbf{1}_j) = 0$, if $i \neq j$ (trivially);

$$((x, y'y'')) = (\chi(x), \chi(y')\chi(y'')) = (r(\chi(x)), \chi(y') \otimes \chi(y''))$$

(b) $$= (\chi \otimes \chi)(r(x), \chi(y') \otimes \chi(y'')) = ((r(x), y' \otimes y''));$$

we have used 13.1.10(a) and the commutativity of the diagram above. By (b) and the symmetry of $((,))$, we obtain

(c) $$((xx', y)) = ((x \otimes x', r(y))).$$

We have $((1,1)) = 1$ (see 13.1.12(f)). Thus, $((x,y))$ satisfies the defining properties of $(,)$ in 1.2.3; hence it coincides with $(,)$. Since $\mathcal{I}$ is defined as the radical of $(,)$ on ${}'\mathbf{f}$, we also get

(d) $$\mathcal{I} = \{x \in {}'\mathbf{f} | (\chi(x), \chi(y)) = 0 \quad \forall y \in {}'\mathbf{f}\}.$$

Hence, if $x \in {}'\mathbf{f}$ satisfies $\chi(x) = 0$, then $x \in \mathcal{I}$, so that Ker $\chi \subset \mathcal{I}$. Conversely, assume that $x \in \mathcal{I}$. Let $z \in \mathbf{k}$. We have $z = \chi(y)$ for some $y \in {}'\mathbf{f}$ (recall that χ is surjective). We have $(\chi(x), z) = (\chi(x), \chi(y)) = 0$ by (d). Thus $\chi(x)$ is in the radical of the form $(,)$ on $\mathbf{k}$. But this radical is zero (see 13.2.8). Hence $\chi(x) = 0$. Thus we have proved that Ker $\chi = \mathcal{I}$. The theorem follows.

CHAPTER 14

The Signed Basis of f

14.1. Cartan Data and Graphs with Automorphisms

14.1.1. There is a very close connection between Cartan data and graphs with automorphisms. Given an admissible automorphism a of a finite graph $(\mathbf{I}, H, h \mapsto [h])$ (see 12.1.1), we define I to be the set of a-orbits on $\mathbf{I}$. For $i, j \in I$, we define $i \cdot j \in \mathbf{Z}$ as follows: if $i \neq j$ in I, then $i \cdot j$ is -1 times the number of edges which join some vertex in the a-orbit i to some vertex in the a-orbit j; $i \cdot i$ is 2 times the number of vertices in the a-orbit i. As shown in 13.2.9, $(I, \cdot)$ is a Cartan datum. Conversely, we have the following result.

Proposition 14.1.2. *Let $(I, \cdot)$ be a Cartan datum. There exists a finite graph $(\mathbf{I}, H, h \mapsto [h])$ and an admissible automorphism a of this graph such that $(I, \cdot)$ is obtained from them by the construction in 14.1.1.*

For each $i \in I$, we consider a set D_i of cardinal $d_i = i \cdot i/2$ and a cyclic permutation $a : D_i \to D_i$. Let $\mathbf{I} = \sqcup_{i \in I} D_i$ and let $a : \mathbf{I} \to \mathbf{I}$ be the permutation whose restriction to each D_i is the permutation $a : D_i \to D_i$ considered above.

For each unordered pair i, j of distinct elements of I, we choose an a-orbit $\wp$ of the permutation $a \times a : D_i \times D_j \to D_i \times D_j$. Then $\wp$ has cardinality equal to the lowest common multiple $l(d_i, d_j)$ of d_i and d_j, which by the definition of a Cartan datum, divides $-i \cdot j$. Hence we may consider a set $H_{i,j}$ which is a disjoint union of $-i \cdot j/l(d_i, d_j)$ copies of $\wp$ with a permutation $a : H_{i,j} \to H_{i,j}$ whose restriction to each copy of $\wp$ is the permutation defined by $a \times a$. We have a natural map $H_{i,j} \to D_i \times D_j$ whose restriction to each copy of $\wp$ is the imbedding $\wp \to D_i \times D_j$.

Let $H = \sqcup H_{i,j}$ (union over the unordered pairs i, j of distinct elements in I). This has a permutation $a : H \to H$ (defined by the permutations $a : H_{i,j} \to H_{i,j}$) and a map $H \to \sqcup D_i \times D_j$ (union over the unordered pairs i, j of distinct elements in I) induced by $H_{i,j} \to D_i \times D_j$. This defines a structure of a graph on $\mathbf{I}, H$. This clearly has the required properties.

G. Lusztig, *Introduction to Quantum Groups*, Modern Birkhäuser Classics,
DOI 10.1007/978-0-8176-4717-9_14, © Springer Science+Business Media, LLC 2010

14.1.3. Remark. In general, the graph with automorphism whose existence is asserted in the previous proposition is not uniquely determined by $(I, \cdot)$. However, if the Cartan datum $(I, \cdot)$ is symmetric, the construction in the previous proposition attaches to $(I, \cdot)$ a graph $(\mathbf{I}, H, h \mapsto [h])$, called *the graph of* $(I, \cdot)$, which is unique up to isomorphism; in this case, $\mathbf{I} = I, H_{i,j}$ is a set with $-i \cdot j$ elements and a is the identity automorphism.

14.1.4. Classification of symmetric Cartan data of affine or finite type. The symmetric Cartan data of affine type are completely described by their graphs. We enumerate below the graphs that appear in this way.

$\tilde{A}_n (n \geq 1)$; a polygon with $n + 1$ vertices; for $n = 1$, this is the graph with two vertices which are joined with two edges.

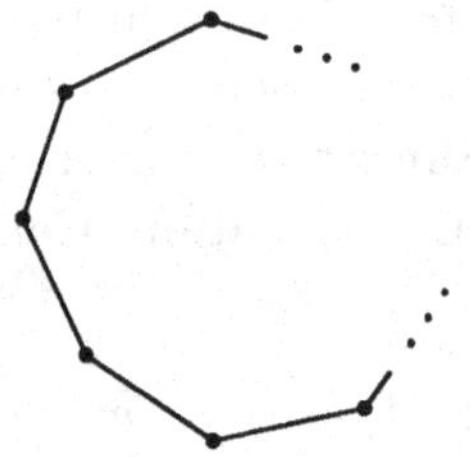

$\tilde{D}_n (n \geq 4)$ (a graph with $n + 1$ vertices):

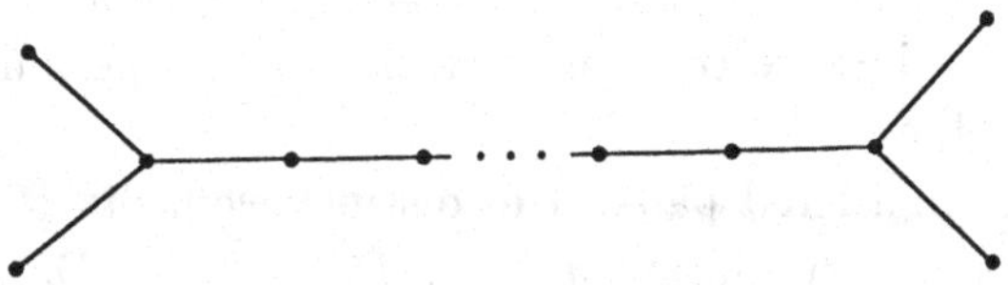

$\tilde{E}_6$:

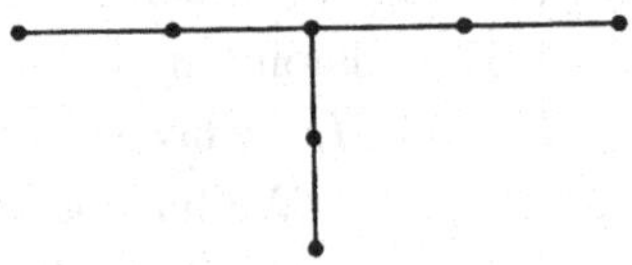

$\tilde{E}_7$:

$\tilde{E}_8$:

According to McKay, these graphs are in 1-1 correspondence with the various finite subgroups of $SL_2(\mathbf{C})$, up to isomorphism.

Certain vertices of these graphs are said to be *special* : namely, all vertices for $\tilde{A}_n$, the four end points for $\tilde{D}_n$, the three end points for $\tilde{E}_6$, the two end points furthest from the branch point for $\tilde{E}_7$, the end point furthest from the branch point for $\tilde{E}_8$.

The group of automorphisms of any of the graphs above acts transitively on the set of special vertices. Therefore, by removing a special vertex from one of the graphs above, we obtain a graph which is independent of the special vertex chosen. The resulting graphs are denoted A_n, D_n, E_6, E_7, E_8. We get in this way the various graphs corresponding to irreducible, simply laced Cartan data of finite type.

14.1.5. Classification of non-symmetric Cartan data of affine type. Let us consider one of the graphs $\tilde{A}_n, \ldots, \tilde{E}_8$, together with an admissible automorphism a of order $\mathbf{n} > 1$, which has at least one fixed vertex. We enumerate the various possibilities (up to isomorphism).

(a) $\tilde{A}_n$ ($n \geq 3$, odd), $\mathbf{n} = 2$ and a has 2 fixed vertices:

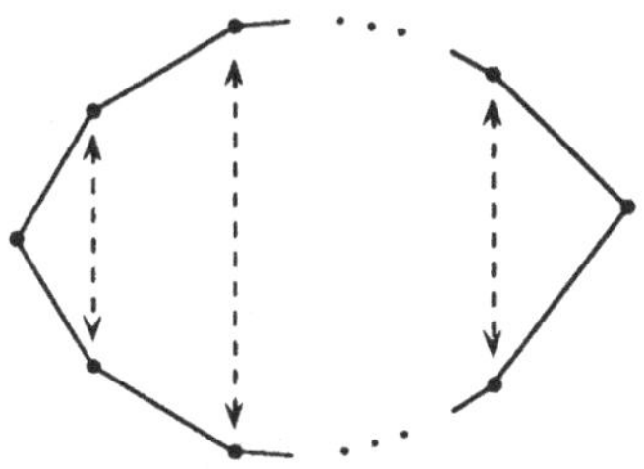

(b) $\tilde{D}_n$, $\mathbf{n} = 2$ and a has $n-1$ fixed vertices:

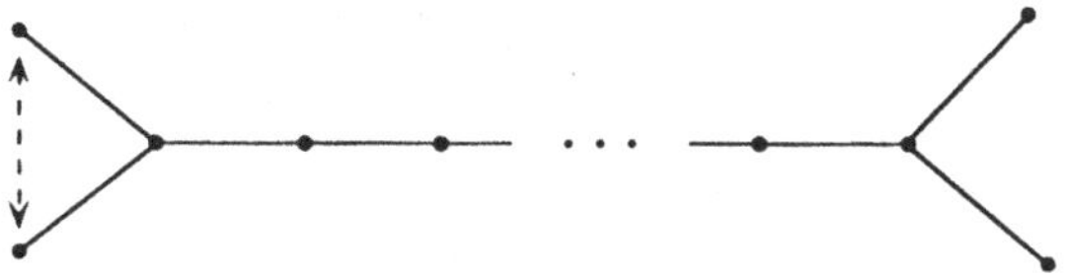

(c) $\tilde{D}_n$, ($n \geq 5$), $\mathbf{n} = 2$ and a has $n-3$ fixed vertices.

(d) $\tilde{D}_n$ (n even), $\mathbf{n} = 2$ and a has 1 fixed vertex:

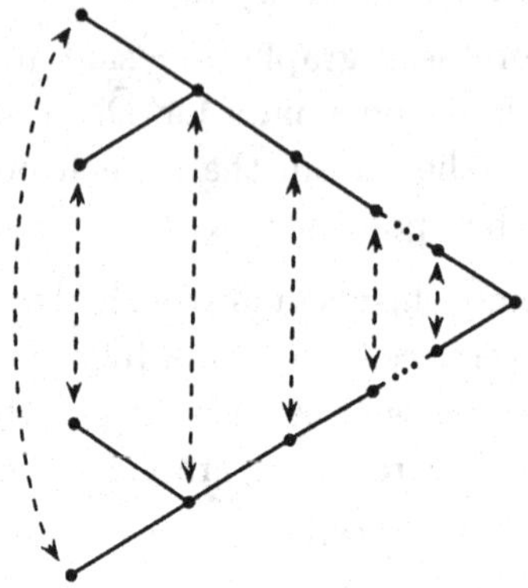

(e) $\tilde{D}_n$ (n even), $\mathbf{n} = 4$ and a has 1 fixed vertex:

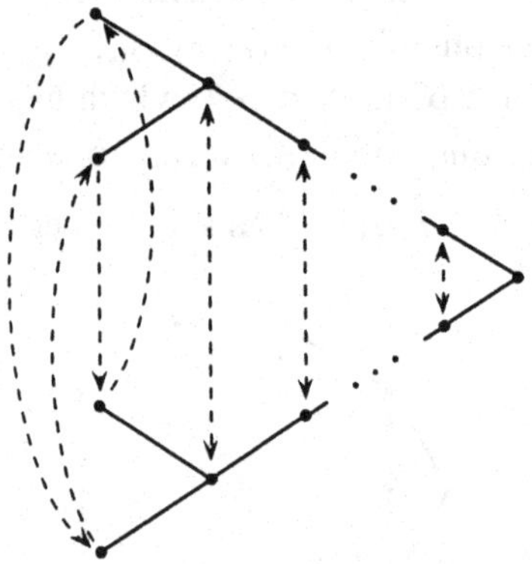

(f) $\tilde{D}_4$, $\mathbf{n} = 3$ and a has 2 fixed vertices:

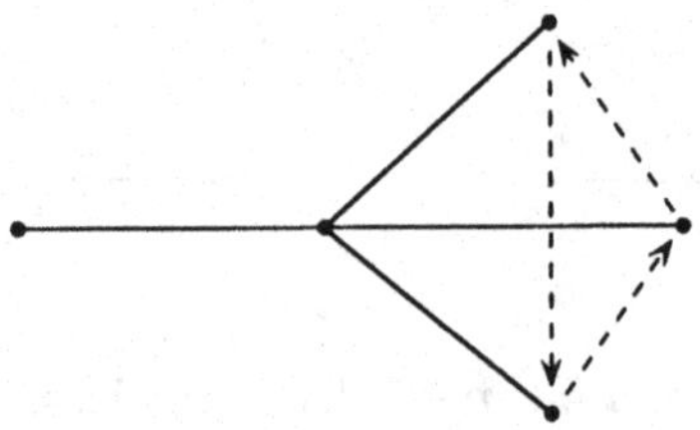

(g) $\tilde{E}_6$, $\mathbf{n} = 2$ and a has 3 fixed vertices:

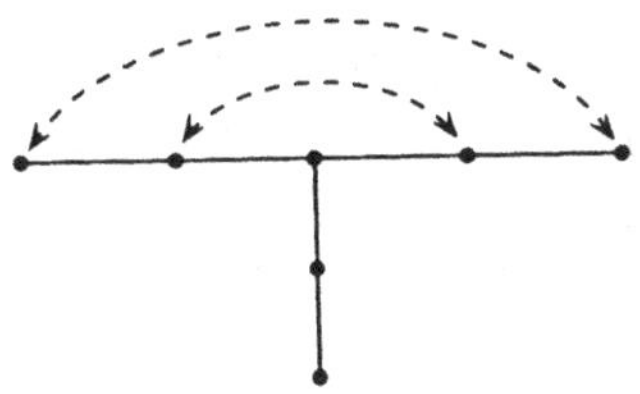

(h) $\tilde{E}_6$, $\mathbf{n} = 3$ and a has 1 fixed vertex:

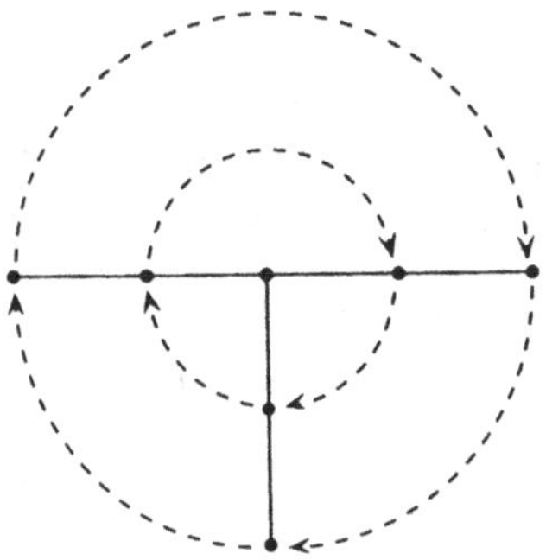

(i) $\tilde{E}_7$, $\mathbf{n} = 2$ and a has 2 fixed vertices:

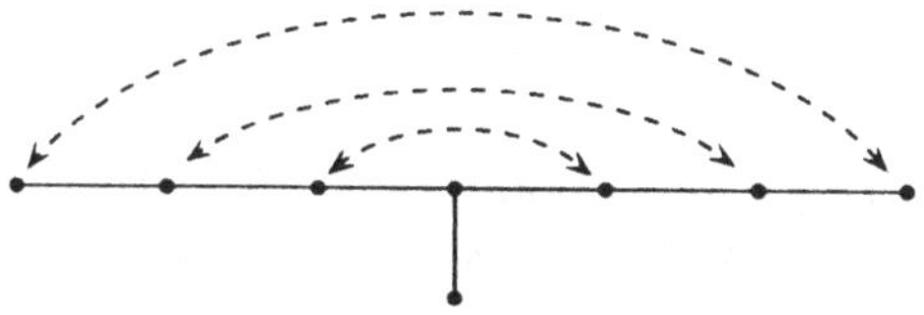

In each case (a)–(i), we may define a Cartan datum as in 14.1.1. We then obtain exactly the various affine non-symmetric Cartan data, up to proportionality (see 1.1.1) which were classified by Kac, Macdonald, Moody and Bruhat-Tits.

14.1.6. Classification of irreducible, non-symmetric Cartan data of finite type. We consider one of the graphs $A_n, \ldots, E_8$, together with an admissible automorphism a of order $\mathbf{n} > 1$.

We enumerate the various possibilities (up to isomorphism).

(a) A_n ($n \geq 3$, odd), $\mathbf{n} = 2$.

(b) D_n, $\mathbf{n} = 2$.
(c) D_4, $\mathbf{n} = 3$.
(d) E_6, $\mathbf{n} = 2$.
In each case (a)–(d), we may define a Cartan datum as in 14.1.1. We then obtain exactly the irreducible non-symmetric Cartan data of finite type, up to proportionality.

14.2. The Signed Basis $\mathcal{B}$

14.2.1. Let V be a $\mathbf{Q}(v)$-vector space with a given basis B and a given symmetric bilinear form $(,) : V \times V \to \mathbf{Q}(v)$. We say that B is *almost orthonormal* for $(,)$ if

(a) $(b, b') \in \delta_{b,b'} + v^{-1}\mathbf{Z}[[v^{-1}]] \cap \mathbf{Q}(v)$ for all $b, b' \in B$.

Let $\mathbf{A} = \mathbf{Q}[[v^{-1}]] \cap \mathbf{Q}(v)$. Let ${}_{\mathcal{A}}V$ be the $\mathcal{A}$-submodule of V generated by B and let $L(V) = \{x \in V | (x, x) \in \mathbf{A}\}$.

Lemma 14.2.2. *In the setup above, the following hold.*

(a) *$L(V)$ is an $\mathbf{A}$-submodule of V and B is a basis of it.*

(b) *Let $x \in {}_{\mathcal{A}}V$ be such that $(x, x) \in 1 + v^{-1}\mathbf{A}$. Then there exists $b \in B$ such that $x = \pm b \mod v^{-1}L(V)$.*

(c) *Let $x \in V$ be such that $(x, x) \in v^{-1}\mathbf{A}$. Then $x \in v^{-1}L(V)$.*

Let $x \in V$. Assume that $x \neq 0$. We can write uniquely $x = \sum_{b \in B} c_b b$ with $c_b \in \mathbf{Q}(v)$. Since only finitely many c_b are non-zero, we can find uniquely $t \in \mathbf{Z}$ and $p_b \in \mathbf{Z}$ (zero for all but finitely many b, but non-zero for some b) such that, for all b, we have $v^{-t}c_b - p_b \in v^{-1}\mathbf{A}$.

We have $(x, x) = (\sum_b p_b^2)v^{2t} \mod v^{2t-1}\mathbf{A}$. Note that $\sum_b p_b^2$ is a rational number > 0. Hence $(x, x) \in \mathbf{A}$ if and only if $t \leq 0$; this is equivalent to the condition that $c_b \in \mathbf{A}$ for all b and (a) follows.

If $(x, x) \in v^{-1}\mathbf{A}$, then we must have $t < 0$; hence $c_b \in v^{-1}\mathbf{A}$ for all b; (c) follows. If $x \in {}_{\mathcal{A}}V$ and $(x, x) \in 1 + v^{-1}\mathbf{A}$, then we must have $t = 0$ and $\sum_b p_b^2 = 1$ with $p_b \in \mathbf{Z}$; hence $p_b = \pm 1$ for some b and $p_b = 0$ for all other b; thus, (b) follows. The lemma is proved.

In the remainder of this chapter we fix a Cartan datum $(I, \cdot)$. Let $\mathbf{f}, {}_{\mathcal{A}}\mathbf{f}$ be defined in terms of $(I, \cdot)$ as in 1.2.5, 1.4.7.

Theorem 14.2.3. *Let $\mathcal{B}$ be the set of all $x \in \mathbf{f}$ such that $x \in {}_{\mathcal{A}}\mathbf{f}$, $\bar{x} = x$ and $(x, x) \in 1 + v^{-1}\mathbf{Z}[[v^{-1}]]$. (The last condition is equivalent to $\{x, x\} \in 1 + v\mathbf{Z}[[v]]$ since $\bar{x} = x$.)*

(a) *$\mathcal{B}$ is a signed basis of the $\mathcal{A}$-module ${}_{\mathcal{A}}\mathbf{f}$ and of the $\mathbf{Q}(v)$-vector space* $\mathbf{f}$.

(b) *If $b, b' \in \mathcal{B}$ and $b' \neq \pm b$, then $(b, b') \in v^{-1}\mathbf{Z}[[v^{-1}]]$ and $\{b, b'\} \in v\mathbf{Z}[[v]]$.*

By 14.1.2, we can find a finite graph $(\mathbf{I}, H, h \mapsto [h])$ and an admissible automorphism a of this graph such that $(I, \cdot)$ is obtained from these data by the construction in 14.1.1. Let $\mathbf{n} \geq 1$ be such that $a^{\mathbf{n}} = 1$. Then, by 13.2.11, $\mathbf{f}$ has a natural isomorphism, say χ, onto the corresponding algebra $\mathbf{k}$ (see 13.2.6). Under χ, the pairings $(,)$ and $(,)$ on $\mathbf{k}$ and $\mathbf{f}$ correspond to each other. This has been seen in the proof of 13.2.11. Moreover, the involutions $D : \mathbf{k} \to \mathbf{k}$ and $^{-} : \mathbf{f} \to \mathbf{f}$ correspond to each other (they both map the generators $\mathbf{1}_i$ and θ_i to themselves).

Note that χ carries $\mathbf{1}_{ni} \in \mathbf{k}$ to $\theta_i^{(n)}$ for any $i \in I$ and any $n \geq 0$ (this follows from 13.1.12(c)); hence it carries the $\mathcal{A}$-subalgebra ${}_{\mathcal{A}}\mathbf{k}$ (which is generated by the $\mathbf{1}_{ni}$) onto ${}_{\mathcal{A}}\mathbf{f}$ (see 1.4.7). Moreover, χ carries the signed basis $\mathcal{B}$ of $\mathbf{k}$ (see 13.1.2) onto a signed basis of $\mathbf{f}$, which we denote by the same letter. By the already known properties of the signed basis of $\mathbf{k}$, it remains to prove the following statement: let $x \in {}_{\mathcal{A}}\mathbf{f}$ be such that $\bar{x} = x$ and $(x, x) \in 1 + v^{-1}\mathbf{Z}[[v^{-1}]]$; then $x \in \mathcal{B}$. Let B be a basis of $\mathbf{f}$ such that $\mathcal{B} = B \cup (-B)$. We can write uniquely $x = \sum_b c_b b$ where b runs over B and $c_b \in \mathcal{A}$ are zero except for finitely many b. Using 14.2.2(b), we see that there is a unique $b \in B$ such that $c_b \in \pm 1 + v^{-1}\mathbf{Z}[v^{-1}]$ and $c_{b'} \in v^{-1}\mathbf{Z}[v^{-1}]$ for $b' \neq b$. From $\bar{x} = x$ and $\bar{b}' = b'$ for all $b' \in B$, it follows that $\bar{c}_{b'} = c_{b'}$ for all $b' \in B$. It follows that $c_b = \pm 1$ and $c_{b'} = 0$ for all $b' \neq b$. Thus, $x \in B \cup (-B)$. The theorem is proved.

14.2.4. Definition. $\mathcal{B}$ is called the *canonical signed basis* of $\mathbf{f}$.

Although the proof of the existence of $\mathcal{B}$ requires a choice of a graph with automorphism, which is not unique in general, the resulting signed basis is independent of any choice, hence the word *canonical.*

14.2.5. The following properties of $\mathcal{B}$ follow immediately from the definitions.

(a) We have $\mathcal{B} = \sqcup_\nu \mathcal{B}_\nu$ where $\mathcal{B}_\nu = \mathcal{B} \cap \mathbf{f}_\nu$.

(b) We have $\theta_i^{(s)} \in \mathcal{B}$ for any $i \in I$ and $s \geq 0$; in particular, $1 \in \mathcal{B}$.

Using 1.2.8(b), we see that

(c) $$\sigma(\mathcal{B}) = \mathcal{B}.$$

Proposition 14.2.6. (a) *r and $\bar{r}$ map ${}_{\mathcal{A}}\mathbf{f}$ into the $\mathcal{A}$-submodule ${}_{\mathcal{A}}\mathbf{f} \otimes_{\mathcal{A}} ({}_{\mathcal{A}}\mathbf{f})$ of $\mathbf{f} \otimes \mathbf{f}$.*

(b) *For any $x, y \in {}_{\mathcal{A}}\mathbf{f}$ we have $(x, y) \in \mathbf{Z}[[v^{-1}]] \cap \mathbf{Q}(v)$ and $\{x, y\} \in \mathbf{Z}[[v]] \cap \mathbf{Q}(v)$.*

This follows immediately from the analogous properties of ${}_{\mathcal{A}}\mathbf{k}$ (see 13.2.2, 13.2.4, 13.2.5) which are already known.

14.3. The Subsets $\mathcal{B}_{i;n}$ of $\mathcal{B}$

14.3.1. Given $i \in I$ and $n \geq 0$, we set $\mathcal{B}_{i;\geq n} = \mathcal{B} \cap \theta_i^n \mathbf{f}$ and ${}^\sigma\mathcal{B}_{i;\geq n} = \mathcal{B} \cap \mathbf{f}\theta_i^n$. Let $\mathcal{B}_{i;n} = \mathcal{B}_{i;\geq n} - \mathcal{B}_{i;\geq n+1}$ and ${}^\sigma\mathcal{B}_{i;n} = {}^\sigma\mathcal{B}_{i;\geq n} - {}^\sigma\mathcal{B}_{i;\geq n+1}$. Thus, we have partitions $\mathcal{B}_{i;\geq n} = \sqcup_{n' \geq n} \mathcal{B}_{i;n'}$ and ${}^\sigma\mathcal{B}_{i;\geq n} = \sqcup_{n' \geq n} {}^\sigma\mathcal{B}_{i;n'}$. Since $\sigma(\mathcal{B}) = \mathcal{B}$ and $\sigma(\theta_i^n \mathbf{f}) = \mathbf{f}\theta_i^n$, we have ${}^\sigma\mathcal{B}_{i;\geq n} = \sigma(\mathcal{B}_{i;\geq n})$ and ${}^\sigma\mathcal{B}_{i;n} = \sigma(\mathcal{B}_{i;n})$.

Theorem 14.3.2. (a) *$\mathcal{B}_{i;\geq n}$ is a signed basis of the $\mathbf{Q}(v)$-vector space $\theta_i^n \mathbf{f}$ and of the $\mathcal{A}$-module $\sum_{n';n' \geq n} \theta_i^{(n')} {}_{\mathcal{A}}\mathbf{f}$.*

(b) *${}^\sigma\mathcal{B}_{i;\geq n}$ is a signed basis of the $\mathbf{Q}(v)$-vector space $\mathbf{f}\theta_i^n$ and of the $\mathcal{A}$-module $\sum_{n';n' \geq n} ({}_{\mathcal{A}}\mathbf{f}\theta_i^{(n')})$.*

(c) *If $b \in \mathcal{B}_{i;0}$, then there is a unique element $b' \in \mathcal{B}_{i;n}$ such that $\theta_i^{(n)} b = b'$ plus an $\mathcal{A}$-linear combination of elements in $\mathcal{B}_{i;\geq n+1}$. Moreover, $b \mapsto b'$ is a bijection $\pi_{i,n} : \mathcal{B}_{i;0} \to \mathcal{B}_{i;n}$.*

(d) *If $b \in {}^\sigma\mathcal{B}_{i;0}$, then there is a unique element $b'' \in {}^\sigma\mathcal{B}_{i;n}$ such that $b\theta_i^{(n)} = b''$ plus an $\mathcal{A}$-linear combination of elements in ${}^\sigma\mathcal{B}_{i;\geq n+1}$. Moreover, $b \mapsto b''$ is a bijection ${}^\sigma\pi_{i,n} : {}^\sigma\mathcal{B}_{i;0} \to {}^\sigma\mathcal{B}_{i;n}$.*

For the proof we place ourselves in the setup considered in the proof of Theorem 14.2.3. Thus i is now regarded as an a-orbit in $\mathbf{I}$. Let $\mathbf{V} \in \mathcal{V}_\nu^a$. For any $n \geq 0$, let $\mathcal{B}_{\mathbf{V};i;n}$ be the set of all $\pm[B, \phi] \in \mathcal{B}_{\mathbf{V}}$ (see 12.6.4) such that $B \in \mathcal{P}_{\mathbf{V};i;n\gamma}$ where $\gamma = \sum_{\mathbf{i} \in i} \mathbf{i}$. By 10.3.3 and 12.5.1, we have a partition $\mathcal{B}_{\mathbf{V}} = \sqcup_{n \geq 0} \mathcal{B}_{\mathbf{V};i;n}$. By our identification $\mathcal{B}_{\mathbf{V}} = \mathcal{B}_\nu$, this becomes a partition $\mathcal{B}_\nu = \sqcup_{n \geq 0} \mathcal{B}_{\nu;i;n}$.

Let $\mathcal{B}'_{i;n} = \cup_\nu \mathcal{B}_{\nu;i;n}$. We will show below that $\mathcal{B}'_{i,n}$ just defined is the same as $\mathcal{B}_{i;n}$ in 14.3.1; see (h) below. We then have a partition $\mathcal{B} = \sqcup_{n \geq 0} \mathcal{B}'_{i;n}$.

Translating the geometric properties of $\mathcal{B}_{\mathbf{V};i;n}$ expressed by 10.3.2(c) we obtain the following property of $\mathcal{B}'_{i;n}$.

(e) For any $n \geq 0$, there is a unique 1-1 correspondence $b \leftrightarrow b'$ between $\mathcal{B}'_{i;0}$ and $\mathcal{B}'_{i;n}$ such that $\theta_i^{(n)} b = b'$ plus an $\mathcal{A}$-linear combination of elements in $\cup_{n' > n} \mathcal{B}'_{i;n'}$.

Let $M_n = \sum_{n':n'\ge n} \theta_i^{(n')}{}_{\mathcal{A}}\mathbf{f}$ and let M'_n be the $\mathcal{A}$-submodule of $\mathbf{f}$ generated by $\cup_{n':n'\ge n}\mathcal{B}'_{i;n'}$. We show that for fixed ν,

(f) any $b \in \mathcal{B}_{\nu;i;n}$ is contained in M_n.

We argue by descending induction on n; note that $\mathcal{B}_{\nu;i;n}$ is empty unless $n \le \nu_{\mathbf{i}}$ for any $\mathbf{i} \in i$. By (e), we have

$$b \in \theta_i^{(n)}{}_{\mathcal{A}}\mathbf{f} + \sum_{n':n'>s} \mathcal{A}\mathcal{B}_{\nu;i,n'};$$

by the induction hypothesis, we have $\mathcal{B}_{\nu;i,n'} \subset M_{n'}$ and it follows that $b \in M_n$. Thus (f) is proved. Thus we have $\mathcal{B}'_{i;n} \subset M_n$. If $n' \ge n$, we have $\mathcal{B}'_{i;n'} \subset M_{n'} \subset M_n$. It follows that, for any $n \ge 0$, we have $M'_n \subset M_n$.

Next we show that

(g) for any $b \in \mathcal{B}_\nu$, we have $\theta_i^{(n)} b \in M'_n$.

We argue by induction on $c(b) = \sum_{\mathbf{i}} \nu_{\mathbf{i}}$. If $b \in \mathcal{B}_{\nu;i;t}$ where $t > 0$, then as we have seen, we have $b \in M_t$; hence b is an $\mathcal{A}$-linear combination of elements $\theta_i^{(m)} b_1$ with $m \ge t$ and with $b_1 \in \mathcal{B}$ such that the induction hypothesis applies to b_1. Then $\theta_i^{(n)} b$ is an $\mathcal{A}$-linear combination of elements $\theta_i^{(m)}\theta_i^{(n)} b_1$ with b_1 as above. By the induction hypothesis, we have $\theta_i^{(n)} b_1 \in M'_n$. Hence $\theta_i^{(n)} b \in \sum_m \theta_i^{(m)} M'_n \subset M'_n$, as required.

Next we assume that $b \in \mathcal{B}_{\nu;i;0}$. Then $\theta_i^{(n)} b \in M'_n$ by (e). Thus, (g) is proved. It follows that, for any $n \ge 0$, we have

$$M_n = \sum_{n':n'\ge n} \theta_i^{(n')}{}_{\mathcal{A}}\mathbf{k} \subset \sum_{n':n'\ge n} M'_{n'} \subset M'_n.$$

We have proved that $M_n \subset M_{n'} \subset M_n$. Thus, $M'_n = M_n$. It follows that (a) holds and $\mathcal{B}_{i;\ge n} = \cup_{n':n'\ge n}\mathcal{B}'_{i;n'}$. In particular, we have

(h) $\mathcal{B}_{i;n} = \mathcal{B}'_{i;n}$.

We now prove (c). The existence of b' asserted in (c) follows immediately from (e). We now prove uniqueness. Assume that $b'_1 \in \mathcal{B}_{i;n}$ has the same property as that asserted for b' in (c). Then $b' - b'_1$ is on the one hand a linear combination of elements in $\cup_{n'>n}\mathcal{B}_{i;n'}$ and on the other hand it is a linear combination of elements in $\mathcal{B}_{i;n}$. It follows that $b' = b'_1$ and (c) is proved.

Now (b) and (d) follow from (a),(c) by applying σ. The theorem is proved.

14.3.3. By 12.5.1(c), we have

$$\mathcal{B} - \{\pm 1\} = \cup_{i \in I; n > 0} \mathcal{B}_{i;n}.$$

14.4. The Canonical Basis B of f

14.4.1. We would like to find in a natural way a basis of $\mathbf{f}$ contained in its canonical signed basis. If $(I, \cdot)$ is symmetric, such a basis is given by geometry. In this case, a is the identity automorphism of our graph and we can take $\mathbf{n} = 1$. Hence we have $\mathcal{O}' = \mathcal{A}$ and $\mathcal{K}(\mathcal{Q}_{\mathbf{V}}) = {}_{\mathcal{A}}\mathcal{K}(\mathcal{Q}_{\mathbf{V}})$ (for $\mathbf{V} \in \mathcal{V}$) has not only a natural signed basis, but a natural basis consisting of the elements $[B, 1]$ where B is a simple object of $\mathcal{P}_{\mathbf{V}}$ and 1 is the identity isomorphism $1 : B \cong B$.

14.4.2. In the general case, the descent from a signed basis to a basis will be non-geometric. We lay the groundwork with some definitions.

For any $\nu \in \mathbf{N}[I]$ we define a subset $\mathbf{B}_\nu$ of $\mathcal{B}_\nu$ by induction on $\mathrm{tr}\,\nu$ as follows. If $\nu = 0$, we set $\mathbf{B}_\nu = \{1\}$. If $\mathrm{tr}\,\nu > 0$, we set $\mathbf{B}_\nu = \cup_{i \in I, n > 0; \nu_i \geq n} \pi_{i,n}(\mathbf{B}_{\nu - ni} \cap \mathcal{B}_{i;0})$.

Let $\mathbf{B} = \sqcup_\nu \mathbf{B}_\nu \subset \mathcal{B}$. By 14.3.3, we have that $\mathcal{B} = \mathbf{B} \cup (-\mathbf{B})$. We can now state the following result.

Theorem 14.4.3. *Let* $\nu \in \mathbf{N}[I]$. *Then*

(a) $\mathbf{B}_\nu \cap (-\mathbf{B}_\nu) = \emptyset$;

(b) $\mathbf{B}_\nu \cap (-\sigma(\mathbf{B}_\nu)) = \emptyset$;

(c) $\sigma(\mathbf{B}_\nu) = \mathbf{B}_\nu$.

(d) $\mathbf{B}$ *is a basis of the* $\mathcal{A}$*-module* ${}_{\mathcal{A}}\mathbf{f}$ *and a basis of the* $\mathbf{Q}(v)$*-vector space* $\mathbf{f}$.

(e) *For any* ν, $\mathbf{B}_\nu$ *is a basis of the* $\mathcal{A}$*-module* ${}_{\mathcal{A}}\mathbf{f}_\nu$ *and a basis of the* $\mathbf{Q}(v)$*-vector space* $\mathbf{f}_\nu$.

14.4.4. Proof of the theorem, assuming that $(I, \cdot)$ is symmetric. In this case, as in the proof of Theorem 14.2.3, we have a natural choice for the graph (with identity automorphism a), see Remark 14.1.3. Moreover since a=1, the corresponding algebra $\mathbf{k}$ has a natural basis inside its natural signed basis, defined as in 14.4.1. From the definitions, it is clear that this basis (transferred to $\mathbf{f}$) coincides with $\mathbf{B}$ and has all the required properties. This completes the proof (in the symmetric case).

14.4.5. The proof of Theorem 14.4.3 in the general case will be given in 19.2.3; in the remainder of this section we shall assume that the theorem is known in general.

14.4.6. Definition. $\mathbf{B}$ is called the *canonical basis* of $\mathbf{f}$.

We shall use the following notation: $\mathbf{B}_{i;n} = \mathcal{B}_{i;n} \cap \mathbf{B}$ for any $i \in I$ and $n \in \mathbf{N}$; note that $\pi_{i,n}$ defines a bijection $\mathbf{B}_{i;0} \cong \mathbf{B}_{i;n}$. Set ${}^\sigma\mathbf{B}_{i;n} = \sigma(\mathbf{B}_{i;n})$. Then ${}^\sigma\pi_{i,n}$ defines a bijection ${}^\sigma\mathbf{B}_{i;0} \cong {}^\sigma\mathbf{B}_{i;n}$.

14.4.7. We can regard $\mathbf{B}$ as the set of vertices of a graph colored by $I \times \{1, 2, \dots\}$ in which b, b' are joined by an edge of color (i, n) if $b \in \mathbf{B}_{i;0}, b' \in \mathbf{B}_{i;n}$ and $b' = \pi_{i,n}(b)$. This is called the *left graph* on $\mathbf{B}$.

Similarly, we can regard $\mathbf{B}$ as the set of vertices of a graph colored by $I \times \{1, 2, \dots\}$ in which b, b'' are joined by an edge of color (i, n) if $b \in {}^\sigma\mathbf{B}_{i;0}, b'' \in {}^\sigma\mathbf{B}_{i;n}$ and $b'' = {}^\sigma\pi_{i,n}(b)$. This is called the *right graph* on $\mathbf{B}$.

14.4.8. Let us choose a finite graph $(\mathbf{I}, H, h \mapsto [h])$ and an admissible automorphism a of this graph such that $(I, \cdot)$ is obtained from them by the construction in 14.1.1. We define a new (symmetric) Cartan datum $(\tilde{I}, \cdot)$ associated to the same graph and to its identity automorphism, as in 14.1.3. More precisely, we have $\tilde{I} = \mathbf{I}$, $\mathbf{i} \cdot \mathbf{i} = 2$ and, for $\mathbf{i} \neq \mathbf{j} \in \mathbf{I}$, we have that $\mathbf{i} \cdot \mathbf{j}$ is -1 times the number of edges joining $\mathbf{i}$ to $\mathbf{j}$.

Let $\tilde{\mathbf{f}}$ be the algebra defined like $\mathbf{f}$, in terms of the Cartan datum $(\tilde{I}, \cdot)$ and let $\tilde{\mathbf{B}} \subset \tilde{\mathbf{f}}$ be its canonical basis. Similarly, let $\pi_{\mathbf{i},n} : \tilde{\mathbf{B}}_{\mathbf{i};0} \to \tilde{\mathbf{B}}_{\mathbf{i};n}$ and ${}^\sigma\pi_{\mathbf{i},n} : {}^\sigma\tilde{\mathbf{B}}_{\mathbf{i};0} \to {}^\sigma\tilde{\mathbf{B}}_{\mathbf{i};n}$ be the bijections analogous to $\pi_{i,n} : \mathbf{B}_{i;0} \to \mathbf{B}_{i;n}$ and ${}^\sigma\pi_{i,n} : {}^\sigma\mathbf{B}_{i;0} \to {}^\sigma\mathbf{B}_{i;n}$ in 14.4.6. Now $a : \mathbf{I} \to \mathbf{I}$ induces an algebra automorphism $a : \tilde{\mathbf{f}} \to \tilde{\mathbf{f}}$ which restricts to a bijection $a : \tilde{\mathbf{B}} \to \tilde{\mathbf{B}}$ whose fixed point set is denoted by $\tilde{\mathbf{B}}^a$.

The $\mathbf{I} \times \{1, 2, \dots\}$-colored left graph structure on $\tilde{\mathbf{B}}$ (as in 14.4.7) defines a $I \times \{1, 2, \dots\}$-colored graph structure on the subset $\tilde{\mathbf{B}}^a$ as follows. We say that $b, b' \in \tilde{\mathbf{B}}^a$ are joined by an edge of color (i, n) if they can be joined in the left graph on $\tilde{\mathbf{B}}$ by a sequence of edges of colors $(\mathbf{i}_1, n), (\mathbf{i}_2, n), \dots, (\mathbf{i}_s, n)$ where $\mathbf{i}_1, \mathbf{i}_2, \dots, \mathbf{i}_s$ is an enumeration of the elements of i in some order. This is called the left graph on $\tilde{\mathbf{B}}^a$. By replacing "left" by "right" we obtain a $I \times \{1, 2, \dots\}$-colored graph structure on $\tilde{\mathbf{B}}^a$, called the right graph on $\tilde{\mathbf{B}}^a$.

Theorem 14.4.9. *There is a unique bijection $\chi : \mathbf{B} \to \tilde{\mathbf{B}}^a$ compatible with the structures of $I \times \{1, 2, \dots\}$-colored left (resp. right) graphs and such that $\chi(1) = 1$. The two bijection corresponding to "left" and "right" coincide.*

The inverse bijection is obtained geometrically by attaching to a pair (B,ϕ) where B is a simple object of a suitable $\mathcal{P}_{\mathbf{V}}$ and ϕ is an isomorphism $a^*B \cong B$, the simple object B without specifying ϕ. The fact that this bijection is compatible with the colored graph structures is also clear geometrically (using, in particular, 12.5.1(a)).

14.4.10. Remark. This theorem shows that to describe the left or right graph structure on $\mathbf{B}$ it is enough to do the same in the case where the Cartan datum is symmetric.

Theorem 14.4.11. *Assume that the root datum is Y-regular. Let $\lambda \in X^+$ and let $\Lambda_\lambda = \mathbf{f}/\sum_i \mathbf{f}\theta^{\langle i,\lambda\rangle+1}$ be the $\mathbf{U}$-module defined in 3.5.6. As in 3.5.7, let $\eta \in \Lambda_\lambda$ be the image of 1. Let $\mathbf{B}(\lambda) = \cap_{i\in I}(\cup_{n;0\le n\le\langle i,\lambda\rangle}{}^\sigma\mathbf{B}_{i,n})$.*

(a) *The map $b \to b^-\eta$ define a bijection of $\mathbf{B}(\lambda)$ onto a basis $\mathbf{B}(\Lambda_\lambda)$ of Λ_λ.*

(b) *If $b \in \mathbf{B} - \mathbf{B}(\lambda)$, then $b^-\eta = 0$.*

An equivalent statement is that

$$\cup_{i,n;n\ge\langle i,\lambda\rangle+1}{}^\sigma\mathbf{B}_{i,n}$$

is a basis of $\sum_i \mathbf{f}\theta^{\langle i,\lambda\rangle+1}$. This follows immediately from Theorem 14.3.2.

14.4.12. Definition. $\mathbf{B}(\Lambda_\lambda)$ is called the *canonical basis* of Λ_λ.

Theorem 14.4.13 (Positivity). *Assume that $(I,\cdot)$ is symmetric.*

(a) *For any $b, b' \in \mathbf{B}$, we have*

$$bb' = \sum_{b''\in\mathbf{B};n\in\mathbf{Z}} c_{b,b',b'',n}v^n b''$$

where $c_{b,b',b'',n} \in \mathbf{N}$ are zero except for finitely many b'', n.

(b) *For any $b \in \mathbf{B}$ we have*

$$r(b) = \sum_{b',b''\in\mathbf{B};n\in\mathbf{Z}} d_{b,b',b'',n}v^n b' \otimes b''$$

where $d_{b,b',b'',n} \in \mathbf{N}$ are zero except for finitely many b', b'', n.

(c) *For any $b, b' \in \mathbf{B}$ we have*

$$(b,b') = \sum_{n\in\mathbf{N}} f_{b,b',n}v^{-n}$$

where $f_{b,b',n} \in \mathbf{N}$.

The theorem asserts the positivity of certain integers; in our definition in the framework of perverse sheaves, these integers are the dimensions of certain $\bar{\mathbf{Q}}_l$-vector spaces. The theorem follows.

14.4.14. Remark. For non-symmetric $(I, \cdot)$, the integers in question are not dimensions, but traces of automorphisms of finite order of certain $\bar{\mathbf{Q}}_l$-vector spaces and it is not clear whether they are positive or not.

14.4.15. In the case where $(I, \cdot)$ is symmetric, the set $\mathbf{B}_\nu$ is (conjecturally) in natural $1-1$ correspondence with the set X_ν of irreducible components of a certain Lagrangian variety naturally attached to $(I, \cdot)$ and to ν (see [9, 13.7]). The union $\sqcup_\nu X_\nu$ has a natural colored graph structure (defined as in [8]) and one can hope that the previous bijection respects the colored graph structures.

14.5. Examples

14.5.1. Assume that $(I, \cdot)$ is a simply laced Cartan datum of finite type. Let $(\mathbf{I}, H, \dots)$ be the graph of $(I, \cdot)$ (see 14.1.3); note that $\mathbf{I} = I$. We choose an orientation of this graph. Let $\mathbf{V} \in \mathcal{V}$ and let $G_\mathbf{V}, \mathbf{E}_\mathbf{V}$ be as in 9.1.2.

From the results in [9], it follows that there is a 1-1 correspondence between the set of orbits of $G_\mathbf{V}$ on $\mathbf{E}_\mathbf{V}$ (a finite set, by Gabriel's theorem) and the set of isomorphism classes of objects of $\mathcal{P}_\mathbf{V}$ (see 9.1.3): to an orbit of $G_\mathbf{V}$ corresponds the $G_\mathbf{V}$-equivariant simple perverse sheaf whose support is the closure of that orbit. This is well-defined since the action of $G_\mathbf{V}$ has connected isotropy groups.

14.5.2. Assume that $(I, \cdot)$ is such that $I = \{i, j\}$ and $i \cdot i = j \cdot j = 2, i \cdot j = j \cdot i = -2$. Then $(I, \cdot)$ is a symmetric Cartan datum of affine type.

Let $(\mathbf{I}, H, \dots)$ be the graph of $(I, \cdot)$ (see 14.1.3); note that $\mathbf{I} = I$ and H has two elements. We orient this graph so that $h' = i$ for both $h \in H$. Let $\mathbf{V} \in \mathcal{V}$ and let $G_\mathbf{V}, \mathbf{E}_\mathbf{V}$ be as in 9.1.2. Note that $\mathbf{E}_\mathbf{V}$ consists of all pairs T, T' of linear maps $\mathbf{V}_i \to \mathbf{V}_j$. Assume that both $\mathbf{V}_i$ and $\mathbf{V}_j$ are n-dimensional and $n \geq 2$. Then $G_\mathbf{V}$ acts on $\mathbf{E}_\mathbf{V}$ with infinitely many orbits.

Let $\boldsymbol{\nu} = (i, j, i, j, \dots)$ ($2n$ terms). Then $\pi_{\boldsymbol{\nu}} : \tilde{\mathcal{F}}_{\boldsymbol{\nu}} \to \mathbf{E}_\mathbf{V}$ (see 9.1.3) is a principal covering with group S_n (the symmetric group) over an open dense subset of $\mathbf{E}_\mathbf{V}$. This gives rise to irreducible local systems over an open dense subset of $\mathbf{E}_\mathbf{V}$, and hence to simple perverse sheaves on $\mathbf{E}_\mathbf{V}$, indexed by the irreducible representations of S_n. These simple perverse sheaves belong to $\mathcal{P}_\mathbf{V}$.

14.5.3. Assume that $(I, \cdot)$ is such that $I = \{i\}$ and $i \cdot i = 2$. The canonical basis $\mathbf{B}$ of $\mathbf{f}$ consists of the elements $\theta_i^{(a)}$ $(a \in \mathbf{N})$.

14.5.4. Assume that $(I,\cdot)$ is such that $I=\{i,j\}$ and $i\cdot i=j\cdot j=2$ and $i\cdot j=j\cdot i=-1$. The canonical basis $\mathbf{B}$ of $\mathbf{f}$ consists of the elements $\theta_i^{(a)}\theta_j^{(b)}\theta_i^{(c)}$ $(a,b,c\in\mathbf{N}, b\ge a+c)$ and of the elements $\theta_j^{(c)}\theta_i^{(b)}\theta_j^{(a)}$ $(a,b,c\in\mathbf{N}, b\ge a+c)$ with the identification $\theta_i^{(a)}\theta_j^{(b)}\theta_i^{(c)}=\theta_j^{(c)}\theta_i^{(b)}\theta_j^{(a)}$ for $b=a+c$.

14.5.5. Assume that $(I,\cdot)$ is as in 14.5.2. The elements of $\mathbf{B}_{i+j}$ are $\theta_i\theta_j,\theta_j\theta_i$.

The elements of $\mathbf{B}_{2i+2j}$ are:

$$\theta_i^{(2)}\theta_j^{(2)},\theta_j^{(2)}\theta_i^{(2)},\theta_i\theta_j^{(2)}\theta_i,\theta_j\theta_i^{(2)}\theta_j,\theta_i\theta_j\theta_i\theta_j-\theta_i^{(2)}\theta_j^{(2)},\theta_j\theta_i\theta_j\theta_i-\theta_j^{(2)}\theta_i^{(2)}.$$

For further examples, see [11].

Notes on Part II

1. The canonical basis of $\mathbf{f}$ has been introduced by the author in [7], assuming that the Cartan datum is symmetric, of finite type. In fact, in [7] two definitions for the canonical basis were given: an elementary algebraic one, involving braid group actions, and a topological one, based on quivers and perverse sheaves. (The elementary definition applies essentially without change to not necessarily symmetric Cartan data of finite type.) The topological definition in [7] was in terms of intersection cohomology of certain singular varieties arising from quivers by a construction reminiscent of that in [4] of the new basis of a Hecke algebra (which used the intersection cohomology of Schubert varieties). One of the main observations of [7] was that the canonical basis of $\mathbf{f}$ gives rise simultaneoulsy to a canonical basis in each $\mathbf{U}$-module Λ_λ, which had rather favourable properties.
2. After [7] became available, Kashiwara announced an elementary algebraic definition of the canonical basis which applied to an arbitrary Cartan datum. Kashiwara's paper [3] contains an inductive construction of the canonical basis, both of $\mathbf{f}$ and of Λ_λ, which advances like a huge spiral. His construction agrees with that in [7], as shown in [8].

   On the other hand, the author [9] extended the topological definition [7] of the canonical basis to arbitrary (symmetric) Cartan data. (The case of not necessarily symmetric Cartan data was only sketched in [9].) The definition of [9] resembles that of character sheaves [5]. While the method of [9] is not elementary, it has the advantage of being more global and to yield positivity results which cannot be obtained by the elementary approach. The agreement of the definitions in [3], [9], was proved in [2].
3. The exposition in Part II essentially follows the treatment in [9], with two main differences. First, in order to include not necessarily symmetric Cartan data in our treatment, we have to take into account the action of a cyclic group, which is a complicating factor, not present in [9], where only symmetric Cartan data were treated. In addition, we make use of the geometric interpretation of the inner product on $\mathbf{f}$ given in [2]; this simplifies somewhat the original proof in [9] and provides the link with [3].
4. The basic reference for the theory of perverse sheaves on algebraic varieties is the work of Beilinson, Bernstein, Deligne and Gabber [1].
5. The representation theory of quivers (which is implicit in the constructions in Chapter 9) has a long history going back to Kronecker. In Ringel's work [12], the connection between the representations of a quiver of finite type over a finite field F_q and the plus part of the corresponding Drinfeld-Jimbo algebra at parameter $\sqrt{q}$ was observed for the first time. This work of Ringel was an important source of inspiration for the author's work on the canonical basis. In particular, the definition of the induction functor in 9.2 was inspired by Ringel's definition of the Hall algebra arising from quivers over F_q. On the other hand, the definition of the restriction functor in 9.2 was inspired by the analogous concept for character sheaves [5].
6. The geometric definition of the inner product in 12.2 is taken from [2] where,

however, the cyclic group action was not present.

7. The idea that the canonical basis can be characterized by an almost orthonormality property for the natural inner product, has originally appeared in Kashiwara's paper [3] and has been later used in [2]. This is analogous to the orthogonality properties of character sheaves [5]; it is a hallmark of intersection cohomology.
8. The description of non-symmetric affine Cartan data given in 14.1.5 is different, as far as I know, from the ones in the literature.
9. The ingredients for the definition of the colored graph in 14.4.7 were introduced in [9]; it turns out that that graph contains the same information as the colored graph defined by Kashiwara (but the two graphs are different).
10. The statement 14.4.13 appeared in [2].
11. The example in 14.5.2 is a special case of the results in [10] where the perverse sheaves which constitute the canonical basis in the affine case are described explicitly. (The results in [10] dealt with symmetric affine Cartan data; but in view of Theorem 14.4.9, the same results can be applied in the case of non-symmetric affine Cartan data.)
12. The geometric method used here to construct canonical bases can be applied, more or less word by word, to quivers in which edges joining a vertex with itself are allowed. (This includes, for example, the classical Hall algebra with its canonical basis.) We have not included this more general case in our discussion (but see [11]).

REFERENCES

1. A. Beilinson, J. Bernstein and P. Deligne, *Faisceaux pervers*, Astérisque **100** (1982), Soc. Math. France.
2. I. Grojnowski and G. Lusztig, *A comparison of bases of quantized enveloping algebras*, Contemp. Math. **153** (1993), 11–19.
3. M. Kashiwara, *On crystal bases of the q-analogue of universal enveloping algebras*, Duke Math. J. **63** (1991), 465-516.
4. D. Kazhdan and G. Lusztig, *Schubert varieties and Poincaré duality*, Proc. Symp. Pure Math. **36** (1980), 185-203, Amer. Math. Soc., Providence, R. I..
5. G. Lusztig, *Character sheaves, I*, Adv. in Math. **56** (1985), 193-237 *II*, Adv. in Math. **57** (1985), 226-265.
6. ———, *Cuspidal local systems and graded Hecke algebras*, Publ. Mathématiques **67** (1988), 145-202.
7. ———, *Canonical bases arising from quantized enveloping algebras*, J. Amer. Math. Soc. **3** (1990), 447-498.
8. ———, *Canonical bases arising from quantized enveloping algebras, II*, Common trends in mathematics and quantum field theories (T. Eguchi et. al., eds.), Progr. Theor. Phys. Suppl., vol. 102, 1990, pp. 175-201.
9. ———, *Quivers, perverse sheaves and quantized enveloping algebras*, J. Amer. Math. Soc. **4** (1991), 365-421.
10. ———, *Affine quivers and canonical bases*, Publ. Math. I. H. E. S. **76** (1992), 111–163.
11. ———, *Tight monomials in quantized enveloping algebras*, Quantum deformations of algebras and representation (S. Shnider, ed.), Israel Math. Conf. Proc. (Amer. Math. Soc.), vol. 7, 1993.
12. C. M. Ringel, *Hall algebras and quantum groups*, Invent. Math. **101** (1990), 583-592.

Part III

KASHIWARA'S OPERATORS AND APPLICATIONS

In the author's elementary algebraic definition [4] of the canonical basis of **f**, there were three main ingredients: (a) the basis was assumed to be integral in a suitable sense; (b) the basis was assumed fixed by the involution $^-$; (c) the basis was assumed to be in a specified $\mathbf{Z}[v^{-1}]$-lattice $\mathcal{L}$ and had prescribed image in $\mathcal{L}/v^{-1}\mathcal{L}$.

Of these three ingredients, the last one is the most subtle; in [4], $\mathcal{L}$ and the basis of $\mathcal{L}/v^{-1}\mathcal{L}$ were defined in terms of a braid group action. This definition does not work for Cartan data of infinite type.

Kashiwara's scheme [2] to define a basis of **f** involves again the ingredients (a),(b),(c) above, but he proposes a quite different way to construct the lattice $\mathcal{L}$ and the basis of $\mathcal{L}/v^{-1}\mathcal{L}$, which makes sense for any Cartan datum. The main ingredients in his definition were certain operators $\tilde{\epsilon}_i, \tilde{\phi}_i : \mathbf{f} \to \mathbf{f}$ and some analogous operators $\tilde{E}_i, \tilde{F}_i$ on any integrable **U**-module. (The last operators were already introduced, in a dual form, in an earlier paper [1].)

Part III gives an account of Kashiwara's approach and its applications. (The results in Part III will be needed in Part IV.) Our exposition differs from that of Kashiwara to some extent. In particular, we will make use of the existence of canonical bases (up to sign) established in Part II, while for Kashiwara, that existence was one of the goals.

The algebra $\mathfrak{U}$ in Chapter 15 is defined in a different way than in [2], but eventually, the two definitions agree. The operators $\tilde{\epsilon}_i, \tilde{\phi}_i, \tilde{E}_i, \tilde{F}_i$ are defined in Chapter 16. Chapter 17 contains a proof of a crucial result of Kashiwara on the behaviour of $\tilde{E}_i, \tilde{F}_i$ in a tensor product. Chapters 18 and 19 are concerned with various properties of the canonical basis of Λ_λ, in particular with the fact that this basis is almost orthonormal for the natural inner product. Chapter 20 deals with bases at ∞ (or crystal bases in Kashiwara's terminology). Chapter 21 deals with the special features which hold in the case where the Cartan datum is of finite type. Chapter 22 contains some new positivity results.

In the remainder of this book we assume that, unless otherwise specified, a Cartan datum $(I, \cdot)$ and a root datum $(Y, X, \dots)$ of type $(I, \cdot)$ have been fixed. The notation $\mathbf{f}, \mathbf{U}$, *etc. will refer to these fixed data.*

CHAPTER 15

The Algebra $\mathfrak{U}$

Lemma 15.1.1. *The algebra homomorphism $\chi : {}'\mathbf{f} \to \mathbf{U}$ given by $\theta_i \mapsto E'_i = (v_i - v_i^{-1})\tilde{K}_{-i}E_i \quad (i \in I)$ factors through an algebra homomorphism $\mathbf{f} \to \mathbf{U}$.*

Let $\chi' : {}'\mathbf{f} \to \mathbf{U}$ be the algebra homomorphism given by $\theta_i \mapsto E_i \quad (i \in I)$. A simple computation shows that, if $f \in {}'\mathbf{f}_\nu$, then

$$\chi(f) = v^N(\prod_i (v_i - v_i^{-1})^{\nu_i})\tilde{K}_{-|f|}\chi'(f)$$

where N depends only on ν and not on f. Hence if f is a homogeneous element in $\mathcal{I}$ (so that $\chi'(f) = 0$), then $\chi(f) = 0$. The lemma is proved.

15.1.2. Let $\tilde{\mathbf{U}}^+$ be the image of χ. Using the previous lemma we see that $E_i \mapsto E'_i$ defines an algebra isomorphism $\mathbf{U}^+ \cong \tilde{\mathbf{U}}^+$.

Let $\tilde{\mathbf{U}}^0$ be the subalgebra of $\mathbf{U}$ generated by the elements $\tilde{K}_{-i}$ for $i \in I$. From the triangular decomposition of $\mathbf{U}$, we can deduce that multiplication defines injective maps $\mathbf{U}^- \otimes \tilde{\mathbf{U}}^0 \otimes \tilde{\mathbf{U}}^+ \to \mathbf{U}$ and $\tilde{\mathbf{U}}^+ \otimes \tilde{\mathbf{U}}^0 \otimes \mathbf{U}^- \to \mathbf{U}$. These maps have the same image, which is a subalgebra $\tilde{\mathbf{U}}$ of $\mathbf{U}$; this follows from the identity $E'_iF_j = v^{i \cdot j}F_jE'_i + \delta_{i,j}(1 - \tilde{K}^2_{-i})$ for all i, j. Note that the elements $\tilde{K}_{-i}$ which are invertible in $\mathbf{U}$ are not invertible in $\tilde{\mathbf{U}}$. The left ideal generated by them in $\tilde{\mathbf{U}}$ coincides with the right ideal generated by them in $\tilde{\mathbf{U}}$. The quotient of $\tilde{\mathbf{U}}$ by this ideal will be denoted by $\mathfrak{U}$. We have obvious algebra homomorphisms $\mathbf{U}^- \to \mathfrak{U}$ and $\tilde{\mathbf{U}}^+ \to \mathfrak{U}$ and it is clear that

(a) multiplication defines isomorphisms of vector spaces $\mathbf{U}^- \otimes \tilde{\mathbf{U}}^+ \cong \mathfrak{U}$ and $\tilde{\mathbf{U}}^+ \otimes \mathbf{U}^- \cong \mathfrak{U}$;

(b) the algebra $\mathfrak{U}$ is defined by the generators $\epsilon_i, \phi_i \quad (i \in I)$ and the relations $\epsilon_i\phi_j = v^{i \cdot j}\phi_j\epsilon_i + \delta_{i,j}$ for all i, j, together with the relations $f(\epsilon_i) = f(\phi_i) = 0$ for any homogeneous $f = f(\theta_i) \in \mathcal{I}$. Here, ϵ_i, ϕ_i are the images of E'_i, F_i in $\mathfrak{U}$.

15.1.3. There is a unique $\mathbf{Q}$-algebra homomorphism $\omega : \mathfrak{U} \to \mathfrak{U}$ such that $\omega(\epsilon_i) = v_i\phi_i, \quad \omega(\phi_i) = -v_i\epsilon_i, \quad \omega(v) = v^{-1}$. We will not use it. Note that $\omega^2 = 1$.

G. Lusztig, *Introduction to Quantum Groups*, Modern Birkhäuser Classics,
DOI 10.1007/978-0-8176-4717-9_15, © Springer Science+Business Media, LLC 2010

Lemma 15.1.4. *For each $i \in I$ we define $\phi_i : \mathbf{f} \to \mathbf{f}$ to be left multiplication by θ_i and $\epsilon_i : \mathbf{f} \to \mathbf{f}$ to be the linear map ${}_ir$ in 1.2.13.*

(a) *ϕ_i, ϵ_i make $\mathbf{f}$ into a left $\mathfrak{U}$-module.*

(b) *$\epsilon_i : \mathbf{f} \to \mathbf{f}$ is locally nilpotent for any $i \in I$.*

The identity $\epsilon_i \phi_j = v^{i \cdot j} \phi_j \epsilon_i + \delta_{i,j}$ (as maps $\mathbf{f} \to \mathbf{f}$) follows from ${}_ir(\theta_j y) = {}_ir(\theta_j)y + v^{j \cdot i}\theta_j({}_ir(y))$. Let $f = f(\theta_i)$ be a homogeneous element of $\mathcal{I}$. From the definition we have $f(\phi_i) = 0$ as a linear map $\mathbf{f} \to \mathbf{f}$. We must show that $f(\epsilon_i) = 0$. Let $f' = \sigma(f) \in \mathcal{I}$. From the definition we have

$$(x, \epsilon_i(y)) = (1 - v_i^{-2})(\phi_i x, y)$$

for all $x, y \in \mathbf{f}$. It follows that $(x, f(\epsilon_i)(y)) = c(f'(\phi_i)(x), y)$ where $c \in \mathbf{Q}(v)$. From $f'(\phi_i)(x) = 0$ we deduce that $(x, f(\epsilon_i)(y)) = 0$. By the non-degeneracy of $(,)$, this implies that $f(\epsilon_i) = 0$ as a linear map $\mathbf{f} \to \mathbf{f}$. Thus the relations 15.1.2(b) of $\mathfrak{U}$ are verified; (a) is proved.

If $x \in \mathbf{f}_\nu$, then $\epsilon_i(x) \in \mathbf{f}_{\nu - i}$ if $\nu_i \geq 1$ and $\epsilon_i(x) = 0$ if $\nu_i = 0$. It follows that $\epsilon_i : \mathbf{f} \to \mathbf{f}$ is locally nilpotent. The lemma is proved.

Lemma 15.1.5. *There is a unique algebra homomorphism $d : \mathfrak{U} \to \mathfrak{U} \otimes \mathbf{U}$ such that $d(\phi_i) = \phi_i \otimes \tilde{K}_{-i} + 1 \otimes F_i$ and $d(\epsilon_i) = \epsilon_i \otimes \tilde{K}_{-i} + (v_i - v_i^{-1}) 1 \otimes \tilde{K}_{-i} E_i$ for all $i \in I$.*

The homomorphism $\Delta : \mathbf{U} \to \mathbf{U} \otimes \mathbf{U}$, satisfies $\Delta(F_i) = F_i \otimes \tilde{K}_{-i} + 1 \otimes F_i$, $\Delta(E_i') = E_i' \otimes \tilde{K}_{-i} + (v_i - v_i^{-1}) 1 \otimes \tilde{K}_{-i} E_i$, and $\Delta(\tilde{K}_{-i}) = \tilde{K}_{-i} \otimes \tilde{K}_{-i}$. Hence Δ restricts to an algebra homomorphism $\tilde{\mathbf{U}} \to \tilde{\mathbf{U}} \otimes \mathbf{U}$ and this induces an algebra homomorphism $d : \mathfrak{U} \to \mathfrak{U} \otimes \mathbf{U}$ which has the required properties.

CHAPTER 16

Kashiwara's Operators in Rank 1

16.1. Definition of the Operators $\tilde{\phi}_i, \tilde{\epsilon}_i$ and $\tilde{F}_i, \tilde{E}_i$

16.1.1. In this chapter we fix $i \in I$. Besides the category $\mathcal{C}'_i$ (see 5.1.1), we shall consider another category $\mathcal{D}_i$ which shares some of the properties of $\mathcal{C}'_i$.

Let $\mathcal{D}_i$ be the category whose objects are $\mathbf{Q}(v)$-vector spaces P provided with two $\mathbf{Q}(v)$-linear maps $\epsilon_i, \phi_i : P \to P$ such that ϵ_i is locally nilpotent and

(a) $\epsilon_i \phi_i = v_i^2 \phi_i \epsilon_i + 1$;

the morphisms in the category are $\mathbf{Q}(v)$-linear maps commuting with ϵ_i, ϕ_i.

For $P \in \mathcal{D}_i$ and $s \in \mathbf{Z}$, let $\phi_i^{(s)} : P \to P$ be defined as $\phi_i^s/[s]_i^!$ if $s \geq 0$ and as 0, if $s < 0$. From (a) we deduce by induction on N:

(b) $\epsilon_i \phi_i^{(N)} = v_i^{2N} \phi_i^{(N)} \epsilon_i + v_i^{N-1} \phi_i^{(N-1)}$ for all N.

For any $t \geq 0$, we consider the operator

$$\Pi_t = \sum_{s \geq 0} (-1)^s v_i^{s(s-1)/2} \phi_i^{(s)} \epsilon_i^{s+t} : P \to P.$$

This is well-defined, since ϵ_i is locally nilpotent. For $N \geq 0$, we define a subspace $P(N)$ of P by $P(0) = \{x \in P | \epsilon_i(x) = 0\}$ and $P(N) = \phi_i^{(N)} P(0)$.

Lemma 16.1.2. (a) *We have* $\epsilon_i \Pi_t = 0$ *for all* $t \geq 0$.

(b) *We have* $\sum_{t \geq 0} v^{-t(t-1)/2} \phi_i^{(t)} \Pi_t = 1$. *The sum is well-defined since, for any* $x \in P$, *we have* $\Pi_t(x) = 0$, *for large* t.

(c) *We have a direct sum decomposition* $P = \oplus_{N \geq 0} P(N)$ *as a vector space. Moreover, for any* $N \geq 0$, *the map* $\phi_i^{(N)}$ *restricts to an isomorphism of vector spaces* $P(0) \cong P(N)$.

(d) $\phi_i : P \to P$ *is injective.*

Using 16.1.1(b), we have

$$\begin{aligned}\epsilon_i \Pi_t &= \sum_{s \geq 0} (-1)^s v_i^{s(s-1)/2} (v_i^{2s} \phi_i^{(s)} \epsilon_i^{s+t+1} + v_i^{s-1} \phi_i^{(s-1)} \epsilon_i^{s+t}) \\ &= \sum_{s \geq 0} (-1)^s \phi_i^{(s)} \epsilon_i^{s+t+1} (v_i^{s(s-1)/2+2s} - v_i^{s(s+1)/2+s}) = 0\end{aligned}$$

G. Lusztig, *Introduction to Quantum Groups*, Modern Birkhäuser Classics,
DOI 10.1007/978-0-8176-4717-9_16, © Springer Science+Business Media, LLC 2010

and (a) is proved. Now (b) follows immediately from 1.3.4.

We prove (c). If $x \in P$, we have by (b): $x = \sum_N \phi_i^{(N)} x_N$ where $x_N = v_i^{-N(N-1)/2}\Pi_N(x)$. By (a), we have $x_N \in P(0)$. It remains to show the uniqueness of the x_N; it is enough to prove the following statement. If $0 = \sum_{N \geq 0} \phi_i^{(N)} x_N$, where $x_N \in P(0)$ are zero for all $N > N_0$ (for some $N_0 \geq 0$), then $x_{N_0} = 0$.

We argue by induction on N_0. For $N_0 = 0$ there is nothing to prove. Assume that $N_0 \geq 1$. Applying ϵ_i and using 16.1.1(b), we obtain $0 = \sum_{N \geq 0} v_i^{(N-1)} \phi_i^{(N-1)} x_N$. The induction hypothesis is applicable to this equation and gives $x_{N_0} = 0$. This proves (c).

(d) follows immediately from (c).

16.1.3. We define linear maps $\tilde{\phi}_i, \tilde{\epsilon}_i : P \to P$ by

$\tilde{\phi}_i(\phi_i^{(N)} y) = \phi_i^{(N+1)} y$ and $\tilde{\epsilon}_i(\phi_i^{(N)} y) = \phi_i^{(N-1)} y$ for all $y \in P(0)$.

Lemma 16.1.4. *Let $M \in \mathcal{C}_i'$ and let $x \in M^t$.*

(a) *We can write uniquely $x = \sum_{s; s \geq 0; s+t \geq 0} F_i^{(s)} x_s$ where $x_s \in \ker(E_i : M^{t+2s} \to M)$ and $x_s = 0$ for large enough s; we can write uniquely $x = \sum_{s; s \geq 0; s-t \geq 0} E_i^{(s)} x'_s$ where $x'_s \in \ker(F_i : M^{t-2s} \to M)$ and $x'_s = 0$ for large enough s.*

(b) *We have $\sum_{s; s \geq 0; s+t \geq 0} F_i^{(s+1)} x_s = \sum_{s; s \geq 0; s-t \geq 0} E_i^{(s-1)} x'_s$. We denote either of these sums by $\tilde{F}_i(x)$.*

(c) *We have $\sum_{s; s \geq 0; s+t \geq 0} F_i^{(s-1)} x_s = \sum_{s; s \geq 0; s-t \geq 0} E_i^{(s+1)} x'_s$. We denote either of these sums by $\tilde{E}_i(x)$.*

This follows from 5.1.5 (we are reduced by 5.1.4 to the case considered there.)

The operators $\tilde{\phi}_i, \tilde{\epsilon}_i$ and $\tilde{F}_i, \tilde{E}_i$ in this and the previous subsection are called *Kashiwara's operators.*

16.1.5. Let $M \in \mathcal{C}_i'$. Consider the $\mathbf{Q}(v)$-linear maps $\tilde{E}_i, \tilde{F}_i : M \to M$ defined in the previous lemma. We have

$$x \in M^n \implies \tilde{E}_i(x) \in M^{n+2}, \tilde{F}_i(x) \in M^{n-2}.$$

16.2. Admissible Forms

16.2.1. We fix $P \in \mathcal{D}_i, M \in \mathcal{C}_i'$. We will study the properties of the operators $\tilde{\phi}_i : P \to P, \tilde{\epsilon}_i : P \to P$ and $\tilde{F}_i : M \to M, \tilde{E}_i : M \to M$ in parallel.

16.2.2. A symmetric bilinear form $(,) : P \times P \to \mathbf{Q}(v)$ is said to be *admissible* if

(a) $(x, \epsilon_i(y)) = (1 - v_i^{-2})(\phi_i x, y)$ for all $x, y \in P$.

A symmetric bilinear form $(,) : M \times M \to \mathbf{Q}(v)$ is said to be *admissible* if

(a′) $(M^n, M^{n'}) = 0$ for $n \neq n'$ and

(b′) $(E_i x, y) = v_i^{n-1}(x, F_i y)$ for all $x \in M^{n-2}, y \in M^n$.

16.2.3. Besides the subrings $\mathcal{A} = \mathbf{Z}[v, v^{-1}]$ and $\mathbf{A} = \mathbf{Q}[[v^{-1}]] \cap \mathbf{Q}(v)$ of $\mathbf{Q}(v)$ we shall need the subrings $\mathbf{A}(\mathbf{Z}) = \mathbf{Z}[[v^{-1}]] \cap \mathbf{Q}(v)$ and $\hat{\mathcal{A}} = \mathbf{Z}((v^{-1})) \cap \mathbf{Q}(v)$ of $\mathbf{Q}((v^{-1}))$.

16.2.4. Let B be a basis of the $\mathbf{Q}(v)$-vector space P (resp. M). We say that B is *integral* if

(a) the $\mathcal{A}$-submodule ${}_{\mathcal{A}}P$ of P generated by B is stable under $\epsilon_i, \phi_i^{(t)} : P \to P$ for all $t \geq 0$ (resp. the $\hat{\mathcal{A}}$-submodule ${}_{\hat{\mathcal{A}}}M$ of M generated by B is stable under $E_i^{(t)}, F_i^{(t)} : M \to M$ for all $t \geq 0$); in the case of M, it is further assumed that $B \cap M^n$ is a basis of M^n for all n.

Assume that we are given an admissible form $(,)$ and an integral basis B for P (resp. M) which is almost orthonormal (see 14.2.1). Let

$$\mathcal{L}(P) = \{x \in {}_{\mathcal{A}}P | (x, x) \in \mathbf{A}\}$$

and

$$\mathcal{L}(M) = \{x \in {}_{\hat{\mathcal{A}}}M | (x, x) \in \mathbf{A}\}.$$

Lemma 16.2.5. (a) *$\mathcal{L}(P)$ is a $\mathbf{Z}[v^{-1}]$-submodule of ${}_{\mathcal{A}}P$ and B is a basis of it.*

(b) *Let $x \in {}_{\mathcal{A}}P$ be such that $(x, x) \in 1 + v^{-1}\mathbf{A}$. Then there exists $b \in B$ such that $x = \pm b \mod v^{-1}\mathcal{L}(P)$.*

(c) *Let $x \in {}_{\mathcal{A}}P$ be such that that $(x, x) \in v^{-1}\mathbf{A}$. Then $x \in v^{-1}\mathcal{L}(P)$.*

(d) *$\mathcal{L}(M)$ is an $\mathbf{A}(\mathbf{Z})$-submodule of ${}_{\hat{\mathcal{A}}}M$ and B is a basis of it.*

(e) *Let $x \in {}_{\hat{\mathcal{A}}}M$ be such that $(x, x) \in 1 + v^{-1}\mathbf{A}$. Then there exists $b \in B$ such that $x = \pm b \mod v^{-1}\mathcal{L}(M)$.*

(f) *Let $x \in {}_{\hat{\mathcal{A}}}M$ be such that that $(x, x) \in v^{-1}\mathbf{A}$. Then $x \in v^{-1}\mathcal{L}(M)$.*

This follows from Lemma 14.2.2.

Lemma 16.2.6. *Let $y \in {}_{\mathcal{A}}P$ (resp. $y \in {}_{\hat{\mathcal{A}}}M \cap M^t$ with $t \geq 0$) be such that $\epsilon_i y = 0$ (resp. $E_i y = 0$); let $n \geq 0$ (resp. $0 \leq n \leq t$). We have $(\phi_i^{(n)} y, \phi_i^{(n)} y) = \pi_n(y, y)$ (resp. $(F_i^{(n)} y, F_i^{(n)} y) = \pi'_n(y, y)$) where $\pi_n \in 1 + v^{-1}\mathbf{Z}[[v^{-1}]]$ (resp. $\pi'_n \in 1 + v^{-1}\mathbf{Z}[[v^{-1}]]$).*

It suffices to show that

$$(\phi_i^{(n+1)} y, \phi_i^{(n+1)} y) = \pi(\phi_i^{(n)} y, \phi_i^{(n)} y)$$

(resp. $(F_i^{(n+1)} y, F_i^{(n+1)} y) = \pi'(F_i^{(n)} y, F_i^{(n)} y)$) where $\pi \in 1 + v^{-1}\mathbf{Z}[[v^{-1}]]$ (resp. $\pi' \in 1 + v^{-1}\mathbf{Z}[[v^{-1}]]$) and $n \geq 0$ (resp. $0 \leq n < t$).

We have

$$\begin{aligned}
&(\phi_i^{(n+1)} y, \phi_i^{(n+1)} y) = ([n+1]_i^{-1} \phi_i \phi_i^{(n)} y, \phi_i^{(n+1)} y) \\
&= (1 - v_i^{-2})^{-1} [n+1]_i^{-1} (\phi_i^{(n)} y, \epsilon_i \phi_i^{(n+1)} y) \\
&= (1 - v_i^{-2})^{-1} [n+1]_i^{-1} v_i^n (\phi_i^{(n)} y, \phi_i^{(n)} y).
\end{aligned}$$

Similarly,

$$\begin{aligned}
&(F_i^{(n+1)} y, F_i^{(n+1)} y) = ([n+1]_i^{-1} F_i F_i^{(n)} y, F_i^{(n+1)} y) \\
&= v_i^{-t+2n+1} [n+1]_i^{-1} (F_i^{(n)} y, E_i F_i^{(n+1)} y) \\
&= v_i^{-t+2n+1} [-n+t]_i [n+1]_i^{-1} (F_i^{(n)} y, F_i^{(n)} y).
\end{aligned}$$

It remains to observe that

$$[n+1]_i^{-1} v_i^n \in 1 + v^{-1}\mathbf{Z}[[v^{-1}]]$$

for $0 \leq n$ and

$$v_i^{-t+2n+1} [-n+t]_i [n+1]_i^{-1} \in 1 + v^{-1}\mathbf{Z}[[v^{-1}]]$$

for $0 \leq n < t$. The lemma follows.

Lemma 16.2.7. (a) *Let $x \in {}_{\mathcal{A}}P$; write $x = \sum_{N \geq 0} y_N$ where $y_N = \phi_i^{(N)} x_N$ and $x_N \in P(0)$ are zero for large N (see 16.1.2(c)). Then $x_N \in {}_{\mathcal{A}}P$ for all N.*

(b) *If $x \in \mathcal{L}(P)$, then each x_N and y_N above is in $\mathcal{L}(P)$. If, in addition, $(x, x) \in 1 + v^{-1}\mathbf{A}$, then there exists $N_0 \geq 0$ and $b \in B$ such that $x_{N_0} = \pm b \mod v^{-1}\mathcal{L}(P)$ and $x_N = 0 \mod v^{-1}\mathcal{L}(P)$, $y_N = 0 \mod v^{-1}\mathcal{L}(P)$ for all $N \neq N_0$.*

(c) *Let* $x \in M^t \cap {}_{\hat{A}}M$. *We write* $x = \sum_{s;s\geq 0;s+t\geq 0} y_s$ *where* $y_s = F_i^{(s)} x_s$ *and* $x_s \in \ker(E_i : M^{t+2s} \to M)$ *are zero for large enough* s; *then* $x_s \in {}_{\hat{A}}M$ *for all* s.

(d) *If* $x \in M^t \cap \mathcal{L}(M)$, *then each* x_s *and* y_s *above is in* $\mathcal{L}(M)$. *If, in addition,* $(x,x) \in 1 + v^{-1}\mathbf{A}$, *then there exists* $s_0 \geq 0$ *and* $b \in B$ *such that* $x_{s_0} = \pm b \mod v^{-1}\mathcal{L}(M)$ *and* $x_s = 0 \mod v^{-1}\mathcal{L}(M)$, $y_s = 0 \mod v^{-1}\mathcal{L}(M)$ *for all* $s \neq s_0$.

We prove (a). We have $x_N = v_i^{-N(N-1)/2}\Pi_N(x)$. Since ${}_{A}P$ is stable under ϵ_i and $\phi_i^{(t)}$ for all t, we see that ${}_{A}P$ is stable under $\Pi_N : P \to P$. Hence $x_N \in {}_{A}P$. This proves (a).

Next we show that the subspaces $\phi^{(N)}P(0), \phi^{(N')}P(0)$ are orthogonal to each other under $(,)$, if $N \neq N'$. We argue by induction on $N + N'$. If $N \geq 1$, we have for $z, z' \in P(0)$ that $(\phi_i^{(N)} z, \phi_i^{(N')} z')$ is equal to a scalar times $(\phi_i^{(N-1)} z, \epsilon_i \phi_i^{(N')} z')$, hence to a scalar times $(\phi_i^{(N-1)} z, \phi_i^{(N'-1)} z)$ so that it is zero by the induction hypothesis. We treat similarly the case where $N' \geq 1$. If $N \leq 0$ and $N' \leq 0$, the result is trivial; our assertion follows.

Now let $x \in {}_{A}P$ be non-zero. We have $(x,x) = \sum_N (y_N, y_N)$. We can find $t \in \mathbf{Z}$ such that $v^{-t} y_N \in \mathcal{L}(P)$ for all N and $v^{-t+1} y_N \notin \mathcal{L}(P)$ for some N. Then there exist integers $a_N \geq 0$, not all equal to 0 such that $v^{-2t}(y_N, y_N) - a_N \in v^{-1}\mathbf{A}$ for all N. Hence

(e) $v^{-2t}(x,x) - \sum_N a_N \in v^{-1}\mathbf{A}$ and $\sum_N a_N > 0$.

If $x \in \mathcal{L}(P)$, then (e) shows that $t \leq 0$; hence $y_N \in \mathcal{L}(P)$ for all N and, using the previous lemma, we see that $x_N \in \mathcal{L}(P)$ for all N. If now $x \in \mathcal{L}(P)$ satisfies $(x,x) \in 1 + v^{-1}\mathbf{A}$, then (e) shows that $t = 0$ and $a_{N_0} = 1$ for some N_0 and $a_N = 0$ for all $N \neq N_0$. In other words, we have $(y_{N_0}, y_{N_0}) \in 1 + v^{-1}\mathbf{A}$ and $(y_N, y_N) \in v^{-1}\mathbf{A}$ for all $N \neq N_0$. Using 16.2.6, we deduce that $(x_{N_0}, x_{N_0}) \in 1 + v^{-1}\mathbf{A}$ and $(x_N, x_N) \in v^{-1}\mathbf{A}$ for all $N \neq N_0$ and the second assertion of (b) follows from 16.2.5.

We prove (c). If $x_s = 0$ for all s, then there is nothing to prove. Hence we may assume that $x_s \neq 0$ for some s and we denote by N the largest index such that $x_N \neq 0$. We have $N \geq 0, N + t \geq 0$. We argue by induction on N. If $N = 0$, there is nothing to prove; hence we may assume that $N > 0$. We have $E_i^{(N)} x = \sum_{s;s\geq 0;s+t\geq 0} E_i^{(N)} F_i^{(s)} x_s = \begin{bmatrix} 2N+t \\ N \end{bmatrix}_i x_N$. Since $E_i^{(N)} {}_{\hat{A}}M \subset {}_{\hat{A}}M$, we have $\begin{bmatrix} 2N+t \\ N \end{bmatrix}_i x_N \in {}_{\hat{A}}M$. We have $\begin{bmatrix} 2N+t \\ N \end{bmatrix}_i^{-1} \in \hat{A}$, hence $x_N \in {}_{\hat{A}}M$. Then $x' = x - F_i^{(N)} x_N \in {}_{\hat{A}}M$. The induction hypothesis is applicable to x' and (c) follows.

Next we show that $(F_i^{(N)}z, F_i^{(N')}z') = 0$ if $N \neq N'$ and z, z' are homogeneous elements in the kernel of E_i. We argue by induction on $N + N'$. If $N \geq 1$, we have that $(F_i^{(N)}z, F_i^{(N')}z')$ is equal to a scalar times $(F_i^{(N-1)}z, E_iF_i^{(N')}z')$, hence to a scalar times $(F_i^{(N-1)}z, F_i^{(N'-1)}z)$ so that it is zero, by the induction hypothesis. We treat similarly the case where $N' \geq 1$. If $N \leq 0$ and $N' \leq 0$, the result is trivial; our assertion follows. The remainder of the proof is entirely similar to that of (b).

Lemma 16.2.8. (a) *$\tilde{\phi}_i, \tilde{\epsilon}_i : P \to P$ map $\mathcal{L}(P)$ into itself.*

(b) *$\tilde{F}_i, \tilde{E}_i : M \to M$ map $\mathcal{L}(M)$ into itself.*

Let $x \in \mathcal{L}(P)$. We must show that $\tilde{\phi}_i x \in \mathcal{L}(P), \tilde{\epsilon}_i x \in \mathcal{L}(P)$. By 16.2.7, we may assume that $x = \phi_i^{(N)} x_N$ for some x_N as in that lemma. But then $\tilde{\phi}_i x = \phi_i^{(N+1)} x_N \in \mathcal{L}(P)$ and $\tilde{\epsilon}_i x = \phi_i^{(N-1)} x_N \in \mathcal{L}(P)$, by 16.2.6. We argue in the same way for M.

16.2.9. For any $N \geq 0$, we denote by $T_N(P)$ the set of all $x \in {}_{\mathcal{A}}P$ such that $x = \phi_i^{(N)} x'$ for some $x' \in P(0) \cap {}_{\mathcal{A}}P$ with $(x', x') = 1 \mod v^{-1}\mathbf{A}$.

For any s, t such that $s \geq 0, s + t \geq 0$, we denote by $T_{s,t}(M)$ the set of all $x \in {}_{\hat{\mathcal{A}}}M$ such that $x = F_i^{(s)} x'$ for some $x' \in \ker(E_i : M^{t+2s} \to M) \cap {}_{\hat{\mathcal{A}}}M$ with $(x', x') = 1 \mod v^{-1}\mathbf{A}$.

From the definitions we see that

(a) $\tilde{\phi}_i(T_N(P)) \subset T_{N+1}(P)$;

(b) $\tilde{\epsilon}_i(T_N(P)) \subset T_{N-1}(P)$ for $N \geq 1$, $\tilde{\epsilon}_i(T_0(P)) = 0$;

(c) if $N \geq 0$, then $\tilde{\phi}_i : T_N(P) \to T_{N+1}(P)$ and $\tilde{\epsilon}_i : T_{N+1}(P) \to T_N(P)$ are inverse bijections;

(d) $\tilde{F}_i(T_{s,t}(M)) \subset T_{s+1,t-2}(M)$ if $s \geq 0, s + t \geq 1$, and $\tilde{F}_i(T_{s,t}(M)) = 0$ if $s \geq 0, s + t = 0$;

(e) $\tilde{E}_i(T_{s,t}(M)) \subset T_{s-1,t+2}(M)$ if $s \geq 1, s + t \geq 0$, and $\tilde{E}_i(T_{s,t}(M)) = 0$ if $s = 0, t \geq 0$;

(f) if $s \geq 0, s + t \geq 1$, then $\tilde{F}_i : T_{s,t}(M) \to T_{s+1,t-2}(M)$ and $\tilde{E}_i : T_{s+1,t-2}(M)) \to T_{s,t}(M)$ are inverse bijections.

Lemma 16.2.10. (a) *Case of P. We have*

$$\pm B + v^{-1}\mathcal{L}(P) = \cup_{N \geq 0} T_N(P) + v^{-1}\mathcal{L}(P).$$

Moreover, the sets $B_N = B \cap (T_N(P) + v^{-1}\mathcal{L}(P))$ $(N \geq 0)$ form a partition of B.

(b) *Case of* M. *We have*

$$\pm B + v^{-1}\mathcal{L}(M) = \cup_{s,t;s\geq 0,s+t\geq 0} T_{s,t}(M) + v^{-1}\mathcal{L}(M).$$

Moreover, the sets $B_{s,t} = B \cap (T_{s,t}(M) + v^{-1}\mathcal{L}(M))$ $(s \geq 0, s+t \geq 0)$ *form a partition of* B.

By 16.2.6 and 16.2.5, we have $T_N(P) \subset \pm B + v^{-1}\mathcal{L}(P)$. Conversely, let $x \in \pm B$. We have $(x,x) \in 1 + v^{-1}\mathbf{A}$. Hence, by 16.2.7, we have $x = y' + y''$ where $y'' \in v^{-1}\mathcal{L}(P)$ and $y' = \phi_i^{(N)} x'$ for some $x' \in P(0) \cap {}_{\mathcal{A}}P$ such that $x' \in \pm B + v^{-1}\mathcal{L}(P)$ and some $N \geq 0$. Thus $x \in T_N(P) + v^{-1}\mathcal{L}(P)$ and the first assertion of (a) follows. To prove the second assertion of (a), it is enough to show that $T_{N_1}(P) \cap (T_{N_2}(P) + v^{-1}\mathcal{L}(P))$ is empty for $N_1 \neq N_2$. Assume that $\phi_i^{(N_1)} x_1 = \phi_i^{(N_2)} x_2 + v^{-1} z$ where $z \in \mathcal{L}(P)$ and $x_1, x_2 \in P(0) \cap ({}_{\mathcal{A}}P)$ satisfy $(x_1, x_1) = 1 \mod v^{-1}\mathbf{A}$ and $(x_2, x_2) = 1 \mod v^{-1}\mathbf{A}$. By 16.2.7, we can write $z = \sum_{N \geq 0} \phi_i^{(N)} z_N$ where $z_N \in \mathcal{L}(P) \cap P(0)$. We have $\phi_i^{(N_1)} x_1 = \phi_i^{(N_2)} x_2 + v^{-1} \sum_{N\geq 0} \phi_i^{(N)} z_N$. This implies, by 16.1.2(c), that $z_N = 0$ for $N \neq N_1, N_2$, $v^{-1} z_{N_1} = x_1$ and $v^{-1} z_{N_2} = -x_2$. From the last equality we deduce that $(x_2, x_2) = v^{-2}(z_{N_2}, z_{N_2}) \in v^{-2}\mathbf{A}$, a contradiction. Thus (a) is proved. The proof of (b) is entirely similar.

16.2.11. Using the previous lemma and the results in 16.2.9, we deduce the following.

In the case of P we have:

(a) $\tilde{\phi}_i(\pm B_N + v^{-1}\mathcal{L}(P)) \subset \pm B_{N+1} + v^{-1}\mathcal{L}(P)$;

(b) $\tilde{\epsilon}_i(\pm B_N + v^{-1}\mathcal{L}(P)) \subset \pm B_{N-1} + v^{-1}\mathcal{L}(P)$ for $N \geq 1$, and $\tilde{\epsilon}_i(\pm B_0 + v^{-1}\mathcal{L}(P)) \subset v^{-1}\mathcal{L}(P)$;

(c) if $N \geq 0$, then $\tilde{\phi}_i : \pm B_N + v^{-1}\mathcal{L}(P) \to \pm B_{N+1} + v^{-1}\mathcal{L}(P)$ and $\tilde{\epsilon}_i : \pm B_{N+1} + v^{-1}\mathcal{L}(P) \to \pm B_N + v^{-1}\mathcal{L}(P)$ are inverse bijections.

In the case of M we have:

(d) $\tilde{F}_i(\pm B_{s,t} + v^{-1}\mathcal{L}(M)) \subset \pm B_{s+1,t-2} + v^{-1}\mathcal{L}(M))$ if $s \geq 0, s+t \geq 1$, and $\tilde{F}_i(\pm B_{s,t} + v^{-1}\mathcal{L}(M)) = 0$ if $s \geq 0, s+t = 0$;

(e) $\tilde{E}_i(\pm B_{s,t} + v^{-1}\mathcal{L}(M)) \subset \pm B_{s-1,t+2} + v^{-1}\mathcal{L}(M))$ if $s \geq 1, s+t \geq 0$, and $\tilde{E}_i(\pm B_{s,t} + v^{-1}\mathcal{L}(M)) = 0$ if $s = 0, t \geq 0$;

(f) if $s \geq 0, s+t \geq 1$, then $\tilde{F}_i : \pm B_{s,t} + v^{-1}\mathcal{L}(M)) \to \pm B_{s+1,t-2} + v^{-1}\mathcal{L}(M))$ and $\tilde{E}_i : \pm B_{s+1,t-2} + v^{-1}\mathcal{L}(M)) \to \pm B_{s,t} + v^{-1}\mathcal{L}(M))$ are inverse bijections.

16.3. Adapted Bases

16.3.1. $P, M, (,)$ are as in the previous section. We say that a basis B of P is *adapted* if there exists a partition $B = \cup_{n\geq 0} B(n)$ and bijections $\pi_n : B(0) \to B(n)$ for all $n \geq 0$ such that

(a) for any $N \geq 0$, $B(N) \cup B(N+1) \cup B(N+2) \cup \dots$ is a basis of $\phi_i^{(N)} P$;

(b) for any $b \in B(0)$ and any $N \geq 0$ we have $\phi_i^{(N)} b - \pi_N(b) \in \phi_i^{(N+1)} P$.

We say that a basis B of M is *adapted* if there exists a partition

$$B = \cup_{s,t;s\geq 0, s+t\geq 0} B(s,t)$$

and bijections

$$\pi_{s,t} : B(0, 2s+t) \to B(s,t)$$

for all s, t as above, such that

(a) $B(s,t) \cup B(s+1,t) \cup B(s+2,t) \cup \dots$ is a basis of $M^t \cap F_i^{(s)} M$;

(b) for any $b \in B(0, 2s+t)$, we have $F_i^{(s)} b - \pi_{s,t}(b) \in F_i^{(s+1)} M$.

In this section it is assumed that B is integral, almost orthonormal (with respect to $(,)$) and adapted.

Lemma 16.3.2. *Let $b \in B$.*

(a) *Case of P. We have $b \in B_0$ if and only if $b \notin \phi_i(P)$.*

(b) *Case of M. We have $b \in \cup_{t\geq 0} B_{0,t}$ if and only if $b \notin F_i M$.*

We prove (a). Assume first that $b \in B_N$ with $N > 0$. Then $b = \phi_i^{(N)} x' + v^{-1} z$ where $z \in \mathcal{L}(P)$ and $x' \in P$. Since B is adapted, we can write $\phi_i^{(N)} x' = \sum c_{b'} b'$ where b' runs over $B \cap \phi_i^{(N)} P$ and $c_{b'} \in \mathbf{Q}(v)$. We can also write $z = \sum_{b''} d_{b''} b''$ where b'' runs over B and $d_{b''} \in \mathbf{Z}[v^{-1}]$. If $b \notin \phi_i^{(N)} P$, then by comparing the coefficients of b, we obtain $1 = v^{-1} d_b$, a contradiction. Thus, we have $b \in \phi_i^{(N)} P$. Since $N > 0$, we have $b \in \phi_i P$. Conversely, assume that $b \in \phi_i P$ and $b \in B_0$. Then $b = x' + v^{-1} z$ where $z \in \mathcal{L}(P), x' \in P(0)$, and $b \in \sum_{N>0} \phi^{(N)} P(0)$; using the equation $(P(0), \phi_i^{(N)} P(0)) = 0$ for $N > 0$, we deduce that $(x', b) = 0$, hence $(b, b) = v^{-1}(z, b) \in v^{-1}\mathbf{A}$, a contradiction. This proves (a). The proof of (b) is entirely similar.

Lemma 16.3.3. (a) *Case of P. Let $b \in B_0, N \geq 0$ and let b' be the unique element of $\pm B$ such that $\tilde{\phi}_i^N(b) = b'$ mod $v^{-1}\mathcal{L}(P)$. Then $b' = \pi_N b$.*

(b) *Case of M. Let $b \in B_{0,s+2t}$ where $s \geq 0$. If $s+t \geq 0$, then there is a unique element $b' \in \pm B$ such that $\tilde{F}_i^s(b) = b'$ mod $v^{-1}\mathcal{L}(M)$ and $b' = \pi_{s,t}b$. If $s+t < 0$, then $\tilde{F}_i^s(b) = 0$ mod $v^{-1}\mathcal{L}(M)$.*

We prove (a). We write $b = x + v^{-1}z$ where $z \in \mathcal{L}(P)$ and $x \in P(0) \cap {}_{\mathcal{A}}P$ satisfies $(x, x) = 1$ mod $v^{-1}\mathbf{A}$. Using 16.2.7, we write $z = \sum_{N'} z_{N'}$ where $z_{N'} \in \mathcal{L}(P) \cap \phi_i^{(N')}P(0)$ for all N'. Replacing x by $x + v^{-1}z_0$ and z by $z - z_0$, we see that we may assume that z satisfies in addition $z \in \phi_i P$. The equalities $\tilde{\phi}_i^N b = \phi_i^{(N)}x + v^{-1}\tilde{\phi}_i^N z$ and $\phi_i^{(N)}b = \phi_i^{(N)}x + v^{-1}\phi_i^{(N)}z$, together with $\tilde{\phi}_i^N z \in \mathcal{L}(P)$ and $\phi_i^{(N)}z \in \phi_i^{(N+1)}P$, imply

$$\tilde{\phi}_i^N b = \phi_i^{(N)}b \mod v^{-1}\mathcal{L}(P) + \phi_i^{(N+1)}P.$$

By assumption we have $\tilde{\phi}_i^N(b) = b'$ mod $v^{-1}\mathcal{L}(P)$. Hence

$$b' = \phi_i^{(N)}b \mod v^{-1}\mathcal{L}(P) + \phi_i^{(N+1)}P.$$

Moreover, we have

$$\phi_i^{(N)}b = b_1 \mod \phi_i^{(N+1)}P$$

where $b_1 = \pi_N b \in B$.

We must prove that $b' = b_1$. We have $b_1 + c_1 = b' + c'$ where $c_1 \in \phi_i^{(N+1)}P$ and $c' \in v^{-1}\mathcal{L}(P)$. We have $b_1 \notin \phi_i^{(N+1)}P$. (Otherwise, we would have $\phi_i^{(N)}b \in \phi_i^{(N+1)}P$; hence $b \in \phi_i P$, contradicting the previous lemma.) Hence, if we express $b_1 + c_1$ as a $\mathbf{Q}(v)$-linear combination of elements of B, the element $b_1 \in B$ will appear with coefficient 1. On the other hand, if we express $b' + c'$ as a $\mathbf{Q}(v)$-linear combination of elements of B, then all coefficients are in $v^{-1}\mathbf{Z}[v^{-1}]$ except that of $\pm b'$.

This forces $b_1 = b'$ or $b_1 = -b'$. If $b_1 = -b'$, then we have $2b_1 + c_1 = c'$ and $\pm b_1$ appears in the left hand side with coefficient 2 and in the right hand side with coefficient in $v^{-1}\mathbf{A}$, a contradiction. Hence we have $b_1 = b'$ and (a) is proved.

The proof of (b) is entirely similar.

16.3.4. The following result shows that the action of the operators $\tilde{\phi}_i, \tilde{\epsilon}_i$ (resp. $\tilde{F}_i, \tilde{E}_i$) on the elements of B is described up to elements in $v^{-1}\mathcal{L}(P)$ (resp. $v^{-1}\mathcal{L}(M)$) in terms of the bijections π_n (resp. $\pi_{s,t}$) in 16.3.1.

Proposition 16.3.5.

(a) *Case of P. Let $b \in B(N)$. Let $b_0 \in B(0)$ be the unique element such that $\pi_N b_0 = b$. We have $\tilde{\phi}_i(b) = \pi_{N+1} b_0 \mod v^{-1}\mathcal{L}(P)$. We have $\tilde{\epsilon}_i(b) = \pi_{N-1} b_0 \mod v^{-1}\mathcal{L}(P)$ if $N \geq 1$ and $\tilde{\epsilon}_i(b) = 0 \mod v^{-1}\mathcal{L}(P)$ if $N = 0$. In particular, we have $B_N = B(N)$ for all N.*

(b) *Case of M. Let $b \in B(s,t)$. Let $b_0 \in B(0, 2s+t)$ be the unique element such that $\pi_{s,t} b_0 = b$. We have $\tilde{F}_i(b) = \pi_{s+1,t-2} b_0 \mod v^{-1}\mathcal{L}(M)$ if $s+t \geq 1$ and $\tilde{F}_i(b) = 0 \mod v^{-1}\mathcal{L}(M)$ if $s+t = 0$. We have $\tilde{E}_i(b) = \pi_{s-1,t+2} b_0 \mod v^{-1}\mathcal{L}(M)$ if $s \geq 1$ and $\tilde{E}_i(b) = 0 \mod v^{-1}\mathcal{L}(M)$ if $s = 0$. In particular, we have $B_{s,t} = B(s,t)$ for all s, t.*

This follows from 16.3.3.

CHAPTER 17

Applications

17.1. First Application to Tensor Products

17.1.1. In this chapter we shall give three applications of Proposition 16.3.5: two to tensor products, and one to $\mathbf{f}$.

17.1.2. Let $\tilde{M}, M \in \mathcal{C}'_i$. Then $\tilde{M} \otimes M$ is an object in $\mathcal{C}'_i$ (see 5.3.1). Now let $P \in \mathcal{D}_i$ and $M \in \mathcal{C}'_i$. We define $\mathbf{Q}(v)$-linear maps $\phi_i, \epsilon_i : P \otimes M \to P \otimes M$ by

$$\phi_i(x \otimes y) = \phi_i(x) \otimes \tilde{K}_i^{-1} y + x \otimes F_i(y)$$
$$\epsilon_i(x \otimes y) = \epsilon_i(x) \otimes \tilde{K}_i^{-1} y + (v_i - v_i^{-1}) x \otimes \tilde{K}_i^{-1} E_i(y)$$

where $\tilde{K}_i : M \to M$ is the linear map given by $\tilde{K}_i y = v_i^n y$ for $y \in M^n$. It is easy to check that $(P \otimes M, \phi_i, \epsilon_i)$ is an object of $\mathcal{D}_i$. (This also follows from 15.1.5.) Hence the linear maps $\tilde{\phi}_i, \tilde{\epsilon}_i : P \otimes M \to P \otimes M$ are well-defined.

From the definitions we deduce (using the quantum binomial formula) that

$$\phi_i^{(t)}(x \otimes y) = \sum v_i^{-t't''} \phi_i^{(t')} x \otimes \tilde{K}_i^{-t'} F_i^{(t'')} y$$

for all $x \in P, y \in M$ and $t \geq 0$; the sum is taken over all $t', t'' \in \mathbf{N}$ such that $t' + t'' = t$.

Lemma 17.1.3. (a) *If* $(,) : P \times P \to \mathbf{Q}(v)$ *and* $(,) : M \times M \to \mathbf{Q}(v)$ *are admissible symmetric bilinear forms in the sense of 16.2.2, then the symmetric bilinear form on* $P \otimes M$ *given by* $(x \otimes y, x' \otimes y') = (x, x')(y, y')$ *is admissible.*

(b) *If* $(,) : \tilde{M} \times \tilde{M} \to \mathbf{Q}(v)$ *and* $(,) : M \times M \to \mathbf{Q}(v)$ *are admissible symmetric bilinear forms in the sense of 16.2.2, then the symmetric bilinear form on* $\tilde{M} \otimes M$ *given by* $(x \otimes y, x' \otimes y') = (x, x')(y, y')$ *is admissible.*

We prove (a). Let $x, x' \in P, y \in M^n, y' \in M^{n'}$. We have

$$(x \otimes y, \epsilon_i(x' \otimes y')) = (x \otimes y, \epsilon_i(x') \otimes \tilde{K}_i^{-1} y' + (v_i - v_i^{-1}) x' \otimes \tilde{K}_i^{-1} E_i(y'))$$
$$= (x, \epsilon_i(x'))(y, \tilde{K}_i^{-1} y') + (v_i - v_i^{-1})(x, x')(y, \tilde{K}_i^{-1} E_i(y'))$$
$$= \delta_{n,n'} v_i^{-n}(1 - v_i^{-2})(\phi_i x, x')(y, y')$$
$$+ \delta_{n-2,n'} v_i^{-n'-2} v_i^{n-1}(v_i - v_i^{-1})(x, x')(F_i(y), y').$$

G. Lusztig, *Introduction to Quantum Groups*, Modern Birkhäuser Classics,
DOI 10.1007/978-0-8176-4717-9_17, © Springer Science+Business Media, LLC 2010

On the other hand, we have

$$\begin{aligned}(\phi_i(x\otimes y),x'\otimes y') &= (\phi_i(x)\otimes \tilde{K}_i^{-1}y + x\otimes F_i(y), x'\otimes y')\\ &= \delta_{n,n'}v_i^{-n}(\phi_i(x),x')(y,y') + \delta_{n-2,n'}(x,x')(F_i(y),y').\end{aligned}$$

This proves (a). The proof of (b) is entirely similar.

17.1.4. We consider the following example. Let P_0 be the $\mathbf{Q}(v)$-vector space with basis $\beta_0,\beta_1,\beta_2,\ldots$. We define $\mathbf{Q}(v)$-linear maps $\phi_i,\epsilon_i : P_0 \to P_0$ by $\phi_i(\beta_s) = [s+1]_i\beta_{s+1}$ for $s\geq 0$ and $\epsilon_i(\beta_s) = v_i^{s-1}\beta_{s-1}$ for $s \geq 0$ (with the convention $\beta_{-1} = 0$). It is easy to check that this makes P_0 into an object of $\mathcal{D}_i$.

We have $\phi_i^{(t)}(\beta_s) = \begin{bmatrix} s+t \\ t\end{bmatrix}_i\beta_{s+t}$ for $s\geq 0, t\geq 0$, and $\epsilon_i^t(\beta_s) = v_i^{-t(t+1)/2+st}\beta_{s-t}$, for $s\geq 0, t\geq 0$, (with the convention $\beta_{-1}=\beta_{-2}=\cdots=0$). Let $(,)$ be the symmetric bilinear form on P_0 given by

$$(\beta_s,\beta_{s'}) = \delta_{s,s'}\prod_{t=1}^{s}(1-v_i^{-2t})^{-1}.$$

It is easy to check that this bilinear form is admissible.

17.1.5. We fix an integer $n\geq 0$. Let M_n be the $\mathbf{Q}(v)$-vector space with basis $b_0, b_1, \ldots, b_n$ with $\mathbf{Z}$-grading such that b_m has degree $n-2m$. It will be convenient to define $b_m = 0$ for $m > n$ and for $m < 0$.

Let $E_i, F_i : M_n \to M_n$ be the linear maps given by $E_i(b_s) = [n-s+1]_i b_{s-1}$ and $F_i(b_s) = [s+1]_i b_{s+1}$ for all s. It is easy to check that in this way M_n is an object of $\mathcal{C}_i'$. Note that for $t\geq 0$, we have $E_i^{(t)}(b_s) = \begin{bmatrix} n-s+t \\ t\end{bmatrix}_i b_{s-t}$ and $F_i^{(t)}(b_s) = \begin{bmatrix} s+t \\ t\end{bmatrix}_i b_{s+t}$ for all s. Let $(,)$ be the bilinear form on M_n given by $(b_s, b_{s'}) = \delta_{s,s'}v_i^{-s(n-s)}\begin{bmatrix} n \\ s\end{bmatrix}_i$ for $0\leq s\leq n$ and $0\leq s'\leq n$. It is easy to check that $(,)$ is an admissible form on M_n.

17.1.6. Let P_0 be as in 17.1.4 and let M_n be as in 17.1.5 ($n\geq 0$). Then, as in 17.1.2, $P = P_0\otimes M_n$ is a well-defined object of $\mathcal{D}_i$. Note that P has a basis $\{b_{s,s'} = \beta_s\otimes b_{s'} | s\geq 0, 0\leq s'\leq n\}$.

For any $t\geq 0$ we have

$$\phi_i^{(t)}b_{s,s'} = \sum v_i^{-t'(n-2s'-t'')}\begin{bmatrix} t'+s \\ t'\end{bmatrix}_i\begin{bmatrix} t''+s' \\ t''\end{bmatrix}_i b_{s+t',s'+t''}$$

where the sum is taken over all $t', t''\in \mathbf{N}$ such that $t'+t''=t$, and

$$\epsilon_i b_{s,s'} = v_i^{-n+2s'+s-1}b_{s-1,s'} + v_i^{-n+2s'-2}(v_i - v_i^{-1})[n-s'+1]_i b_{s,s'-1}.$$

By convention, we set $b_{s,s'} = 0$ if either $s < 0$ or $s' < 0$ or $s' > n$.

For any $m \in \mathbf{Z}$, we define P^m to be the $\mathbf{Q}(v)$-subspace of P spanned by the vectors $b_{s,s'}$ with $s + s' = m$. We have $P = \oplus_m P^m$, $\phi_i^{(t)} P^m \subset P^{m+t}$ and $\epsilon_i P^m \subset P^{m-1}$.

By definition, the operator $\tilde{\phi}_i : P \to P$ (resp. $\tilde{\epsilon}_i : P \to P$) is an (infinite) linear combination of operators $\phi_i^{(t+1)} \epsilon_i^t$ (resp. $\phi_i^{(t)} \epsilon_i^{t+1}$); it follows that

(a) $\tilde{\phi}_i(P^m) \subset P^{m+1}$ and $\tilde{\epsilon}_i(P^m) \subset P^{m-1}$.

17.1.7. Let

$$\zeta_{s,s'} = \sum_{t=0}^{s'} v_i^{-t(n+t-s')} \begin{bmatrix} s+t \\ t \end{bmatrix}_i b_{s+t,s'-t}$$

for $s \geq 0, s' \geq 0, s + s' \leq n$,

$$\zeta_{s,s'} = \sum_{t=0}^{s'} v_i^{-t(s+t)} \begin{bmatrix} n+t-s' \\ t \end{bmatrix}_i b_{s+t,s'-t}$$

for $s \geq 0$, $0 \leq s' \leq n$, $s + s' \geq n$; the two definitions agree if $s + s' = n$. Note that $\zeta_{s,s'} \in P^{s+s'}$.

For $s + s' \geq n$, we have

(a) $$b_{s,s'} = \sum_{t'=0}^{s'} v_i^{-t's-t'} \begin{bmatrix} -1-n+s' \\ t' \end{bmatrix}_i \zeta_{s+t',s'-t'}.$$

Indeed, the right hand side of this equality is, by definition,

$$\sum_{t'=0}^{s'} \sum_{t''=0}^{s'-t'} v_i^{-t''(s+t'+t'')-t's-t'} \begin{bmatrix} n+t''-s'+t' \\ t'' \end{bmatrix}_i \begin{bmatrix} -1-n+s' \\ t' \end{bmatrix}_i b_{s+t'+t'',s'-t'-t''}.$$

The coefficient of $b_{s+t,s'-t}$ (where $0 \leq t \leq s'$) is

$$\sum_{t'+t''=t} v_i^{-t''(s+t'+t'')-t's-t'} \begin{bmatrix} n+t-s' \\ t'' \end{bmatrix}_i \begin{bmatrix} -1-n+s' \\ t' \end{bmatrix}_i.$$

We replace the exponent $-t''(s+t'+t'') - t's - t'$ by $(n+t-s')t' - (-1-n+s')t'' + f$ where $f = t(-n-t-s+s'-1)$ depends on t', t'' only through their sum. Hence the coefficient of $b_{s+t,s'-t}$ is

$$v_i^f \sum_{t'+t''=t} v_i^{(n+t-s')t'-(-1-n+s')t''} \begin{bmatrix} n+t-s' \\ t'' \end{bmatrix}_i \begin{bmatrix} -1-n+s' \\ t' \end{bmatrix}_i$$

$$= v^f \begin{bmatrix} t-1 \\ t \end{bmatrix} = v^f \delta_{t,0} = \delta_{t,0};$$

(a) is proved.

From the definitions and from (a), we see that the subset B of P consisting of the vectors $\zeta_{s,s'}$ (with $s \geq 0$ and $0 \leq s' \leq n$) is a basis of P.

For $m \in \mathbf{Z}$, let $\mathcal{L}^m$ be the $\mathbf{Z}[v^{-1}]$-submodule of P generated by the vectors $b_{s,s'}$ with $s + s' = m$.

Lemma 17.1.8. (a) *For any $s \geq 0$ and $0 \leq s' \leq n$, we have*

$$\zeta_{s,s'} - b_{s,s'} \in v^{-1}\mathcal{L}^{s+s'}.$$

(b) *For $m \geq 0$, $\mathcal{L}^m$ is the $\mathbf{Z}[v^{-1}]$-submodule of P generated by the vectors $\zeta_{s,s'}$ with $s + s' = m$.*

Assume first that $s + s' \leq n$. The coefficient of $b_{s+t,s'-t}$ in $\zeta_{s,s'}$ is in $v_i^{-t(n+t-s')+st}(1 + v^{-1}\mathbf{Z}[v^{-1}])$. Here $t \geq 0$; hence $-t(n + t - s') + st = t(s + s' - n) - t^2 \leq 0$; the inequality becomes an equality only for $t = 0$.

Assume next that $s + s' \geq n$. The coefficient of $b_{s+t,s'-t}$ in $\zeta_{s,s'}$ is in

$$v_i^{-t(s+t)+t(n-s')}(1 + v^{-1}\mathbf{Z}[v^{-1}]).$$

Here $t \geq 0$; hence $-t(s+t)+t(n-s') = t(n-s-s')-t^2 \leq 0$; the inequality becomes an equality only for $t = 0$. This proves (a).

The previous proof also shows that the matrix expressing the vectors $\zeta_{s,s'}$ in terms of the vectors $b_{s,s'}$ (with $s+s' = m$ fixed) is upper triangular, with diagonal entries equal to 1 and with off-diagonal entries in $v^{-1}\mathbf{Z}[v^{-1}]$. This implies (b). The lemma is proved.

Lemma 17.1.9. *The $\mathcal{A}$-submodule ${}_{\mathcal{A}}P$ of P generated by B is stable under $\epsilon_i, \phi_i^{(t)} : P \to P$ for all $t \geq 0$. (In other words, the basis B of P is integral.)*

The formulas in 17.1.6 show that $\epsilon_i(b_{s,s'}) \in {}_{\mathcal{A}}P$ and $\phi_i^{(t)}(b_{s,s'}) \in {}_{\mathcal{A}}P$ for all $t \geq 0$. The lemma follows.

Lemma 17.1.10. *Assume that $0 \leq s \leq n$ and $t \geq 0$. We have*

$$\phi_i^{(t)} b_{s,0} = \zeta_{s,t}$$

if $s + t \leq n$ and

$$\phi_i^{(t)} b_{s,0} = \sum_{u;u\geq 0;u\geq t-n;u\leq s+t-n} \begin{bmatrix} t+s-n \\ u \end{bmatrix}_i \zeta_{s+u,t-u}$$

if $s+t \geq n$.

Assume first that $s+t \leq n$. We have

$$\phi_i^{(t)} b_{s,0} = \sum_{t'=0}^{t} v_i^{-t'(n-t+t')} \begin{bmatrix} t'+s \\ t' \end{bmatrix}_i b_{s+t',t-t'} = \zeta_{s,t}.$$

Assume now that $s+t \geq n$. We have

$$\begin{aligned}
\phi_i^{(t)} b_{s,0} &= \sum_{t';t'\geq 0;t'\leq t;t-t'\leq n} v_i^{-t'(n-t+t')} \begin{bmatrix} t'+s \\ t' \end{bmatrix}_i b_{s+t',t-t'} \\
&= \sum_{t',t''\in\mathbf{N};t'+t''\leq t;t-t'\leq n} v_i^{-t''(s+t')-t''} v_i^{-t'(n-t+t')} \\
&\quad \times \begin{bmatrix} -1-n+t-t' \\ t'' \end{bmatrix}_i \begin{bmatrix} t'+s \\ t' \end{bmatrix}_i \zeta_{s+t'+t'',t-t'-t''} \\
&= \sum_{u=0}^{t} \Big(\sum_{t',t'';t'+t''=u;t'\geq 0;t''\geq 0;t'\geq t-n} v_i^{-t''(s+t')-t''-t'(n-t+t')} \\
&\quad \times \begin{bmatrix} -1-n+t-t' \\ t'' \end{bmatrix}_i \begin{bmatrix} t'+s \\ t' \end{bmatrix}_i \zeta_{s+u,t-u} \Big).
\end{aligned}$$

Since the index t' satisfies $t' \geq t-n$ and $u \geq t'$, the index u must satisfy $u \geq t-n$. We substitute $\left[\begin{smallmatrix} -1-n+t-t' \\ t'' \end{smallmatrix}\right]_i = (-1)^{t''} \left[\begin{smallmatrix} n-t+u \\ t'' \end{smallmatrix}\right]_i$, $\left[\begin{smallmatrix} t'+s \\ t' \end{smallmatrix}\right]_i = (-1)^{t'} \left[\begin{smallmatrix} -s-1 \\ t' \end{smallmatrix}\right]_i$ and $v_i^{-t''(s+t')-t''-t'(n-t+t')} = v_i^{-(n-t+u)t'+(-s-1)t''}$.

The condition on u implies $n-t+u \geq 0$; hence $\left[\begin{smallmatrix} n-t+u \\ t'' \end{smallmatrix}\right]_i$ is automatically zero unless $n-t+u \geq t''$, i.e., if $t' \geq t-n$. Thus the condition $t' \geq t-n$ can be omitted in the summation and we obtain

$$\begin{aligned}
&\sum_{\substack{0\leq u\leq t \\ u\geq t-n}} \sum_{\substack{t',t''\geq 0 \\ t'+t''=u}} (-1)^u v^{-(n-t+u)t'+(-s-1)t''} \\
&\quad \times \begin{bmatrix} n-t+u \\ t'' \end{bmatrix}_i \begin{bmatrix} -s-1 \\ t' \end{bmatrix}_i \zeta_{s+u,t-u} \\
&= \sum_{u;0\leq u\leq t;u\geq t-n} (-1)^u \begin{bmatrix} n-t+u-s-1 \\ u \end{bmatrix}_i \zeta_{s+u,t-u} \\
&= \sum_{u;0\leq u\leq t;u\geq t-n} \begin{bmatrix} -n+t+s \\ u \end{bmatrix}_i \zeta_{s+u,t-u}.
\end{aligned}$$

(We have used 1.3.1(e), 1.3.1(a).) Recall that $s+t \geq n$. It follows that $\left[\begin{smallmatrix} -n+t+s \\ u \end{smallmatrix}\right]_i = 0$ unless $u \leq t+s-n$ and then the condition $u \leq t$ is automatic. Hence our sum becomes $\sum_{u;u\geq 0;u\geq t-n;u\leq s+t-n} \left[\begin{smallmatrix} t+s-n \\ u \end{smallmatrix}\right]_i \zeta_{s+u,t-u}$. The lemma is proved.

Lemma 17.1.11. *We consider the partition of B into the subsets*

$$B(t) = \{\zeta_{s,t} | s + t \leq n\} \cup \{\zeta_{t+2s-n,n-s} | s + t > n\}$$

where $t = 0, 1, 2, \ldots$.

(a) *For any $t \geq 0$, the set $B(t) \cup B(t+1) \cup B(t+2) \cup \cdots$ is a basis of $\phi_i^{(t)}P$.*

(b) *The basis B of P is adapted.*

From 17.1.10, we see that

$$\zeta_{s,t} \in \phi_i^{(t)}P$$

if $s + t \leq n$ and

$$\zeta_{s,t} \in \phi_i^{(2t+s-n)}P$$

if $s + t \geq n$. (The last inclusion is seen by induction on t.) It follows that

(c) $B(t) \cup B(t+1) \cup B(t+2) \cup \cdots \subset \phi_i^{(t)}P$.

Hence $X(t) \subset \phi_i^{(t)}P$ where $X(t)$ is the subspace of P spanned by $B(t) \cup B(t+1) \cup B(t+2) \cup \cdots$. We now prove the inclusion

(d) $\phi_i^{(t)}b \subset X(t)$

for any $b \in B \cap P^m$, by induction on $m \geq 0$.

Note that $B(0) = \{\zeta_{s,0} | 0 \leq s \leq n\} = \{b_{s,0} | 0 \leq s \leq n\}$. If $b \in B(0)$, then (d) follows from 17.1.10. If $m = 0$, then $b = b_{0,0} \in B(0)$, hence (d) holds. Assume now that $m \geq 1$. If $b \in B(0)$, then (d) holds; hence we may assume that $b \in B - B(0)$. Then $b \in X(1)$ and by (c) we have $b = \phi_i y$ where $y \in P^{m-1}$. By the induction hypothesis we have $\phi_i^{(t+1)}y \in X(t+1) \subset X(t)$, hence $\phi_i^{(t)}b \in X(t)$. This proves (d). Thus (a) is proved.

From (a) we see that $\{b \in B | b \notin \phi_i P\} = B(0)$. Let $\pi_t : B_0 \to B_t$ be the bijection given by

(e) $\pi_t \zeta_{s,0} = \zeta_{s,t}$ if $s + t \leq n$ and $\pi_t \zeta_{s,0} = \zeta_{t+2s-n,n-s}$ if $s + t \geq n$.

From 17.1.10, we see that $\phi_i^{(t)}\zeta_{s,0} - \pi_t \zeta_{s,0} \in X(t+1)$, hence

(f) $\phi_i^{(t)}\zeta_{s,0} = \pi_t \zeta_{s,0} \mod \phi_i^{(t+1)}P$.

The lemma is proved.

Lemma 17.1.12. *Consider the admissible form $(,)$ on P defined as in 17.1.3, in terms of the admissible forms 17.1.4, 17.1.5, on M_n and P_0. Then B is almost orthonormal with respect to $(,)$.*

From the definition it is clear that the basis $(b_{s,s'})$ of P is almost orthonormal (actually different elements in this basis are orthogonal to each other). Since B is related to this basis as described in 17.1.8, it follows that B is also almost orthonormal.

We now see that the hypotheses of 16.3.5 are verified in our case. Applying Proposition 16.3.5 to B, and taking into account 17.1.11(e),(f), we obtain the following result.

Proposition 17.1.13. *We have*

$\tilde{\phi}_i(\zeta_{s,s'}) = \zeta_{s,s'+1} \mod v^{-1}\mathcal{L}(P)$ *if* $s+s' < n$,

$\tilde{\phi}_i(\zeta_{s,s'}) = \zeta_{s+1,s'} \mod v^{-1}\mathcal{L}(P)$ *if* $s+s' \geq n$,

$\tilde{\epsilon}_i(\zeta_{s,s'}) = \zeta_{s,s'-1} \mod v^{-1}\mathcal{L}(P)$ *if* $s+s' \leq n$ *and* $s' \geq 1$,

$\tilde{\epsilon}_i(\zeta_{s,s'}) = \zeta_{s-1,s'} \mod v^{-1}\mathcal{L}(P)$ *if* $s+s' > n$,

$\tilde{\epsilon}_i(\zeta_{s,0}) = 0 \mod v^{-1}\mathcal{L}(P)$ *if* $s \leq n$.

Using Lemma 17.1.8, we can restate the proposition as follows.

Corollary 17.1.14. $\tilde{\phi}_i(b_{s,s'}) = b_{s,s'+1} \mod v^{-1}\mathcal{L}(P) \cap P^{s+s'+1}$ *if* $s+s' < n$,

$\tilde{\phi}_i(b_{s,s'}) = b_{s+1,s'} \mod v^{-1}\mathcal{L}(P) \cap P^{s+s'+1}$ *if* $s+s' \geq n$,

$\tilde{\epsilon}_i(b_{s,s'}) = b_{s,s'-1} \mod v^{-1}\mathcal{L}(P) \cap P^{s+s'-1}$ *if* $s+s' \leq n$,

$\tilde{\epsilon}_i(b_{s,s'}) = b_{s-1,s'} \mod v^{-1}\mathcal{L}(P) \cap P^{s+s'-1}$ *if* $s+s' > n$.

What we actually get are the statements of the corollary with $\mathcal{L}(P) \cap P^{s+s'\pm 1}$ replaced by $\mathcal{L}(P)$. But $b_{s,s'} \in P^{s+s'}$; hence from 17.1.6(a), $\tilde{\phi}_i(b_{s,s'}) \in P^{s+s'+1}$ and $\tilde{\epsilon}_i(b_{s,s'}) \in P^{s+s'-1}$. The corollary follows.

Corollary 17.1.15. *Let* $P \in \mathcal{D}_i$, $M \in \mathcal{C}'_i$ *and let* $(P \otimes M, \phi_i, \epsilon_i) \in \mathcal{D}_i$ *be defined as in 17.1.2. Let* $x \in P$ *and* $y \in M^n$ *be such that* $\epsilon_i x = 0, E_i y = 0$. *(Then* $n \geq 0$.*) For any* $m \geq 0$, *let* $\mathcal{L}_m$ *be the* $\mathbf{Z}[v^{-1}]$*-submodule of* $P \otimes M$ *generated by the vectors* $\phi_i^{(s)} x \otimes F_i^{(s')} y$ *with* $s+s' = m$. *We set* $\mathcal{L}_{-1} = 0$. *We have*

$\tilde{\phi}_i(\phi_i^{(s)} x \otimes F_i^{(s')} y) = \phi_i^{(s)} x \otimes F_i^{(s'+1)} y \mod v^{-1}\mathcal{L}_{s+s'+1}$ *if* $s+s' < n$;

$\tilde{\phi}_i(\phi_i^{(s)} x \otimes F_i^{(s')} y) = \phi_i^{(s+1)} x \otimes F_i^{(s')} y \mod v^{-1}\mathcal{L}_{s+s'+1}$ *if* $s+s' \geq n$;

$\tilde{\epsilon}_i(\phi_i^{(s)} x \otimes F_i^{(s')} y) = \phi_i^{(s)} x \otimes F_i^{(s'-1)} y \mod v^{-1}\mathcal{L}_{s+s'-1}$ *if* $s+s' \leq n$;

$\tilde{\epsilon}_i(\phi_i^{(s)} x \otimes F_i^{(s')} y) = \phi_i^{(s-1)} x \otimes F_i^{(s')} y \mod v^{-1}\mathcal{L}_{s+s'-1}$ *if* $s+s' > n$.

We may identify $P_0 \otimes M_n$ with the subspace of $P \otimes M$ spanned by the vectors $\phi_i^{(s)} x \otimes F_i^{(s')} y$ with $s \geq 0$ and $0 \leq s' \leq n$. It is in fact a subobject in $\mathcal{D}_i$. Therefore the result follows from the previous corollary.

17.2. Second Application to Tensor Products

17.2.1. We consider two integers $p \geq 0$ and $n \geq 0$ and form the tensor product $M = M_p \otimes M_n$. This is again an object of $\mathcal{C}'_i$; hence the operators

$\tilde{E}_i, \tilde{F}_i : M_p \otimes M_n \to M_p \otimes M_n$ are well-defined. Now M_n has a basis $\{b_{s'} | 0 \le s' \le n\}$ as in 17.1.5; similarly, M_p has a basis $\{b_s | 0 \le s \le p\}$ as in 17.1.5; Then

$$\{b_{s,s'} = b_s \otimes b_{s'} | 0 \le s \le p, 0 \le s' \le n\}$$

is a basis of M. As in 17.1.7, we define

$$\zeta_{s,s'} = \sum_{t=0}^{s'} v_i^{-t(n+t-s')} \begin{bmatrix} s+t \\ t \end{bmatrix}_i b_{s+t,s'-t}$$

for $0 \le s \le p$, $s' \ge 0$, $s + s' \le n$,

$$\zeta_{s,s'} = \sum_{t=0}^{s'} v_i^{-t(s+t)} \begin{bmatrix} n+t-s' \\ t \end{bmatrix}_i b_{s+t,s'-t}$$

for $0 \le s \le p$, $0 \le s' \le n$, $s + s' \ge n$; the two definitions agree if $s + s' = n$.

The vectors $\zeta_{s,s'}$ just described form a basis B of the vector space M, which is related to the basis $(b_{s,s'})$ by a matrix with entries in $\mathbf{Z}[v^{-1}]$ whose constant terms form the identity matrix. (This is seen as in 17.1.8 or can be deduced from that lemma, using the natural surjective map $P \to M$ which takes $b_{s,s'}$ to $b_{s,s'}$ if $s \le p$ and to zero if $s > p$; that map also takes $\zeta_{s,s'}$ to $\zeta_{s,s'}$ if $s \le p$ and to zero if $s > p$.) Hence the $\mathbf{A}(\mathbf{Z})$-submodule of M, generated by the elements $(b_{s,s'})$, coincides with the $\mathbf{A}(\mathbf{Z})$-submodule generated by the elements $(\zeta_{s,s'})$; we denote it by $\mathcal{L}(M)$.

As in 17.1.9, we see that the $\hat{\mathcal{A}}$-submodule of M generated by B is stable under $E_i^{(n)}, F_i^{(n)}$; hence B is an integral basis. As in 17.1.12, we see that B is almost orthonormal with respect to the form $(,)$ on M defined as in 17.1.3 in terms of the forms $(,)$ on M_p, M_n (see 17.1.5). As in 17.1.11, we see that the basis B of M is adapted. (Again, this could be deduced from the corresponding result for P.)

We now see that the hypotheses of 16.3.5 are verified in our case. Applying Proposition 16.3.5 to B, we obtain the following result, analogous to 17.1.13.

Proposition 17.2.2. *We have*

$\tilde{F}_i(\zeta_{s,s'}) = \zeta_{s,s'+1} \mod v^{-1}\mathcal{L}(M)$ *if* $s + s' < n$;

$\tilde{F}_i(\zeta_{s,s'}) = \zeta_{s+1,s'} \mod v^{-1}\mathcal{L}(M)$ *if* $s < p$ *and* $s + s' \ge n$;

$\tilde{F}_i(\zeta_{p,s'}) = 0 \mod v^{-1}\mathcal{L}(M)$ *if* $s + s' \ge n$;

$\tilde{E}_i(\zeta_{s,s'}) = \zeta_{s,s'-1} \mod v^{-1}\mathcal{L}(M)$ *if* $s + s' \le n$ *and* $s' \ge 1$;

$\tilde{E}_i(\zeta_{s,s'}) = \zeta_{s-1,s'} \mod v^{-1}\mathcal{L}(M)$ *if* $s + s' > n$;

$\tilde{E}_i(\zeta_{s,0}) = 0 \mod v^{-1}\mathcal{L}(M)$ *if* $s \le n$.

As in 17.1.4, we can restate the proposition as follows.

Corollary 17.2.3. *$\tilde{F}_i(b_{s,s'}) = b_{s,s'+1} \mod v^{-1}\mathcal{L}(M)$ if $s+s' < n$;*

$\tilde{F}_i(b_{s,s'}) = b_{s+1,s'} \mod v^{-1}\mathcal{L}(M)$ if $s < p$ and $s+s' \geq n$;

$\tilde{F}_i(b_{p,s'}) = 0 \mod v^{-1}\mathcal{L}(M)$ if $s+s' \geq n$;

$\tilde{E}_i(b_{s,s'}) = b_{s,s'-1} \mod v^{-1}\mathcal{L}(M)$ if $s+s' \leq n$ and $s' \geq 1$;

$\tilde{E}_i(b_{s,s'}) = b_{s-1,s'} \mod v^{-1}\mathcal{L}(M)$ if $s+s' > n$;

$\tilde{E}_i(b_{s,0}) = 0 \mod v^{-1}\mathcal{L}(M)$ if $s \leq n$.

Corollary 17.2.4. *Let $\tilde{M} \in \mathcal{C}'_i$, $M \in \mathcal{C}'_i$ and let $\tilde{M} \otimes M \in \mathcal{C}'_i$ be defined as in 5.3.1. Let $x \in \tilde{M}^p$ and $y \in M^n$ be such that $E_i x = 0, E_i y = 0$. (Then $p \geq 0, n \geq 0$.) For any $m \geq 0$, let $\mathcal{L}$ be the $\mathbf{A}(\mathbf{Z})$-submodule of $\tilde{M} \otimes M$ generated by the vectors $F_i^{(s)}x \otimes F_i^{(s')}y$. We have*

$$\tilde{F}_i(F_i^{(s)}x \otimes F_i^{(s')}y) = F_i^{(s)}x \otimes F_i^{(s'+1)}y \mod v^{-1}\mathcal{L} \text{ if } s+s' < n;$$

$$\tilde{F}_i(F_i^{(s)}x \otimes F_i^{(s')}y) = F_i^{(s+1)}x \otimes F_i^{(s')}y \mod v^{-1}\mathcal{L} \text{ if } s+s' \geq n;$$

$$\tilde{E}_i(F_i^{(s)}x \otimes F_i^{(s')}y) = F_i^{(s)}x \otimes F_i^{(s'-1)}y \mod v^{-1}\mathcal{L} \text{ if } s+s' \leq n;$$

$$\tilde{E}_i(F_i^{(s)}x \otimes F_i^{(s')}y) = F_i^{(s-1)}x \otimes F_i^{(s')}y \mod v^{-1}\mathcal{L} \text{ if } s+s' > n.$$

We may identify $M_p \otimes M_n$ with the subspace of $\tilde{M} \otimes M$ spanned by the vectors $F_i^{(s)}x \otimes F_i^{(s')}y$ with $0 \leq s \leq p$ and $0 \leq s' \leq n$. It is in fact a subobject in $\mathcal{C}'_i$. Therefore the result follows from the previous corollary.

17.3. The Operators $\tilde{\phi}_i, \tilde{\epsilon}_i : \mathbf{f} \to \mathbf{f}$

17.3.1. We shall regard $\mathbf{f}$ as a $\mathfrak{U}$-module as in 15.1.4. Thus, for each $i \in I$, $\phi_i : \mathbf{f} \to \mathbf{f}$ acts as left multiplication by θ_i and $\epsilon_i : \mathbf{f} \to \mathbf{f}$ is the linear map ${}_ir$ in 1.2.13. For any $i \in I$, $\mathbf{f}$ with the operators $\phi_i, \epsilon_i : \mathbf{f} \to \mathbf{f}$ is then an object of $\mathcal{D}_i$. (See 16.1.1.) Hence the operators $\tilde{\phi}_i, \tilde{\epsilon}_i : \mathbf{f} \to \mathbf{f}$ (see 16.1.3) are well-defined.

Note that the form $(,)$ on $\mathbf{f}$ is admissible in the sense of 16.2.2 for any i.

17.3.2. For a fixed $i \in I$, we define a $\mathbf{Q}(v)$-basis B^i of $\mathbf{f}$ as follows. By definition, $B^i = \sqcup_{t \geq 0} B^i(t)$ where $B^i(0)$ is any subset of $\mathcal{B}_{i;0}$ such that $\mathcal{B}_{i;0} = B^i(0) \sqcup (-B^i(0))$ and, for $t > 0$, $B^i(t)$ is the image of $B^i(0)$ under $\pi_{i,t} : \mathcal{B}_{i;0} \cong \mathcal{B}_{i;t}$ (see 14.3.2(c)). By definition, we have $\mathcal{B} = B^i \cup (-B^i)$ and B^i is *adapted* (in the sense of 16.3.1) to $\mathbf{f} \in \mathcal{D}_i$.

By definition of $\mathcal{B}$, we see that B^i is almost orthonormal for $(,)$ and the $\mathcal{A}$-module it generates is ${}_{\mathcal{A}}\mathbf{f}$.

17.3.3. Let $\mathcal{L}(\mathbf{f}) = \{x \in {}_{\mathcal{A}}\mathbf{f} | (x, x) \in \mathbf{A}\}$. From Theorem 14.2.3 and Lemma 14.2.2, it follows that $\mathcal{L}(\mathbf{f})$ is the $\mathbf{Z}[v^{-1}]$-submodule of $\mathbf{f}$ generated by B^i.

Lemma 17.3.4. (a) *${}_{\mathcal{A}}\mathbf{f}$ is stable under the operators $\epsilon_i, \phi_i^{(t)} : \mathbf{f} \to \mathbf{f}$, for any $i \in I$.*

(b) *$\mathcal{L}(\mathbf{f})$ is stable under the operators $\tilde{\phi}_i, \tilde{\epsilon}_i : \mathbf{f} \to \mathbf{f}$, for any $i \in I$.*

The stability under $\phi_i^{(t)}$ is clear from definitions. The stability under ϵ_i follows from 13.2.4. This gives (a). Now (b) follows from Lemma 16.2.8(a) applied to $\mathbf{f}, (,)$ and B^i.

Applying Proposition 16.3.5 to our case, we see that the following holds.

Proposition 17.3.5. *Let $b \in B^i(t)$. Let $b_0 \in B^i(0)$ be the unique element such that $\pi_{i,t} b_0 = b$. We have $\tilde{\phi}_i(b) = \pi_{i,t+1} b_0 \mod v^{-1}\mathcal{L}(\mathbf{f})$. We have $\tilde{\epsilon}_i(b) = \pi_{i,t-1} b_0 \mod v^{-1}\mathcal{L}(\mathbf{f})$ if $t \geq 1$ and $\tilde{\epsilon}_i(b) = 0 \mod v^{-1}\mathcal{L}(\mathbf{f})$ if $t = 0$.*

17.3.6. The following result shows that the endomorphisms of the $\mathbf{Z}$-module $\mathcal{L}(\mathbf{f})/v^{-1}\mathcal{L}(\mathbf{f})$ induced by $\tilde{\phi}_i, \tilde{\epsilon}_i$ act, with respect to the signed basis given by the image of $\mathcal{B}$, in a very simple way, described in terms of the bijections $\pi_{i,n}$ in 14.3.2(c).

Corollary 17.3.7. *Let $i \in I$ and let $b \in \mathcal{B}_{i;t}$. Let $b_0 \in \mathcal{B}_{i;0}$ be the unique element such that $\pi_{i,t} b_0 = b$. We have*

(a) *$\tilde{\phi}_i(b) = \pi_{i,t+1} b_0 \mod v^{-1}\mathcal{L}(\mathbf{f})$;*

(b) *$\tilde{\epsilon}_i(b) = \pi_{i,t-1} b_0 \mod v^{-1}\mathcal{L}(\mathbf{f})$ if $t \geq 1$ and $\tilde{\epsilon}_i(b) = 0 \mod v^{-1}\mathcal{L}(\mathbf{f})$ if $t = 0$.*

(c) *If $i \in I$ and $b \in \mathcal{B}$, then we have $\tilde{\phi}_i(b) = b' \mod v^{-1}\mathcal{L}(\mathbf{f})$ for a unique $b' \in \mathcal{B}$. Moreover, $\tilde{\epsilon}_i b' = b \mod v^{-1}\mathcal{L}(\mathbf{f})$.*

(d) *If $i \in I$ and $b \in \mathcal{B}_{i;n}$ for some $n > 0$, then we have $\tilde{\epsilon}_i(b) = b'' \mod v^{-1}\mathcal{L}(\mathbf{f})$ for a unique $b'' \in \mathcal{B}$. Moreover, $\tilde{\phi}_i b'' = b \mod v^{-1}\mathcal{L}(\mathbf{f})$.*

We apply 17.3.5 to b if $b \in B_t^i$ or to $-b$ if $-b \in B_t^i$. This gives (a) and (b).

Let $b' = \pi_{i,n+1} b_0 \in \mathcal{B}_{i;n+1}$. We have $\tilde{\phi}_i(b) = b' \mod v^{-1}\mathcal{L}(\mathbf{f})$ by (a) and $\tilde{\epsilon}_i(b') = b \mod v^{-1}\mathcal{L}(\mathbf{f})$ by (b). This proves (c).

Assume now that $b \in \mathcal{B}_{i;n}$ with $n > 0$. Let $b'' = \pi_{i,n-1} b_0 \in \mathcal{B}_{i;n-1}$. We have $\tilde{\epsilon}_i(b) = b'' \mod v^{-1}\mathcal{L}(\mathbf{f})$ by (b) and $\tilde{\phi}_i(b'') = b \mod v^{-1}\mathcal{L}(\mathbf{f})$ by (a). This proves (d).

CHAPTER 18

Study of the Operators $\tilde{F}_i, \tilde{E}_i$ on Λ_λ

18.1. Preliminaries

18.1.1. In this chapter we assume that the root datum is Y-regular. Let $\lambda \in X^+$. As in 3.5.6, we set $\Lambda_\lambda = \mathbf{f}/\sum_i \mathbf{f}\theta_i^{\langle i,\lambda\rangle+1}$. Since λ will be fixed in this chapter, we shall write Λ instead of Λ_λ. As in 3.5.7, we denote the image of $1 \in \mathbf{f}$ by $\eta \in \Lambda$.

Recall that there is a unique $\mathbf{U}$-module structure on Λ such that $E_i\eta = 0$ for all $i \in I$, $K_\mu\eta = v^{\langle\mu,\lambda\rangle}\eta$ for all $\mu \in Y$, and F_i acts by the map obtained from left multiplication by θ_i on $\mathbf{f}$. From the triangular decomposition for $\mathbf{U}$, we see that Λ can be naturally identified with the $\mathbf{U}$-module

(a) $$\mathbf{U}/(\sum_i \mathbf{U}E_i + \sum_\mu \mathbf{U}(K_\mu - v^{\langle\mu,\lambda\rangle}) + \sum_i \mathbf{U}F_i^{\langle i,\lambda\rangle+1})$$

by the unique isomorphism which makes η correspond to the image of $1 \in \mathbf{U}$.

For any $\nu \in \mathbf{N}[I]$, we denote by $(\Lambda)_\nu$ the image of $\mathbf{f}_\nu$ under the canonical map $\mathbf{f} \to \Lambda$. We have a direct sum decomposition $\Lambda = \oplus_\nu(\Lambda)_\nu$. Note that $(\Lambda)_\nu$ is contained in the $(\lambda - \nu)$-weight space $\Lambda^{\lambda-\nu}$ (the containment may be strict if the root datum is not X-regular).

18.1.2. By Theorem 14.3.2(b), the subset $\cup_{i,n;n\ge\langle i,\lambda\rangle+1}{}^\sigma\mathcal{B}_{i;n}$ of $\mathcal{B}$ is a signed basis of the $\mathbf{Q}(v)$-subspace $\sum_i \mathbf{f}\theta_i^{\langle i,\lambda\rangle+1}$ of $\mathbf{f}$. Hence the natural projection $\mathbf{f} \to \Lambda$ maps this subset to zero and maps its complement $\mathcal{B}(\lambda) = \cap_{i\in I}(\cup_{n;0\le n\le\langle i,\lambda\rangle}{}^\sigma\mathcal{B}_{i,n})$ bijectively onto a signed basis of the $\mathbf{Q}(v)$-vector space Λ. Thus $\{b^-\eta | b \in \mathcal{B}(\lambda)\}$ is a signed basis of Λ.

18.1.3. We shall regard $\mathbf{f}$ as an object of $\mathcal{D}_i$, for any $i \in I$ as in 17.3.1. Since $\mathbf{f}$ is a $\mathfrak{U}$-module (see 15.1.4), the tensor product $\mathbf{f} \otimes \Lambda$ is a $\mathfrak{U}$-module with

$$\phi_i(x \otimes y) = \phi_i(x) \otimes \tilde{K}_i^{-1}y + x \otimes F_i(y)$$

and

$$\epsilon_i(x \otimes y) = \epsilon_i(x) \otimes \tilde{K}_i^{-1}y + (v_i - v_i^{-1})x \otimes \tilde{K}_i^{-1}E_i(y)$$

for all $x \in \mathbf{f}$ and $y \in \Lambda$. (See 15.1.5.) Hence for each $i \in I$, we have $\mathbf{f} \otimes \Lambda \in \mathcal{D}_i$.

G. Lusztig, *Introduction to Quantum Groups*, Modern Birkhäuser Classics,
DOI 10.1007/978-0-8176-4717-9_18, © Springer Science+Business Media, LLC 2010

Lemma 18.1.4. *There is a unique* $\mathbf{Q}(v)$*-linear map* $\Xi : \mathbf{f} \to \mathbf{f} \otimes \Lambda$ *such that*

(a) $\Xi(1) = 1 \otimes \eta$;

(b) $\Xi(\phi_i x) = \phi_i(\Xi(x))$ *for all* $x \in \mathbf{f}$ *and all* $i \in I$;

(c) $\Xi(\epsilon_i x) = \epsilon_i(\Xi(x))$ *for all* $x \in \mathbf{f}$ *and all* $i \in I$.

By 3.1.4, there is a unique algebra homomorphism $\mathbf{f} \to \mathbf{f} \otimes \mathbf{U}$ such that $\theta_i \mapsto \theta_i \otimes \tilde{K}_{-i} + 1 \otimes F_i$ for all $i \in I$. Composing this with the linear map $\mathbf{f} \otimes \mathbf{U} \to \mathbf{f} \otimes \Lambda$ (identity on the first factor, the map $u \mapsto u\eta$ on the second factor) we obtain a linear map $\Xi : \mathbf{f} \to \mathbf{f} \otimes \Lambda$ which clearly satisfies (a) and (b). We show that it satisfies (c). For $x = 1$, (c) is trivial. Since the algebra $\mathbf{f}$ is generated by the various θ_j, it is enough to show that (c) holds for $x = \theta_j x'$, assuming that it holds for x'. We have

$$\Xi(\epsilon_i x) = \Xi(\epsilon_i \phi_j x') = \Xi(v^{i \cdot j} \phi_j \epsilon_i x' + \delta_{i,j} x') = v^{i \cdot j} \phi_j \epsilon_i \Xi(x') + \delta_{i,j} \Xi(x')$$

and

$$\epsilon_i(\Xi(x)) = \epsilon_i(\Xi(\phi_j x')) = \epsilon_i \phi_j(\Xi(x'));$$

hence (c) holds for x. This proves the existence of Ξ. The uniqueness of Ξ (assuming only (a),(b)) is clear since $\mathbf{f}$ is generated by the θ_j as an algebra.

18.1.5. Let $\mathcal{L}(\mathbf{f})$ be as in 17.3.3. We have $\mathcal{L}(\mathbf{f}) = \oplus_\nu \mathcal{L}(\mathbf{f})_\nu$ (sum over all $\nu \in \mathbf{N}[I]$) where $\mathcal{L}(\mathbf{f})_\nu$ is the $\mathbf{Z}[v^{-1}]$-submodule of $\mathbf{f}$ generated by $\mathcal{B}_\nu$.

Lemma 18.1.6. (a) *If* $b \in \mathcal{B}$ *is not equal to* ± 1, *then there exist* $i \in I$ *and* $b'' \in \mathcal{B}$ *such that* $b - \tilde{\phi}_i b'' \in v^{-1}\mathcal{L}(\mathbf{f})$.

(b) *If* $\nu \in \mathbf{N}[I]$ *is non-zero, then*

$$\mathcal{L}(\mathbf{f})_\nu = \sum_{i;\nu_i > 0} \tilde{\phi}_i(\mathcal{L}(\mathbf{f})_{\nu - i}).$$

We prove (a). According to 14.3.3, if b is as in (a), then there exist $i \in I$ and $n > 0$ such that $b \in \mathcal{B}_{i;n}$. By 17.3.7, we then have $\tilde{\phi}_i b'' - b \in v^{-1}\mathcal{L}(\mathbf{f})$ for some $b'' \in \mathcal{B}$.

We prove (b). The sum $\sum_{i;\nu_i>0} \tilde{\phi}_i(\mathcal{L}(\mathbf{f})_{\nu-i})$ is a $\mathbf{Z}[v^{-1}]$-submodule of $\mathcal{L}(\mathbf{f})_\nu$ (by 17.3.4) and the corresponding quotient module is annihilated by v^{-1} (by (a)). By Nakayama's lemma, this quotient is zero; therefore (b) holds. The lemma is proved.

Proposition 18.1.7. *Let $\nu \in \mathbf{N}[I]$.*

(a) *$\mathcal{L}(\mathbf{f})_\nu$ coincides with the $\mathbf{Z}[v^{-1}]$-submodule of $\mathbf{f}$ generated by the elements $\tilde{\phi}_{i_1}\tilde{\phi}_{i_2}\cdots\tilde{\phi}_{i_t}1$ for various sequences $i_1, i_2, \ldots, i_t$ in I in which i appears exactly ν_i times for each $i \in I$.*

(b) *The subset of $\mathcal{L}(\mathbf{f})_\nu/v^{-1}\mathcal{L}(\mathbf{f})_\nu$ consisting of the images of the elements $\pm\tilde{\phi}_{i_1}\tilde{\phi}_{i_2}\cdots\tilde{\phi}_{i_t}1$ for various $i_1, i_2, \ldots, i_t$, as above, coincides with the image of $\mathcal{B}_\nu$ in $\mathcal{L}(\mathbf{f})_\nu/v^{-1}\mathcal{L}(\mathbf{f})_\nu$.*

This follows immediately from the previous lemma.

18.1.8. We will denote by $L(\Lambda)$ the $\mathbf{A}$-submodule of Λ generated by the signed basis $\{b^-\eta | b \in \mathcal{B}(\lambda)\}$ of Λ. We have a direct sum decomposition $L(\Lambda) = \oplus_\nu L(\Lambda)_\nu$ (ν runs over $\mathbf{N}[I]$) where $L(\Lambda)_\nu$ is the $\mathbf{A}$-submodule of Λ generated by the elements $\{b^-\eta | b \in \mathcal{B}(\lambda) \cap \mathcal{B}_\nu\}$. We have $L(\Lambda)_\nu \subset (\Lambda)_\nu$.

Since Λ is integrable (see 3.5.6), Λ belongs to the category $\mathcal{C}'_i$ for any $i \in I$; hence the operators $\tilde{E}_i, \tilde{F}_i : \Lambda \to \Lambda$ (see 16.1.4) are well-defined. For any $\nu \in \mathbf{N}[I]$, we will denote by $L'(\Lambda)_\nu$ the $\mathbf{A}$-submodule of Λ generated by the elements $\tilde{F}_{i_1}\tilde{F}_{i_2}\cdots\tilde{F}_{i_t}\eta$ for various sequences $i_1, i_2, \ldots, i_t$ in I in which i appears exactly ν_i times for each $i \in I$.

Let $L'(\Lambda) = \sum_\nu L'(\Lambda)_\nu \subset \Lambda$. We have $L'(\Lambda)_\nu \subset (\Lambda)_\nu$.

18.2. A General Hypothesis and Some Consequences

Until the end of 18.3.6, we shall make the following

General hypothesis 18.2.1. *N is a fixed integer ≥ 1 such that, for any $\nu \in \mathbf{N}[I]$ with tr $\nu < N$, we have*

(a) *$L(\Lambda)_\nu = L'(\Lambda)_\nu$;*

(b) *if i is such that $\nu_i > 0$, then $\tilde{F}_i(x^-\eta) = (\tilde{\phi}_i x)^-\eta \mod v^{-1}L(\Lambda)$, for all $x \in \mathcal{L}(\mathbf{f})_{\nu-i}$;*

(c) *if i is such that $\nu_i > 0$, then $\tilde{E}_i(b^-\eta) = (\tilde{\epsilon}_i b)^-\eta \mod v^{-1}L(\Lambda)$, for all $b \in \mathcal{B}_\nu$ such that $b^-\eta \neq 0$; in particular, $\tilde{E}_i(L(\Lambda)_\nu) \subset L(\Lambda)_{\nu-i}$.*

In this section and the next we will derive various consequences of the general hypothesis; we will eventually show that this is not only a hypothesis, but a theorem (see 18.3.8).

Lemma 18.2.2. *Let ν be such that tr $\nu < N$, let $i \in I$ and let $x \in L'(\Lambda)_\nu$. Write $x = \sum_{r=0}^{\nu_i} F_i^{(r)} x_r$ where the $x_r \in (\Lambda)_{\nu-ri}$ satisfy $E_i x_r = 0$ for all r and $x_r = 0$ unless $r + \langle i, \lambda - \nu\rangle \geq 0$ (see 16.1.4). Then*

(a) *$x_r \in L'(\Lambda)_{\nu-ri}$ for all r.*

(b) *If, in addition,* $x = b^-\eta \mod v^{-1}L(\Lambda)$ *for some* $b \in \mathcal{B}_\nu \cap \mathcal{B}(\lambda)$, *then there exist* $r_0 \in [0, \nu_i]$ *and* $b_0 \in \mathcal{B}_{\nu - r_0 i} \cap \mathcal{B}(\lambda)$ *such that* $x_{r_0} = b_0^-\eta$ $\mod v^{-1}L'(\Lambda)_{\nu - r_0 i}$, *and* $x_r \in v^{-1}L'(\Lambda)_{\nu - ri}$ *for all* $r \neq r_0$.

Let t be an integer such that $0 \leq t \leq \nu_i$ and $x_r = 0$ for $r > t$. We prove (a) by induction on t. If $t = 0$, then (a) is obvious. Assume now that $t \geq 1$. We have $\tilde{E}_i x = \sum_{r=0}^{t-1} F_i^{(r)} x_{r+1}$ and $x_{r+1} = 0$ unless $r + \langle i, \lambda - \nu + i' \rangle \geq 0$. (If we had simultaneously $x_{r+1} \neq 0$ and $r + \langle i, \lambda - \nu + i' \rangle < 0$ then $r + 1 + \langle i, \lambda - \nu \rangle < 0$, a contradiction.) By the general hypothesis, we have $\tilde{E}_i x \in L(\Lambda)_{\nu - i}$. By the induction hypothesis applied to $\tilde{E}_i x$, we have $x_r \in L'(\Lambda)_{\nu - ri}$ for all $r > 0$. Hence $F_i^{(r)} x_r = \tilde{F}_i^r x_r \in L'(\Lambda)_\nu$ for all $r > 0$. Since $x \in L'(\Lambda)_\nu$, it follows that $x_0 \in L'(\Lambda)_\nu$. This proves (a).

We prove (b) by induction on t as above. If $t = 0$, then (b) is obvious. Assume now that $t \geq 1$. By the general hypothesis, we have $\tilde{E}_i x = (\tilde{\epsilon}_i b)^- \eta$ $\mod v^{-1}L(\Lambda)$. By 17.3.7, we have that $\tilde{\epsilon}_i b$ is equal modulo $v^{-1}\mathcal{L}(\mathbf{f})$ to either 0 or to b' for some $b' \in \mathcal{B}_{\nu - i}$.

If the first alternative occurs, or if the second alternative occurs with $(b')^- \eta = 0$, then $\tilde{E}_i(b^-\eta) \in v^{-1}L(\Lambda)$; applying (a) to $v\tilde{E}_i(b^-\eta)$ we see that $x_r \in v^{-1}L(\Lambda)$ for all $r > 0$. We then have $x_0 = b^-\eta \mod v^{-1}L(\Lambda)$, as required.

Hence we may assume that

(c) $\tilde{\epsilon}_i b = b' \mod v^{-1}\mathcal{L}(\mathbf{f})$, where $b' \in \mathcal{B}_{\nu - i} \cap \mathcal{B}(\lambda)$.

We have, by assumption,

(d) $\tilde{E}_i x = \tilde{E}_i(b^-\eta) \mod v^{-1}\tilde{E}_i\mathcal{L}(\mathbf{f})_\nu$.

By the general hypothesis, we have $\tilde{E}_i\mathcal{L}(\mathbf{f})_\nu \subset \mathcal{L}(\mathbf{f})_{\nu - i}$ and $\tilde{E}_i(b^-\eta) = (\tilde{\epsilon}_i b)^-\eta \mod v^{-1}L(\Lambda)_{\nu - i}$ (we have $b^-\eta \neq 0$, by assumption). Introducing this in (d), and using (c), we obtain

$$\tilde{E}_i x = b'^-\eta \mod v^{-1}\mathcal{L}(\mathbf{f})_{\nu - i}.$$

By the induction hypothesis applied to $\tilde{E}_i x$, we see that there exist $r_0 \in [1, \nu_i]$ and $b_0 \in \mathcal{B}_{\nu - r_0 i} \cap \mathcal{B}(\lambda)$ such that $x_{r_0} = b_0^-\eta \mod v^{-1}L'(\Lambda)_{\nu - r_0 i}$, and $x_r \in v^{-1}L'(\Lambda)_{\nu - ri}$ for all r such that $r > 0$ and $r \neq r_0$. It follows that

$$\tilde{E}_i x = \tilde{F}_i^{r_0 - 1} x_{r_0} \mod v^{-1}L(\Lambda).$$

By the general hypothesis, we have

$$\tilde{F}_i\tilde{E}_i x = \tilde{F}_i\tilde{E}_i(b^-\eta) = \tilde{F}_i((\tilde{\epsilon}_i b)^-\eta) = (\tilde{\phi}_i\tilde{\epsilon}_i b)^-\eta$$

(equalities modulo $v^{-1}L(\Lambda)$.) Since $\tilde{\epsilon}_i b = b'$ mod $v^{-1}\mathcal{L}(\mathbf{f})$, we see from 17.3.7 that $\tilde{\phi}_i\tilde{\epsilon}_i b = b$ mod $v^{-1}\mathcal{L}(\mathbf{f})$. It follows that $\tilde{F}_i\tilde{E}_i x = b^-\eta = x$ (equalities modulo $v^{-1}L(\Lambda)$). We deduce that $x = \tilde{F}_i(\tilde{F}_i^{r_0-1}x_{r_0}) = \tilde{F}_i^{r_0}x_{r_0}$ (equalities modulo $v^{-1}L(\Lambda)$.) Since $\tilde{F}_i^r x_r \in v^{-1}L(\Lambda)$ for all $r > 0$, $r \neq r_0$ and $x = \sum_r \tilde{F}_i^r x_r$, we deduce that $x_0 \in v^{-1}L(\Lambda)$. This completes the proof.

Lemma 18.2.3. *Let $i \in I$ and let $x \in \mathcal{L}(\mathbf{f})$. Write $x = \sum_{r\geq 0}\phi_i^{(r)}x_r$ where $x_r \in \mathbf{f}$ are 0 for all but finitely many r and $\epsilon_i x_r = 0$ for all r (see 16.1.2(c)). Then $x_r \in \mathcal{L}(\mathbf{f})$ for all r.*

This is a special case of Lemma 16.2.7(b).

18.2.4. If H, H' are two subsets of $\mathbf{f}, \Lambda$ respectively, we denote by $H \odot H'$ the subgroup of $\mathbf{f}\otimes\Lambda$ generated by the vectors $h\otimes h'$ with $h \in H$, $h' \in H'$.

Lemma 18.2.5. *Assume that $tr\,\nu < N$ and let $i \in I$. Then*

$$\tilde{\phi}_i(\mathcal{L}(\mathbf{f}) \odot L'(\Lambda)_\nu) \subset \mathcal{L}(\mathbf{f}) \odot L'(\Lambda)$$

and

$$\tilde{\epsilon}_i(\mathcal{L}(\mathbf{f}) \odot L'(\Lambda)_\nu) \subset \mathcal{L}(\mathbf{f}) \odot L'(\Lambda).$$

By Lemmas 18.2.2(a) and 18.2.3, the $\mathbf{A}$-module $\mathcal{L}(\mathbf{f}) \odot L'(\Lambda)_\nu$ is generated by elements $\phi^{(a)}x\otimes F_i^{(a')}x'$ where $x \in \mathcal{L}(\mathbf{f})$ and $x' \in L'(\Lambda)_{\nu-a'i}$ satisfy $\epsilon_i(x) = 0$ and $E_i(x') = 0$. The image of such elements under $\tilde{\phi}_i$ or $\tilde{\epsilon}_i$ is contained in $\mathcal{L}(\mathbf{f}) \odot L'(\Lambda)$ by Corollary 17.1.15. The lemma follows.

Lemma 18.2.6. *Let $x \in L'(\Lambda)_\nu$ where $tr\,\nu < N$. Assume that there exists $b \in \mathcal{B}_\nu \cap \mathcal{B}(\lambda)$ such that $x = b^-\eta$ mod $v^{-1}L'(\Lambda)$. Assume also that $\tilde{F}_i x \notin v^{-1}L'(\Lambda)$. Then $\tilde{\phi}_i(1\otimes x) = 1\otimes\tilde{F}_i x$ mod $v^{-1}\mathcal{L}(\mathbf{f})\odot L'(\Lambda)$.*

By 18.2.2, we may assume that $x = F_i^{(s)}x'$ where $x' \in L'(\Lambda)_{\nu-si}$ satisfies $E_i x' = 0$. Since $x' \neq 0$ and $E_i x' = 0$, we have $n = \langle i, \lambda - \nu + si'\rangle \in \mathbf{N}$; moreover, $F_i^{(n+1)}x' = 0$. By Corollary 17.1.15, $\tilde{\phi}_i(1\otimes F_i^{(s)}x')$ is equal modulo $v^{-1}\mathcal{L}(\mathbf{f})\odot L'(\Lambda)$ to $1\otimes F_i^{(s+1)}x'$ (if $s < n$) or to $\theta_i\otimes F_i^{(s)}x'$ (if $s \geq n$). If the second alternative occurs, then $\tilde{F}_i x = \tilde{F}_i^{s+1}x' = 0$, contradicting our assumptions. Thus the first alternative occurs and the lemma is proved.

Lemma 18.2.7. *For any ν such that $tr\,\nu \leq N$, we have $\Xi(\mathcal{L}(\mathbf{f})_\nu) \subset \mathcal{L}(\mathbf{f}) \odot L'(\Lambda)$.*

We argue by induction on $\mathrm{tr}\,\nu$. If $\nu = 0$, the result is obvious. Assume that $\nu \neq 0$ and that the result is known when ν is replaced by ν' with $\mathrm{tr}\,\nu' < \mathrm{tr}\,\nu$. By 18.1.6(b), the $\mathbf{Z}[v^{-1}]$-module $\mathcal{L}(\mathbf{f})_\nu$ is spanned by vectors $\tilde{\phi}_i x$ with $i \in I$ and $x \in \mathcal{L}(\mathbf{f})_{\nu-i}$, so it suffices to show that for such i, x, we have $\Xi(\tilde{\phi}_i x) \in \mathcal{L}(\mathbf{f}) \odot L'(\Lambda)$. Since Ξ is a morphism in $\mathcal{D}_i$, we have $\Xi(\tilde{\phi}_i x) = \tilde{\phi}_i(\Xi(x))$. By the induction hypothesis, we have $\Xi(x) \in \mathcal{L}(\mathbf{f}) \odot L'(\Lambda)$.

From the definition of Ξ we see immediately that

$$\Xi(\mathbf{f}_{\nu'}) \subset \sum_{\nu'';\ \mathrm{tr}\,\nu'' \leq\ \mathrm{tr}\,\nu'} \mathbf{f} \otimes \mathbf{U}^-_{\nu''}\eta.$$

Combining this with the previous inclusion, we see that

$$\Xi(x) \in \sum_{\nu'';\ \mathrm{tr}\,\nu'' < N} \mathcal{L}(\mathbf{f}) \odot L'(\Lambda)_{\nu''}.$$

Hence it is enough to show that

$$\tilde{\phi}_i(\mathcal{L}(\mathbf{f}) \odot L'(\Lambda)_{\nu''}) \subset \mathcal{L}(\mathbf{f}) \odot L'(\Lambda)$$

whenever $\mathrm{tr}\,\nu'' < N$.

Now the $\mathbf{A}$-module $\mathcal{L}(\mathbf{f}) \odot L'(\Lambda)_{\nu''}$ is spanned by vectors of the form $\phi^{(a)} x \otimes F_i^{(c)} y$ where $x \in \mathcal{L}(\mathbf{f}), y \in L'(\Lambda)_{\nu''-ci}$ satisfy $\epsilon_i x = 0, E_i y = 0$ (see Lemmas 18.2.2, 18.2.3). Hence it suffices to show that $\tilde{\phi}_i(\phi_i^{(a)} x \otimes F_i^{(c)} y)$ belongs to the $\mathbf{Z}[v^{-1}]$-submodule generated by the vectors $\phi_i^{(a')} x \otimes F_i^{(c')} y$ for various $a', c' \geq 0$. But this follows from Corollary 17.1.15. The lemma is proved.

18.2.8. Consider the linear form $\mathbf{f} \to \mathbf{Q}(v)$ which takes $\mathbf{f}_\nu$ to zero for all $\nu \neq 0$ and takes 1 to 1; tensoring it with the identity map of Λ, we obtain a $\mathbf{Q}(v)$-linear map $pr : \mathbf{f} \otimes \Lambda \to \Lambda$.

From the definitions, we see easily that

(a) $pr(\Xi(x)) = x^- \eta$ for all $x \in \mathbf{f}$.

Lemma 18.2.9. (a) *We have $pr(\mathcal{L}(\mathbf{f}) \odot L'(\Lambda)) \subset L'(\Lambda)$.*

(b) *Let $i \in I$. Let $y \in \mathcal{L}(\mathbf{f}) \odot L'(\Lambda)_\nu$ where $tr\,\nu < N$. We have $pr(\tilde{\phi}_i(y)) = \tilde{F}_i(pr(y)) \mod v^{-1} L'(\Lambda)_{\nu+i}$.*

Let $x \in \mathcal{L}(\mathbf{f})_\nu$ and let $x' \in L'(\Lambda)_{\nu'}$. If $\nu \neq 0$, we have $pr(x \otimes x') = 0$; if $\nu = 0$, we have $x = f1$ where $f \in \mathbf{Z}[v^{-1}]$ and $pr(x \otimes x') = fx'$. Thus (a) holds.

We prove (b). By Lemmas 18.2.2, 18.2.3, we may assume that $y = \phi_i^{(a)} z \otimes F_i^{(a')} z'$ where $z \in \mathcal{L}(\mathbf{f})$ (homogeneous) and $z' \in L'(\Lambda)_{\nu'}$ satisfy $\epsilon_i(z) = 0$, $E_i(z') = 0$ and $a, a' \in \mathbf{N}$. We may assume that $z' \neq 0$. Let n be the smallest integer ≥ 0 such that $F_i^{n+1} z' = 0$. By Corollary 17.1.15, we have that $\tilde{\phi}_i(y)$ is equal to

(c) $\phi_i^{(a+1)} z \otimes F_i^{(a')} z'$ modulo $v^{-1}\mathcal{L}(\mathbf{f}) \odot L'(\Lambda)$, if $a + a' \geq n$, and to

(d) $\phi_i^{(a)} z \otimes F_i^{(a'+1)} z'$ modulo $v^{-1}\mathcal{L}(\mathbf{f}) \odot L'(\Lambda)$, if $a + a' < n$.

If $a > 0$ or $z \notin \mathbf{f}_0$, then y and both vectors (c),(d) are in the kernel of pr, by the definition of pr; on the other hand, by (a), we have $pr(v^{-1}\mathcal{L}(\mathbf{f}) \odot L'(\Lambda)) \subset v^{-1}L'(\Lambda)$. Hence in this case the lemma holds for y. Hence we may assume that $a = 0$ and $z = 1$. We then have $pr(y) = F_i^{(a')} z'$; moreover, by the previous argument:

$$pr(\tilde{\phi}_i(y)) = F_i^{(a'+1)}(z') \mod v^{-1}L'(\Lambda)$$

if $a' < n$ and

$$pr(\tilde{\phi}_i(y)) = 0 \mod v^{-1}L'(\Lambda)$$

if $a' \geq n$.

On the other hand, by the definition of $\tilde{F}_i$, we have

$$\tilde{F}_i(pr(y)) = \tilde{F}_i(F_i^{(a')}(z')) = F_i^{(a'+1)}(z').$$

It remains to observe that $F_i^{(a'+1)}(z') = 0$ if $a' \geq n$ (by the definition of n). The lemma is proved.

Lemma 18.2.10. *Let $x \in \mathcal{L}(\mathbf{f})_\nu$ with $\operatorname{tr} \nu < N$. We have $(\tilde{\phi}_i x)^- \eta = \tilde{F}_i(x^- \eta) \mod v^{-1}L'(\Lambda)$.*

Using 18.2.8(a) and the commutation of Ξ with $\tilde{\phi}_i$, we have

$$(\tilde{\phi}_i x)^- \eta = pr(\Xi(\tilde{\phi}_i x)) = pr(\tilde{\phi}_i(\Xi(x))).$$

Using again 18.2.8(a), we have

$$\tilde{F}_i(x^- \eta) = \tilde{F}_i(pr(\Xi(x))).$$

It remains to show that

$$pr(\tilde{\phi}_i(\Xi(x))) = \tilde{F}_i(pr(\Xi(x))) \mod v^{-1}L'(\Lambda).$$

This follows from Lemma 18.2.9(b) applied to $y = \Xi(x)$. (We have $\Xi(x) \in \mathcal{L}(\mathbf{f}) \odot L'(\Lambda)$ by 18.2.7, and $\Xi(x) \in \sum_{\nu''; \operatorname{tr} \nu'' \leq \operatorname{tr} \nu} \mathbf{f} \otimes \mathbf{U}^-_{\nu''}\eta$; hence

$$\Xi(x) \in \sum_{\nu''; \operatorname{tr} \nu'' < N} \mathcal{L}(\mathbf{f}) \odot L'(\Lambda)_{\nu''}$$

so that Lemma 18.2.9(b) is applicable.) The lemma is proved.

Lemma 18.2.11. *If* $tr\ \nu = N$, *we have* $L(\Lambda)_\nu \subset L'(\Lambda)_\nu$.

By definition, $L(\Lambda)_\nu$ consists of the vectors of form $x^-\eta$ where $x \in \mathcal{L}(\mathbf{f})_\nu$. Since $\nu \neq 0$, the $\mathbf{Z}[v^{-1}]$-module $\mathcal{L}(\mathbf{f})_\nu$ is equal to $\sum_{i;\nu_i>0} \tilde{\phi}_i(\mathcal{L}(\mathbf{f})_{\nu-i})$ (see 18.1.6). Hence it suffices to show that $(\tilde{\phi}_i(x))^-\eta \in L'(\Lambda)_\nu$ for any $i \in I$ such that $\nu_i > 0$ and any $x \in \mathcal{L}(\mathbf{f})_{\nu-i}$. By 18.2.10, we have $(\tilde{\phi}_i x)^-\eta = \tilde{F}_i(x^-\eta)$ mod $v^{-1}L'(\Lambda)_\nu$. Hence it suffices to show that $\tilde{F}_i(x^-\eta) \in L'(\Lambda)_\nu$.

By the definition of $L(\Lambda)_{\nu-i}$, we have $x^-\eta \in L(\Lambda)_{\nu-i}$. Using our general hypothesis, we deduce that $x^-\eta \in L'(\Lambda)_{\nu-i}$. It remains to observe that $\tilde{F}_i(L'(\Lambda)_{\nu-i}) \subset L'(\Lambda)_\nu$ (from the definitions). The lemma is proved.

Lemma 18.2.12. *If* $tr\ \nu = N$, *we have* $L'(\Lambda)_\nu \subset L(\Lambda)_\nu + v^{-1}L'(\Lambda)_\nu$.

We have

$$\begin{aligned} L'(\Lambda)_\nu &= \sum_{i;\nu_i>0} \tilde{F}_i L'(\Lambda)_{\nu-i} = \sum_{i;\nu_i>0} \tilde{F}_i L(\Lambda)_{\nu-i} \\ &= \sum_{i;\nu_i>0} \mathbf{A}\tilde{F}_i(\mathcal{L}(\mathbf{f})^-_{\nu-i}\eta). \end{aligned}$$

The first and third equalities are by definition; the second one follows from our general hypothesis. Hence it suffices to show that

$$\tilde{F}_i(x^-\eta) \in L(\Lambda)_\nu + v^{-1}L'(\Lambda)_\nu$$

for all $x \in \mathcal{L}(\mathbf{f})_{\nu-i}$ (where $\nu_i > 0$).

By 18.2.10, we have $\tilde{F}_i(x^-\eta) = (\tilde{\phi}_i x)^-\eta \mod v^{-1}L'(\Lambda)_\nu$. On the other hand, we have $\tilde{\phi}_i x \in \mathcal{L}(\mathbf{f})_\nu$ (see 17.3.4); hence we have $(\tilde{\phi}_i x)^-\eta \in L(\Lambda)_\nu$. The lemma is proved.

Lemma 18.2.13. *If* $tr\ \nu = N$, *we have* $L(\Lambda)_\nu = L'(\Lambda)_\nu$.

By 18.2.11, $L(\Lambda)_\nu$ is an $\mathbf{A}$-submodule of $L'(\Lambda)_\nu$. The corresponding quotient module is annihilated by v^{-1}, see Lemma 18.2.12. This quotient is then zero by Nakayama's lemma. The lemma is proved.

18.3. Further Consequences of the General Hypothesis

Lemma 18.3.1. *Let* $y = \tilde{F}_{i_1}\tilde{F}_{i_2}\cdots\tilde{F}_{i_t}\eta \in \Lambda$ *where* $i_1 + i_2 + \cdots + i_t = \nu$ *and* $t = tr\ \nu \leq N$. *Let* $x = \tilde{\phi}_{i_1}\tilde{\phi}_{i_2}\cdots\tilde{\phi}_{i_t}(1 \otimes \eta) \in \mathbf{f} \otimes \Lambda$. *Assume that* $y \notin v^{-1}L'(\Lambda)$. *Then* $x = 1 \otimes y \mod v^{-1}\mathcal{L}(\mathbf{f}) \odot L'(\Lambda)$.

We argue by induction on t. If $t = 0$, there is nothing to prove. Assume now that $t > 0$ and that the result is known for $t - 1$.

Let $y' = \tilde{F}_{i_2} \cdots \tilde{F}_{i_t}\eta \in \Lambda$ and let $x' = \tilde{\phi}_{i_2} \cdots \tilde{\phi}_{i_t}(1 \otimes \eta) \in \mathbf{f} \otimes \Lambda$. Since $\tilde{F}_{i_1}(v^{-1}L'(\Lambda)) \subset v^{-1}L'(\Lambda)$, and $y \notin v^{-1}L'(\Lambda)$, we have $y' \notin v^{-1}L'(\Lambda)$. By the induction hypothesis, we have $x' = 1 \otimes y' \mod v^{-1}\mathcal{L}(\mathbf{f}) \odot L'(\Lambda)$. Applying $\tilde{\phi}_{i_1}$ and using 18.2.5, we deduce that $x = \tilde{\phi}_{i_1}x' = \tilde{\phi}_{i_1}(1 \otimes y') \mod v^{-1}\mathcal{L}(\mathbf{f}) \odot L'(\Lambda)$.

By our general hypothesis, we have $y' = (\tilde{\phi}_{i_2} \cdots \tilde{\phi}_{i_t}1)^{-}\eta \mod v^{-1}L(\Lambda)$; hence, by 18.1.7(b), we have $y' = b^{-}\eta \mod v^{-1}L(\Lambda)_{\nu - i_1}$ for some $b \in \mathcal{B}_{\nu - i_1}$. Since $y' \notin v^{-1}L'(\Lambda)_{\nu - i_1}$, we have $b \in \mathcal{B}(\lambda)$. Applying Lemma 18.2.6, we see that $\tilde{\phi}_{i_1}(1 \otimes y') = 1 \otimes \tilde{F}_{i_1}y' = 1 \otimes y \mod v^{-1}\mathcal{L}(\mathbf{f}) \odot L'(\Lambda)$. (That lemma is applicable since $\tilde{F}_{i_1}y' = y \notin v^{-1}L'(\Lambda)$.) Hence $x = 1 \otimes y \mod v^{-1}\mathcal{L}(\mathbf{f}) \odot L'(\Lambda)$. The lemma is proved.

Lemma 18.3.2. *If* $\operatorname{tr}\nu \le N$, *then* $\tilde{E}_i(L(\Lambda)_\nu) \subset L'(\Lambda)$.

We will prove, for any $n \ge 0$, that $\tilde{E}_i(L(\Lambda)_\nu) \subset v^n L'(\Lambda)$, by descending induction on n. This is obvious for n large since $L(\Lambda)_\nu$ is a finitely generated **A**-module. Hence it is enough to prove the following statement.

(a) Assume that $n \ge 1$ and $\tilde{E}_i(L(\Lambda)_\nu) \subset v^n L'(\Lambda)$; then $\tilde{E}_i(L(\Lambda)_\nu) \subset v^{n-1}L'(\Lambda)$.

We first show that

(b) $\tilde{\epsilon}_i(\mathcal{L}(\mathbf{f})_{\nu'} \odot L(\Lambda)_{\nu''}) \subset v^n\mathcal{L}(\mathbf{f}) \odot L'(\Lambda)$

provided that $\nu' + \nu'' = \nu$. In the case where $\operatorname{tr}\nu'' < N$, this follows from 18.2.5. Assume now that $\operatorname{tr}\nu'' = N$; then $\nu' = 0$. It suffices to show that $\tilde{\epsilon}_i(1 \otimes x) \in v^n\mathcal{L}(\mathbf{f}) \odot L'(\Lambda)$ for any $x \in L(\Lambda)_\nu$. We write $x = \sum_{r \ge 0} F_i^{(r)}x_r$ where $x_r = 0$ unless $r + \langle i, \lambda - \nu\rangle \ge 0$ and $E_i x_r = 0$. By the assumption of (a), we have $\sum_{r \ge 1} F_i^{(r-1)}x_r \in v^n L'(\Lambda)$ and by 18.2.2, we deduce that $F_i^{(r-1)}x_r \in v^n L'(\Lambda)$ for $r \ge 1$, or equivalently, $\tilde{F}_i^{r-1}x_r \in v^n L'(\Lambda)$ for $r \ge 1$. Using the general hypothesis $(r-1)$ times, we have $\tilde{E}_i^{r-1}(\tilde{F}_i^{r-1}x_r) \subset v^n L'(\Lambda)$; hence $x_r \in v^n L'(\Lambda)$ for all $r \ge 1$. We have $\tilde{\epsilon}_i(1 \otimes x) = \sum_{r \ge 1} \tilde{\epsilon}_i(1 \otimes F^{(r)}x_r)$ since $\epsilon_i(1 \otimes x_0) = 0$. By 17.1.15, this belongs to the $\mathbf{Z}[v^{-1}]$-submodule generated by the elements $\theta_i^{(r_1)} \otimes F^{(r_2)}x_r$ with $r \ge 1$ and $r_1 + r_2 = r - 1$ and these elements belong to $v^n\mathcal{L}(\mathbf{f}) \odot L'(\Lambda)$. Thus (b) is proved.

To prove (a), it suffices to show that $\tilde{E}_i(y) \in v^{n-1}L'(\Lambda)$ for all y of the form $y = \tilde{F}_{i_1}\tilde{F}_{i_2} \ldots \tilde{F}_{i_t}\eta$ where $i_1 + i_2 + \cdots + i_t = \nu$. If $y \in v^{-1}L'(\Lambda)$, then our inductive assumption shows that $\tilde{E}_i(vy) \in v^n L'(\Lambda)$; hence $\tilde{E}_i(y) \in v^{n-1}L'(\Lambda)$, as desired. Thus we may assume that $y \notin v^{-1}L'(\Lambda)$. Using now Lemma 18.3.1, we see that

(c) $\tilde{\phi}_{i_1}\tilde{\phi}_{i_2} \cdots \tilde{\phi}_{i_t}(1 \otimes \eta) = (1 \otimes y) + v^{-1}z$

where $z \in \mathcal{L}(\mathbf{f}) \odot L'(\Lambda)$. From the definition of the operators $\tilde{\phi}_i$ on $\mathbf{f} \otimes \Lambda$ we see that we have necessarily

$$z \in \sum_{\nu'+\nu''=\nu} \mathcal{L}(\mathbf{f})_{\nu'} \odot L'(\Lambda)_{\nu''}.$$

Hence, by (b), we have

$$\tilde{\epsilon}_i(z) \in v^n \mathcal{L}(\mathbf{f}) \odot L'(\Lambda).$$

Thus applying $\tilde{\epsilon}_i$ to (c), we obtain

(d) $\tilde{\epsilon}_i \tilde{\phi}_{i_1} \tilde{\phi}_{i_2} \cdots \tilde{\phi}_{i_t}(1 \otimes \eta) = \tilde{\epsilon}_i(1 \otimes y) \mod v^{n-1}\mathcal{L}(\mathbf{f}) \odot L'(\Lambda)$.

We have (using 18.2.7)

$$\tilde{\epsilon}_i \tilde{\phi}_{i_1} \tilde{\phi}_{i_2} \cdots \tilde{\phi}_{i_t}(1 \otimes \eta) = \tilde{\epsilon}_i \tilde{\phi}_{i_1} \tilde{\phi}_{i_2} \cdots \tilde{\phi}_{i_t}(\Xi(1)) = \Xi(\tilde{\epsilon}_i \tilde{\phi}_{i_1} \tilde{\phi}_{i_2} \cdots \tilde{\phi}_{i_t} 1)$$
$$\subset \Xi(\mathcal{L}(\mathbf{f})_\nu) \subset \mathcal{L}(\mathbf{f}) \odot L'(\Lambda) \subset v^{n-1}\mathcal{L}(\mathbf{f}) \odot L'(\Lambda).$$

Hence from (d) we deduce that

(e) $\tilde{\epsilon}_i(1 \otimes y) \in v^{n-1}\mathcal{L}(\mathbf{f}) \odot L'(\Lambda)$.

We write $y = \sum_{r \geq 0} F_i^{(r)} y_r$, where the $y_r \in (\Lambda)_{\nu - ri}$ satisfy $E_i y_r = 0$ for all r and $y_r = 0$ unless $r + \langle i, \lambda - \nu \rangle \geq 0$.

By our inductive assumption, we have $\tilde{E}_i y = \sum_{r \geq 1} F_i^{(r-1)} y_r \in v^n L(\Lambda)$. Using 18.2.2, we deduce that $y_r \in v^n L(\Lambda)$ for $r \geq 1$. We have $\tilde{\epsilon}_i(1 \otimes y) = \sum_{r \geq 1} \tilde{\epsilon}_i(1 \otimes F_i^{(r)} y_r)$, since $\epsilon_i(1 \otimes y_0) = 0$. By 17.1.15, we have for any $r \geq 1$, $\tilde{\epsilon}_i(1 \otimes F_i^{(r)} y_r) = 1 \otimes F_i^{(r-1)} y_r$ plus a linear combination with coefficients in $v^{-1}\mathbf{Z}[v^{-1}]$ of terms $\theta_i^{(r_1)} \otimes F_i^{(r_2)} y_r$ where $r_1 + r_2 = r - 1$. The last linear combination is in $v^{n-1}\mathcal{L}(\mathbf{f}) \odot L'(\Lambda)$ since $y_r \in v^n L(\Lambda)$ for $r \geq 1$. Taking sum over $r \geq 1$ we obtain

$$\tilde{\epsilon}_i(1 \otimes y) = 1 \otimes \tilde{E}_i y \mod v^{n-1}\mathcal{L}(\mathbf{f}) \odot L'(\Lambda).$$

Using this and (e), we deduce that $1 \otimes \tilde{E}_i y \in v^{n-1}\mathcal{L}(\mathbf{f}) \odot L'(\Lambda)$. Applying pr, we obtain

$$pr(1 \otimes \tilde{E}_i y) \in v^{n-1} pr(\mathcal{L}(\mathbf{f}) \odot L'(\Lambda));$$

hence $\tilde{E}_i y \in v^{n-1} L'(\Lambda)$. The lemma is proved.

Lemma 18.3.3. *Let ν be such that $\operatorname{tr} \nu \leq N$, let $i \in I$ and let $x \in L'(\Lambda)_\nu$. Write $x = \sum_{r=0}^{\nu_i} F_i^{(r)} x_r$ where the $x_r \in (\Lambda)_{\nu - ri}$ satisfy $E_i x_r = 0$ for all r and $x_r = 0$ unless $r + \langle i, \lambda - \nu \rangle \geq 0$ (see 16.1.4). Then $x_r \in L'(\Lambda)_{\nu - ri}$ for all r.*

When $\operatorname{tr} \nu < N$, this is just Lemma 18.2.2(a). In the case where $\operatorname{tr} \nu = N$, we can use the same proof since the inclusion $\tilde{\epsilon}_i(L(\Lambda)_\nu) \subset L'(\Lambda)$ is now known for $\operatorname{tr} \nu = N$ by the previous lemma.

Lemma 18.3.4. *Let $x \in L'(\Lambda)_\nu$ where tr $\nu \leq N$. We have*

$$\tilde{\epsilon}_i(1 \otimes x) = 1 \otimes \tilde{E}_i x \mod v^{-1}\mathcal{L}(\mathbf{f}) \odot L'(\Lambda).$$

Using the previous lemma we can assume that $x = F_i^{(r)} x'$ where $x' \in L'(\Lambda)_{\nu - ri}$ and $E_i x' = 0$. We can assume that $x' \neq 0$. Then

$$n = \langle i, \lambda - \nu + ri' \rangle \in \mathbf{N};$$

we have $F_i^{(n+1)} x' = 0$. By 17.1.15, $\tilde{\epsilon}_i(1 \otimes F_i^{(s)} x') = 1 \otimes F_i^{(s-1)} x' \mod v^{-1}\mathcal{L}(\mathbf{f}) \odot L'(\Lambda)$. The lemma follows.

Lemma 18.3.5. *Assume that tr $\nu \leq N$. Then*

$$\tilde{\epsilon}_i(\mathcal{L}(\mathbf{f}) \odot L'(\Lambda)_\nu) \subset \mathcal{L}(\mathbf{f}) \odot L'(\Lambda).$$

When tr $\nu < N$ this is shown in 18.2.5. When tr $\nu \leq N$, the same proof applies since Lemma 18.3.3 is now available.

Lemma 18.3.6. *Assume that tr $\nu = N$. Let i be such that $\nu_i > 0$. Then $\tilde{E}_i(b^-\eta) = (\tilde{\epsilon}_i b)^-\eta \mod v^{-1}L'(\Lambda)$, for all $b \in \mathcal{B}_\nu$ such that $b^-\eta \neq 0$.*

We can find $i_1, i_2, \ldots, i_N$ with $i_1 + i_2 + \cdots + i_N = \nu$ such that

$$b = \tilde{\phi}_{i_1}\tilde{\phi}_{i_2} \cdots \tilde{\phi}_{i_N} 1 \mod v^{-1}\mathcal{L}(\mathbf{f})$$

(see 18.1.7(b)). Then

$$y = \tilde{F}_{i_1}\tilde{F}_{i_2} \cdots \tilde{F}_{i_N}\eta \in L(\Lambda)_\nu$$

satisfies $b^-\eta = y \mod v^{-1}L'(\Lambda)_\nu$ (see Lemma 18.2.10) and $y \notin v^{-1}L'(\Lambda)$. By 18.3.1, we have $\tilde{\phi}_{i_1}\tilde{\phi}_{i_2} \cdots \tilde{\phi}_{i_N}(1 \otimes \eta) = 1 \otimes y = 1 \otimes b^-\eta$ up to elements in $v^{-1}\sum_{\mathrm{tr}\, \nu' \leq N} \mathcal{L}(\mathbf{f}) \odot L'(\Lambda)_{\nu'}$. From this we deduce using Lemma 18.3.5, that

$$\tilde{\epsilon}_i\tilde{\phi}_{i_1}\tilde{\phi}_{i_2} \cdots \tilde{\phi}_{i_N}(1 \otimes \eta) = \tilde{\epsilon}_i(1 \otimes b^-\eta) \mod v^{-1}\mathcal{L}(\mathbf{f}) \odot L'(\Lambda).$$

We have

$$\begin{aligned}\tilde{\epsilon}_i\tilde{\phi}_{i_1}\tilde{\phi}_{i_2} \cdots \tilde{\phi}_{i_N}(1 \otimes \eta) &= \tilde{\epsilon}_i\tilde{\phi}_{i_1}\tilde{\phi}_{i_2} \cdots \tilde{\phi}_{i_N}(\Xi(1)) \\ &= \Xi(\tilde{\epsilon}_i\tilde{\phi}_{i_1}\tilde{\phi}_{i_2} \cdots \tilde{\phi}_{i_N} 1) \\ &= \Xi(\tilde{\epsilon}_i(b))\end{aligned}$$

modulo $v^{-1}\mathcal{L}(\mathbf{f}) \odot L'(\Lambda)$ (using Lemma 18.2.7). Using Lemma 18.3.4, we have $\tilde{\epsilon}_i(1 \otimes b^-\eta) = 1 \otimes \tilde{E}_i(b^-\eta) \mod v^{-1}\mathcal{L}(\mathbf{f}) \odot L'(\Lambda)$. We deduce that $\Xi(\tilde{\epsilon}_i(b)) = 1 \otimes \tilde{E}_i(b^-\eta) \mod v^{-1}\mathcal{L}(\mathbf{f}) \odot L'(\Lambda)$. Apply *pr* to this congruence and use $pr(\Xi(x)) = x^-\eta$ for all $x \in \mathbf{f}$. We deduce that $(\tilde{\epsilon}_i(b))^-\eta = \tilde{E}_i(b^-\eta) \mod v^{-1}L'(\Lambda)$. The lemma is proved.

18.3.7. From the lemmas above, we see that, if we assume the general hypothesis 18.2.1 for N, then the properties (a),(b),(c) in 18.2.1 also hold when N is replaced by $N+1$. Since they are obvious for $N=1$, we see that we have proved by induction the following result.

Theorem 18.3.8. *Let $\nu \in \mathbf{N}[I]$. We have*

(a) $L(\Lambda)_\nu = L'(\Lambda)_\nu$;

(b) *for any i we have* $\tilde{F}_i(x^-\eta) = (\tilde{\phi}_i x)^-\eta \mod v^{-1}L(\Lambda)$, *for all* $x \in \mathcal{L}(\mathbf{f})_\nu$;

(c) *if i is such that $\nu_i > 0$, then* $\tilde{E}_i(b^-\eta) = (\tilde{\epsilon}_i b)^-\eta \mod v^{-1}L(\Lambda)$, *for all $b \in \mathcal{B}_\nu$ such that $b^-\eta \neq 0$; in particular,* $\tilde{E}_i(L(\Lambda)_\nu) \subset L(\Lambda)_{\nu-i}$.

From now on, we shall not distinguish between $L(\Lambda)$ and $L'(\Lambda)$.

CHAPTER 19

Inner Product on Λ

19.1. First Properties of the Inner Product

19.1.1. In this chapter, we preserve the setup of the previous chapter. In particular, we write $\Lambda = \Lambda_\lambda$ where $\lambda \in X^+$ is fixed, except in subsections 19.2.3, 19.3.6 and 19.3.7.

Let $\rho_1 : \mathbf{U} \to \mathbf{U}$ be the algebra isomorphism given by

$$\rho_1(E_i) = -v_i F_i, \quad \rho_1(F_i) = -v_i^{-1} E_i, \quad \rho_1(K_\mu) = K_{-\mu}.$$

Let $\rho : \mathbf{U} \to \mathbf{U}^{opp}$ be the algebra isomorphism given by the composition $S\rho_1$ where $S : \mathbf{U} \to \mathbf{U}^{opp}$ is the antipode. We have

$$\rho(E_i) = v_i \tilde{K}_i F_i, \quad \rho(F_i) = v_i \tilde{K}_{-i} E_i, \quad \rho(K_\mu) = K_\mu.$$

It is clear that $\rho^2 = 1$.

Proposition 19.1.2. *There is a unique bilinear form* $(,) : \Lambda \times \Lambda \to \mathbf{Q}(v)$ *such that*

(a) $(\eta, \eta) = 1$;

(b) $(ux, y) = (x, \rho(u)y)$ *for all* $x, y \in \Lambda$ *and* $u \in \mathbf{U}$.

This bilinear form is symmetric. If $x \in (\Lambda)_\nu, y \in (\Lambda)_{\nu'}$ *with* $\nu \neq \nu'$, *then* $(x, y) = 0$.

For any $u \in \mathbf{U}$, we consider the linear map of the dual space $\Lambda^* = \mathrm{Hom}(\Lambda, \mathbf{Q}(v))$ into itself, given by $\xi \mapsto u(\xi)$ where $u(\xi)(x) = \xi(\rho(u)x)$ for all $x \in \Lambda$. This defines a $\mathbf{U}$-module structure on Λ^*, since $\rho : \mathbf{U} \to \mathbf{U}^{opp}$ is an algebra homomorphism. Let $\xi_0 \in \Lambda^*$ be the unique linear form such that $\xi_0(\eta) = 1$ and ξ_0 is zero on $(\Lambda)_\nu$ for $\nu \neq 0$. It is clear that $E_i \xi_0 = 0$ for all $i \in I$ and $K_\mu \xi_0 = v^{\langle \mu, \lambda \rangle} \xi_0$ for all $\mu \in Y$. We show that $F_i^{\langle i, \lambda \rangle + 1} \xi_0 = 0$. It is enough to show that, for any x in a weight space of Λ, the vector $E_i^{\langle i, \lambda \rangle + 1} x$ cannot be a non-zero multiple of η. This follows from Lemma 5.1.6, since $E_i \eta = 0$ and $F_i^{\langle i, \lambda \rangle + 1} \eta = 0$. From the description 18.1.1(a) of

G. Lusztig, *Introduction to Quantum Groups*, Modern Birkhäuser Classics,
DOI 10.1007/978-0-8176-4717-9_19,

Λ, we now see that there is a unique homomorphism of $\mathbf{U}$-modules $\Lambda \to \Lambda^*$ which takes η to ξ_0.

Now it is clear that there is a 1-1 correspondence between homomorphisms of $\mathbf{U}$-modules $\Lambda \to \Lambda^*$ which take η to ξ_0 and bilinear forms $(,)$ on Λ which satisfy (a) and (b). The existence and uniqueness of the form $x, y \mapsto (x, y)$ follow. The form $x, y \mapsto (y, x)$ satisfies the defining properties of the form $x, y \mapsto (x, y)$, since $\rho^2 = 1$. By the uniqueness, we see that these two forms coincide; hence $(,)$ is symmetric.

Proposition 19.1.3. *Let $\nu \in \mathbf{N}[I]$.*

(a) *We have $(L(\Lambda)_\nu, L(\Lambda)_\nu) \in \mathbf{A}$.*

(b) *For any $i \in I$ such that $\nu_i > 0$, and any $x \in L(\Lambda)_{\nu-i}, x' \in L(\Lambda)_\nu$ we have $(\tilde{F}_i x, x') = (x, \tilde{E}_i x')$ modulo $v^{-1}\mathbf{A}$.*

When $\operatorname{tr} \nu = 0$, (a) and (b) are trivial. We may therefore assume that $\operatorname{tr} \nu = N > 0$ and that both (a),(b) are already known for ν' with $\operatorname{tr} \nu' < N$. Since $L(\Lambda)_\nu = \sum_{i;\nu_i>0} \tilde{F}_i L(\Lambda)_{\nu-i}$, and $\tilde{E}_i(L(\Lambda)_\nu) \subset L(\Lambda)_{\nu-i}$ whenever $\nu_i > 0$, we see that (a) for ν follows from the induction hypothesis. (We use (a) and (b) for $\operatorname{tr} \nu' = N - 1$.)

We now prove (b) for ν. Let i, x, x' be as in (b). By 18.2.2, we may assume that $x = F_i^{(s)} y$ and $x' = F_i^{(s')} y'$ where $y \in L(\Lambda)_{\nu-i-si}$, $y' \in L(\Lambda)_{\nu-s'i}$, $E_i y = E_i y' = 0$ and $s \geq 0, s' \geq 0$, $s + \langle i, \lambda - \nu + i' \rangle \geq 0, s' + \langle i, \lambda - \nu \rangle \geq 0$. We must show that

(c) $(F_i^{(s+1)} y, F_i^{(s')} y') = (F_i^{(s)} y, F_i^{(s'-1)} y') \mod v^{-1}\mathbf{A}$.

For any r, r' we have from the definitions

$$\begin{aligned}(F_i^{(r)} y, F_i^{(r')} y') &= (y, v_i^{r^2} \tilde{K}_{-ri} E_i^{(r)} F_i^{(r')} y') \\ &= v_i^{r^2} \begin{bmatrix} r - r' + \langle i, \lambda - \nu + s'i' \rangle \\ r \end{bmatrix}_i (y, K_{-ri} F_i^{(r'-r)} y').\end{aligned}$$

This is zero unless $r' \geq r$. By symmetry, it is also zero unless $r \geq r'$. Thus

$$\begin{aligned}(F_i^{(r)} y, F_i^{(r')} y') &= \delta_{r,r'} v_i^{r^2} \begin{bmatrix} \langle i, \lambda - \nu + s'i' \rangle \\ r \end{bmatrix}_i (y, K_{-ri} y') \\ &= \delta_{r,r'} v_i^{r^2 - r\langle i, \lambda - nu + s'i' \rangle} \begin{bmatrix} \langle i, \lambda - \nu + s'i' \rangle \\ r \end{bmatrix}_i (y, y').\end{aligned}$$

Hence (c) is equivalent to

(d)
$$\delta_{s+1,s'} v_i^{-(s+1)((s+1)+\langle i,\lambda-\nu\rangle)} \begin{bmatrix} 2(s+1)+\langle i,\lambda-\nu\rangle \\ s+1 \end{bmatrix}_i (y,y')$$
$$= \delta_{s,s'-1} v_i^{-s(s+2+\langle i,\lambda-\nu\rangle)} \begin{bmatrix} 2(s+1)+\langle i,\lambda-\nu\rangle \\ s \end{bmatrix}_i (y,y') \mod v^{-1}\mathbf{A}.$$

We may therefore assume that $s+1 = s'$. Now y, y' are contained in $L(\Lambda)_{\nu-i-si}$ and $\operatorname{tr}(\nu - i - si) < N$; hence, by the induction hypothesis, we have $(y,y') \in \mathbf{A}$. We see that to prove (d) it suffices to prove

(e)
$$v_i^{-(s+1)((s+1)+\langle i,\lambda-\nu\rangle)} \begin{bmatrix} 2(s+1)+\langle i,\lambda-\nu\rangle \\ s+1 \end{bmatrix}_i$$
$$= v_i^{-s(s+2+\langle i,\lambda-\nu\rangle)} \begin{bmatrix} 2(s+1)+\langle i,\lambda-\nu\rangle \\ s \end{bmatrix}_i \mod v^{-1}\mathbf{Z}[v^{-1}].$$

From the inequality $s+1+\langle i,\lambda-\nu\rangle \geq 0$, we deduce $2(s+1)+\langle i,\lambda-\nu\rangle \geq s+1$ and $2(s+1)+\langle i,\lambda-\nu\rangle \geq s$. Since $v_i^{-pq}\left[\begin{smallmatrix} p+q \\ p \end{smallmatrix}\right]_i \in 1+v^{-1}\mathbf{Z}[v^{-1}]$ for $p \geq 0, q \geq 0$, it follows that we have

$$v_i^{-(s+1)((s+1)+\langle i,\lambda-\nu\rangle)} \begin{bmatrix} 2(s+1)+\langle i,\lambda-\nu\rangle \\ s+1 \end{bmatrix}_i \in 1+v^{-1}\mathbf{Z}[v^{-1}]$$

and

$$v_i^{-s(s+2+\langle i,\lambda-\nu\rangle)} \begin{bmatrix} 2(s+1)+\langle i,\lambda-\nu\rangle \\ s \end{bmatrix}_i \in 1+v^{-1}\mathbf{Z}[v^{-1}]$$

and (e) follows. The proposition is proved.

Lemma 19.1.4. *Let $b, b' \in \mathcal{B}(\lambda)$.*

(a) *If $b' \neq \pm b$ then $(b^-\eta, b'^-\eta) \in v^{-1}\mathbf{A}$.*

(b) *We have $(b^-\eta, b^-\eta) \in 1+v^{-1}\mathbf{A}$.*

We may assume that $b, b' \in \mathcal{B}_\nu$ for some ν. We argue by induction on $N = \operatorname{tr}\nu$. The case where $N = 0$ is trivial. Assume now that $N \geq 1$. We can find $i \in I$ such that $\nu_i > 0$ and $b_1 \in \mathcal{B}_{\nu-i}$ such that $\tilde{\phi}_i b_1 = b$ and $\tilde{\epsilon}_i b = b_1$ (both equalities are modulo $v^{-1}\mathcal{L}(\mathbf{f})$). By 18.3.8, we have $\tilde{E}_i(b^-\eta) = b_1^-\eta$ and $\tilde{F}_i(b_1^-\eta) = b^-\eta$ (both equalities are modulo $v^{-1}L(\Lambda)$). It follows that $b_1^-\eta \neq 0$. (From $b_1^-\eta = 0$ we could deduce by applying $\tilde{F}_i$ that $b^-\eta \in v^{-1}L(\Lambda)$ which contradicts $b \in \mathcal{B}(\lambda)$.) Thus $b_1 \in \mathcal{B}(\lambda)$. Using again 18.3.8, we have $\tilde{E}_i(b'^-\eta) = (\tilde{\epsilon}_i b')^-\eta \mod v^{-1}L(\Lambda)$.

Using the previous proposition, we have

(c) $(b^-\eta, b'^-\eta) = (\tilde{F}_i(b_1^-\eta), b'^-\eta) = (b_1^-\eta, \tilde{E}_i(b'^-\eta)) = (b_1^-\eta, (\tilde{\epsilon}_i b')^-\eta)$

equalities modulo $v^{-1}\mathbf{A}$. Assume first that $b = b'$. Then $(\tilde{\epsilon}_i b')^-\eta = b_1^-\eta$ mod $v^{-1}L(\Lambda)$ and (c) becomes $(b^-\eta, b^-\eta) = (b_1^-\eta, b_1^-\eta)$ mod $v^{-1}L(\Lambda)$; by the induction hypothesis, we have $(b_1^-\eta, b_1^-\eta) \in 1 + v^{-1}\mathbf{A}$ so that $(b^-\eta, b^-\eta) \in 1 + v^{-1}\mathbf{A}$, as required.

Assume next that $b' \neq \pm b$. There are two cases: we have either $\tilde{\epsilon}_i b' = b_2$ mod $v^{-1}\mathcal{L}(\mathbf{f})$ for some $b_2 \in \mathcal{B}$ or $\tilde{\epsilon}_i b' \in v^{-1}\mathcal{L}(\mathbf{f})$. If the second alternative occurs, then $(b_1^-\eta, (\tilde{\epsilon}_i b')^-\eta) \in v^{-1}\mathbf{A}$ by 19.1.3(a); hence $(b^-\eta, b'^-\eta) \in v^{-1}\mathbf{A}$, by (c). If the first alternative occurs, then $b_2 \in \mathcal{B}(\lambda)$ (by the same argument as the one showing that $b_1 \in \mathcal{B}(\lambda)$) and we have $b_2 \neq \pm b_1$ (if we had $b_2 = \pm b_1$, then by applying $\tilde{\phi}_i$ we would deduce $b' = \pm b$ mod $v^{-1}\mathcal{L}(\mathbf{f})$; hence $b' = \pm b$).

From (c) we have $(b^-\eta, b'^-\eta) = (b_1^-\eta, b_2^-\eta)$ mod $v^{-1}\mathbf{A}$ and from the induction hypothesis we have $(b_1^-\eta, b_2^-\eta) \in v^{-1}\mathbf{A}$. It follows that $(b^-\eta, b'^-\eta) \in v^{-1}\mathbf{A}$. The lemma is proved.

19.2. Normalization of Signs

19.2.1. Let $\mathbf{B}_\nu$ be as in 14.4.2. From 17.3.7 and 18.1.7, we see that the following two conditions for an element $x \in \mathbf{f}_\nu$ are equivalent:

(a) $x \in \mathbf{B}_\nu + v^{-1}\mathcal{L}(\mathbf{f})_\nu$

(b) $x = \tilde{\phi}_{i_1}\tilde{\phi}_{i_2}\cdots\tilde{\phi}_{i_t}1$ mod $v^{-1}\mathcal{L}(\mathbf{f})_\nu$,

for some sequence $i_1, i_2, \dots, i_t$ in I such that $i_1 + i_2 + \cdots + i_t = \nu$.

For the proof of Theorem 14.4.3, we shall need the following result. We regard $\mathbf{f} \otimes \Lambda$ as an $\mathbf{f}$-module with θ_i acting as ϕ_i (see 18.1.3.)

Lemma 19.2.2. (a) *Let $b \in \mathbf{B}_\nu$ and $b' \in \mathbf{B}_{\nu'}$. The vector $\tilde{\phi}_i(b \otimes b'^-\eta)$ is equal modulo $v^{-1}\mathcal{L}(\mathbf{f}) \odot L(\Lambda)$ to $\tilde{\phi}_i(b) \otimes b'^-\eta$ or to $b \otimes \tilde{F}_i(b'^-\eta)$.*

(b) *Let $b_0 \in \mathbf{B}_\nu$. The vector $b_0(1 \otimes \eta)$ is equal modulo $v^{-1}\mathcal{L}(\mathbf{f}) \odot L(\Lambda)$ to $b_1 \otimes b_2^-\eta$ for some $b_1 \in \mathbf{B}_{\nu_1}$, $b_2 \in \mathbf{B}_{\nu_2}$ with $\nu_1 + \nu_2 = \nu$.*

We prove (a). By 18.2.2, (which is now known to be valid unconditionally) there exists $r_0 \geq 0$ such that $\nu'_i \geq r_0$ and $b'^-\eta = F_i^{(r_0)}x'$ mod $v^{-1}L(\Lambda)$ where $x' \in L(\Lambda)_{\nu'-r_0 i}$, $E_i x' = 0$, $x' \neq 0$.

By 16.2.7(b), there exists $r_1 \geq 0$ such that $\nu_i \geq r_1$ and $b = \phi_i^{(r_1)}x$ mod $v^{-1}\mathcal{L}(\mathbf{f})$ where $x \in \mathcal{L}(\mathbf{f})_{\nu - r_1 i}, \epsilon_i x = 0, x \neq 0$.

By 18.2.5 (which is now known to hold unconditionally), we have $\tilde{\phi}_i(b \otimes b'^-\eta) = \tilde{\phi}_i(\phi_i^{(r_1)}x \otimes F_i^{(r_0)}x')$ mod $v^{-1}\mathcal{L}(\mathbf{f}) \odot L(\Lambda)$.

By 17.1.15, $\tilde{\phi}_i(\phi_i^{(r_1)}x \otimes F_i^{(r_0)}x')$ is equal modulo $v^{-1}\mathcal{L}(\mathbf{f}) \odot L(\Lambda)$ to $\tilde{\phi}_i^{r_1+1}x \otimes \tilde{F}_i^{r_0}x'$ or to $\tilde{\phi}_i^{r_1}x \otimes \tilde{F}_i^{r_0+1}x'$ or equivalently to $\tilde{\phi}_i(b) \otimes b'^-\eta$ or $b \otimes \tilde{F}_i(b'^-\eta)$. This proves (a).

We prove (b). From 19.2.1, it follows that

$$b_0 = \tilde{\phi}_{i_1}\tilde{\phi}_{i_2}\cdots\tilde{\phi}_{i_t}1 \mod v^{-1}\mathcal{L}(\mathbf{f}),$$

for some sequence $i_1, i_2, \ldots, i_t$ in I such that $i_1+i_2+\cdots+i_t = \nu$. We have

$$\begin{aligned} b_0(1\otimes\eta) &= b_0\Xi(1) = \Xi(b_0) = \Xi(\tilde{\phi}_{i_1}\tilde{\phi}_{i_2}\cdots\tilde{\phi}_{i_t}1) \\ &= \tilde{\phi}_{i_1}\tilde{\phi}_{i_2}\cdots\tilde{\phi}_{i_t}(\Xi(1)) = \tilde{\phi}_{i_1}\tilde{\phi}_{i_2}\cdots\tilde{\phi}_{i_t}(1\otimes\eta). \end{aligned}$$

The third equality is modulo $v^{-1}\mathcal{L}(\mathbf{f}) \odot L(\Lambda)$.

It remains to show that $\tilde{\phi}_{i_1}\tilde{\phi}_{i_2}\cdots\tilde{\phi}_{i_t}(1 \otimes \eta)$ is equal to $b_1 \otimes b_2^-\eta$ mod $v^{-1}\mathcal{L}(\mathbf{f}) \odot L(\Lambda)$ for some $b_1 \in \mathbf{B}_{\nu_1}$ and $b_2 \in \mathbf{B}_{\nu_2}$ with $\nu_1 + \nu_2 = \nu$. We show that this holds for any sequence $i_1, i_2, \ldots, i_t$, by induction on t. The case where $t = 0$ is trivial. We assume that $t \geq 1$. Using the induction hypothesis and (a), we have that $\tilde{\phi}_{i_1}\tilde{\phi}_{i_2}\cdots\tilde{\phi}_{i_t}(1\otimes\eta)$ is equal to $\tilde{\phi}_{i_1}b\otimes b'^-\eta$ or to $b\otimes\tilde{F}_{i_1}(b'^-\eta)$ modulo $v^{-1}\mathcal{L}(\mathbf{f})\odot L(\Lambda)$, for some $b \in \mathbf{B}_{\nu_1}, b' \in \mathbf{B}_{\nu_2}$ such that $\nu_1+\nu_2 = \nu-i_1$. We have $\tilde{\phi}_{i_1}b = b_1 \mod v^{-1}\mathcal{L}(\mathbf{f})$ for some $b_1 \in \mathbf{B}_{\nu_1+i_1}$ and $\tilde{F}_{i_1}(b'^-\eta) = (\tilde{\phi}_{i_1}b')^-\eta = b_2^-\eta \mod v^{-1}\mathcal{L}(\mathbf{f})$ for some $b_2 \in \mathbf{B}_{\nu_2+i_1}$; (b) follows.

19.2.3. Proof of Theorem 14.4.3. The theorem is obvious when $\nu = 0$. Thus we may assume that $\operatorname{tr}\nu = N > 0$ and that the result is true when N is replaced by $N' \in [0, N-1]$.

We first prove part (b) of the theorem. Recall that $\sigma(\mathcal{B}_\nu) = \mathcal{B}_\nu$. (See 14.2.5(c).) Assume that $b, b' \in \mathbf{B}_\nu$ satisfy $\sigma(b) = -b'$. We will show that this leads to a contradiction.

We can find $i \in I, n > 0$ such that $\nu_i \geq n$, and $b'' \in \mathbf{B}_{\nu-ni}\cap\mathcal{B}_{i;0}$ such that $b = \pi_{i,n}b''$. By 17.3.7, we have $\tilde{\phi}_i^n b'' = b \mod v^{-1}\mathcal{L}(\mathbf{f})$. Since $b'' \in \mathcal{B}_{i;0}$, we can find $\beta \in \mathcal{L}(\mathbf{f})$ such that $\epsilon_i(\beta) = 0$ and $b'' = \beta \mod v^{-1}\mathcal{L}(\mathbf{f})$. This is a special case of the equality $B_N = B(N)$ in 16.3.5(a). By definition, we have $\tilde{\phi}_i^n\beta = \phi_i^{(n)}\beta$. Since $\tilde{\phi}_i^n$ preserves $v^{-1}\mathcal{L}(\mathbf{f})$, we have $\tilde{\phi}_i^n\beta = \tilde{\phi}_i^n b''$ mod $v^{-1}\mathcal{L}(\mathbf{f})$; hence $b = \phi_i^{(n)}\beta \mod v^{-1}\mathcal{L}(\mathbf{f})$.

For any $j \in I$, we define $c_j \in \mathbf{N}$ by $b'' \in \mathcal{B}_{j;c_j}$. Thus, we have $c_i = 0$. Since the root datum is assumed to be Y-regular, we can find $\lambda \in X$ such that $\langle i, \lambda\rangle = 0$ and $\langle j, \lambda\rangle \geq c_j$ for all $j \in I - \{i\}$. These inequalities show, using the definition of the c_j, that $\sigma(b'')\eta \neq 0$, where $\eta \in \Lambda = \Lambda_\lambda$ is as in 3.5.7.

In the **f**-module $\mathbf{f} \otimes \Lambda$, we have $\theta_i^{(n)}(1 \otimes \eta) = \theta_i^{(n)} \otimes \eta$ since $F_i\eta = 0$ and $\tilde{K}_{-i}\eta = 0$ (recalling that $\langle i, \lambda \rangle = 0$). By definition of the **f**-module structure on $\mathbf{f} \otimes \Lambda$, we have

$$\sigma(\beta)\theta_i^{(n)}(1 \otimes \eta) = \sigma(\beta)(\theta_i^{(n)} \otimes \eta) = \theta_i^{(n)} \otimes \sigma(\beta)(\eta) + z$$

where z is in the kernel of the obvious projection $pr_n : \mathbf{f}\otimes\Lambda \to \mathbf{f}_{ni}\otimes\Lambda$. Since $b = \theta_i^{(n)}\beta \mod v^{-1}\mathcal{L}(\mathbf{f})$, we have $-b' = \sigma(b) = \sigma(\beta)\theta_i^{(n)} \mod v^{-1}\mathcal{L}(\mathbf{f})$; hence

(a) $$-b'(1 \otimes \eta) = \theta_i^{(n)} \otimes \sigma(\beta)(\eta) + z \mod v^{-1}\mathcal{L}(\mathbf{f}) \odot L(\Lambda).$$

We have used that $x \in \mathcal{L}(\mathbf{f}) \implies x(1 \otimes \eta) \in \mathcal{L}(\mathbf{f}) \odot L(\Lambda)$; since $x(1 \otimes \eta) = \Xi(x)$, this follows from Lemma 18.2.7, which is now known unconditionally.

By 19.2.2(b), we have $b'(1 \otimes \eta) = b_1 \otimes b_2^- \eta \mod v^{-1}\mathcal{L}(\mathbf{f}) \odot L(\Lambda)$ where $b_1 \in \mathbf{B}_{\nu_1}, B_2 \in \mathbf{B}_{\nu_2}$ and $\nu_1 + \nu_2 = \nu$. Comparing with (a), we deduce that

$$\theta_i^{(n)} \otimes \sigma(\beta)^-(\eta) + b_1 \otimes b_2^- \eta + z \in v^{-1}\mathcal{L}(\mathbf{f}) \odot L(\Lambda).$$

Since $\beta = b'' \mod v^{-1}\mathcal{L}(\mathbf{f})$, we have $\sigma(\beta) = \sigma(b'') \mod v^{-1}\mathcal{L}(\mathbf{f})$. By the induction hypothesis, we have $\sigma(b'') \in \mathbf{B}_{\nu-ni}$. Recall also that $\sigma(b'')\eta \neq 0$. Thus we have

$$\theta_i^{(n)} \otimes \sigma(b'')\eta + b_1 \otimes b_2^- \eta + z \in v^{-1}\mathcal{L}(\mathbf{f}) \odot L(\Lambda).$$

By the definition of z, this implies that

$$\sigma(b'')^- \eta + b_2^- \eta \in v^{-1}L(\Lambda)$$

if $b_2 \in \mathbf{B}_{\nu-ni}$ and $b_2^- \eta \neq 0$ and

$$\sigma(b'')^- \eta \in v^{-1}L(\Lambda)$$

if $b_2 \notin \mathbf{B}_{\nu-ni}$ or $b_2^- \eta = 0$.

Both alternatives are impossible, since, by the induction hypothesis, $\sigma(b'')^- \eta, b_2^- \eta$ (in the first case) and $\sigma(b'')^- \eta$ (in the second case) are a part of an **A**-basis of $L(\Lambda)$; by the induction hypothesis, we cannot have $\sigma(b'') + b_2 = 0$. This proves part (b) of the theorem.

We now prove part (a) of the theorem. Assume that $b, b' \in \mathbf{B}_\nu$ satisfy $b' = -b$. Since $\sigma(b) \in \mathcal{B}_\nu = \mathbf{B}_\nu \cup (-\mathbf{B}_\nu)$, we have either $\sigma(b) = b_1$ with $b_1 \in \mathbf{B}_\nu$ or $\sigma(b) = -b_2$ with $b_2 \in \mathbf{B}_\nu$. The second alternative cannot occur, by part (b). Thus the first alternative holds. But then $b_1 = -\sigma(b')$ and this again contradicts part (b). This proves part (a).

We prove part (c). Let $b \in \mathbf{B}_\nu$. We have $\sigma(b) \in \mathbf{B}_\nu \cup (-\mathbf{B}_\nu)$ and $\sigma(b) \notin (-\mathbf{B}_\nu)$ by (b), hence $\sigma(b) \in \mathbf{B}_\nu$. This proves part (c). Clearly, parts (d) and (e) follow from part (a) since $\mathcal{B}_\nu$ is a signed basis of $\mathbf{f}_\nu$. The theorem is proved.

19.3. Further Properties of the Inner Product

19.3.1. We shall denote by ${}_{\mathcal{A}}\Lambda_\lambda$ the image of the canonical map ${}_{\mathcal{A}}\mathbf{f} \to \Lambda_\lambda$. The canonical basis $\mathbf{B}(\Lambda_\lambda)$ of Λ_λ (see 14.4.11) is clearly an $\mathcal{A}$-basis of ${}_{\mathcal{A}}\Lambda_\lambda$. For any $\nu \in \mathbf{N}[I]$, let $({}_{\mathcal{A}}\Lambda_\lambda)_\nu$ be the image of ${}_{\mathcal{A}}\mathbf{f}_\nu$ under the canonical map $\mathbf{f} \to \Lambda_\lambda$. We have a direct sum decomposition ${}_{\mathcal{A}}\Lambda_\lambda = \oplus_\nu ({}_{\mathcal{A}}\Lambda_\lambda)_\nu$.

Proposition 19.3.2. *${}_{\mathcal{A}}\Lambda_\lambda$ is stable under the operators $x^-, x^+ : \Lambda_\lambda \to \Lambda_\lambda$, for any $x \in {}_{\mathcal{A}}\mathbf{f}$.*

For x^-, this is obvious. To prove the assertion about x^+, we may assume that $x = \theta_i^{(n)}$ for some i, n. Let $y \in ({}_{\mathcal{A}}\Lambda_\lambda)_\nu$. We show that $E_i^{(n)} y \in {}_{\mathcal{A}}\Lambda_\lambda$ by induction on $N = \text{tr}\ \nu$. If $N = 0$, the result is obvious. Assume that $N \geq 1$. We may assume that $y = F_j^{(t)} y'$ where $1 \leq t \leq \nu_j$ and $y' \in ({}_{\mathcal{A}}\Lambda_\lambda)_{\nu - tj}$. By 3.4.2(b), the operator $E_i^{(n)} F_j^{(t)}$ on Λ_λ is an $\mathcal{A}$-linear combination of operators $F_j^{(t')} E_i^{(n')}$. By the induction hypothesis, we have $E_i^{(n')} y' \in {}_{\mathcal{A}}\Lambda_\lambda$; hence $F_j^{(t')} E_i^{(n')} y' \in {}_{\mathcal{A}}\Lambda_\lambda$ so that $E_i^{(n)} y = E_i^{(n)} F_j^{(t)} y' \in {}_{\mathcal{A}}\Lambda_\lambda$. This completes the proof.

The following result is a strengthening of Lemma 19.1.4.

Proposition 19.3.3. *Let $b, b' \in \mathcal{B}(\lambda)$.*

(a) *If $b' \neq \pm b$ then $(b^- \eta, b'^- \eta) \in v^{-1}\mathbf{Z}[v^{-1}]$.*

(b) *We have $(b^- \eta, b^- \eta) \in 1 + v^{-1}\mathbf{Z}[v^{-1}]$.*

We shall prove by induction on $\text{tr}\ \nu$ that

(c) $(x, y) \in \mathcal{A}$

for any $x, y \in ({}_{\mathcal{A}}\Lambda_\lambda)_\nu$. When $\text{tr}\ \nu = 0$, (c) is trivial. We may therefore assume that $\text{tr}\ \nu = N > 0$ and that (c) is already known for ν' with $\text{tr}\ \nu' < N$. We may assume that $x = F_i^{(r)} x'$ where $0 < r \leq \nu_i$ and $x' \in ({}_{\mathcal{A}}\Lambda_\lambda)_{\nu - ri}$. From the definitions we have

(d) $(F_i^{(r)} x', y) = (x', v_i^{r^2} \tilde{K}_{-ri} E_i^{(r)} y)$.

By 19.3.2, we have $\tilde{K}_{-ri} E_i^{(r)} y \in ({}_{\mathcal{A}}\Lambda_\lambda)_{\nu - ri}$; hence the right hand side of (d) is in $\mathcal{A}$, by the induction hypothesis. This proves (c). The proposition follows by combining (c) with Lemma 19.1.4, since $\mathbf{A} \cap \mathcal{A} = \mathbf{Z}[v^{-1}]$.

19.3.4. From the description 18.1.1(a) of Λ_λ, we see that there is a unique **Q**-linear isomorphism $^- : \Lambda_\lambda \to \Lambda_\lambda$ such that $\overline{u\eta_\lambda} = \bar{u}\eta_\lambda$ for all $u \in \mathbf{U}$. It has square equal to 1.

Theorem 19.3.5. *Let $b \in \Lambda_\lambda$. We have $b \in \mathbf{B}(\Lambda_\lambda)$ if and only if*

(1) $b \in {}_{\mathcal{A}}\Lambda_\lambda$, $\bar{b} = b$ and

(2) there exists a sequence $i_1, i_2, \dots, i_p$ in I such that $b = \tilde{F}_{i_1}\tilde{F}_{i_2}\cdots\tilde{F}_{i_p}\eta_\lambda$ mod $v^{-1}L(\Lambda_\lambda)$.

We have $b \in \pm\mathbf{B}(\Lambda_\lambda)$ if and only if b satisfies (1) and

(3) $(b, b) = 1 \mod v^{-1}\mathbf{Z}[v^{-1}]$.

If $b \in \pm\mathbf{B}(\Lambda_\lambda)$, then b obviously satisfies (1); it satisfies (3) by 19.3.3.

Assume now that b satisfies (1) and (3). Since the canonical basis is almost orthonormal, from Lemma 14.2.2 it follows that there exists $b' \in \mathbf{B}(\Lambda_\lambda)$ such that $b = \pm b' \mod v^{-1}L(\Lambda_\lambda)$. Since $\overline{b - (\pm b')} = b - (\pm b')$, it follows that $b - (\pm b') = 0$.

Assume now that $b \in \mathbf{B}(\Lambda_\lambda)$. We show that b satisfies (2). We have $b = \beta^-\eta_\lambda$ for some $\beta \in \mathbf{B}$. We can find a sequence $i_1, i_2, \dots, i_p$ in I such that $\beta = \tilde{\phi}_{i_1}\tilde{\phi}_{i_2}\cdots\tilde{\phi}_{i_p}1 \mod v^{-1}\mathcal{L}(\mathbf{f})$. Using 18.3.8, it follows that $b = \tilde{F}_{i_1}\tilde{F}_{i_2}\cdots\tilde{F}_{i_p}\eta_\lambda \mod v^{-1}L(\Lambda_\lambda)$, as required.

Finally, assume that b satisfies (1),(2). Let $i_1, i_2, \dots, i_p$ in I be as in (2). Using again 18.3.8, we see that $b = (\tilde{\phi}_{i_1}\tilde{\phi}_{i_2}\cdots\tilde{\phi}_{i_p}1)^-\eta_\lambda \mod v^{-1}L(\Lambda_\lambda)$. Let β be the unique element of $\mathbf{B}$ such that

$$\beta = \tilde{\phi}_{i_1}\tilde{\phi}_{i_2}\cdots\tilde{\phi}_{i_p}1 \mod v^{-1}\mathcal{L}(\mathbf{f}).$$

We have $b = \beta^-\eta_\lambda \mod v^{-1}L(\Lambda_\lambda)$. Let $b' = \beta^-\eta_\lambda$. Then $b - b' \in {}_{\mathcal{A}}\Lambda_\lambda$, $\overline{b - b'} = b - b'$ and $b - b' \in v^{-1}L(\Lambda_\lambda)$. It follows that $b - b' = 0$. Thus, $b \in \mathbf{B}(\Lambda_\lambda)$. The theorem is proved.

19.3.6. We will now investigate the relation between the inner product $(,)$ on $\mathbf{f}$ (see 1.2.5) and the inner product $(,)$ on Λ_λ, which we now denote by $(,)_\lambda$ since $\lambda \in X^+$ will vary.

Proposition 19.3.7. *Let $x, y \in \mathbf{f}$. When $\lambda \in X^+$ tends to ∞ (in the sense that $\langle i, \lambda\rangle$ tends to ∞ for all i), then the inner product $(x^-\eta_\lambda, y^-\eta_\lambda)_\lambda \in \mathbf{Q}(v)$ converges in $\mathbf{Q}((v^{-1}))$ to (x, y).*

We may assume that both x and y belong to $\mathbf{f}_\nu$ for some ν. We prove the proposition by induction on $N = \operatorname{tr} \nu$. When $N = 0$, the result is trivial. Assume now that $N \geq 1$. We may assume that $\nu_i > 0$ and $x = \theta_i x'$ for some i and some $x' \in \mathbf{f}_{\nu-i}$. We have

$$(x^- \eta_\lambda, y^- \eta_\lambda)_\lambda = (F_i x'^- \eta_\lambda, y^- \eta_\lambda)_\lambda = (x'^- \eta_\lambda, v_i \tilde{K}_{-i} E_i y^- \eta_\lambda)_\lambda.$$

Using the commutation formula 3.1.6(b) and the equality $E_i \eta_\lambda = 0$ we see that the last inner product is equal to

$$\begin{aligned}
&(v_i - v_i^{-1})^{-1}(x'^- \eta_\lambda, v_i \tilde{K}_{-i}(-r_i(y)^- \tilde{K}_{-i} + \tilde{K}_i({}_i r(y)^-))\eta_\lambda)_\lambda \\
&= -(v_i - v_i^{-1})^{-1} v_i^{-2\langle i,\lambda\rangle + \langle i,|y|\rangle - 1}(x'^- \eta_\lambda, r_i(y)^- \eta_\lambda)_\lambda \\
&+ (1 - v_i^{-2})^{-1}(x'^- \eta_\lambda, {}_i r(y)^- \eta_\lambda)_\lambda.
\end{aligned}$$

Using the induction hypothesis, we see that in the last expression, the first term converges to 0 for $\lambda \to \infty$ (note that $v_i^{-\langle i,\lambda\rangle}$ converges to 0) and the second term converges to $(1 - v_i^{-2})^{-1}(x', {}_i r(y))$ which by 1.2.13(a), is equal to (x, y). The proposition is proved.

CHAPTER 20

Bases at ∞

20.1. The Basis at ∞ of Λ_λ

20.1.1. Let M be an object of $\mathcal{C}'$. We define a *basis at* ∞ of M to be a pair consisting of

(a) a free $\mathbf{A}$-submodule L of M such that $M = \mathbf{Q}(v) \otimes_{\mathbf{A}} L = M$ and

(b) a basis $\mathbf{b}$ of the $\mathbf{Q}$-vector space $L/v^{-1}L$;

it is required that the properties (c)–(f) below are satisfied.

(c) L is stable under the operators $\tilde{E}_i, \tilde{F}_i : M \to M$ for all i; thus, $\tilde{E}_i, \tilde{F}_i$ act on $L/v^{-1}L$;

(d) $\tilde{F}_i(\mathbf{b}) \subset \mathbf{b} \cup \{0\}$ and $\tilde{E}_i(\mathbf{b}) \subset \mathbf{b} \cup \{0\}$ for all i;

(e) we have $L = \oplus L^\lambda$ where $L^\lambda = L \cap M^\lambda$ and $\mathbf{b} = \sqcup \mathbf{b}^\lambda$ where $\mathbf{b}^\lambda = \mathbf{b} \cap (L^\lambda / v^{-1} L^\lambda)$;

(f) given $b, b' \in \mathbf{b}$ and $i \in I$, we have $\tilde{E}_i b = b'$ if and only if $\tilde{F}_i b' = b$.

The definition given above of bases at ∞ is due to Kashiwara who calls them *crystal bases.*

Lemma 20.1.2. *Let $x \in L \cap M^\lambda$ and let $i \in I$. Let $t = \langle i, \lambda \rangle$. Write $x = \sum_{s; s \geq 0; s+t \geq 0} F_i^{(s)} x_s$ where $x_s \in \ker(E_i : M^{\lambda + si'} \to M)$ and $x_s = 0$ for large enough s. (See 16.1.4.)*

(a) *For all $s \geq 0$ we have $x_s \in L$.*

(b) *If $x \mod v^{-1}L$ belongs to $\mathbf{b}$, then there exists s_0 such that $x_s \in v^{-1}L$ for $s \neq s_0$, $x_{s_0} \mod v^{-1}L$ belongs to $\mathbf{b}$ and $x = F_i^{(s_0)} x_{s_0} \mod v^{-1}L$.*

We prove (a) by induction on $N \geq 0$ such that $x_s = 0$ for $s > N$. For $N = 0$, the result is clear. Assume now that $N \geq 1$. We have $\tilde{E}_i x = \sum_{s; s \geq 1; s+t \geq 0} F_i^{(s-1)} x_s$ where $x_s = 0$ for $s > N$. By definition, we have $\tilde{E}_i x \in L \cap M^{\lambda + i'}$. Hence if $t' = \langle i, \lambda + i' \rangle = t + 2$, we have $\tilde{E}_i x = \sum_{s'; s' \geq 0; s'+t' \geq 1} F_i^{(s')} x_{s'+1}$ and $x_{s'+1} = 0$ for $s' \geq N$. By the induction hypothesis, we have $x_s \in L$ for all $s \geq 1$. Since L is stable under $\tilde{F}_i$ and

G. Lusztig, *Introduction to Quantum Groups*, Modern Birkhäuser Classics,
DOI 10.1007/978-0-8176-4717-9_20, © Springer Science+Business Media, LLC 2010

$F_i^{(s)}x_s = \tilde{F}_i^s x_s$, it follows that $F_i^{(s)}x_s \in L$ for all $s \geq 1$. Since $x \in L$, we deduce that $x_0 \in L$. This proves (a).

We prove (b) by induction on $N \geq 0$ as above. If $N = 0$, there is nothing to prove. Assume that $N \geq 1$. If $\tilde{E}_i x \in v^{-1}L$, then $v\sum_{s';s'\geq 0;s'+t'\geq 1} F_i^{(s')}x_{s'+1} \in L$ and by (a) we have $vx_{s'+1} \in L$ for all $s' \geq 0$. Hence $x_s \in v^{-1}L$ for all $s \geq 1$. As before we then have $F_i^{(s)}x_s \in v^{-1}L$ for $s \geq 1$ and $x = x_0 \mod v^{-1}L$. If $\tilde{E}_i x \notin v^{-1}L$, then $\tilde{E}_i x \mod v^{-1}L$ belongs to $\mathbf{b}$. By the induction hypothesis, there exists $s_0 \geq 1$ such that $x_s \in v^{-1}L$ for $s \neq s_0$ and $s \geq 1$. Therefore we have $\tilde{E}_i x = F_i^{(s_0-1)}x_{s_0} \mod v^{-1}L$. Equivalently, we have $\tilde{E}_i x = \tilde{F}_i^{s_0-1}x_{s_0} \mod v^{-1}L$. Applying $\tilde{F}_i$ to this and using 20.1.1(f), we obtain $x = \tilde{F}_i\tilde{E}_i x = \tilde{F}_i^{s_0}x_{s_0} = F_i^{(s_0)}x_{s_0} \mod v^{-1}L$. The lemma is proved.

20.1.3. In the next theorem we assume that the root datum is Y-regular. Let $\lambda \in X^+$. Let L be the $\mathbf{A}$-submodule of Λ_λ generated by the canonical basis $\mathbf{B}(\Lambda_\lambda)$ and let $\mathbf{b}$ be the image of the canonical basis in $L/v^{-1}L$.

Theorem 20.1.4. *$(L, \mathbf{b})$ is a basis at ∞ of Λ_λ.*

Property 20.1.1(c) follows from Theorem 18.3.8. We prove that property 20.1.1(d) is satisfied. Let $b \in \mathbf{b}$. There exists $\beta \in \mathbf{B}$ such that b is $\beta^-\eta_\lambda \mod v^{-1}L$. From Theorem 18.3.8 we see that $\tilde{F}_i b$ is $\tilde{\phi}_i(\beta)^-\eta_\lambda \mod v^{-1}L$ and $\tilde{E}_i b$ is $\tilde{\epsilon}_i(\beta)^-\eta_\lambda \mod v^{-1}L$ or zero. By 17.3.7, we have $\tilde{\phi}_i(\beta) = \beta' \mod v^{-1}\mathcal{L}(\mathbf{f})$ for some $\beta' \in \mathcal{B}$ and this is necessarily in $\mathbf{B}$. Then $\tilde{\phi}_i(\beta)^-\eta_\lambda = \beta'^-\eta_\lambda \mod v^{-1}L$ so that $\tilde{F}_i b$ is $\beta'^-\eta_\lambda \mod v^{-1}L$ and $\beta'^-\eta_\lambda \mod v^{-1}L$ is in $\mathbf{b} \cup \{0\}$.

By 17.3.7, we have either $\tilde{\epsilon}_i(\beta) = \beta'' \mod v^{-1}\mathcal{L}(\mathbf{f})$ for some $\beta'' \in \mathcal{B}$ (which is necessarily in $\mathbf{B}$) or $\tilde{\epsilon}_i(\beta) = 0 \mod v^{-1}\mathcal{L}(\mathbf{f})$. Then $\tilde{\epsilon}_i(\beta)^-\eta_\lambda = \beta''^-\eta_\lambda \mod v^{-1}L$ or $\tilde{\epsilon}_i(\beta)^-\eta_\lambda = 0 \mod v^{-1}L$ so that $\tilde{E}_i b$ is $\beta''^-\eta_\lambda \mod v^{-1}L$ or 0. Now $\beta''^-\eta_\lambda \mod v^{-1}L$ is in $\mathbf{b}\cup\{0\}$. This proves property 20.1.1(d).

Property 20.1.1(e) is clearly satisfied. We prove that property 20.1.1(f) is satisfied. Let $b, b' \in \mathbf{b}$. We have $b = \beta^-\eta_\lambda \mod v^{-1}L$ and $b' = \beta'^-\eta_\lambda \mod v^{-1}L$ where $\beta, \beta' \in \mathbf{B}$.

By 18.3.8, we have $\tilde{E}_i b = b'$ if and only if $(\tilde{\epsilon}_i\beta)^-\eta_\lambda = \beta'^-\eta_\lambda \mod v^{-1}L$. This is equivalent to the condition that

(a) $\tilde{\epsilon}_i\beta = \beta' \mod v^{-1}\mathcal{L}(\mathbf{f})$.

Similarly, the condition that $\tilde{F}_i b' = b$ is equivalent to the condition that

(b) $\tilde{\phi}_i\beta' = \beta \mod v^{-1}\mathcal{L}(\mathbf{f})$.

Now conditions (a) and (b) are equivalent by 17.3.7. The theorem is proved.

20.2. Basis at ∞ in a Tensor Product

20.2.1. Let $M, M' \in \mathcal{C}'$. Assume that M and M' have finite dimensional weight spaces. Assume that $(L, \mathbf{b})$ (resp. $(L', \mathbf{b}')$) is a given basis at ∞ of M (resp. M'). Consider the tensor product $M \otimes M' \in \mathcal{C}'$.

Theorem 20.2.2. *The free* $\mathbf{A}$*-submodule* $L \otimes_{\mathbf{A}} L'$ *of* $M \otimes M'$ *and the* $\mathbf{Q}$*-basis* $\mathbf{b} \otimes \mathbf{b}'$ *of* $(L \otimes_{\mathbf{A}} L')/v^{-1}(L \otimes_{\mathbf{A}} L') = (L/v^{-1}L) \otimes_{\mathbf{Q}} (L'/v^{-1}L')$ *define a basis at* ∞ *of* $M \otimes M'$.

Only properties (c),(d),(f) in the definition 20.1.1 need to be verified. In verifying these properties, we shall fix $i \in I$ and write L^t for the sum $\oplus L^\lambda$ over all λ such that $\langle i, \lambda \rangle = t$. The notation L'^t has a similar meaning.

Let G^t be the set of all $z \in L^t$ such that $z \mod v^{-1}L$ belongs to $\mathbf{b}$ and such that $E_i z = 0$. Let G'^t be the set of all $z' \in L'^t$ such that $z' \mod v^{-1}L'$ belongs to $\mathbf{b}'$ and such that $E_i z' = 0$. From the definitions, all elements of the form $F_i^{(s)} z \quad (z \in G^t, s \in [0, t])$ belong to $\mathbf{b}$ modulo $v^{-1}L$ and according to 20.1.2, all elements of $\mathbf{b}$ are obtained in this way.

Similarly, all elements of the form $F_i^{(s')} z' \quad (z' \in G'^{t'}, s' \in [0, t'])$ belong to $\mathbf{b}'$ modulo $v^{-1}L'$ and all elements of $\mathbf{b}'$ are obtained in this way.

Using Nakayama's lemma, which is applicable since the weight spaces are assumed to be finite dimensional, we deduce that the elements $F_i^{(s)} z \quad (z \in G^t, s \in [0, t])$ generate the $\mathbf{A}$-module L; similarly, the elements $F_i^{(s')} z' \ (z' \in G'^{t'}, s' \in [0, t'])$ generate the $\mathbf{A}$-module L'.

Let $z \in G^t, z' \in G'^{t'}, s \in [0, t], s' \in [0, t']$. According to 17.2.4, we have

$$\tilde{F}_i(F_i^{(s)} z \otimes F_i^{(s')} z') = F_i^{(s)} z \otimes F_i^{(s'+1)} z' \mod v^{-1}(L \otimes_{\mathbf{A}} L')$$

if $s + s' < t'$;

$$\tilde{F}_i(F_i^{(s)} z \otimes F_i^{(s')} z') = F_i^{(s+1)} z \otimes F_i^{(s')} z' \mod v^{-1}(L \otimes_{\mathbf{A}} L')$$

if $s + s' \geq t'$;

$$\tilde{E}_i(F_i^{(s)} z \otimes F_i^{(s')} z') = F_i^{(s)} z \otimes F_i^{(s'-1)} z' \mod v^{-1}(L \otimes_{\mathbf{A}} L')$$

if $s + s' \leq t'$;

$$\tilde{E}_i(F_i^{(s)} z \otimes F_i^{(s')} z') = F_i^{(s-1)} z \otimes F_i^{(s')} z' \mod v^{-1}(L \otimes_{\mathbf{A}} L')$$

if $s + s' > t'$.

It follows that $\tilde{E}_i, \tilde{F}_i$ map a set of generators of the $\mathbf{A}$-module $L \otimes_{\mathbf{A}} L'$ into $L \otimes_{\mathbf{A}} L'$; hence they map $L \otimes_{\mathbf{A}} L'$ into $L \otimes_{\mathbf{A}} L'$. This verifies property 20.1.1(c) of a basis at ∞. Properties 20.1.1(e),(f) of a basis at ∞ are also clear from the previous formulas. The theorem is proved.

20.2.3. Assume that $z \in G^t, z' \in G'^{t'}$ and that $s \in [0,t], s' \in [0,t']$ are such that $t + t' = 2(s+s')$. By the formulas in 20.2.2, the condition that $\tilde{F}_i(F_i^{(s)}z \otimes F_i^{(s')}z') \in v^{-1}(L \otimes_{\mathbf{A}} L')$ is that either $s' = t'$ and $s + s' < t'$, or $s = t$ and $s + s' \geq t'$. The first case cannot occur since $s \geq 0$. Hence the condition is that $s = t$ and $s + s' \geq t'$. But if $s = t$ then $t' = s + 2s'$ hence $s + s' \geq s + 2s'$ so that $s' = 0$. Thus the condition is $s = t = t', s' = 0$. We can reformulate this as follows.

Proposition 20.2.4. *Let* $(M, L, \mathbf{b}), (M', L', \mathbf{b}')$ *be as above. Let* $b \in \mathbf{b}, b' \in \mathbf{b}'$. *Assume that* $b \in \mathbf{b}^\lambda$ *and* $b' \in \mathbf{b}'^{\lambda'}$ *and* $\langle i, \lambda \rangle + \langle i, \lambda' \rangle = 0$. *Then the following two conditions are equivalent:*

(a) $\tilde{F}_i(b \otimes b') = 0$ *in* $(L \otimes_{\mathbf{A}} L')/v^{-1}(L \otimes_{\mathbf{A}} L')$;

(b) $\tilde{F}_i(b) = 0$ *in* $L/v^{-1}L$ *and* $\tilde{E}_i(b') = 0$ *in* $L'/v^{-1}L'$.

CHAPTER 21

Cartan Data of Finite Type

21.1.1. In this chapter we assume that the Cartan datum is of finite type; then the root datum is automatically Y-regular and X-regular.

Let $\lambda' = -w_0(\lambda)$. Then $\lambda' \in X^+$ and we may consider the $\mathbf{U}$-module ${}^\omega\Lambda_{\lambda'}$ as in 3.5.7. Since ${}^\omega\Lambda_{\lambda'} = \Lambda_{\lambda'}$ as a vector space, the canonical basis $\mathbf{B}(\Lambda_{\lambda'})$ of $\Lambda_{\lambda'}$ may be regarded as a basis of ${}^\omega\Lambda_{\lambda'}$.

Proposition 21.1.2. *There is a unique isomorphism of* $\mathbf{U}$*-modules* $\chi : \Lambda_\lambda \to {}^\omega\Lambda_{\lambda'}$ *such that* χ *maps* $\mathbf{B}(\Lambda_\lambda)$ *onto* $\mathbf{B}(\Lambda_{\lambda'})$.

By 6.3.4, $\Lambda_{\lambda'}$ is a finite dimensional simple object of $\mathcal{C}'$. Hence ${}^\omega\Lambda_{\lambda'}$ is a finite dimensional simple object of $\mathcal{C}'$. By definition, its $(-\lambda')$-weight space is one dimensional and the $(-\lambda' - i')$-weight space is zero for any i. By Weyl group invariance (5.2.7), it follows that the λ-weight space is one dimensional and the $(\lambda + i')$-weight space is zero for any i (we have $\lambda = w_0(-\lambda')$).

Let x be the unique element of $\mathbf{B}(\Lambda_{\lambda'})$ in this λ-weight space. Then $E_i x = 0$ for all i. By Lemma 3.5.8, there is a unique morphism (in $\mathcal{C}'$) $\chi : \Lambda_\lambda \to {}^\omega\Lambda_{\lambda'}$ which carries η to x. Since χ is a non-zero morphism between simple objects, it is an isomorphism. We can regard χ as an isomorphism of vector spaces $\Lambda_\lambda \cong \Lambda_{\lambda'}$ such that $\chi(uy) = \omega(u)\chi(y)$ for all $u \in \mathbf{U}$ and $y \in \Lambda_\lambda$ and such that $\chi(\eta) \in \mathbf{B}(\Lambda_{\lambda'})$.

We have

(a) $\chi({}_{\mathcal{A}}\Lambda_\lambda) \subset {}_{\mathcal{A}}\Lambda_{\lambda'}$.

Indeed, let $y \in {}_{\mathcal{A}}\Lambda_\lambda$. Then $y = g^-\eta$ for some $g \in {}_{\mathcal{A}}\mathbf{f}$; hence $\chi(y) = g^+\chi(\eta)$. It remains to use the fact that ${}_{\mathcal{A}}\Lambda_{\lambda'}$ is stable under g^+ (see 19.3.2).

We have

(b) $\overline{\chi(x)} = \chi(\bar{x})$ for all $x \in \Lambda_\lambda$.

Indeed, we can write $x = u\eta$ with $u \in \mathbf{U}$. We have

$$\overline{\chi(u\eta)} = \overline{\omega(u)\chi(\eta)} = \overline{\omega(u)\chi(\eta)} = \omega(\bar{u})\chi(\eta) = \chi(\bar{u}\eta) = \chi(\bar{x}),$$

as required.

G. Lusztig, *Introduction to Quantum Groups*, Modern Birkhäuser Classics,
DOI 10.1007/978-0-8176-4717-9_21, © Springer Science+Business Media, LLC 2010

We have

(c) $(x,x') = (\chi(\eta),\chi(\eta))^{-1}(\chi(x),\chi(x'))$ for $x,x' \in \Lambda_\lambda$.

Indeed, if we set $((x,x')) = (\chi(\eta),\chi(\eta))^{-1}(\chi(x),\chi(x'))$ we obtain a form satisfying the defining properties of (x,x'), hence equal to it.

Let $b \in \mathbf{B}(\Lambda_\lambda)$. Let $b' = \chi(b)$. Using (a),(b), we see that $b' \in {}_{\mathcal{A}}\Lambda_{\lambda'}$ and $\bar{b}' = b'$. Using (c) and the fact that (b,b) and $(\chi(\eta),\chi(\eta))$ are in $1+v^{-1}\mathbf{Z}[v^{-1}]$, we see that $(b',b') \in 1+v^{-1}\mathbf{A}$. Since $b' \in {}_{\mathcal{A}}\Lambda_{\lambda'}$, we have also $(b',b') \in \mathcal{A}$; hence $(b',b') \in 1+v^{-1}\mathbf{Z}[v^{-1}]$. Using Theorem 19.3.5, it follows that $\pm b' \in \mathbf{B}(\Lambda_{\lambda'})$. This argument also shows that, if $(L,\mathrm{b}),(L',\mathrm{b}')$ are the bases at ∞ of $\Lambda_\lambda,\Lambda_{\lambda'}$ defined in 20.1.4, then $\chi(L) = L'$.

We can find a sequence $i_1,i_2,\dots,i_t$ in I such that $b = \tilde{F}_{i_1}\tilde{F}_{i_2}\cdots\tilde{F}_{i_t}\eta$ mod $v^{-1}L$. From the definitions we have that $\chi\tilde{F}_i = \tilde{E}_i\chi$ for all i. It follows that $b' = \tilde{E}_{i_1}\tilde{E}_{i_2}\cdots\tilde{E}_{i_t}\chi(\eta)$ mod $v^{-1}L'$, so that b' mod $v^{-1}L'$ belongs to b'. Since $\pm b' \in \mathbf{B}(\Lambda_{\lambda'})$, it follows that $b' \in \mathbf{B}(\Lambda_{\lambda'})$. The proposition is proved.

21.1.3. We shall identify the $\mathbf{U}$-modules Λ_λ and ${}^\omega\Lambda_{\lambda'}$ via χ. In particular, the generator $\xi = \xi_{-\lambda'}$ of ${}^\omega\Lambda_{\lambda'}$ (see 3.5.7) is now regarded as a vector in the $w_0(\lambda)$-weight space of Λ_λ, which belongs to the canonical basis of Λ_λ.

CHAPTER 22

Positivity of the Action of F_i, E_i in the Simply-Laced Case

22.1.1. In this chapter, the root datum is assumed to be Y-regular. We fix $\lambda \in X^+$ and we set $\Lambda = \Lambda_\lambda$. The main result of this chapter is Theorem 22.1.7, which asserts, in the simply laced case, that the matrices of the linear maps E_i and F_i of Λ into itself, with respect to the canonical basis of Λ, have as entries polynomials with integer, ≥ 0 coefficients.

Using Theorem 18.3.8, we see that Lemma 18.2.7 is true unconditionally. We restate it here as follows.

Theorem 22.1.2. *We have* $\Xi(\mathcal{L}(\mathbf{f})) \subset \mathcal{L}(\mathbf{f}) \odot L(\Lambda)$.

22.1.3. In the following corollary we shall use the notation

$$\nu \circ \lambda = \sum_{i \in I} \nu_i \langle i, \lambda \rangle (i \cdot i/2)$$

for any $\nu \in \mathbf{Z}[I], \lambda \in X$.

Corollary 22.1.4. *Let* $b \in \mathbf{B}$. *Write* $r(b) = \sum h_{b;b_1,b_2} b_1 \otimes b_2$ *and* $\bar{r}(b) = \sum g_{b;b_1,b_2} b_1 \otimes b_2$ *where* $h_{b;b_1,b_2} \in \mathcal{A}$ *and* $g_{b;b_1,b_2} \in \mathcal{A}$; *here* b_1, b_2 *run over* $\mathbf{B}$. *Thus,* $g_{b;b_1,b_2} = \overline{h_{b;b_1,b_2}}$.

If $b_2 \in \mathbf{B}(\lambda)$ *and* $b, b_1 \in \mathbf{B}$, *we have*

$$v^{-|b_1| \circ (\lambda - |b_2|)} g_{b;b_1,b_2} \in \mathbf{Z}[v^{-1}] \text{ and } v^{|b_1| \circ (\lambda - |b_2|)} h_{b;b_1,b_2} \in \mathbf{Z}[v].$$

It suffices to prove the statement about $g_{b;b_1,b_2}$.

By 3.1.5, we have $\Delta(b^-) = \sum g_{b;b_1,b_2} b_1^- \otimes \tilde{K}_{-|b_1|} b_2^-$. By the definition of Ξ, we have

$$\Xi(b) = \sum g_{b;b_1,b_2} b_1 \otimes \tilde{K}_{-|b_1|} b_2^- \eta = \sum v^{-|b_1| \circ (\lambda - |b_2|)} g_{b;b_1,b_2} b_1 \otimes b_2^- \eta.$$

By the previous theorem, if $b_2 \in \mathbf{B}(\lambda)$, the coefficient of $b_1 \otimes b_2^- \eta$ is in $\mathbf{A}$. This coefficient is clearly in $\mathcal{A}$; hence it is in $\mathbf{Z}[v^{-1}]$. The corollary follows.

G. Lusztig, *Introduction to Quantum Groups*, Modern Birkhäuser Classics,
DOI 10.1007/978-0-8176-4717-9_22, © Springer Science+Business Media, LLC 2010

Corollary 22.1.5. *Let $i \in I$. Let $b \in \mathbf{B}$. Let us write ${}_ir(b) = \sum_{b';n\in\mathbf{Z}} d_{b,\theta_i,b',n} v^n b'$ where b' runs over $\mathbf{B}$ and $d_{b,\theta_i,b',n}$ are integers.*

(a) *Let $b' \in \mathbf{B}(\lambda)$ and let $n \in \mathbf{Z}$ be such that $d_{b,\theta_i,b',n} \neq 0$. Then*

$$i \circ (\lambda - |b'|) + n \geq 0.$$

(b) *We have*

$$E_i(b^- \eta) = \sum_{b';n\in\mathbf{Z}} d_{b,\theta_i,b',n} \frac{v^{i\circ(\lambda-|b'|)+n} - v^{-i\circ(\lambda-|b'|)-n}}{v_i - v_i^{-1}} b'^- \eta$$

where b' runs over $\mathbf{B}(\lambda)$.

We apply the previous corollary to $h_{b;b_1,b_2}$ with $b_1 = \theta_i$; we obtain

$$\sum_n v^{i\circ(\lambda-|b'|)+n} d_{b,\theta_i,b',n} \in \mathbf{Z}[v]$$

for any $b' \in \mathbf{B}(\lambda)$; (a) follows. We now prove (b). By 3.1.6(b), we have

$$E_i(b^-\eta) = (v_i - v_i^{-1})^{-1}(-r_i(b)^- \tilde{K}_{-i}\eta + v^{-|b|\cdot i + i\cdot i}{}_ir(b)^- \tilde{K}_i \eta)$$

since $E_i\eta = 0$. By 1.2.14, we have $r_i(b) = v^{|b|\cdot i - i\cdot i}\overline{{}_ir(b)}$, since $\bar{b} = b$. Note also that $\tilde{K}_{\pm i}\eta = v_i^{\pm\langle i,\lambda\rangle}\eta$. Thus,

$$E_i(b^-\eta) = \sum_{b';n} d_{b,\theta_i,b',n} \frac{v^{-|b|\cdot i + i\cdot i + n} v_i^{\langle i,\lambda\rangle} - v^{|b|\cdot i - i\cdot i - n} v_i^{-\langle i,\lambda\rangle}}{v_i - v_i^{-1}} b'^- \eta.$$

Using now $|b| = |b'| + i$, we obtain (b).

22.1.6. Let $b \in \mathbf{B}(\Lambda)$. For any $i \in I$, we set $F_i b = \sum_{b',n} f_{b,b',i,n} v^n b'$, $E_i b = \sum_{b',n} \tilde{f}_{b,b',i,n} v^n b'$ where b' runs over $\mathbf{B}(\Lambda)$ and n runs over $\mathbf{Z}$; the coefficients $f_{b,b',i,n}, \tilde{f}_{b,b',i,n}$ are integers.

Theorem 22.1.7. *Assume that the Cartan datum is simply laced. Then $f_{b,b',i,n} \in \mathbf{N}$ and $\tilde{f}_{b,b',i,n} \in \mathbf{N}$ for any b, b', i, n.*

If $\beta, \beta' \in \mathbf{B}$ are such that $\beta\eta = b, \beta'\eta = b'$, then with the notation of Theorem 14.4.13, we have

$$f_{b,b',i,n} = c_{\theta_i,\beta,\beta',n}$$

and

$$\sum_n \tilde{f}_{b,b',i,n} v^n = \sum_n [i \circ (\lambda - |b'|) + n] d_{b,\theta_i,b',n}$$

(we have used Corollary 22.1.5(b) and the equality $v_i = v$). By Theorem 14.4.13, the integers $c_{\theta_i,\beta,\beta',n}$ are ≥ 0. Hence $f_{b,b',i,n} \in \mathbf{N}$.

Again by Theorem 14.4.13, the integers $d_{b,\theta_i,b',n}$ are ≥ 0 and, by Corollary 22.1.15(a), we have $i \circ (\lambda - |b'|) + n \geq 0$ for any n such that $d_{b,\theta_i,b',n} \neq 0$. Since $[N]$ is a sum of powers of v if $N \geq 0$, we deduce that $\tilde{f}_{b,b',i,n} \in \mathbf{N}$. The theorem is proved.

Notes on Part III

1. Most results in Part III are due to Kashiwara [2]. An exception is Theorem 22.1.7, which is new.
2. Although Theorem 22.1.2 does not appear explicitly in Kashiwara's papers, it is close to results which do appear; the same applies to the results in 17.1. The proofs in 17.2 are quite different from Kashiwara's.
3. The proof in 19.2.3 is an adaptation of arguments in [3].

References

1. M. Kashiwara, *Crystallizing the q-analogue of universal enveloping algebras*, Comm. Math. Phys. **133** (1990), 249–260.
2. ———, *On crystal bases of the q-analogue of universal enveloping algebras*, Duke Math. J. **63** (1991), 465–516.
3. ———, *Crystal base and Littelman's refined Demazure character formula*, Duke Math. J. **71** (1993), 839–858.
4. G. Lusztig, *Canonical bases arising from quantized enveloping algebras*, J. Amer. Math. Soc. **3** (1990), 447–498.

Part IV

CANONICAL BASIS OF $\dot{\mathbf{U}}$

The algebra $\dot{\mathbf{U}}$ is a modified form of $\mathbf{U}$ in which $\mathbf{U}^0$ is replaced by the direct sum of infinitely many one-dimensional algebras, one for each element of the lattice X of weights. This is an algebra without unit; it has instead many orthogonal idempotents by means of which one can approximate the unit element. The $\mathbf{U}$-modules in the category $\mathcal{C}$ can be regarded naturally as modules for $\dot{\mathbf{U}}$; on the other hand, more exotic $\mathbf{U}$-modules, like those without weight decomposition, cannot be regarded as $\dot{\mathbf{U}}$-modules. Thus, $\dot{\mathbf{U}}$ is an algebra more appropriate than $\mathbf{U}$ for the study of objects of $\mathcal{C}$. One of its main virtues is that it has a canonical basis $\dot{\mathbf{B}}$ with remarkable properties.

In Chapter 23, we define and study the algebra $\dot{\mathbf{U}}$ and its $\mathcal{A}$-form ${}_{\mathcal{A}}\dot{\mathbf{U}}$.

In Chapter 24, we prove an integrality property for the quasi-$\mathcal{R}$-matrix and we use it to define a canonical basis in any tensor product ${}^{\omega}\Lambda_\lambda \otimes \Lambda_{\lambda'}$, which is fixed by an antilinear involution obtained by composing the obvious involution $^-$ with the action of the quasi-$\mathcal{R}$-matrix.

In Chapter 25, we show that these bases have some rather nice stability properties with respect to certain transition maps (see 25.1.5). The proof is based on results in Part III. From this we deduce (Theorem 25.2.1) that these bases can be "glued together" to form a single basis $\dot{\mathbf{B}}$ of $\dot{\mathbf{U}}$. In Theorem 25.2.5, we show that $\dot{\mathbf{B}}$ is simultaneously compatible with many subspaces of $\dot{\mathbf{U}}$ and in Proposition 25.2.6, we show that $\dot{\mathbf{B}}$ is a generalization of $\mathbf{B}$.

In Chapter 26 we show that $\dot{\mathbf{B}}$ is characterized (up to signs) by an almost orthonormality property with respect to an inner product. This is in the same spirit as Kashiwara's idea of characterizing $\mathbf{B}$ by an almost orthonormality property (see Notes on Part II). As an application, we show that $\dot{\mathbf{B}}$ is stable (up to signs) under σ and ω. The same should be true without signs.

In Chapter 27 we assume that the Cartan datum is of finite type; we show that a tensor product of form $\Lambda_{\lambda_1} \otimes \Lambda_{\lambda_2} \cdots \otimes \Lambda_{\lambda_n}$ has a canonical basis and that this induces a canonical basis on the space of coinvariants. This last basis has an invariance property with respect to cyclic permutations, as shown in Chapter 28.

In Chapter 29 we prove a refinement of the Peter-Weyl theorem (in $\dot{\mathbf{U}}$ instead of the coordinate algebra) and we discuss an extension of the theory of cells in Weyl groups to the case of $\dot{\mathbf{U}}$.

In Chapter 30 we define a canonical topological basis of (a completion of) $\mathbf{U}^- \otimes \mathbf{U}^+$ (in finite type) and show that this basis gives rise simultaneously to the canonical bases of the various tensor products $\Lambda_\lambda \otimes {}^\omega\Lambda_{\lambda'}$.

CHAPTER 23

The Algebra $\dot{\mathbf{U}}$

23.1. Definition and First Properties of $\dot{\mathbf{U}}$

23.1.1. The objects of $\mathcal{C}$ may be regarded as modules over a certain $\mathbf{Q}(v)$-algebra $\dot{\mathbf{U}}$, in general without 1, which is a modified form of $\mathbf{U}$. We prepare the ground for the definition of $\dot{\mathbf{U}}$.

If $\lambda', \lambda'' \in X$, we set

(a) $$_{\lambda'}\mathbf{U}_{\lambda''} = \mathbf{U}/(\sum_{\mu \in Y}(K_\mu - v^{\langle \mu, \lambda' \rangle})\mathbf{U} + \sum_{\mu \in Y}\mathbf{U}(K_\mu - v^{\langle \mu, \lambda'' \rangle})).$$

Let $\pi_{\lambda',\lambda''} : \mathbf{U} \to {}_{\lambda'}\mathbf{U}_{\lambda''}$ be the canonical projection. Let

(b) $$\dot{\mathbf{U}} = \oplus_{\lambda',\lambda'' \in X}({}_{\lambda'}\mathbf{U}_{\lambda''}).$$

Consider the direct sum decomposition $\mathbf{U} = \oplus_\nu \mathbf{U}(\nu)$, where ν runs through $\mathbf{Z}[I]$, defined by the conditions $\mathbf{U}(\nu')\mathbf{U}(\nu'') \subset \mathbf{U}(\nu' + \nu''), K_\mu \in \mathbf{U}(0)$, $E_i \in \mathbf{U}(i), F_i \in \mathbf{U}(-i)$ for all $\nu', \nu'' \in \mathbf{Z}[I]$, $i \in I$, $\mu \in Y$. We have $\mathbf{U}^+_\nu \subset \mathbf{U}(\nu), \mathbf{U}^-_\nu \subset \mathbf{U}(-\nu)$.

There is a natural associative $\mathbf{Q}(v)$-algebra structure on $\dot{\mathbf{U}}$ inherited from that of $\mathbf{U}$. It is defined by the following requirement: for any $\lambda'_1, \lambda''_1, \lambda'_2, \lambda''_2$ and any $t \in \mathbf{U}(\lambda'_1 - \lambda''_1)$, $s \in \mathbf{U}(\lambda'_2 - \lambda''_2)$, the product $\pi_{\lambda'_1,\lambda''_1}(t)\pi_{\lambda'_2,\lambda''_2}(s)$ is equal to $\pi_{\lambda'_1,\lambda''_2}(ts)$ if $\lambda''_1 = \lambda'_2$ and is zero otherwise.

The elements $1_\lambda = \pi_{\lambda,\lambda}(1) \quad (\lambda \in X)$ of $\dot{\mathbf{U}}$ satisfy

(c) $$1_\lambda 1_{\lambda'} = \delta_{\lambda,\lambda'} 1_\lambda.$$

We have

(d) $$_{\lambda'}\mathbf{U}_{\lambda''} = 1_{\lambda'}\dot{\mathbf{U}}1_{\lambda''}.$$

The algebra $\dot{\mathbf{U}}$ does not generally have 1, since the infinite sum $\sum_{\lambda \in X} 1_\lambda$ does not belong to $\dot{\mathbf{U}}$ (if $X \neq 0$); however the family $(1_\lambda)_{\lambda \in X}$ is in some sense a substitute for the unit element.

G. Lusztig, *Introduction to Quantum Groups*, Modern Birkhäuser Classics,
DOI 10.1007/978-0-8176-4717-9_23, © Springer Science+Business Media, LLC 2010

23.1.2. A finer decomposition of $\dot{\mathbf{U}}$. We shall need a direct sum decomposition of $\dot{\mathbf{U}}$ which is slightly different from the one in 23.1.1(b).

The images of the summands $\mathbf{U}(\nu)$ under $\pi_{\lambda',\lambda''}$ form the direct sum decomposition ${}_{\lambda'}\mathbf{U}_{\lambda''} = \oplus_\nu({}_{\lambda'}\mathbf{U}_{\lambda''}(\nu))$. We set $\dot{\mathbf{U}}(\nu) = \oplus_{\lambda',\lambda''}({}_{\lambda'}\mathbf{U}_{\lambda''}(\nu))$. We then have direct sums decompositions

$$\dot{\mathbf{U}} = \oplus_\nu \dot{\mathbf{U}}(\nu) \text{ and } \dot{\mathbf{U}} = \oplus_{\lambda',\lambda'',\nu}(1_{\lambda'}\dot{\mathbf{U}}(\nu)1_{\lambda''})$$

where λ',λ'' run through X and ν runs through $\mathbf{Z}[I]$. Actually, the summand $1_{\lambda'}\dot{\mathbf{U}}(\nu)1_{\lambda''} = {}_{\lambda'}\mathbf{U}_{\lambda''}(\nu)$ is zero, unless $\lambda' - \lambda'' = \nu$ in X.

23.1.3. U-bimodule structure. Note that $\dot{\mathbf{U}}$ is naturally a $\mathbf{U}$-bimodule by the following rule: if $t' \in \mathbf{U}(\nu'), s \in \mathbf{U}, t'' \in \mathbf{U}(\nu'')$, we have by definition, $t'\pi_{\lambda',\lambda''}(s)t'' = \pi_{\lambda'+\nu',\lambda''-\nu''}(t'st'')$ for all $\lambda',\lambda'' \in X$.

For all $i \in I$ and $\lambda \in X$, we have the following identities in $\dot{\mathbf{U}}$:

$$E_i 1_\lambda = 1_{\lambda+i'}E_i, \quad F_i 1_\lambda = 1_{\lambda-i'}F_i;$$

$$(E_iF_j - F_jE_i)1_\lambda = \delta_{i,j}[\langle i,\lambda\rangle]_i 1_\lambda;$$

and more generally,

$$E_i^{(a)}1_\lambda = 1_{\lambda+ai'}E_i^{(a)}, \quad F_i^{(a)}1_\lambda = 1_{\lambda-ai'}F_i^{(a)};$$

$$E_i^{(a)}F_j^{(b)}1_\lambda = F_j^{(b)}E_i^{(a)}1_\lambda \quad \text{if} \quad i \neq j;$$

$$E_i^{(a)}1_{-\lambda}F_i^{(b)} = \sum_{t\geq 0}\begin{bmatrix} a+b-\langle i,\lambda\rangle \\ t \end{bmatrix}_i F_i^{(b-t)}1_{-\lambda+(a+b-t)i'}E_i^{(a-t)};$$

$$F_i^{(b)}1_\lambda E_i^{(a)} = \sum_{t\geq 0}\begin{bmatrix} a+b-\langle i,\lambda\rangle \\ t \end{bmatrix}_i E_i^{(a-t)}1_{\lambda-(a+b-t)i'}F_i^{(b-t)}.$$

These identities follow from 3.1.9.

23.1.4. Unital modules. A $\dot{\mathbf{U}}$-module M is said to be *unital* if

(a) for any $m \in M$ we have $1_\lambda m = 0$ for all but finitely many $\lambda \in X$;

(b) for any $m \in M$ we have $\sum_{\lambda\in X} 1_\lambda m = m$.

If M is a unital $\dot{\mathbf{U}}$-module, then we regard it as an object of $\mathcal{C}$ as follows. The decomposition $M = \oplus M^\lambda$ is given by $M^\lambda = 1_\lambda M$; the action of $u \in \mathbf{U}$ is given by $um = (u1_\lambda)m$ for any $\lambda \in X$ and any $m \in M^\lambda$. Here $u1_\lambda$ is regarded as an element of $\dot{\mathbf{U}}$ as in 23.1.3. In this way, we see that to give a unital $\dot{\mathbf{U}}$-module is the same as to give an object of $\mathcal{C}$.

23.1.5. The comultiplication of $\mathbf{U}$ induces a similar structure on $\dot{\mathbf{U}}$. More precisely, for any $\lambda'_1, \lambda'_2, \lambda''_1, \lambda''_2 \in X$, there is a well-defined linear map

$$\Delta_{\lambda'_1,\lambda''_1,\lambda'_2,\lambda''_2} : {}_{\lambda'_1+\lambda'_2}\mathbf{U}_{\lambda''_1+\lambda''_2} \to ({}_{\lambda'_1}\mathbf{U}_{\lambda''_1}) \otimes ({}_{\lambda'_2}\mathbf{U}_{\lambda''_2})$$

such that

$$\Delta_{\lambda'_1,\lambda''_1,\lambda'_2,\lambda''_2}(\pi_{\lambda'_1+\lambda'_2,\lambda''_1+\lambda''_2}(x)) = (\pi_{\lambda'_1,\lambda''_1} \otimes \pi_{\lambda'_2,\lambda''_2})(\Delta(x))$$

for all $x \in \mathbf{U}$. This collection of maps is called the comultiplication of $\dot{\mathbf{U}}$, and may be regarded as a single linear map from $\dot{\mathbf{U}}$ to the direct product

$$\prod_{\lambda'_1,\lambda''_1,\lambda'_2,\lambda''_2 \in X} ({}_{\lambda'_1}\mathbf{U}_{\lambda''_1}) \otimes ({}_{\lambda'_2}\mathbf{U}_{\lambda''_2}).$$

23.1.6. The algebra automorphism $\omega : \mathbf{U} \to \mathbf{U}$ induces, for each λ', λ'', a linear isomorphism ${}_{\lambda'}\mathbf{U}_{\lambda''} \to {}_{-\lambda'}\mathbf{U}_{-\lambda''}$. Taking direct sums, we obtain an algebra automorphism $\omega : \dot{\mathbf{U}} \to \dot{\mathbf{U}}$ such that $\omega(1_\lambda) = 1_{-\lambda}$ for all λ and $\omega(uxx'u') = \omega(u)\omega(x)\omega(x')\omega(u')$ for all $u, u' \in \mathbf{U}$ and $x, x' \in \dot{\mathbf{U}}$. It has square 1.

The map $\sigma : \mathbf{U} \to \mathbf{U}$ induces, for each λ', λ'', a linear isomorphism ${}_{\lambda'}\mathbf{U}_{\lambda''} \to {}_{-\lambda''}\mathbf{U}_{-\lambda'}$. Taking direct sums, we obtain a linear isomorphism $\sigma : \dot{\mathbf{U}} \to \dot{\mathbf{U}}$ such that $\sigma(1_\lambda) = 1_{-\lambda}$ for all $\lambda \in X$, and $\sigma(uxx'u') = \sigma(u')\sigma(x')\sigma(x)\sigma(u)$ for all $u, u' \in \mathbf{U}$ and $x, x' \in \dot{\mathbf{U}}$.

The antipode $S : \mathbf{U} \to \mathbf{U}$ (resp. its inverse S') induces, for each λ', λ'', a linear isomorphism ${}_{\lambda'}\mathbf{U}_{\lambda''} \to {}_{-\lambda''}\mathbf{U}_{-\lambda'}$. Taking direct sums, we obtain a linear isomorphism $S : \dot{\mathbf{U}} \to \dot{\mathbf{U}}$ (resp. $S' : \dot{\mathbf{U}} \to \dot{\mathbf{U}}$) such that $S(1_\lambda) = 1_{-\lambda}$ (resp. $S'(1_\lambda) = 1_{-\lambda}$) for all $\lambda \in X$, and $S(uxx'u') = S(u')S(x')S(x)S(u)$ (resp. $S'(uxx'u') = S'(u')S'(x')S'(x)S'(u)$) for all $u, u' \in \mathbf{U}$ and $x, x' \in \dot{\mathbf{U}}$.

These maps are related to σ as follows.

Lemma 23.1.7. *Let $x \in 1_{\lambda'}\dot{\mathbf{U}}(\nu)1_{\lambda''}$. We have $S(x) = (-1)^{\operatorname{tr}(\nu)} v^p \sigma(x)$ and $S'(x) = (-1)^{\operatorname{tr}(\nu)} v^{-p}\sigma(x)$, where $\nu \in \mathbf{Z}[I]$ is such that $\nu = \lambda' - \lambda''$ in X and $p = \sum_i \nu_i \langle i, \lambda' + \lambda'' \rangle (i \cdot i/4) - \sum_i \nu_i (i \cdot i/2) \in \mathbf{Z}$.*

This follows easily from the definitions.

23.1.8. The $\mathbf{Q}$-algebra homomorphism $^- : \mathbf{U} \to \mathbf{U}$ induces, for any $\lambda, \lambda' \in X$, a $\mathbf{Q}$-linear map $^- : {}_\lambda\mathbf{U}_{\lambda'} \to {}_\lambda\mathbf{U}_{\lambda'}$. Taking the direct sum of these maps, we obtain a $\mathbf{Q}$-linear map $^- : \dot{\mathbf{U}} \to \dot{\mathbf{U}}$ with square 1, which respects the multiplication of $\dot{\mathbf{U}}$, maps each 1_λ into itself and satisfies $\overline{tst'} = \bar{t}\bar{s}\bar{t'}$ for $t, t' \in \mathbf{U}$ and $s \in \dot{\mathbf{U}}$.

23.2. Triangular Decomposition, $\mathcal{A}$-form for $\dot{\mathbf{U}}$

23.2.1. The $\mathbf{U}$-bimodule structure on $\dot{\mathbf{U}}$ gives us a $\mathbf{f}\otimes\mathbf{f}^{opp}$-module structure $(x\otimes x'): u\mapsto x^+ux'^-$ on $\dot{\mathbf{U}}$. Similarly, the $\mathbf{U}$-bimodule structure on $\dot{\mathbf{U}}$ gives us a $\mathbf{f}\otimes\mathbf{f}^{opp}$-module structure $(x\otimes x'): u\mapsto x^-ux'^+$ on $\dot{\mathbf{U}}$. From the triangular decomposition for $\mathbf{U}$, we see that $\dot{\mathbf{U}}$ is a free $\mathbf{f}\otimes\mathbf{f}^{opp}$-module with basis $(1_\lambda)_{\lambda\in X}$ for either of the two $\mathbf{f}\otimes\mathbf{f}^{opp}$-module structures. In other words,

(a) the elements $b^+1_\lambda b'^-$ $(b,b'\in\mathbf{B},\lambda\in X)$ form a basis of the $\mathbf{Q}(v)$-vector space $\dot{\mathbf{U}}$;

(b) the elements $b^-1_\lambda b'^+$ $(b,b'\in\mathbf{B},\lambda\in X)$ form a basis of the $\mathbf{Q}(v)$-vector space $\dot{\mathbf{U}}$.

This is called the *triangular decomposition* of $\dot{\mathbf{U}}$.

Lemma 23.2.2. (a) *The $\mathcal{A}$-submodule of $\dot{\mathbf{U}}$ spanned by the elements $x^+1_\lambda x'^-$ (with $x,x'\in{}_{\mathcal{A}}\mathbf{f}$) coincides with the $\mathcal{A}$-submodule of $\dot{\mathbf{U}}$ spanned by the elements $x^-1_\lambda x'^+$ (with $x,x'\in{}_{\mathcal{A}}\mathbf{f}$). We denote it by ${}_{\mathcal{A}}\dot{\mathbf{U}}$.*

(b) *The elements in* 23.2.1(*a*) *form an $\mathcal{A}$-basis of ${}_{\mathcal{A}}\dot{\mathbf{U}}$. The elements in* 23.2.1(*b*) *form an $\mathcal{A}$-basis of ${}_{\mathcal{A}}\dot{\mathbf{U}}$.*

(c) ${}_{\mathcal{A}}\dot{\mathbf{U}}$ is an $\mathcal{A}$-subalgebra of $\dot{\mathbf{U}}$. This algebra is generated by the elements $E_i^{(n)}1_\lambda, F_i^{(n)}1_\lambda$ for various $i\in I$, $n\geq 0$ and $\lambda\in X$.

(a) follows immediately from the commutation formulas 23.1.3. (b) follows from (a) and the fact that $\mathbf{B}$ is an $\mathcal{A}$-basis of ${}_{\mathcal{A}}\mathbf{f}$. (c) follows from the commutation formulas 23.1.3.

23.2.3. From the integrality properties of the maps $r,\bar{r}$, it follows easily that the maps $\Delta_{\lambda'_1,\lambda''_1,\lambda'_2,\lambda''_2}: 1_{\lambda'_1+\lambda'_2}\dot{\mathbf{U}}1_{\lambda''_1+\lambda''_2}\to(1_{\lambda'_1}\dot{\mathbf{U}}1_{\lambda''_1})\otimes(1_{\lambda'_2}\dot{\mathbf{U}}1_{\lambda''_2})$ restrict to $\mathcal{A}$-linear maps

$$1_{\lambda'_1+\lambda'_2}({}_{\mathcal{A}}\dot{\mathbf{U}})1_{\lambda''_1+\lambda''_2}\to(1_{\lambda'_1}({}_{\mathcal{A}}\dot{\mathbf{U}})1_{\lambda''_1})\otimes_{\mathcal{A}}(1_{\lambda'_2}({}_{\mathcal{A}}\dot{\mathbf{U}})1_{\lambda''_2}).$$

23.2.4. From the definitions we see that $^-:\dot{\mathbf{U}}\to\dot{\mathbf{U}}$ leaves stable the $\mathcal{A}$-subalgebra ${}_{\mathcal{A}}\dot{\mathbf{U}}$ of $\dot{\mathbf{U}}$. The same holds for ω and σ.

23.2.5. Let $(Y',X',\dots)$ be the simply connected root datum of type $(I,\cdot)$ and let $f:Y'\to Y, g:X\to X'$ be the unique morphism of root data. We have an induced algebra homomorphism $\phi:\mathbf{U}'\to\mathbf{U}$ between the corresponding Drinfeld-Jimbo algebras (see 3.1.2). Assume that we are given $\zeta\in X$; let $\zeta'=g(\zeta)\in X'$.

Now ϕ maps the left ideal $\sum_{\mu'\in Y'} \mathbf{U}'(K_{\mu'} - v^{\langle \mu',\zeta'\rangle})$ of $\mathbf{U}'$ into the left ideal $\sum_{\mu \in Y} \mathbf{U}(K_\mu - v^{\langle \mu,\zeta\rangle})$ of $\mathbf{U}$; hence it induces a linear map on the quotients:

$$\dot{\phi} : \dot{\mathbf{U}}'1_{\zeta'} \to \dot{\mathbf{U}}1_\zeta.$$

From the triangular decomposition in $\mathbf{U}'$ and $\mathbf{U}$, it follows that the linear map $\dot{\phi}$ is an isomorphism. Using the definitions, we see that $\dot{\phi}$ restricts to an isomorphism of $\mathcal{A}$-modules

$$_{\mathcal{A}}\dot{\phi} : {}_{\mathcal{A}}\dot{\mathbf{U}}'1_{\zeta'} \cong {}_{\mathcal{A}}\dot{\mathbf{U}}1_\zeta.$$

23.3. $\dot{\mathbf{U}}$ and Tensor Products

23.3.1. Let $\lambda, \lambda' \in X$. The $\mathbf{U}$-module ${}^{\omega}M_\lambda \otimes M_{\lambda'} = \mathbf{f} \otimes \mathbf{f}$ belongs to $\mathcal{C}$; hence it is naturally a unital $\dot{\mathbf{U}}$-module.

From 23.2.1, we see that the elements $b^+b'^-1_{\lambda''}$ with $b, b' \in \mathbf{B}$ and $\lambda'' \in X$ form a $\mathbf{Q}(v)$-basis of $\dot{\mathbf{U}}$. Such an element (with $\lambda'' = \lambda' - \lambda$) acts on $1 \otimes 1 \in \mathbf{f} \otimes \mathbf{f}$ as follows:

(a) $(b^+b'^-1_{\lambda'-\lambda})(1 \otimes 1) = b^+(1 \otimes b') = b \otimes b' + \sum c_{b_1,b'_1} b_1 \otimes b'_1$

where $c_{b_1,b'_1} \in \mathbf{Q}(v)$, and the sum is taken over elements b_1, b'_1 in $\mathbf{B}$ such that $\operatorname{tr}|b_1| < \operatorname{tr}|b|$, $\operatorname{tr}|b'_1| < \operatorname{tr}|b'|$ and b'_1 belongs to the $\dot{\mathbf{U}}$-submodule of $M_{\lambda'} = \mathbf{f}$ generated by b'.

Similarly, the elements $b'^-b^+1_{\lambda''}$ with $b, b' \in \mathbf{B}$ and $\lambda'' \in X$ form a $\mathbf{Q}(v)$-basis of $\dot{\mathbf{U}}$. Such an element (with $\lambda'' = \lambda' - \lambda$) acts on $1 \otimes 1 \in \mathbf{f} \otimes \mathbf{f}$ as follows:

(b) $(b'^-b^+1_{\lambda'-\lambda})(1 \otimes 1) = b'^-(b \otimes 1) = b \otimes b' + \sum c'_{b_1,b'_1} b_1 \otimes b'_1$

where $c'_{b_1,b'_1} \in \mathbf{Q}(v)$, and the sum is taken over elements b_1, b'_1 in $\mathbf{B}$ such that $\operatorname{tr}|b_1| < \operatorname{tr}|b|$, $\operatorname{tr}|b'_1| < \operatorname{tr}|b'|$ and b_1 belongs to the $\dot{\mathbf{U}}$-submodule of ${}^{\omega}M_\lambda = \mathbf{f}$ generated by b.

Using either (a) or (b), we see that

(c) the $\mathbf{Q}(v)$-linear map $\pi' : \dot{\mathbf{U}}1_{\lambda'-\lambda} \to {}^{\omega}M_\lambda \otimes M_{\lambda'}$ given by $u \mapsto u(1 \otimes 1)$ is an isomorphism.

23.3.2. For any $\lambda \in X^+$, we define a ${}_{\mathcal{A}}\dot{\mathbf{U}}$-submodule ${}_{\mathcal{A}}M_\lambda$ of the Verma module M_λ as follows. As an $\mathcal{A}$-module, it is the $\mathcal{A}$-submodule ${}_{\mathcal{A}}\mathbf{f}$ of $\mathbf{f} = M_\lambda$. This submodule is obviously stable under the action of the operators $F_i^{(n)}1_\zeta \in \dot{\mathbf{U}}$ on M_λ; using repeatedly the commutation formulas 23.1.3, we see that it is also stable under the operators $E_i^{(n)}1_\zeta \in \dot{\mathbf{U}}$ on M_λ; hence it is a ${}_{\mathcal{A}}\dot{\mathbf{U}}$-submodule of M_λ. The same $\mathcal{A}$-submodule is an ${}_{\mathcal{A}}\dot{\mathbf{U}}$-submodule of ${}^\omega M_\lambda$, denoted ${}^\omega_{\mathcal{A}}M_\lambda$. The argument used to prove 23.3.1(c) can be repeated word for word for $\mathcal{A}$ instead of $\mathbf{Q}(v)$ and gives the following result.

(a) Given $\lambda, \lambda' \in X$, the map $u \mapsto u(1 \otimes 1)$ defines an $\mathcal{A}$-linear isomorphism π' of ${}_{\mathcal{A}}\dot{\mathbf{U}}1_{\lambda'-\lambda}$ onto the $\mathcal{A}$-submodule ${}^\omega_{\mathcal{A}}M_\lambda \otimes_{\mathcal{A}} ({}_{\mathcal{A}}M_{\lambda'})$ of ${}^\omega M_\lambda \otimes M_{\lambda'}$.

23.3.3. Let $\zeta \in X$ and let $a = \sum_i a_i i \in \mathbf{N}[I], a' = \sum_i a'_i i \in \mathbf{N}[I]$ be such that $\langle i, \zeta \rangle = a'_i - a_i$ for all $i \in I$.

Let $P(\zeta, a, a')$ be the left ideal $\sum_{i,n>a'_i} \dot{\mathbf{U}}F_i^{(n)}1_\zeta + \sum_{i,n>a_i} \dot{\mathbf{U}}E_i^{(n)}1_\zeta$ of $\dot{\mathbf{U}}$.

Let ${}_{\mathcal{A}}P(\zeta, a, a')$ be the left ideal $\sum_{i,n>a'_i} {}_{\mathcal{A}}\dot{\mathbf{U}}F_i^{(n)}1_\zeta + \sum_{i,n>a_i} {}_{\mathcal{A}}\dot{\mathbf{U}}E_i^{(n)}1_\zeta$ of ${}_{\mathcal{A}}\dot{\mathbf{U}}$.

23.3.4. Let $(Y', X', \dots)$ be the simply connected root datum of type $(I, \cdot)$ and let $f : Y' \to Y, g : X \to X'$ be the unique morphism of root data. Let $\dot{\mathbf{U}}'$ be the algebra defined like $\dot{\mathbf{U}}$, in terms of $(Y', X', \dots)$. Let $\zeta \in X, a, a'$ be as in 23.3.3, and let $\zeta' = g(\zeta)$.

The natural isomorphism $\dot{\mathbf{U}}'1_{\zeta'} \to \dot{\mathbf{U}}1_\zeta$ (see 23.2.5) carries, for any i and any $n \geq 0$, the subspace $\dot{\mathbf{U}}'F_i^{(n)}1_{\zeta'}$ isomorphically onto $\dot{\mathbf{U}}F_i^{(n)}1_\zeta$ and the subspace $\dot{\mathbf{U}}'E_i^{(n)}1_{\zeta'}$ isomorphically onto $\dot{\mathbf{U}}E_i^{(n)}1_\zeta$. Hence it carries the left ideal $P(\zeta', a, a')$ isomorphically onto the left ideal $P(\zeta, a, a')$.

The same argument shows that the isomorphism ${}_{\mathcal{A}}\dot{\mathbf{U}}'1_{\zeta'} \to {}_{\mathcal{A}}\dot{\mathbf{U}}1_\zeta$ (see 23.2.5) carries the left ideal ${}_{\mathcal{A}}P(\zeta', a, a')$ isomorphically onto the left ideal ${}_{\mathcal{A}}P(\zeta, a, a')$.

23.3.5. In the remainder of this section we assume that the root datum $(Y, X, \dots)$ is Y-regular. Let $\lambda, \lambda' \in X^+$. We set $\langle i, \lambda \rangle = a_i, \langle i, \lambda' \rangle = a'_i$ for all i and $\zeta = \lambda' - \lambda$. Let $a = \sum_i a_i i, a' = \sum_i a'_i i$.

Let $\mathcal{T}$ (resp. $\mathcal{T}'$) be the kernel of the canonical homomorphism of $\mathbf{U}$-modules $\mathbf{f} = {}^\omega M_\lambda \to {}^\omega \Lambda_\lambda$ (resp. $\mathbf{f} = M_{\lambda'} \to \Lambda_{\lambda'}$). By 14.4.11, $\mathcal{T}$ (resp. $\mathcal{T}'$) is the subspace of $\mathbf{f}$ generated by a subset of the basis $\mathbf{B}$, namely by $\mathbf{D} = \cup_{i,n;n>a_i} {}^\sigma\mathbf{B}_{i,n}$ (resp. $\mathbf{D}' = \cup_{i,n;n>a'_i} {}^\sigma\mathbf{B}_{i,n}$).

Taking tensor products, we obtain a surjective homomorphism of $\mathbf{U}$-modules, or $\dot{\mathbf{U}}$-modules

$$\chi : {}^\omega M_\lambda \otimes M_{\lambda'} \to {}^\omega \Lambda_\lambda \otimes \Lambda_{\lambda'}$$

and we see that

(a) the kernel of χ is the subspace $\mathcal{T} \otimes M_{\lambda'} + M_\lambda \otimes \mathcal{T}'$ of $\mathbf{f} \otimes \mathbf{f}$.

Using 23.2.1(a),(b), and the description of $\mathcal{T}', \mathcal{T}$ given above, we see that π' (see 23.3.1) maps

(b) the subspace $\sum_{b\in\mathbf{B};b'\in\mathbf{D}'}\mathbf{Q}(v)b^+b'^-1_\zeta = \sum_{i,n>\langle i,\lambda'\rangle}\dot{\mathbf{U}}F_i^{(n)}1_\zeta$ of $\dot{\mathbf{U}}1_\zeta$ onto the subspace $\sum_{b\in\mathbf{B};b'\in\mathbf{D}'}\mathbf{Q}(v)b\otimes b' = M_\lambda\otimes\mathcal{T}'$ of $\mathbf{f}\otimes\mathbf{f}$ and

(c) the subspace $\sum_{b\in\mathbf{D};b'\in\mathbf{B}}\mathbf{Q}(v)b'^-b^+1_\zeta = \sum_{i,n>\langle i,\lambda\rangle}\dot{\mathbf{U}}E_i^{(n)}1_\zeta$ of $\dot{\mathbf{U}}1_\zeta$ onto the subspace $\sum_{b\in\mathbf{D};b'\in\mathbf{B}}\mathbf{Q}(v)b\otimes b' = \mathcal{T}\otimes M_{\lambda'}$ of $\mathbf{f}\otimes\mathbf{f}$.

Combining (a),(b),(c), we see that π' maps the subspace $P(\zeta,a,a')$ of $\dot{\mathbf{U}}1_\zeta$ onto the kernel of χ. This fact, together with 23.3.1(c), implies the following result.

Proposition 23.3.6. *In the setup above, the assignment $u\mapsto u(\xi_{-\lambda}\otimes\eta_{\lambda'})$ defines a surjective linear map $\pi:\dot{\mathbf{U}}1_\zeta\to{}^\omega\Lambda_\lambda\otimes\Lambda_{\lambda'}$ with kernel equal to $P(\zeta,a,a')$.*

23.3.7. Assume that $\lambda\in X^+$. As in 19.3.1, we denote by ${}_\mathcal{A}\Lambda_\lambda$ the $\mathcal{A}$-submodule of Λ_λ generated by its canonical basis. Using 19.3.2, we see that ${}_\mathcal{A}\Lambda_\lambda$ is an ${}_\mathcal{A}\dot{\mathbf{U}}$-submodule of Λ_λ. The same $\mathcal{A}$-submodule is an ${}_\mathcal{A}\dot{\mathbf{U}}$-submodule of ${}^\omega\Lambda_\lambda$, denoted by ${}^\omega_\mathcal{A}\Lambda_\lambda$. From Theorem 14.3.2(b), we see that the kernel of the the $\mathcal{A}$-linear map ${}_\mathcal{A}\mathbf{f}\to{}_\mathcal{A}\Lambda_\lambda$ given by $x\mapsto x^-\eta_\lambda$ is the left ideal ${}_\mathcal{A}\mathcal{T}$ of ${}_\mathcal{A}\mathbf{f}$ generated by the elements $\theta_i^{(n)}$ for various $i\in I$ and $n>\langle i,\lambda\rangle$.

We can now repeat word for word the proof of Proposition 23.3.6 for $\mathcal{A}$ instead of $\mathbf{Q}(v)$ and we obtain the following result, where $\lambda,\lambda'\in X^+$.

Proposition 23.3.8. *The assignment $u\mapsto u(\xi_{-\lambda}\otimes\eta_{\lambda'})$ defines a surjective $\mathcal{A}$-linear map π of ${}_\mathcal{A}\dot{\mathbf{U}}1_\zeta$ onto the $\mathcal{A}$-submodule ${}^\omega_\mathcal{A}\Lambda_\lambda\otimes_\mathcal{A}({}_\mathcal{A}\Lambda_{\lambda'})$ of ${}^\omega\Lambda_\lambda\otimes\Lambda_{\lambda'}$, with kernel equal to ${}_\mathcal{A}P(\zeta,a,a')$.*

Corollary 23.3.9. *${}^\omega_\mathcal{A}\Lambda_\lambda\otimes_\mathcal{A}({}_\mathcal{A}\Lambda_{\lambda'})$ is an ${}_\mathcal{A}\dot{\mathbf{U}}$-submodule of the $\dot{\mathbf{U}}$-module ${}^\omega\Lambda_\lambda\otimes\Lambda_{\lambda'}$.*

Proposition 23.3.10. *Let $M\in\mathcal{C}'$ and let $m\in M^\zeta$. There exist $\lambda,\lambda'\in X^+$ such that $\lambda'-\lambda=\zeta$ and a morphism $f:{}^\omega\Lambda_\lambda\otimes\Lambda_{\lambda'}\to M$ in $\mathcal{C}'$ such that $f(\xi_{-\lambda}\otimes\eta_{\lambda'})=m$.*

Since M is integrable, we can find integers $a_i,a_i'\in\mathbf{N}$ such that $E_i^{(a)}m=0$ for all i and all $a>a_i$ and $F_i^{(a')}m=0$ for all i and all $a'>a_i'$. Since the root datum is Y-regular, we can find $\lambda\in X$ such that $\langle i,\lambda\rangle\ge a_i$ and $\langle i,\lambda+\zeta\rangle\ge a_i'$ for all i. Let $\lambda'=\lambda+\zeta$. Then $\langle i,\lambda'\rangle\ge a_i'$ for all i.

The existence of f follows now from 23.3.6.

CHAPTER 24

Canonical Bases in Certain Tensor Products

24.1. Integrality Properties of the Quasi-$\mathcal{R}$-Matrix

24.1.1. Let $M, M' \in \mathcal{C}$ be such that either ${}^{\omega}M \in \mathcal{C}^{hi}$ or $M' \in \mathcal{C}^{hi}$ (see 3.4.7). We regard $M \otimes M'$ naturally as a $\mathbf{U} \otimes \mathbf{U}$-module and we define a linear map $\Theta : M \otimes M' \to M \otimes M'$ by $\Theta(m \otimes m') = \sum_\nu \Theta_\nu(m \otimes m')$ (notation of 4.1.2.) This is well-defined since only finitely many terms of the sum are non-zero.

Lemma 24.1.2. *Let M, M' be as above. We have*

(a) $\Delta(u)\Theta(m \otimes m') = \Theta(\bar{\Delta}(u)(m \otimes m'))$.

(b) *Assume that we are given* $\mathbf{Q}$*-linear maps* $^- : M \to M$ *and* $^- : M' \to M'$ *such that* $\overline{um} = \bar{u}\bar{m}$ *and* $\overline{um'} = \bar{u}\bar{m}'$ *for all* $u \in \mathbf{U}, m \in M, m' \in M'$*. Let* $^- = {}^- \otimes {}^- : M \otimes M' \to M \otimes M'$*. Then* $\Delta(u)\Theta(m \otimes m') = \Theta(\overline{\Delta(\bar{u})(\bar{m} \otimes \bar{m}')})$ *for any* $m \in M, m' \in M'$ *and any* $u \in \mathbf{U}$*.*

The set of u for which (a) holds is clearly a subalgebra of $\mathbf{U}$ containing all K_μ. Hence it suffices to check (a) in the special case where u is one of the algebra generators E_i, F_i. Applying both sides of the equalities 4.2.5(c),(d) (with large p) to $m \otimes m' \in M \otimes M'$, we obtain

$$(E_i \otimes 1 + \tilde{K}_i \otimes E_i)\Theta(m \otimes m') = \Theta(E_i \otimes 1 + \tilde{K}_{-i} \otimes E_i)(m \otimes m')$$

$$(1 \otimes F_i + F_i \otimes \tilde{K}_{-i})\Theta(m \otimes m') = \Theta(1 \otimes F + F_i \otimes \tilde{K}_i)(m \otimes m').$$

This proves (a). Now (b) is a consequence of (a). The lemma follows.

24.1.3. The following property is just a reformulation of the property in Lemma 24.1.2(b). Let M, M' be as in that lemma. Then for any $m \in M, m' \in M', u \in \dot{\mathbf{U}}$ we have

(a) $$u\Theta(m \otimes m') = \Theta(\overline{\bar{u}(\bar{m} \otimes \bar{m}')}).$$

G. Lusztig, *Introduction to Quantum Groups*, Modern Birkhäuser Classics,
DOI 10.1007/978-0-8176-4717-9_24, © Springer Science+Business Media, LLC 2010

Proposition 24.1.4. *Let $\lambda, \lambda' \in X$. Consider the Verma modules $M_\lambda, M_{\lambda'}$. Note that $M = {}^\omega M_\lambda \in \mathcal{C}$ and $M' = M_{\lambda'} \in \mathcal{C}^{hi}$; hence*

$$\Theta : M \otimes M' \to M \otimes M'$$

is well-defined. Then Θ leaves stable the $\mathcal{A}$-submodule ${}^\omega_{\mathcal{A}} M_\lambda \otimes_{\mathcal{A}} ({}_{\mathcal{A}}M_{\lambda'})$.

Since the ambient space of M and M' is $\mathbf{f}$, we may regard $^- : \mathbf{f} \to \mathbf{f}$ as maps $^- : M \to M, ^- : M' \to M'$. Using the definition of Verma modules, we may identify $M' = \mathbf{U}/\sum_i \mathbf{U}E_i + \sum_\mu \mathbf{U}(K_\mu - v^{\langle \mu, \lambda' \rangle}1)$ as a $\mathbf{U}$-module so that $^- : M' \to M'$ is induced by $^- : \mathbf{U} \to \mathbf{U}$. It follows that $\overline{um'} = \bar{u}\bar{m}'$ for all $u \in \mathbf{U}$ and $m' \in M'$. Similarly, we have $\overline{um} = \bar{u}\bar{m}$ for all $u \in \mathbf{U}$ and $m \in M$.

As in Lemma 24.1.2, we set $^- = {}^- \otimes {}^- : M \otimes M' \to M \otimes M'$ and we have

$$u\Theta(m \otimes m') = \Theta(\overline{\bar{u}(\bar{m} \otimes \bar{m}')})$$

for all $u \in \dot{\mathbf{U}}, m \in M, m' \in M'$. In particular, taking $m = 1 = \bar{1}, m' = 1 = \bar{1}$, we obtain

(a) $u(1 \otimes 1) = \Theta(\overline{\bar{u}(1 \otimes 1)})$ for all $u \in \dot{\mathbf{U}}$,

since $1 = \bar{1}, 1 = \bar{1}$, and $\Theta(1 \otimes 1) = 1 \otimes 1$. Let $x \in {}^\omega_{\mathcal{A}} M_\lambda \otimes_{\mathcal{A}} ({}_{\mathcal{A}}M_{\lambda'})$. Then $x = \bar{x}'$ where $x' \in {}^\omega_{\mathcal{A}} M_\lambda \otimes_{\mathcal{A}} ({}_{\mathcal{A}}M_{\lambda'})$, since the involution $^- \otimes {}^- : \mathbf{f} \to \mathbf{f}$ leaves ${}_{\mathcal{A}}\mathbf{f} \otimes_{\mathcal{A}} ({}_{\mathcal{A}}\mathbf{f})$ stable.

With the notation of 23.3.2(a), we have $x' = \pi'(u')$ for some $u' \in {}_{\mathcal{A}}\dot{\mathbf{U}}1_{\lambda'-\lambda}$. Since ${}_{\mathcal{A}}\dot{\mathbf{U}}1_{\lambda'-\lambda}$ is stable under the involution $^- : \dot{\mathbf{U}} \to \dot{\mathbf{U}}$, we have $u' = \bar{u}$ for some $u \in {}_{\mathcal{A}}\dot{\mathbf{U}}1_{\lambda'-\lambda}$. Hence $x = \bar{x}' = \overline{\bar{u}(1 \otimes 1)}$. Using (a), we have $\Theta(x) = \Theta(\overline{\bar{u}(1 \otimes 1)}) = u(1 \otimes 1) = \pi'(u)$. Using again 23.3.2(a), we have $\pi'(u) \in {}^\omega_{\mathcal{A}} M_\lambda \otimes_{\mathcal{A}} ({}_{\mathcal{A}}M_{\lambda'})$; hence $\Theta(x) \in {}^\omega_{\mathcal{A}} M_\lambda \otimes_{\mathcal{A}} ({}_{\mathcal{A}}M_{\lambda'})$. The proposition is proved.

Corollary 24.1.5. *Assume that the root datum is Y-regular. Let $\lambda, \lambda' \in X^+$. The map $\Theta : {}^\omega\Lambda_\lambda \otimes \Lambda_{\lambda'} \to {}^\omega\Lambda_\lambda \otimes \Lambda_{\lambda'}$ leaves stable the $\mathcal{A}$-submodule ${}^\omega_{\mathcal{A}}\Lambda_\lambda \otimes_{\mathcal{A}} ({}_{\mathcal{A}}\Lambda_{\lambda'})$.*

This follows immediately from the previous proposition since ${}^\omega_{\mathcal{A}}\Lambda_\lambda \otimes_{\mathcal{A}} ({}_{\mathcal{A}}\Lambda_{\lambda'})$ is the image of ${}^\omega_{\mathcal{A}} M_\lambda \otimes_{\mathcal{A}} ({}_{\mathcal{A}}M_{\lambda'})$ under the natural map ${}^\omega M_\lambda \otimes M_{\lambda'} \to {}^\omega\Lambda_\lambda \otimes \Lambda_{\lambda'}$.

Corollary 24.1.6. *Assume that $(I, \cdot)$ is of finite type. Write*

$$\Theta = \sum_\nu \sum_{b,b' \in \mathbf{B}_\nu} p_{b,b'} b^- \otimes b'^+ \quad (p_{b,b'} \in \mathbf{Q}(v)).$$

For any ν^0 and any $b^0, b'^0 \in \mathbf{B}_{\nu^0}$, we have $p_{b^0,b'^0} \in \mathcal{A}$.

We can find $\lambda, \lambda' \in X^+$ such that $b^0 \in \mathbf{B}(-w_0(\lambda)), b'^0 \in \mathbf{B}(-w_0(\lambda'))$. By 24.1.5, $\Theta : {}^\omega\Lambda_\lambda \otimes \Lambda_{\lambda'} \to {}^\omega\Lambda_\lambda \otimes \Lambda_{\lambda'}$ maps the $\mathcal{A}$-submodule ${}^\omega_{\mathcal{A}}\Lambda_\lambda \otimes_{\mathcal{A}} ({}_{\mathcal{A}}\Lambda_{\lambda'})$ into itself. By 21.1.2, we may canonically identify as $\mathbf{U}$-modules ${}^\omega\Lambda_\lambda$ with $\Lambda_{-w_0(\lambda)}$ and $\Lambda_{\lambda'}$ with ${}^\omega\Lambda_{-w_0(\lambda')}$ respecting the canonical bases. It follows that $\Theta : \Lambda_{-w_0(\lambda)} \otimes {}^\omega\Lambda_{-w_0(\lambda')} \to \Lambda_{-w_0(\lambda)} \otimes {}^\omega\Lambda_{-w_0(\lambda')}$ maps the $\mathcal{A}$-submodule ${}_{\mathcal{A}}\Lambda_{-w_0(\lambda)} \otimes_{\mathcal{A}} ({}^\omega_{\mathcal{A}}\Lambda_{-w_0(\lambda')})$ into itself. In particular, this submodule contains the vector $\Theta(\eta \otimes \xi) = \sum_\nu \sum_{b,b'} p_{b,b'} b^- \eta \otimes b'^+ \xi$. Here, $\eta = \eta_{-w_0(\lambda)}$ and $\xi = \xi_{w_0(\lambda')}$; b runs over $\mathbf{B}_\nu \cap \mathbf{B}(-w_0(\lambda))$ and b' runs over $\mathbf{B}_\nu \cap \mathbf{B}(-w_0(\lambda'))$. Since the elements $b^- \eta \otimes b'^+ \xi$ (for all indices (ν, b, b') as in the sum) are a part of an $\mathcal{A}$-basis of ${}_{\mathcal{A}}\Lambda_{-w_0(\lambda)} \otimes_{\mathcal{A}} ({}^\omega_{\mathcal{A}}\Lambda_{-w_0(\lambda')})$, and (ν^0, b^0, b'^0) is an index in the sum, it follows that $p_{b^0,b'^0} \in \mathcal{A}$. The corollary is proved.

24.2. A Lemma on Systems of Semi-Linear Equations

Lemma 24.2.1. *Let H be a set with a partial order $\le$ such that for any $h \le h'$ in H, the set $\{h'' | h \le h'' \le h'\}$ is finite. Assume that for each $h \le h'$ in H we are given an element $r_{h,h'} \in \mathcal{A}$ such that*

(a) $r_{h,h} = 1$ *for all h;*

(b) $\sum_{h''; h \le h'' \le h'} \bar{r}_{h,h''} r_{h'',h'} = \delta_{h,h'}$ *for all $h \le h'$ in H.*

Then there is a unique family of elements $p_{h,h'} \in \mathbf{Z}[v^{-1}]$ defined for all $h \le h'$ in H such that

(c) $p_{h,h} = 1$ *for all $h \in H$;*

(d) $p_{h,h'} \in v^{-1}\mathbf{Z}[v^{-1}]$ *for all $h < h'$ in H;*

(e) $p_{h,h'} = \sum_{h''; h \le h'' \le h'} \bar{p}_{h,h''} r_{h'',h'}$ *for all $h \le h'$ in H.*

For $h \le h'$ in H, we denote by $d(h, h')$ the maximum length of a chain $h = h_0 < h_1 < h_2 < \cdots < h_p = h'$ in H. Note that $d(h, h') < \infty$ by our assumption. For any $n \ge 0$, we consider the statement P_n which is the assertion of the lemma restricted to elements $h \le h'$ such that $d(h, h') \le n$. Note that property (e) is meaningful for this statement. We prove P_n by induction on n. The case $n = 0$ is trivial. Assume now that $n \ge 1$. Let $h \le h'$. If $d(h, h') < n$, then $p_{h,h'}$ is defined by P_{n-1}. If $d(h, h') = n$, we note that $q = \sum_{h''; h \le h'' < h'} \bar{p}_{h,h''} r_{h'',h'}$ is defined. We show that $\bar{q} + q = 0$.

Indeed, using P_{n-1} and (a),(b), we have

$$\begin{aligned}\bar{q}+q &= \sum_{h'';h\le h''<h'} p_{h,h''}\bar{r}_{h'',h'} + \sum_{h_1;h\le h_1<h'} \bar{p}_{h,h_1}r_{h_1,h'}\\ &= \sum_{h'',h_1;h\le h_1\le h''<h'} \bar{p}_{h,h_1}r_{h_1,h''}\bar{r}_{h'',h'} + \sum_{h'',h_1;h\le h_1<h''=h'} \bar{p}_{h,h_1}r_{h_1,h'}\bar{r}_{h'',h'}\\ &= \sum_{h'',h_1;h\le h_1\le h''\le h';h_1<h'} \bar{p}_{h,h_1}r_{h_1,h''}\bar{r}_{h'',h'} = \sum_{h_1;h\le h_1<h'} \bar{p}_{h,h_1}\delta_{h_1,h'} = 0,\end{aligned}$$

as required. Since $q \in \mathcal{A}$ satisfies $\bar{q}+q=0$, there is a unique element $q' \in v^{-1}\mathbf{Z}[v^{-1}]$ such that $q'-\bar{q}'=q$. We set $p_{h,h'}=q'$. Then properties (c),(d),(e) are clearly satisfied as far as P_n is concerned. This proves the existence in P_n. The previous proof also shows uniqueness. The lemma is proved.

24.3. The Canonical Basis of ${}^\omega\Lambda_\lambda \otimes \Lambda_{\lambda'}$

24.3.1. In this section we assume that the root datum is Y-regular.

Let $\lambda, \lambda' \in X^+$. We shall consider the following partial order on the set $\mathbf{B}\times\mathbf{B}$: we say that $(b_1,b_1')\le(b_2,b_2')$ if $\operatorname{tr}|b_1| - \operatorname{tr}|b_1'| = \operatorname{tr}|b_2| - \operatorname{tr}|b_2'|$ and if we have either

$$\operatorname{tr}|b_1| < \operatorname{tr}|b_2| \text{ and } \operatorname{tr}|b_1'| < \operatorname{tr}|b_2'|,$$

or

$$b_1 = b_2 \text{ and } b_1' = b_2'.$$

This induces, for given $\lambda,\lambda'\in X^+$, a partial order on the set $\mathbf{B}(\lambda)\times\mathbf{B}(\lambda')$.

As in 19.3.4, let $^- : \Lambda_{\lambda'} \to \Lambda_{\lambda'}$ be the unique $\mathbf{Q}$-linear involution such that $\overline{u\eta_{\lambda'}} = \bar{u}\eta_{\lambda'}$ for all $u \in \mathbf{U}$; similarly, let $^- : {}^\omega\Lambda_\lambda \to {}^\omega\Lambda_\lambda$ be the unique $\mathbf{Q}$-linear involution such that $\overline{u\xi_{-\lambda}} = \bar{u}\xi_{-\lambda}$ for all $u\in\mathbf{U}$. Let $^- = {}^-\otimes{}^- : {}^\omega\Lambda_\lambda\otimes\Lambda_{\lambda'} \to {}^\omega\Lambda_\lambda\otimes\Lambda_{\lambda'}$. The elements $b^+\xi_{-\lambda}\otimes b'^-\eta_{\lambda'}$ with $b\in\mathbf{B}(\lambda)$ and $b'\in\mathbf{B}(\lambda')$ form a $\mathbf{Q}(v)$-basis of ${}^\omega\Lambda_\lambda\otimes\Lambda_{\lambda'}$. They generate a $\mathbf{Z}[v^{-1}]$-submodule $\mathcal{L}=\mathcal{L}_{\lambda,\lambda'}$ and an $\mathcal{A}$-submodule ${}_\mathcal{A}\mathcal{L}$.

24.3.2. Now $\Theta : {}^\omega\Lambda_\lambda\otimes\Lambda_{\lambda'} \to {}^\omega\Lambda_\lambda\otimes\Lambda_{\lambda'}$ is well-defined (see 24.1.1). Let $\Psi : {}^\omega\Lambda_\lambda\otimes\Lambda_{\lambda'} \to {}^\omega\Lambda_\lambda\otimes\Lambda_{\lambda'}$ be given by $\Psi(x)=\Theta(\bar{x})$. Since Θ and $^- : {}^\omega\Lambda_\lambda\otimes\Lambda_{\lambda'} \to {}^\omega\Lambda_\lambda\otimes\Lambda_{\lambda'}$ leave ${}_\mathcal{A}\mathcal{L}$ stable (see 24.1.5), we have $\Psi({}_\mathcal{A}\mathcal{L}) \subset {}_\mathcal{A}\mathcal{L}$. From 24.1.2 and 4.1.3, it follows that $\Psi^2=1$. We clearly have $\Psi(fx) = \bar{f}\Psi(x)$ for all $f \in \mathcal{A}$ and all x. From the definition we have for all $b_1 \in \mathbf{B}(\lambda), b'_1 \in \mathbf{B}(\lambda')$:

$$\Psi(b_1^+\xi_{-\lambda} \otimes b_1'^-\eta_{\lambda'}) = \sum_{b_2 \in \mathbf{B}(\lambda), b'_2 \in \mathbf{B}(\lambda')} \rho_{b_1,b'_1;b_2,b'_2} b_2^+\xi_{-\lambda} \otimes b_2'^-\eta_{\lambda'}$$

where $\rho_{b_1,b'_1;b_2,b'_2} \in \mathcal{A}$, and $\rho_{b_1,b'_1;b_2,b'_2} = 0$ unless $(b_1, b'_1) \geq (b_2, b'_2)$; hence the last sum is finite.

Note also that $\rho_{b_1,b'_1;b_1,b'_1} = 1$ and

$$\sum_{b_2 \in \mathbf{B}(\lambda), b'_2 \in \mathbf{B}(\lambda')} \bar{\rho}_{b_1,b'_1;b_2,b'_2} \rho_{b_2,b'_2;b_3,b'_3} = \delta_{b_1,b_3} \delta_{b'_1,b'_3},$$

for any $b_1, b_3 \in \mathbf{B}(\lambda), b'_1, b'_3 \in \mathbf{B}(\lambda')$; the last condition follows from $\Psi^2 = 1$. Applying Lemma 24.2.1 to the set $H = \mathbf{B}(\lambda) \times \mathbf{B}(\lambda')$, we see that there is a unique family of elements $\pi_{b_1,b'_1;b_2,b'_2} \in \mathbf{Z}[v^{-1}]$ defined for $b_1, b_2 \in \mathbf{B}(\lambda), b'_1, b'_2 \in \mathbf{B}(\lambda')$ such that

$\pi_{b_1,b'_1;b_1,b'_1} = 1$;

$\pi_{b_1,b'_1;b_2,b'_2} \in v^{-1}\mathbf{Z}[v^{-1}]$ if $(b_1, b'_1) \neq (b_2, b'_2)$;

$\pi_{b_1,b'_1;b_2,b'_2} = 0$ unless $(b_1, b'_1) \geq (b_2, b'_2)$;

$\pi_{b_1,b'_1;b_2,b'_2} = \sum_{b_3,b'_3} \bar{\pi}_{b_1,b'_1;b_3,b'_3} \rho_{b_3,b'_3;b_2,b'_2}$ for all $(b_1, b'_1) \geq (b_2, b'_2)$.

We have the following result.

Theorem 24.3.3. (a) *For any $(b_1, b'_1) \in \mathbf{B}(\lambda) \times \mathbf{B}(\lambda')$, there is a unique element $(b_1 \diamondsuit b'_1)_{\lambda,\lambda'} \in \mathcal{L}$ such that*

$\Psi((b_1 \diamondsuit b'_1)_{\lambda,\lambda'}) = (b_1 \diamondsuit b'_1)_{\lambda,\lambda'}$ *and* $(b_1 \diamondsuit b'_1)_{\lambda,\lambda'} - b_1^+\xi_{-\lambda} \otimes b_1'^-\eta_{\lambda'} \in v^{-1}\mathcal{L}$.

(b) *The element $(b_1 \diamondsuit b'_1)_{\lambda,\lambda'}$ in (a) is equal to $b_1^+\xi_{-\lambda} \otimes b_1'^-\eta_{\lambda'}$ plus a linear combination of elements $b_2^+\xi_{-\lambda} \otimes b_2'^-\eta_{\lambda'}$ with $(b_2, b'_2) \in \mathbf{B}(\lambda) \times \mathbf{B}(\lambda')$, $(b_2, b'_2) < (b_1, b'_1)$ and with coefficients in $v^{-1}\mathbf{Z}[v^{-1}]$.*

(c) *The elements $(b_1 \diamondsuit b'_1)_{\lambda,\lambda'}$ with b_1, b'_1 as above form a $\mathbf{Q}(v)$-basis of ${}^\omega\Lambda_\lambda \otimes \Lambda_{\lambda'}$, an $\mathcal{A}$-basis of ${}_{\mathcal{A}}\mathcal{L}$ and a $\mathbf{Z}[v^{-1}]$-basis of $\mathcal{L}$.*

(d) *The natural homomorphism $\mathcal{L} \cap \Psi(\mathcal{L}) \to \mathcal{L}/v^{-1}\mathcal{L}$ is an isomorphism.*

The element $(b_1 \diamondsuit b'_1)_{\lambda,\lambda'} = \sum_{b_2,b'_2} \pi_{b_1,b'_1;b_2,b'_2} b_2^+\xi_{-\lambda} \otimes b_2'^-\eta_{\lambda'}$ (see 24.3.2) satisfies the requirements of (a). This shows existence in (a). It is also clear that the elements $(b_1 \diamondsuit b'_1)_{\lambda,\lambda'}$ just defined satisfy the requirements of (b),(c),(d). It remains to show the uniqueness in (a). It is enough to show that an element $x \in v^{-1}\mathcal{L}$, such that $\bar{x} = x$, is necessarily 0. But this follows from (d).

24.3.4. The basis $(b_1 \diamondsuit b_1')_{\lambda,\lambda'}$ in 24.3.3(c) is called the *canonical basis* of ${}^\omega\Lambda_\lambda \otimes \Lambda_{\lambda'}$.

24.3.5. Let $\lambda, \tilde{\lambda} \in X^+$. Let $(Y', X', \dots)$ be the simply connected root datum and let $f : Y' \to Y, g : X \to X'$ be the unique morphism of root data. Let $\dot{\mathbf{U}}'$ be the algebra defined like $\dot{\mathbf{U}}$, in terms of $(Y', X', \dots)$. Let $\lambda', \tilde{\lambda}' \in X'^+$ be defined by $\lambda' = g(\lambda), \tilde{\lambda}' = g(\tilde{\lambda})$. Then ${}^\omega\Lambda_\lambda \otimes \Lambda_{\tilde{\lambda}}$, defined in terms of $\dot{\mathbf{U}}$, has the same ambient space as ${}^\omega\Lambda_{\lambda'} \otimes \Lambda_{\tilde{\lambda}'}$, defined in terms of $\dot{\mathbf{U}}'$. We have a priori two definitions of the canonical basis of this space, one in terms of $\dot{\mathbf{U}}$, one in terms of $\dot{\mathbf{U}}'$. From the definitions, we easily see that these two bases coincide.

CHAPTER 25

The Canonical Basis $\dot{\mathbf{B}}$ of $\dot{\mathbf{U}}$

25.1. Stability Properties

25.1.1. In this section, the root datum is assumed to be Y-regular.

Proposition 25.1.2. *Let λ, λ' be dominant elements of X. We write $\eta = \eta_\lambda, \eta' = \eta_{\lambda'}, \eta'' = \eta_{\lambda+\lambda'}$.*

(a) *There is a unique homomorphism of $\mathbf{U}$-modules $\chi : \Lambda_{\lambda+\lambda'} \to \Lambda_\lambda \otimes \Lambda_{\lambda'}$ such that $\chi(\eta'') = \eta \otimes \eta'$.*

(b) *Let $b \in \mathbf{B}(\lambda + \lambda')$. We have $\chi(b^- \eta'') = \sum_{b_1,b_2} f(b; b_1, b_2) b_1^- \eta \otimes b_2^- \eta'$, sum over $b_1 \in \mathbf{B}(\lambda), b_2 \in \mathbf{B}(\lambda')$, with $f(b; b_1, b_2) \in \mathbf{Z}[v^{-1}]$.*

(c) *If $b^- \eta' \neq 0$, then $f(b; 1, b) = 1$ and $f(b; 1, b_2) = 0$ for any $b_2 \neq b$. If $b^- \eta' = 0$, then $f(b; 1, b_2) = 0$ for any b_2.*

The vector $\eta \otimes \eta' \in \Lambda_\lambda \otimes \Lambda_{\lambda'}$ satisfies $E_i(\eta \otimes \eta') = 0$, $K_\mu(\eta \otimes \eta') = v^{\langle \mu, \lambda \rangle + \langle \mu, \lambda' \rangle}$. This implies (a) (by 3.5.8). By the definition of comultiplication in $\mathbf{U}$, we can write $\chi(b^- \eta'') = \sum_{b_1,b_2} f(b; b_1, b_2) b_1^- \eta \otimes b_2^- \eta'$, sum over $b_1 \in \mathbf{B}(\lambda), b_2 \in \mathbf{B}(\lambda')$, with $f(b; b_1, b_2) \in \mathbf{Q}(v)$ satisfying $f(b; 1, b_2) = 1$ if $b = b_2$ and $f(b; 1, b_2) = 0$ if $b \neq b_2$. This proves (c).

By 23.2.3, we have $f(b; b_1, b_2) \in \mathcal{A}$ for all b_1, b_2. Hence to prove (b), it suffices to show that $f(b; b_1, b_2) \in \mathbf{A}$ for all b_1, b_2. We have a commutative diagram

$$\begin{array}{ccc} \mathbf{f} & \xrightarrow{\Xi} & \mathbf{f} \otimes \Lambda_{\lambda'} \\ \downarrow & & \downarrow \\ \Lambda_{\lambda+\lambda'} & \xrightarrow{\chi} & \Lambda_\lambda \otimes \Lambda_{\lambda'} \end{array}$$

where Ξ is as in 18.1.4, the left vertical map is given by $x \mapsto x^- \eta''$ and the right vertical map is given by $x \otimes y \to (x^- \eta') \otimes y$. The commutativity of the diagram follows from the definitions. Now our assertion on $f(b; b_1, b_2)$ follows from 22.1.2.

G. Lusztig, *Introduction to Quantum Groups*, Modern Birkhäuser Classics,
DOI 10.1007/978-0-8176-4717-9_25, © Springer Science+Business Media, LLC 2010

Proposition 25.1.3. *Let λ, λ' be dominant elements of X. We write $\xi = \xi_{-\lambda}, \xi' = \xi_{-\lambda'}, \xi'' = \xi_{-\lambda-\lambda'}$.*

(a) *There is a unique homomorphism of $\mathbf{U}$-modules $\chi' : {}^{\omega}\Lambda_{\lambda+\lambda'} \to {}^{\omega}\Lambda_{\lambda'} \otimes {}^{\omega}\Lambda_{\lambda}$ such that $\chi'(\xi'') = \xi' \otimes \xi$.*

(b) *Let $b \in \mathbf{B}(\lambda+\lambda')$. We have $\chi'(b^+\xi'') = \sum_{b_1,b_2} f(b; b_1, b_2) b_2^+\xi' \otimes b_1^+\xi$, sum over $b_1 \in \mathbf{B}(\lambda), b_2 \in \mathbf{B}(\lambda')$, with $f(b; b_1, b_2) \in \mathbf{Z}[v^{-1}]$. If $b^+\xi' \neq 0$, then $f(b; 1, b) = 1$ and $f(b; 1, b_2) = 0$ for any $b_2 \neq b$. If $b^+\xi' = 0$, then $f(b; 1, b_2) = 0$ for any b_2.*

We have a commutative diagram

$$\begin{array}{ccc} \mathbf{U} & \xrightarrow{\Delta} & \mathbf{U} \otimes \mathbf{U} \\ {\scriptstyle \omega}\downarrow & & \downarrow{\scriptstyle \omega\otimes\omega} \\ \mathbf{U} & \xrightarrow{{}^t\Delta} & \mathbf{U} \otimes \mathbf{U} \end{array}$$

(${}^t\Delta$ as in 3.3.4.) Indeed, both compositions in the diagram are algebra homomorphisms; to check that they are equal, it suffices to check that they agree on the generators E_i, F_i, K_μ and that is immediate. Using this commutative diagram, we see immediately that the proposition follows from the previous proposition.

Proposition 25.1.4. *Let $\eta \in \Lambda_\lambda, \xi \in {}^{\omega}\Lambda_\lambda$ be as above.*

(a) *There is a unique homomorphism of $\mathbf{U}$-modules $\delta_\lambda : {}^{\omega}\Lambda_\lambda \otimes \Lambda_\lambda \to \mathbf{Q}(v)$, where $\mathbf{Q}(v)$ is a $\mathbf{U}$-module via the co-unit $\mathbf{U} \to \mathbf{Q}(v)$, such that $\delta_\lambda(\xi \otimes \eta) = 1$.*

(b) *Let $b, b' \in \mathbf{B}(\lambda)$. Then $\delta_\lambda(b^+\xi \otimes b'^-\eta)$ is equal to 1 if $b = b' = 1$ and is in $v^{-1}\mathbf{Z}[v^{-1}]$ otherwise.*

The following statement is equivalent to (a). There is a unique bilinear pairing $[,] : \Lambda_\lambda \times \Lambda_\lambda \to \mathbf{Q}(v)$ such that

$$[\eta, \eta] = 1$$

and

$$[F_i x, y] = -[\tilde{K}_{-i} x, E_i y], [x, F_i y] = -[E_i x, \tilde{K}_{-i} y], [K_{-\mu} x, K_\mu y] = [x, y]$$

for all $x, y \in \Lambda$, all $i \in I$ and all $\mu \in Y$. We then have $\delta_\lambda(x \otimes y) = [x, y]$.

This is also equivalent to the following statement. There is a unique bilinear pairing $[,] : \Lambda_\lambda \times \Lambda_\lambda \to \mathbf{Q}(v)$ such that

$$[\eta, \eta] = 1$$

and

$$[ux, y] = [x, \tilde{\rho}(u)y]$$

for all $x, y \in \Lambda$, and all $u \in \mathbf{U}$, where $\tilde{\rho} : \mathbf{U} \to \mathbf{U}^{opp}$ is the algebra isomorphism given by the composition $S\omega$ (S is the antipode).

This is proved exactly as in 19.1.2. It follows from the definition that $[x, y] = 0$ if $x \in (\Lambda)_\nu, y \in (\Lambda)_{\nu'}$ and $\nu \neq \nu'$. Let $(,)$ be as in 19.1.2. We show by induction on $\operatorname{tr} \nu \geq 0$, where $\nu \in \mathbf{N}[I]$, that

(c) $$[x, y] = (-1)^{\operatorname{tr} \nu} v_{-|\nu|}(x, y)$$

for all $x, y \in (\Lambda)_\nu$. This is obvious for $\nu = 0$. We assume that $\operatorname{tr} \nu \geq 1$. We can assume that $x = F_i x'$ for some i such that $\nu_i > 0$ and some $x' \in (\Lambda)_{\nu - i}$. Then

$$[x, y] = [F_i x', y] = -[\tilde{K}_{-i} x', E_i y] = -[x', \tilde{K}_{-i} E_i y]$$

and

$$(x, y) = (F_i x, y) = v_i (x', \tilde{K}_{-i} E_i y).$$

By the induction hypothesis, we have

$$-[x', \tilde{K}_{-i} E_i y] = (-1)^{\operatorname{tr} \nu} v_{-|\nu - i|}(x', \tilde{K}_{-i} E_i y);$$

hence $[x, y] = (-1)^{\operatorname{tr} \nu} v_{-|\nu|}(x, y)$, which completes the induction.

Now let b, b' be as in (b). We must show that $[b^- \eta, b'^- \eta]$ is in $v^{-1}\mathbf{Z}[v^{-1}]$ unless $b = b' = 1$. We may assume that there exists ν such that $b^- \eta, b'^- \eta$ both belong to $(\Lambda)_\nu$. The result then follows from (c) since, by 19.3.3, we have $(b^- \eta, b'^- \eta) \in \mathbf{Z}[v^{-1}]$.

25.1.5. Let $\lambda, \lambda', \lambda''$ be dominant elements of X. We define a homomorphism of $\mathbf{U}$-modules

$$t : {}^\omega\Lambda_{\lambda + \lambda'} \otimes \Lambda_{\lambda' + \lambda''} \to {}^\omega\Lambda_{\lambda'} \otimes \Lambda_{\lambda''}$$

as the composition of

$$\chi' \otimes \chi : {}^\omega\Lambda_{\lambda + \lambda'} \otimes \Lambda_{\lambda' + \lambda''} \to {}^\omega\Lambda_\lambda \otimes {}^\omega\Lambda_{\lambda'} \otimes \Lambda_{\lambda'} \otimes \Lambda_{\lambda''},$$

where χ', χ are as in 25.1.3, 25.1.2, with

$$1 \otimes \delta_{\lambda'} \otimes 1 : {}^\omega\Lambda_\lambda \otimes {}^\omega\Lambda_{\lambda'} \otimes \Lambda_{\lambda'} \otimes \Lambda_{\lambda''} \to {}^\omega\Lambda_\lambda \otimes \Lambda_{\lambda''}.$$

Lemma 25.1.6.

(a) *Let* $b \in \mathbf{B}(\lambda)$, $b'' \in \mathbf{B}(\lambda'')$. *We have*

$$t(b^+\xi_{-\lambda-\lambda'} \otimes b''^-\eta_{\lambda'+\lambda''}) = b^+\xi_{-\lambda} \otimes b''^-\eta_{\lambda''} \mod v^{-1}\mathcal{L}_{\lambda,\lambda''}$$

(notation of 24.3.1*).*

(b) *Let* $b \in \mathbf{B}(\lambda+\lambda')$, $b'' \in \mathbf{B}(\lambda'+\lambda'')$. *Assume that either* $b \notin \mathbf{B}(\lambda)$, *or* $b'' \notin \mathbf{B}(\lambda'')$. *We have* $t(b^+\xi_{-\lambda-\lambda'} \otimes b''^-\eta_{\lambda'+\lambda''}) = 0 \mod v^{-1}\mathcal{L}_{\lambda,\lambda''}$.

(c) t *is surjective.*

In this proof we shall use the symbol $\equiv$ to denote congruences modulo v^{-1} times a $\mathbf{Z}[v^{-1}]$-submodule spanned by the natural basis.

Using 25.1.2, 25.1.3, 25.1.4, we see that if b, b'' are as in (a), we have

$$\begin{aligned} t(b^+\xi_{-\lambda-\lambda'} \otimes b''^-\eta_{\lambda'+\lambda''}) &\equiv (1 \otimes \delta_{\lambda'} \otimes 1)(b^+\xi_{-\lambda} \otimes \xi_{-\lambda'} \otimes \eta_{\lambda'} \otimes b''^-\eta_{\lambda''}) \\ &= b^+\xi_{-\lambda} \otimes b''^-\eta_{\lambda''}. \end{aligned}$$

Using again 25.1.2, 25.1.3, 25.1.4, we see that if b, b'' are as in (b), we have

$$t(b^+\xi_{-\lambda-\lambda'} \otimes b''^-\eta_{\lambda'+\lambda''}) \equiv 0.$$

Now (c) follows from the fact that $\xi_{-\lambda} \otimes \eta_{\lambda'}$ is in the image of t and it generates the $\mathbf{U}$-module ${}^{\omega}\Lambda_\lambda \otimes \Lambda_{\lambda'}$ (see 23.3.6).

Using the definition 24.3.3 of the elements $(b\diamondsuit b')_{\lambda,\lambda'}$, we can reformulate the previous lemma as follows.

Lemma 25.1.7. (a) *Let* $b \in \mathbf{B}(\lambda)$, $b'' \in \mathbf{B}(\lambda'')$. *We have*

$$t(b\diamondsuit b'')_{\lambda+\lambda',\lambda'+\lambda''} = (b\diamondsuit b'')_{\lambda,\lambda''} \mod v^{-1}\mathcal{L}_{\lambda,\lambda''}.$$

(b) *Let* $b \in \mathbf{B}(\lambda+\lambda')$, $b'' \in \mathbf{B}(\lambda'+\lambda'')$. *Assume that either* $b \notin \mathbf{B}(\lambda)$, *or* $b'' \notin \mathbf{B}(\lambda'')$. *We have*

$$t(b\diamondsuit b'')_{\lambda+\lambda',\lambda'+\lambda''} = 0 \mod v^{-1}\mathcal{L}_{\lambda,\lambda''}.$$

25.1.8. In the following result we show that the maps

$$\Psi : {}^{\omega}\Lambda_{\lambda+\lambda'} \otimes \Lambda_{\lambda'+\lambda''} \to {}^{\omega}\Lambda_{\lambda+\lambda'} \otimes \Lambda_{\lambda'+\lambda''}$$

and

$$\Psi : {}^{\omega}\Lambda_{\lambda} \otimes \Lambda_{\lambda''} \to {}^{\omega}\Lambda_{\lambda} \otimes \Lambda_{\lambda''},$$

defined as in 24.3.2, are compatible with t.

Lemma 25.1.9. *We have $t\Psi = \Psi t$.*

We write ξ, η instead of $\xi_{-\lambda-\lambda'}, \eta_{\lambda'+\lambda''}$.

Since any element of ${}^\omega\Lambda_{\lambda+\lambda'} \otimes \Lambda_{\lambda'+\lambda''}$ is of the form $u(\xi \otimes \eta)$ for some $u \in \dot{\mathbf{U}}$ (see 23.3.6), it is enough to check that

$$t\Theta(\overline{u(\xi \otimes \eta)}) = \Theta\overline{t(u(\xi \otimes \eta))}$$

for all $u \in \dot{\mathbf{U}}$. Using the definition of Θ and its property 24.1.3(a), we have

$$t\Theta(\overline{u(\xi \otimes \eta)}) = t\bar{u}(\Theta(\xi \otimes \eta)) = t\bar{u}(\xi \otimes \eta) = \bar{u}t(\xi \otimes \eta) = \bar{u}(\xi_{-\lambda} \otimes \eta_{\lambda''})$$

and

$$\Theta\overline{t(u(\xi \otimes \eta))} = \Theta\overline{u(t(\xi \otimes \eta))} = \bar{u}\Theta(\xi_{-\lambda} \otimes \eta_{\lambda''}) = \bar{u}(\xi_{-\lambda} \otimes \eta_{\lambda''}).$$

The lemma is proved.

Proposition 25.1.10. (a) *Let $b \in \mathbf{B}(\lambda)$, $b'' \in \mathbf{B}(\lambda'')$. We have*

$$t(b\diamondsuit b'')_{\lambda+\lambda',\lambda'+\lambda''} = (b\diamondsuit b'')_{\lambda,\lambda''}.$$

(b) *Let $b \in \mathbf{B}(\lambda + \lambda')$, $b'' \in \mathbf{B}(\lambda' + \lambda'')$. Assume that either $b \notin \mathbf{B}(\lambda)$, or $b'' \notin \mathbf{B}(\lambda'')$. We have*

$$t(b\diamondsuit b'')_{\lambda+\lambda',\lambda'+\lambda''} = 0.$$

The difference of the two sides of the equality in (a) is in $v^{-1}\mathcal{L}_{\lambda,\lambda''}$ (by 25.1.7) and is fixed by $\Psi : {}^\omega\Lambda_\lambda \otimes \Lambda_{\lambda''} \to {}^\omega\Lambda_\lambda \otimes \Lambda_{\lambda''}$, using the definitions and Lemma 25.1.9; hence that difference is zero, by 24.3.3(d). Thus the equality in (a) holds. Exactly the same proof shows (b).

25.2. Definition of the Basis $\dot{\mathbf{B}}$ of $\dot{\mathbf{U}}$

Theorem 25.2.1. *Assume that the root datum is Y-regular. Let $\zeta \in X$ and let $b, b'' \in \mathbf{B}$.*

(a) *There is a unique element $u = b\diamondsuit_\zeta b'' \in {}_{\mathcal{A}}\dot{\mathbf{U}}1_\zeta$ such that*

$$u(\xi_{-\lambda} \otimes \eta_{\lambda''}) = (b\diamondsuit b'')_{\lambda,\lambda''}$$

for any $\lambda, \lambda'' \in X^+$ such that $b \in \mathbf{B}(\lambda), b'' \in \mathbf{B}(\lambda'')$ and $\lambda'' - \lambda = \zeta$.

(b) *If $\lambda, \lambda'' \in X^+$ are such that $\lambda'' - \lambda = \zeta$ and either $b \notin \mathbf{B}(\lambda)$ or $b'' \notin \mathbf{B}(\lambda'')$, then $u(\xi_{-\lambda} \otimes \eta_{\lambda''}) = 0$ (u as in (a)).*

(c) *The element* u *in (a) satisfies* $\bar{u} = u$.

(d) *The elements* $b\diamondsuit_\zeta b''$, *for various* ζ, b, b'' *as above, form a* $\mathbf{Q}(v)$*-basis of* $\dot{\mathbf{U}}$ *and an* $\mathcal{A}$*-basis of* ${}_{\mathcal{A}}\dot{\mathbf{U}}$.

Since the root datum is Y-regular, we can find $\lambda, \lambda'' \in X^+$ such that $b \in \mathbf{B}(\lambda), b'' \in \mathbf{B}(\lambda'')$ and $\lambda'' - \lambda = \zeta$.

For any integers N_1, N_2, let $P(N_1, N_2)$ be the $\mathcal{A}$-submodule of ${}_{\mathcal{A}}\dot{\mathbf{U}}$ spanned by the elements $b_1^+ b_2^- 1_\zeta$ where b_1, b_2 run through the set of pairs of elements of $\mathbf{B}$ such that $\operatorname{tr} |b_1| \le N_1$, $\operatorname{tr} |b_2| \le N_2$ and $|b_1| - |b_2| = |b| - |b''|$.

By arguments such as in 23.3.1 or 23.3.2, we see that any element of ${}^{\omega}M_\lambda \otimes M_{\lambda''}$, or ${}^{\omega}\Lambda_\lambda \otimes \Lambda_{\lambda''}$, of the form $\beta^+ \xi_{-\lambda} \otimes \beta'^- \eta_{\lambda''}$, with $\beta, \beta' \in \mathbf{B}$, is equal to $u_1(\xi_{-\lambda} \otimes \eta_{\lambda''})$ for some $u_1 \in P(\operatorname{tr} |\beta|, \operatorname{tr} |\beta'|)$; moreover, u_1 can be taken to be equal to $\beta^+ \beta'^- 1_\zeta$ plus an element in $P(\operatorname{tr} |\beta| - 1, \operatorname{tr} |\beta'| - 1)$. From this we deduce that

(e) $(b\diamondsuit b'')_{\lambda,\lambda''} \in {}^{\omega}\Lambda_\lambda \otimes \Lambda_{\lambda''}$ is of the form $u(\xi_{-\lambda} \otimes \eta_{\lambda''})$ for some $u \in P(\operatorname{tr} |b|, \operatorname{tr} |b''|)$; moreover, u can be taken to be equal to $b^+ b''^- 1_\zeta$ plus an element in $P(\operatorname{tr} |b| - 1, \operatorname{tr} |b''| - 1)$.

Assume that u is such an element and that u' is another element with the same properties as u. Then $(u - u')(\xi_{-\lambda} \otimes \eta_{\lambda''}) = 0$; hence, by 23.3.8, we have

$$u - u' \in \sum_{i, n > \langle i, \lambda'' \rangle} {}_{\mathcal{A}}\dot{\mathbf{U}} F_i^{(n)} 1_\zeta + \sum_{i, n > \langle i, \lambda \rangle} {}_{\mathcal{A}}\dot{\mathbf{U}} E_i^{(n)} 1_\zeta.$$

Since $u - u' \in P(\operatorname{tr} |b|, \operatorname{tr} |b''|)$ we deduce that, if $\langle i, \lambda \rangle$ and $\langle i, \lambda'' \rangle$ are large enough (for all i), then we must have $u = u'$. Thus, for such λ, λ'' the element u above is uniquely determined. We denote it by $u_{\lambda,\lambda''}$.

Assume now that $\lambda, \lambda'' \in X^+$ satisfy $b \in \mathbf{B}(\lambda), b'' \in \mathbf{B}(\lambda'')$ and $\lambda'' - \lambda = \zeta$. Let $\lambda' \in X^+$ be such that $\langle i, \lambda' \rangle$ is large enough for all i, so that $u' = u_{\lambda+\lambda', \lambda'+\lambda''}$ is defined.

We show that $u'(\xi_{-\lambda} \otimes \eta_{\lambda''}) = (b\diamondsuit b'')_{\lambda,\lambda''}$. Indeed, if t is as in 25.1.5, we have

$$\begin{aligned} u'(\xi_{-\lambda} \otimes \eta_{\lambda''}) &= u'(t(\xi_{-\lambda-\lambda'} \otimes \eta_{\lambda'+\lambda''})) = t(u'(\xi_{-\lambda-\lambda'} \otimes \eta_{\lambda'+\lambda''})) \\ &= t((b\diamondsuit b'')_{\lambda+\lambda',\lambda'+\lambda''}) = (b\diamondsuit b'')_{\lambda,\lambda''} \end{aligned}$$

where the last equality follows from 25.1.10. It follows that $u_{\lambda,\lambda''}$ is independent of λ, λ'', provided that $\langle i, \lambda \rangle$ and $\langle i, \lambda'' \rangle$ are large enough (for all i); hence we can denote it as u, without specifying λ, λ''. It also follows that u satisfies the requirements of (a). This proves the existence part of (a). The previous proof shows also uniqueness. Thus (a) is proved.

Now let λ, λ'' be as in (b). Let $\lambda' \in X^+$ be such that $\langle i, \lambda' \rangle$ is large enough for all i, so that $u_{\lambda+\lambda', \lambda'+\lambda''}$ is defined (hence it is u). We have

$$\begin{aligned} u(\xi_{-\lambda} \otimes \eta_{\lambda''}) &= u(t(\xi_{-\lambda-\lambda'} \otimes \eta_{\lambda'+\lambda''})) \\ &= t(u(\xi_{-\lambda-\lambda'} \otimes \eta_{\lambda'+\lambda''})) \\ &= t((b\diamondsuit b'')_{\lambda+\lambda', \lambda'+\lambda''}) = 0, \end{aligned}$$

where the last equality follows from 25.1.10. This proves (b).

We prove (c). Let u, λ, λ'' be as in (a). We have

$$\begin{aligned} \bar{u}(\xi_{-\lambda} \otimes \eta_{\lambda''}) &= \bar{u}\Theta(\xi_{-\lambda} \otimes \eta_{\lambda''}) \\ &= \Theta\overline{u(\xi_{-\lambda} \otimes \eta_{\lambda''})} \\ &= \Theta\overline{(b\diamondsuit b'')_{\lambda, \lambda''}} \\ &= (b\diamondsuit b'')_{\lambda, \lambda''}. \end{aligned}$$

Thus $\bar{u}$ satisfies the defining property of u. By uniqueness, we have $\bar{u} = u$. This proves (c).

We prove (d). From (e) we see that, for fixed ζ, we have

$$b\diamondsuit_\zeta b'' = b^+ b''^- 1_\zeta \mod P(\text{ tr } |b| - 1, \text{ tr } |b''| - 1).$$

Since the elements $b^+ b''^- 1_\zeta$ form an $\mathcal{A}$-basis of ${}_{\mathcal{A}}\dot{\mathbf{U}}$, we see that (d) follows. The theorem is proved.

25.2.2. We now drop the assumption that the root datum $(Y, X, \dots)$ is Y-regular. Assume that we are given $\zeta \in X$. Let $(Y', X', \dots)$ be the simply connected root datum of type $(I, \cdot)$ and let $f : Y' \to Y, g : X \to X'$ be the unique morphism of root data. Let $\dot{\mathbf{U}}'$ be the algebra, defined like $\dot{\mathbf{U}}$, in terms of $(Y', X', \dots)$. Let $\zeta' = g(\zeta)$. By 23.2.5, we have a natural isomorphism

(a) $\dot{\mathbf{U}}' 1_{\zeta'} \cong \dot{\mathbf{U}} 1_\zeta$,

defined by $u^+ u'^- 1_{\zeta'} \mapsto u^+ u'^- 1_\zeta$ for all $u, u' \in \mathbf{f}$. For each $b, b'' \in \mathbf{B}$, we denote by $b\diamondsuit_\zeta b''$ the element of $\dot{\mathbf{U}} 1_\zeta$ corresponding under (a) to $b\diamondsuit_{\zeta'} b'' \in \dot{\mathbf{U}}' 1_{\zeta'}$ (which is defined by the previous theorem.)

Corollary 25.2.3. *The elements $b\diamondsuit_\zeta b''$ for various $b, b'' \in \mathbf{B}$ and various $\zeta \in X$ form an $\mathcal{A}$-basis of ${}_{\mathcal{A}}\dot{\mathbf{U}}$ and a $\mathbf{Q}(v)$-basis of $\dot{\mathbf{U}}$. They are all fixed by the involution $^-: \dot{\mathbf{U}} \to \dot{\mathbf{U}}$.*

25.2.4. Remark. The basis of $\dot{\mathbf{U}}$ just defined is called the *canonical basis* of $\dot{\mathbf{U}}$. We denote it by $\dot{\mathbf{B}}$. In the case where the root datum $(Y, X, \dots)$ is Y-regular, this canonical basis coincides with the one defined in 25.2.1. This follows immediately from definitions, using 23.2.5 and 23.3.4.

From the proof we see that any element of $\dot{\mathbf{B}}$ is contained in one of the summands in the direct sum decomposition 23.1.2 of $\dot{\mathbf{U}}$.

Theorem 25.2.5. *Let $\zeta \in X$ and let a, a' be as in 23.3.3. Let $P(\zeta, a, a')$, ${}_{\mathcal{A}}P(\zeta, a, a')$ be as in 23.3.3. Then $\dot{\mathbf{B}} \cap P(\zeta, a, a')$ is an $\mathcal{A}$-basis of ${}_{\mathcal{A}}P(\zeta, a, a')$ and a $\mathbf{Q}(v)$-basis of $P(\zeta, a, a')$.*

Using the definitions and 23.3.4, we are reduced to the case where the root datum is simply connected. In that case, the result follows immediately from Theorem 25.2.1.

We now show that $\dot{\mathbf{B}}$ is a generalization of $\mathbf{B}$.

Proposition 25.2.6. *Let $b \in \mathbf{B}$ and let $\zeta \in X$. Then $b^{-}1_{\zeta} \in \dot{\mathbf{B}}$ and $b^{+}1_{\zeta} \in \dot{\mathbf{B}}$.*

We can assume that the root datum is simply connected. Choose $\lambda, \lambda' \in X^{+}$ such that $\lambda' - \lambda = \zeta$ and such that $b \in \mathbf{B}(\lambda')$. We have $b^{-}1_{\zeta}(\xi_{-\lambda} \otimes \eta_{\lambda'}) = \xi_{-\lambda} \otimes b^{-}\eta_{\lambda'}$. Using the definitions, we see that $\xi_{-\lambda} \otimes b^{-}\eta_{\lambda'}$ satisfies the defining properties of $(1 \diamondsuit b)_{\lambda, \lambda'}$; hence it is equal to $(1 \diamondsuit b)_{\lambda, \lambda'}$. It follows that $b^{-}1_{\zeta} = 1 \diamondsuit_{\zeta} b$. A similar argument shows that $b^{+}1_{\zeta} = b \diamondsuit_{\zeta} 1$. The proposition is proved.

25.3. Example: Rank 1

25.3.1. In this section we assume that $I = \{i\}$ and $X = Y = \mathbf{Z}$ with $i = 1 \in Y, i' = 2 \in X$. Consider the following elements of $\dot{\mathbf{U}}$:

(a) $E_i^{(a)} 1_{-n} F_i^{(b)} \quad (a, b, n \in \mathbf{N}, n \geq a + b)$

(b) $F_i^{(b)} 1_n E_i^{(a)} \quad (a, b, n \in \mathbf{N}, n \geq a + b)$.

Note that

(c) $E_i^{(a)} 1_{-n} F_i^{(b)} = F_i^{(b)} 1_n E_i^{(a)}$ for $n = a + b$.

Proposition 25.3.2. *The canonical basis of $\dot{\mathbf{U}}$ consists of the elements 25.3.1(a) and (b), with the identification 25.3.1(c). More precisely, if $n \geq a + b$, we have*

$$E_i^{(a)} 1_{-n} F_i^{(b)} = \theta_i^{(a)} \diamondsuit_{-n+2b} \theta_i^{(b)}$$

and

$$F_i^{(b)} 1_n E_i^{(a)} = \theta_i^{(a)} \diamondsuit_{n-2a} \theta_i^{(b)}.$$

We compute the image of the elements 25.3.1(a),(b) under the map $\dot{\mathbf{U}} \to {}^{\omega}\Lambda_p \otimes \Lambda_q$, with $p, q \geq 0$, given by $u \mapsto u(\xi_{-p} \otimes \eta_q)$. The image of the element 25.3.1(a) is zero unless $-n + 2b = q - p$, in which case it is

$$\begin{aligned}
&E_i^{(a)} F_i^{(b)} (\xi_{-p} \otimes \eta_q) = E_i^{(a)} (\xi_{-p} \otimes F_i^{(b)} \eta_q) \\
&= \sum_{a'+a''=a} v_i^{a'a''-a''p} E_i^{(a')} \xi_{-p} \otimes E_i^{(a'')} F_i^{(b)} \eta_q \\
&= \sum_{a'+a''=a} \sum_{t \geq 0} v_i^{a'a''-a''p} E_i^{(a')} \xi_{-p} \otimes \begin{bmatrix} a'' - b + q \\ t \end{bmatrix}_i F_i^{(b-t)} E_i^{(a''-t)} \eta_q \\
&= \sum_{a'+a''=a} v_i^{a'a''-a''p} \begin{bmatrix} a'' - b + q \\ a'' \end{bmatrix}_i E_i^{(a')} \xi_{-p} \otimes F_i^{(b-a'')} \eta_q \\
&= \sum_{s \geq 0; s \leq a, s \leq b} v_i^{s(a-s-p)} \begin{bmatrix} s - b + q \\ s \end{bmatrix}_i E_i^{(a-s)} \xi_{-p} \otimes F_i^{(b-s)} \eta_q.
\end{aligned}$$

This element is fixed by the involution Ψ of ${}^{\omega}\Lambda_p \otimes \Lambda_q$, since the element 25.3.1(a) is fixed by $^- : \dot{\mathbf{U}} \to \dot{\mathbf{U}}$. Using the definitions, we see that this element is $(\theta_i^{(a)} \diamond \theta_i^{(b)})_{p,q}$. Hence the element 25.3.1(a) is $\theta_i^{(a)} \diamond_{-n+2b} \theta_i^{(b)}$.

The image of the element 25.3.1(b) is zero unless $n - 2a = q - p$, in which case it is

$$\begin{aligned}
&F_i^{(b)} E_i^{(a)} (\xi_{-p} \otimes \eta_q) = F_i^{(b)} (E_i^{(a)} \xi_{-p} \otimes \eta_q) \\
&= \sum_{b'+b''=b} v_i^{b'b''-b'q} F_i^{(b')} E_i^{(a)} \xi_{-p} \otimes F_i^{(b'')} \eta_q \\
&= \sum_{b'+b''=b} \sum_{t \geq 0} v_i^{b'b''-b'q} \begin{bmatrix} -a + b' + p \\ t \end{bmatrix}_i E_i^{(a-t)} F_i^{(b'-t)} \xi_{-p} \otimes F_i^{(b'')} \eta_q \\
&= \sum_{s \geq 0; s \leq a, s \leq b} v_i^{s(b-s-q)} \begin{bmatrix} s - a + p \\ s \end{bmatrix}_i E_i^{(a-s)} \xi_{-p} \otimes F_i^{(b-s)} \eta_q.
\end{aligned}$$

As before, we see that this is equal to $(\theta_i^{(a)} \diamond \theta_i^{(b)})_{p,q}$. The proposition follows.

25.4. Structure Constants

25.4.1. For any triplet $a, b, c \in \dot{\mathbf{B}}$, we define elements $m_{ab}^c \in \mathcal{A}$ by $ab = \sum_c m_{ab}^c c$ (ab is the product in $\dot{\mathbf{U}}$). We also define elements $\hat{m}_c^{ab} \in \mathcal{A}$ by the following requirement: for any $\lambda'_1, \lambda''_1, \lambda'_2, \lambda''_2 \in X$ and any $c \in \dot{\mathbf{B}} \cap ({}_{\lambda'_1+\lambda'_2}\mathbf{U}_{\lambda''_1+\lambda''_2})$ we have

$$\Delta_{\lambda'_1,\lambda''_1,\lambda'_2,\lambda''_2}(c) = \sum_{a,b} \hat{m}_c^{ab} a \otimes b;$$

in the last formula, $\Delta_{\lambda'_1,\lambda''_1,\lambda'_2,\lambda''_2}$ is as in 23.1.5; a runs over $\dot{\mathbf{B}} \cap ({}_{\lambda'_1}\mathbf{U}_{\lambda''_1})$; b runs over $\dot{\mathbf{B}} \cap ({}_{\lambda'_2}\mathbf{U}_{\lambda''_2})$. If $a \in \dot{\mathbf{B}} \cap ({}_{\lambda'_1}\mathbf{U}_{\lambda''_1}), b \in \dot{\mathbf{B}} \cap ({}_{\lambda'_2}\mathbf{U}_{\lambda''_2})$ and $c \in \dot{\mathbf{B}} \cap ({}_{\lambda'_3}\mathbf{U}_{\lambda''_3})$, and either $\lambda'_3 \neq \lambda'_1 + \lambda'_2$ or $\lambda''_3 \neq \lambda''_1 + \lambda''_2$, then $\hat{m}_c^{ab}$ is defined to be 0.

The elements $m_{ab}^c, \hat{m}_{ab}^c$ are called the *structure constants* of $\dot{\mathbf{U}}$. They satisfy the following identities, for all $a, b, d, e \in \dot{\mathbf{B}}$ and $\lambda \in X$:

(a) $\sum_c m_{ab}^c m_{cd}^e = \sum_c m_{ac}^e m_{bd}^c$;

(b) $\sum_c \hat{m}_c^{ab} \hat{m}_e^{cd} = \sum_c \hat{m}_e^{ac} \hat{m}_c^{bd}$;

(c) $\sum_c m_{ab}^c \hat{m}_c^{ed} = \sum_{a',b',c',d'} \hat{m}_a^{a'b'} \hat{m}_b^{c'd'} m_{a'c'}^e m_{b'd'}^d$.

(d) $\hat{m}_{1_\lambda}^{ab} = 1$ if $a = 1_{\lambda'}, b = 1_{\lambda''}, \lambda' + \lambda'' = \lambda$ and $\hat{m}_{1_\lambda}^{ab} = 0$ otherwise.

In each sum, all but finitely many terms are zero. The identity (a) expresses the associativity of multiplication in $\dot{\mathbf{U}}$; (b) is a consequence of the coassociativity of comultiplication in $\mathbf{U}$; (c),(d) are consequences of the fact that the comultiplication $\Delta : \mathbf{U} \to \mathbf{U} \otimes \mathbf{U}$ is an algebra homomorphism preserving 1.

Conjecture 25.4.2. *If the Cartan datum is symmetric, then the structure constants* $m_{ab}^c, \hat{m}_c^{ab}$ *are in* $\mathbf{N}[v, v^{-1}]$.

This would generalize the positivity theorem 14.4.13. For the proof an interpretation of $(\dot{\mathbf{U}}, \dot{\mathbf{B}})$ in terms of perverse sheaves, generalizing that of $(\mathbf{f}, \mathbf{B})$ will be required.

CHAPTER 26

Inner Product on $\dot{\mathbf{U}}$

26.1. First Definition of the Inner Product

26.1.1. In the following theorem, $\rho : \mathbf{U} \to \mathbf{U}$ is as in 19.1.1, and $\bar{\rho} : \mathbf{U} \to \mathbf{U}$ is defined by $\bar{\rho}(u) = \overline{\rho(\bar{u})}$.

Theorem 26.1.2. *There exists a unique $\mathbf{Q}(v)$-bilinear pairing $(,) : \dot{\mathbf{U}} \times \dot{\mathbf{U}} \to \mathbf{Q}(v)$ such that (a), (b), (c) below hold.*

(a) $(1_{\lambda_1} x 1_{\lambda_2}, 1_{\lambda'_1} x' 1_{\lambda'_2})$ *is zero for all $x, x' \in \dot{\mathbf{U}}$, unless $\lambda_1 = \lambda'_1$ and $\lambda_2 = \lambda'_2$;*

(b) $(ux, y) = (x, \rho(u)y)$ *for all $x, y \in \dot{\mathbf{U}}$ and $u \in \mathbf{U}$;*

(c) $(x^- 1_\lambda, x'^- 1_\lambda) = (x, x')$ *for all $x, x' \in \mathbf{f}$ and all λ (here (x, x') is the inner product as in 1.2.5).*

(d) *We have $(x, y) = (y, x)$ for all $x, y \in \dot{\mathbf{U}}$.*

Let $\zeta \in X$. If B is a basis of $\mathbf{f}$ consisting of homogeneous elements, the elements $\rho(b^-)b'^- 1_\zeta$, with $b, b' \in B$, form a basis of $\dot{\mathbf{U}}1_\zeta$. (We use the triangular decomposition of $\dot{\mathbf{U}}$.) Hence there is a unique $\mathbf{Q}(v)$-linear map $p : \dot{\mathbf{U}}1_\zeta \to \mathbf{Q}(v)$ such that $p(\rho(x^-)x'^- 1_\zeta) = (x, x')$ for all $x, x' \in \mathbf{f}$; here, (x, x') is as in 1.2.5. The properties of (x, x') imply that for homogeneous x, x', we have $p(\rho(x^-)x'^- 1_\zeta) = 0$ unless x, x' have the same homogeneity, in which case $\rho(x^-)x'^- 1_\zeta = 1_\zeta \rho(x^-)x'^-$. Thus, for $\zeta' \in X$ different from ζ, we have $p(1_{\zeta'}\dot{\mathbf{U}}1_\zeta) = 0$. It follows that, for any $\mu \in Y$, we have $p((K_\mu - v^{\langle \mu, \zeta \rangle})\dot{\mathbf{U}}1_\zeta) = 0$. We now define a pairing $f_\zeta : \dot{\mathbf{U}}1_\zeta \times \dot{\mathbf{U}}1_\zeta \to \mathbf{Q}(v)$ by $f_\zeta(u1_\zeta, u'1_\zeta) = p(\rho(u)u'1_\zeta)$ where $u, u' \in \mathbf{U}$. To show independence of u, u', we must check that $p(\rho(u)u'1_\zeta) = 0$ if either u or u' is in the left ideal of $\mathbf{U}$ generated by $(K_\mu - v^{\langle \mu, \zeta \rangle})$ for some $\mu \in Y$; in the case of u, this follows from the previous sentence, while in the case of u', this is obvious. Thus, f_ζ is well-defined. We define the bilinear pairing $(,)$ on $\dot{\mathbf{U}}$ by $(x, y) = f_\zeta(x, y)$ if $x, y \in \dot{\mathbf{U}}1_\zeta$ and by $(x, y) = 0$ if $x \in \dot{\mathbf{U}}1_\zeta$ and $y \in \dot{\mathbf{U}}1_{\zeta'}$ with $\zeta \neq \zeta'$. It is clear that this pairing satisfies (a),(b),(c); the uniqueness is also clear from the proof above.

The pairing $x, y \mapsto (y, x)$ satisfies the defining properties of (x, y) (since $\rho^2 = 1$), and hence coincides with it. This proves (d).

G. Lusztig, *Introduction to Quantum Groups*, Modern Birkhäuser Classics,
DOI 10.1007/978-0-8176-4717-9_26, © Springer Science+Business Media, LLC 2010

Proposition 26.1.3. *We have $(xu, y) = (x, y\bar{\rho}(u))$ for all $x, y \in \dot{\mathbf{U}}$ and $u \in \mathbf{U}$.*

We may assume that u is one of the standard generators of $\mathbf{U}$. Thus, we must verify that

$$(xE_i, y) = (x, v_i y F_i \tilde{K}_{-i}), (xF_i, y) = (x, v_i y E_i \tilde{K}_i), (xK_{-\mu}, y) = (x, yK_{-\mu})$$

for all $x, y \in \dot{\mathbf{U}}$ and $i \in I, \mu \in Y$.

We may assume that $x = u' 1_\zeta$ where $u' \in \mathbf{U}$ and $\zeta \in X$. Using 26.1.2(b), and setting $\rho(u')y = y'$, we see that the previous equalities are consequences of

$$(1_\zeta E_i, y') = (1_\zeta, v_i y' F_i \tilde{K}_{-i}), (1_\zeta F_i, y') = (1_\zeta, v_i y' E_i \tilde{K}_i)$$

and

$$(1_\zeta K_{-\mu}, y') = (1_\zeta, y' K_{-\mu})$$

for all $y' \in \dot{\mathbf{U}}$ and $i \in I, \mu \in Y$.

We can assume that $y' = \rho(y_1^-) y_2^- 1_{\zeta'}$ where $y_1, y_2 \in \mathbf{f}$ are homogeneous. Then the equalities to be proved can be rewritten as follows:

(a) $(y_1^- E_i 1_{\zeta - i'}, y_2^- 1_{\zeta'}) = v_i^{1 - \langle i, \zeta' + i' \rangle} (y_1^- 1_\zeta, y_2^- F_i 1_{\zeta' + i'})$

(b) $(y_1^- F_i 1_{\zeta + i'}, y_2^- 1_{\zeta'}) = v_i^{1 + \langle i, \zeta' - i' \rangle} (y_1^- 1_\zeta, y_2^- E_i 1_{\zeta' - i'})$

(c) $(y_1^- 1_\zeta K_{-\mu}, y_2^- 1_{\zeta'}) = (y_1^- 1_\zeta, y_2^- 1_{\zeta'} K_{-\mu})$.

Now (c) is obvious and (b) follows from (a), using 26.1.2(d). It remains to prove (a). We may assume that $\zeta' = \zeta - i'$. We substitute

$$y_1^- E_i 1_{\zeta'} = E_i y_1^- 1_{\zeta'} + \frac{r_i(y_1)^- \tilde{K}_{-i} - \tilde{K}_i ({}_i r(y_1)^-)}{v_i - v_i^{-1}} 1_{\zeta'}$$

and note that

$$\begin{aligned}(E_i y_1^- 1_{\zeta'}, y_2^- 1_{\zeta'}) &= (y_1^- 1_{\zeta'}, v_i \tilde{K}_i F_i y_2^- 1_{\zeta'}) \\ &= (y_1^- 1_{\zeta'}, v_i^{-1 + \langle i, \zeta' - |y_2| \rangle} F_i y_2^- 1_{\zeta'}) \\ &= v_i^{-1 + \langle i, \zeta' - |y_2| \rangle} (y_1, \theta_i y_2),\end{aligned}$$

$$\left(\frac{r_i(y_1)^- \tilde{K}_{-i} - \tilde{K}_i ({}_i r(y_1)^-)}{v_i - v_i^{-1}} 1_{\zeta'}, y_2^- 1_{\zeta'}\right) =$$

$$\frac{v_i^{-\langle i, \zeta' \rangle} (r_i(y_1), y_2) - v_i^{\langle i, \zeta' - |y_1| + i' \rangle} ({}_i r(y_1), y_2)}{v_i - v_i^{-1}},$$

$$v_i^{1 - \langle i, \zeta' + i' \rangle} (y_1^- 1_\zeta, y_2^- F_i 1_{\zeta' + i'}) = v_i^{-1 - \langle i, \zeta' \rangle} (y_1, y_2 \theta_i),$$

(see 26.1.2(b), (c)). We see that (a) is equivalent to:

$$v_i^{-1+\langle i,\zeta'-|y_2|\rangle}(y_1,\theta_i y_2)+\frac{v_i^{-\langle i,\zeta'\rangle}(r_i(y_1),y_2)-v_i^{\langle i,\zeta'-|y_1|+i'\rangle}({}_ir(y_1),y_2)}{v_i-v_i^{-1}}$$

$$=v_i^{-1-\langle i,\zeta'\rangle}(y_1,y_2\theta_i).$$

But this follows from the known equalities

$$(y_1,y_2\theta_i)=\frac{(r_i(y_1),y_2)}{1-v_i^{-2}},(y_1,\theta_i y_2)=\frac{({}_ir(y_1),y_2)}{1-v_i^{-2}}$$

(see 1.2.13(a)). The proposition is proved.

Proposition 26.1.4. *We have $(\sigma(x),\sigma(y))=(x,y)$ for all $x,y\in\dot{\mathbf{U}}$.*

We must show that the pairing $x,y\mapsto(\sigma(x),\sigma(y))$ on $\dot{\mathbf{U}}$ satisfies the defining properties of $(,)$. Property 26.1.2(a) is obvious and property 26.1.2(c) follows from 1.2.8(b). Since $\sigma\rho=\bar{\rho}\sigma:\mathbf{U}\to\mathbf{U}$, we see that 26.1.2(b) for the pairing $x,y\mapsto(\sigma(x),\sigma(y))$ is equivalent to the identity in the previous proposition. The proposition follows.

Lemma 26.1.5. *$(x^+1_\lambda,x'^+1_\lambda)=(x,x')$ for all $x,x'\in\mathbf{f}$ and all λ; here (x,x') is as in 1.2.5.*

We may assume that x,x' are homogeneous; moreover, using 26.1.2(a), we may assume that they both belong to $\mathbf{f}_\nu$. Using 26.1.2(b), we have

$$(x^+1_\lambda,x'^+1_\lambda)=(1_\lambda,\rho(x^+)x'^+1_\lambda).$$

Using the previous proposition, we see that the last expression equals

$$\begin{aligned}(\sigma(1_\lambda),\sigma(\rho(x^+)x'^+1_\lambda))&=(1_{-\lambda},1_{-\lambda}\sigma(x'^+)\bar{\rho}(\sigma(x^+)))\\&=(1_{-\lambda},\sigma(x'^+)\bar{\rho}(\sigma(x^+))1_{-\lambda}).\end{aligned}$$

Using 26.1.2(b) and the fact that $\rho^2=1$, we see that the last expression equals

$$(\rho(\sigma(x'^+))1_{-\lambda},\bar{\rho}(\sigma(x^+))1_{-\lambda}).$$

Using the definitions, we see that there exists an integer N depending only on ν and λ such that for any $z\in\mathbf{f}_\nu$ we have $\rho(\sigma(z^+))1_\lambda=v^Nz^-1_\lambda$ and $\bar{\rho}(\sigma(z^+))1_\lambda=v^{-N}z^-1_\lambda$. Hence the last inner product is equal to

$$(v^Nx'^-1_{-\lambda},v^{-N}x^-1_{-\lambda})=(x'^-1_{-\lambda},x^-1_{-\lambda})=(x',x)=(x,x').$$

The lemma is proved.

Proposition 26.1.6. *We have $(\omega(x), \omega(y)) = (x, y)$ for all $x, y \in \dot{\mathbf{U}}$.*

We must show that the pairing $x, y \mapsto (\omega(x), \omega(y))$ on $\dot{\mathbf{U}}$ satisfies the defining properties of $(,)$. Property 26.1.2(a) is obvious and property 26.1.2(c) follows from the previous lemma. Since $\omega\rho = \rho\omega : \mathbf{U} \to \mathbf{U}$, we see that 26.1.2(b) for the pairing $x, y \mapsto (\omega(x), \omega(y))$ is equivalent to the corresponding identity for (x, y). The proposition follows.

26.2. Definition of the Inner Product as a Limit

26.2.1 In this section we assume that the root datum is Y-regular. Let $\zeta \in X$ and let $\lambda, \lambda' \in X^+$ be such that $\lambda' - \lambda = \zeta$. We consider the bilinear pairing $(,)_{\lambda,\lambda'}$ on ${}^\omega\Lambda_\lambda \otimes \Lambda_{\lambda'}$, defined by $(x \otimes x', y \otimes y') = (x, y)_\lambda (x', y')_{\lambda'}$. Here $(,)_{\lambda'}$ is the pairing on $\Lambda_{\lambda'}$ defined in 19.1.2, and $(,)_\lambda$ is the analogous pairing on Λ_λ which has the same ambient space as ${}^\omega\Lambda_\lambda$.

Lemma 26.2.2. *If $x_1, x_2 \in {}^\omega\Lambda_\lambda \otimes \Lambda_{\lambda'}$ and $u \in \mathbf{U}$, we have*

$$(ux_1, x_2)_{\lambda,\lambda'} = (x_1, \rho(u)x_2)_{\lambda,\lambda'}.$$

It is enough to check this in the case where u is one of the algebra generators E_i, F_i, K_μ of $\mathbf{U}$. The case where $u = K_\mu$ is trivial. We now fix i and regard the $\mathbf{U}$-module ${}^\omega\Lambda_\lambda \otimes \Lambda_{\lambda'}$ as an object of $\mathcal{C}'_i$ (by restriction). It is enough to show that the form $(,)$ on this object is admissible in the sense of 16.2.2. Using 17.1.3(b), we see that this would follow if we knew that the forms $(,)_\lambda$ and $(,)_{\lambda'}$ on ${}^\omega\Lambda_\lambda$ and $\Lambda_{\lambda'}$ (regarded as objects of $\mathcal{C}'_i$) are admissible. For $(,)_{\lambda'}$ this follows from the definition. The same holds for $(,)_\lambda$ on Λ_λ. One checks easily that applying ω to an object of $\mathcal{C}'_i$ with an admissible form gives a new object of $\mathcal{C}'_i$ for which the same form is admissible. In particular, $(,)_{\lambda'}$ is admissible for ${}^\omega\Lambda_\lambda$. The lemma is proved.

Proposition 26.2.3. *Let $x, y \in \dot{\mathbf{U}}1_\zeta$. When the pair λ, λ' tends to ∞ (in the sense that $\langle i, \lambda\rangle$ tends to ∞ for all i, or equivalently, $\langle i, \lambda'\rangle$ tends to ∞ for all i, the difference $\lambda' - \lambda$ being fixed and equal to ζ), the inner product $(x(\xi_{-\lambda} \otimes \eta_{\lambda'}), y(\xi_{-\lambda} \otimes \eta_{\lambda'}))_{\lambda,\lambda'} \in \mathbf{Q}(v)$ converges in $\mathbf{Q}((v^{-1}))$ to (x, y).*

Assume first that $x = x_1^- 1_\zeta, y = y_1^- 1_\zeta$ where $x_1, y_1 \in \mathbf{f}$. In this case we have

$$x(\xi_{-\lambda} \otimes \eta_{\lambda'}) = \xi_{-\lambda} \otimes x_1^- \eta_{\lambda'}$$

and

$$y(\xi_{-\lambda} \otimes \eta_{\lambda'}) = \xi_{-\lambda} \otimes y_1^- \eta_{\lambda'};$$

hence

$$(x(\xi_{-\lambda}\otimes\eta_{\lambda'}),y(\xi_{-\lambda}\otimes\eta_{\lambda'}))_{\lambda,\lambda'}=(x_1^-\eta_{\lambda'},y_1^-\eta_{\lambda'})_{\lambda'}$$

and, by 19.3.7, this converges to (x_1,y_1) when $\lambda\to\infty$. Since $(x_1,y_1)=(x_1^-1_\zeta,y_1^-1_\zeta)$, the proposition holds in this case.

Next, we prove the proposition in the case where $x=1_\zeta$ and y is arbitrary. We may assume that $y=\rho(x_1^-)y_1^-1_\zeta$ where $x_1,y_1\in\mathbf{f}$. Using the previous lemma, we have

$$(1_\zeta(\xi_{-\lambda}\otimes\eta_{\lambda'}),y(\xi_{-\lambda}\otimes\eta_{\lambda'}))_{\lambda,\lambda'}=(x_1^-1_\zeta(\xi_{-\lambda}\otimes\eta_{\lambda'}),y_1^-1_\zeta(\xi_{-\lambda}\otimes\eta_{\lambda'}))_{\lambda,\lambda'}$$

and by the earlier part of the proof, this converges to $(x_1^-1_\zeta,y_1^-1_\zeta)$ when $\lambda\to\infty$. Since $(x_1^-1_\zeta,y_1^-1_\zeta)=(1_\zeta,y)$, the proposition holds in this case.

We now consider the general case. We can write $x=u1_\zeta$ where $u\in\mathbf{U}$. Using the previous lemma, we have

$$(u1_\zeta(\xi_{-\lambda}\otimes\eta_{\lambda'}),y(\xi_{-\lambda}\otimes\eta_{\lambda'}))_{\lambda,\lambda'}=(1_\zeta(\xi_{-\lambda}\otimes\eta_{\lambda'}),\rho(u)y(\xi_{-\lambda}\otimes\eta_{\lambda'}))_{\lambda,\lambda'}$$

and by the case previously considered, this converges to $(1_\zeta,\rho(u)y)$ when $\lambda\to\infty$. Since $(1_\zeta,\rho(u)y)=(u1_\zeta,y)$, the proposition is proved.

26.3. A Characterization of $\dot{\mathbf{B}}\sqcup(-\dot{\mathbf{B}})$

In the following result there is no assumption on the root datum.

Theorem 26.3.1. (a) *Let $b,b',b_1,b_1'\in\mathbf{B}$ and let $\zeta,\zeta_1\in X$. We have*

$$(b\diamondsuit_\zeta b',b_1\diamondsuit_{\zeta_1}b_1')=\delta_{b,b_1}\delta_{b',b_1'}\delta_{\zeta,\zeta_1}\mod v^{-1}\mathbf{A}.$$

In particular, the canonical basis $\dot{\mathbf{B}}$ of $\dot{\mathbf{U}}$ is almost orthonormal for $(,)$.

(b) *Let $\beta\in\dot{\mathbf{U}}$. We have $\beta\in\dot{\mathbf{B}}\sqcup(-\dot{\mathbf{B}})$ if and only if β satisfies the following three conditions: $\beta\in{}_\mathcal{A}\dot{\mathbf{U}}$, $\bar\beta=\beta$ and $(\beta,\beta)\in1+v^{-1}\mathbf{A}$.*

Note that (a) is trivial when $\zeta\neq\zeta_1$. Hence we may assume that $\zeta=\zeta_1$. In this case, using the definitions we are immediately reduced to the case where the root datum is simply connected. Then 26.2.3 is applicable. Hence, using the definition, we see that it is enough to prove that

(c) $((b\diamondsuit b')_{\lambda,\lambda'},(b_1\diamondsuit b_1')_{\lambda,\lambda'})_{\lambda,\lambda'}=\delta_{b,b_1}\delta_{b',b_1'}\mod v^{-1}\mathbf{A}$

for any $\lambda,\lambda'\in X^+$ such that $\lambda'-\lambda=\zeta$ and such that $b\in\mathbf{B}(\lambda),b'\in\mathbf{B}(\lambda')$.

Since

$$(b^+\xi_{-\lambda}\otimes b'^-\eta_\lambda,b_1^+\xi_{-\lambda}\otimes b_1'^-\eta_\lambda)_{\lambda,\lambda'}=(b^-\eta_\lambda,b_1^-\eta_\lambda)_\lambda(b'^-\eta_{\lambda'},b_1'^-\eta_{\lambda'})_{\lambda'}$$
$$\in(\delta_{b,b_1}+v^{-1}\mathbf{A})(\delta_{b',b_1'}+v^{-1}\mathbf{A})=\delta_{b,b_1}\delta_{b',b_1'}+v^{-1}\mathbf{A},$$

we see that (c) follows from Theorem 24.3.3(b).

We prove (b). If $\beta \in \pm\dot{\mathbf{B}}$ then, by (a) and 25.2.1, it satisfies the three conditions listed. Conversely, if $\beta \in \dot{\mathbf{U}}$ satisfies the three conditions in (b) then, using (a) and Lemma 14.2.2(b), we see that there exists $\beta' \in \dot{\mathbf{B}}$ such that $\beta - (\pm\beta')$ is a linear combination of elements in $\dot{\mathbf{B}}$ with coefficients in $v^{-1}\mathbf{Z}[v^{-1}]$. These coefficients are necessarily 0, since $\beta - (\pm\beta')$ is fixed by $^{-} : \dot{\mathbf{U}} \to \dot{\mathbf{U}}$. The theorem is proved.

Corollary 26.3.2. *If $\beta \in \dot{\mathbf{B}}$, then $\sigma(\beta) \in \pm\dot{\mathbf{B}}$ and $\omega(\beta) \in \pm\dot{\mathbf{B}}$.*

σ and ω commute with $^{-} : \dot{\mathbf{U}} \to \dot{\mathbf{U}}$, preserve the lattice ${}_{\mathcal{A}}\dot{\mathbf{U}}$ and preserve the inner product $(,)$ (see 26.1.4, 26.1.6). Hence the corollary follows from Theorem 26.3.1(b).

CHAPTER 27

Based Modules

27.1. Isotypical Components

27.1.1. In this chapter we assume that $(I, \cdot)$ is of finite type.

Let $M \in \mathcal{C}$. We assume that M is finite dimensional over $\mathbf{Q}(v)$. For any $\lambda \in X^+$, we denote by $M[\lambda]$ the sum of simple subobjects of M that are isomorphic to Λ_λ. Then $M = \oplus_\lambda M[\lambda]$. We also define for any $\lambda \in X^+$:

$$M[\geq \lambda] = \oplus_{\lambda' \in X^+; \lambda' \geq \lambda} M[\lambda']$$

and

$$M[> \lambda] = \oplus_{\lambda' \in X^+; \lambda' > \lambda} M[\lambda'].$$

Clearly, $M[> \lambda]$ is a subobject of $M[\geq \lambda]$ and $M[\lambda] \oplus M[> \lambda] = M[\geq \lambda]$ as objects in $\mathcal{C}$.

27.1.2. A *based module* is an object M of $\mathcal{C}$, of finite dimension over $\mathbf{Q}(v)$ with a given $\mathbf{Q}(v)$-basis B such that

(a) $B \cap M^\zeta$ is a basis of M^ζ, for any $\zeta \in X$;

(b) the $\mathcal{A}$-submodule ${}_\mathcal{A}M$ generated by B is stable under ${}_\mathcal{A}\dot{\mathbf{U}}$;

(c) the $\mathbf{Q}$-linear involution $\bar{}: M \to M$ defined by $\overline{fb} = \bar{f}b$ for all $f \in \mathbf{Q}(v)$ and all $b \in B$ is compatible with the $\mathbf{U}$-module structure in the sense that $\overline{um} = \bar{u}\bar{m}$ for all $u \in \mathbf{U}, m \in M$;

(d) the $\mathbf{A}$-submodule $L(M)$ generated by B, together with the image of B in $L(M)/v^{-1}L(M)$, forms a basis at ∞ for M (see 20.1.1).

We say that $\bar{}: M \to M$ in (c) is the *associated involution* of (M, B). The direct sum of two based modules (M, B) and (M', B') is again a based module $(M \oplus M', B \sqcup B')$.

27.1.3. The based modules form the objects of a category $\tilde{\mathcal{C}}$; a morphism from the based module (M, B) to the based module (M', B') is by definition a morphism $f : M \to M'$ in $\mathcal{C}$ such that

(a) for any $b \in B$ we have $f(b) \in B' \cup \{0\}$ and

(b) $B \cap \ker f$ is a basis of $\ker f$.

G. Lusztig, *Introduction to Quantum Groups*, Modern Birkhäuser Classics,
DOI 10.1007/978-0-8176-4717-9_27, © Springer Science+Business Media, LLC 2010

27.1.4. Let (M, B) be a based module and let M' be a $\dot{\mathbf{U}}$-submodule of M such that M' is spanned as a $\mathbf{Q}(v)$-subspace of M by a subset B' of B. Then (M', B') is a based module; moreover, M/M' together with the image of $B - B'$ is a based module.

For any $\lambda \in X^+$, Λ_λ together with its canonical basis, is a based module. (See 19.3.4, 23.3.7, 20.1.4.)

27.1.5. Let (M, B) be a based module with associated involution $^-$ and let $m \in M$ be an element such that $\bar{m} = m, m \in {}_{\mathcal{A}}M$ and $m \in B + v^{-1}L(M)$ (resp. $m \in v^{-1}L(M)$). Then we have $m \in B$ (resp. $m = 0$). Indeed, we can write $m = \sum_{b \in B} c_b b$ with $c_b \in \mathcal{A}$. By our assumption, we have $c_b \in \mathbf{A}$ for all b. Hence $c_b \in \mathbf{Z}[v^{-1}]$ for all b. We have $\bar{c}_b = c_b$ for all b. Hence $c_b \in \mathbf{Z}$ for all b. Moreover, by our assumption, we have $c_b \in v^{-1}\mathbf{A}$ for all b, except possibly for a single b for which we have $c_b = 0$ or 1 mod $v^{-1}\mathbf{A}$. It follows that $c_b = 0$ for all b, except possibly for a single b for which we have $c_b = 0$ or 1. Our assertion follows.

27.1.6. Let (M, B) be a based module. Assume that $M \neq 0$. Let $\lambda_1 \in X^+$ be such that $M^{\lambda_1} \neq 0$ and such that λ_1 is maximal with this property. Let $B_1 = B \cap M^{\lambda_1}$. It is a non-empty set. Let $M' = \oplus_{b \in B_1} \Lambda_{\lambda_1, b} \in \mathcal{C}$. Here $\Lambda_{\lambda_1, b}$ is a copy of Λ_{λ_1} corresponding to b; we denote its canonical generator η_{λ_1} by η_b.

For any $b \in B_1$, we have $E_i b = 0$ for all $i \in I$ by the maximality of λ_1. Hence there is a unique homomorphism $\phi : M' \to M$ of objects in $\mathcal{C}$ whose restriction to any summand $\Lambda_{\lambda_1, b}$ carries η_b to b. Let B' be the basis of M' given by the union of the canonical bases of the various summands $\Lambda_{\lambda_1, b}$.

Proposition 27.1.7. *In the setup above, $B \cap M[\lambda_1]$ is a basis of $M[\lambda_1]$ and ϕ defines an isomorphism $M' \cong M[\lambda_1]$ carrying B' onto $B \cap M[\lambda_1]$. Thus ϕ is an isomorphism of based modules $(M', B') \cong (M[\lambda_1], B \cap M[\lambda_1])$.*

Let $^- : M' \to M'$ be the $\mathbf{Q}$-linear involution whose restriction to each summand $\Lambda_{\lambda_1, b}$ is the canonical involution $^- : \Lambda_{\lambda_1, b} \to \Lambda_{\lambda_1, b}$. The involution $^- : M' \to M'$ is compatible under ϕ with that of M. Indeed, both involutions are the identity on B_1. (We regard B_1 as a subset of M' by $b \mapsto \eta_b$.)

Let $b' \in B' \cap \Lambda_{\lambda_1, b}$. We have $\bar{b}' = b'$; hence $\overline{\phi(b')} = \phi(\bar{b}') = \phi(b')$. Thus $\phi(b')$ is fixed by $^- : M \to M$.

We know from 19.3.5 that there exists a sequence $i_1, i_2, \dots, i_p$ in I such that b' is equal to $\tilde{F}_{i_1} \tilde{F}_{i_2} \cdots \tilde{F}_{i_p} \eta_b$ plus a $v^{-1}\mathbf{A}$-linear combination of elements of the same kind. Now the action of $\tilde{F}_i$ on M' is compatible with the action of $\tilde{F}_i$ on M. Hence $\phi(b')$ is equal to $\tilde{F}_{i_1}\tilde{F}_{i_2}\dots\tilde{F}_{i_p}b$ plus a linear combination with coefficients in $v^{-1}\mathbf{A}$ of elements of the same kind. By property 27.1.2(d) of B, we see that either $\phi(b') \in B + v^{-1}L(M)$ or $\phi(b') \in v^{-1}L(M)$.

On the other hand, by the definition of the canonical basis of M', we have that b' belongs to the ${}_{\mathcal{A}}\dot{\mathbf{U}}$-submodule of M' generated by η_b; hence $\phi(b')$ belongs to the ${}_{\mathcal{A}}\dot{\mathbf{U}}$-submodule of M generated by b; by the property 27.1.2(b), we then have $\phi(b') \in {}_{\mathcal{A}}M$. These properties of $\phi(b')$ imply that $\phi(b') \in B$ or $\phi(b') = 0$ (see 27.1.5). The second alternative does not occur: indeed, the restriction of ϕ to the summand $\Lambda_{\lambda_1,b}$ is injective since $\Lambda_{\lambda_1,b}$ is simple. Thus we have $\phi(b') \in B$. We see that ϕ defines a bijection of the canonical basis of $\Lambda_{\lambda_1,b}$ with a subset $B(b)$ of B.

Next we consider an element $\tilde{b} \in B_1$ distinct from b. We show that $B(\tilde{b})$ is disjoint from $B(b)$. Indeed, assume that $b_1 \in B$ belongs to $B(b) \cap B(\tilde{b})$. Then we have

$$b_1 = \tilde{F}_{i_1}\tilde{F}_{i_2}\cdots\tilde{F}_{i_p}b \mod v^{-1}L(M)$$

and

$$b_1 = \tilde{F}_{j_1}\tilde{F}_{j_2}\cdots\tilde{F}_{j_q}\tilde{b} \mod v^{-1}L(M)$$

for some sequences $i_1, i_2, \dots, i_p$ and $j_1, j_2, \dots, j_q$ in I. By property 27.1.2(d), we then have

$$\tilde{b} = \tilde{E}_{j_q}\tilde{E}_{j_{q-1}}\cdots\tilde{E}_{j_1}\tilde{F}_{i_1}\tilde{F}_{i_2}\cdots\tilde{F}_{i_p}b \mod v^{-1}L(M).$$

Hence $\tilde{b}$ is equal to some element in $B(b)$ plus an element of $v^{-1}L(M)$. It follows that $\tilde{b} \in B(b)$.

In particular, we have $\tilde{b} = \phi(\tilde{b}')$ for some $\tilde{b}' \in \Lambda_{\lambda_1,b}$. Since $\tilde{b} \neq b$, we have $\tilde{b}' \neq \eta_b$; hence $\tilde{b}' \in \Lambda^{\lambda'}_{\lambda_1,b}$ with $\lambda' < \lambda_1$. It follows that $\tilde{b} \in M^{\lambda'}$ with $\lambda' < \lambda_1$. This contradicts the assumption that $\tilde{b} \in B_1$. We have proved therefore that $B(\tilde{b})$ is disjoint from $B(b)$.

Since B' is the disjoint union of the canonical bases of the various $\Lambda_{\lambda_1,b}$ and these subsets are carried by ϕ injectively onto disjoint subsets of B, it follows that ϕ restricts to an injective map $B' \to B$. Since B' is a basis of M', it follows that $\phi : M' \to M$ is injective. Thus we may identify M' with a $\dot{\mathbf{U}}$-submodule of M (via ϕ) in such a way that B' becomes a subset of B. This submodule is clearly equal to $M[\lambda_1]$. The proposition follows.

Proposition 27.1.8. *Let (M, B) be a based module and let $\lambda \in X^+$. Then*

(a) *$B \cap M[\geq \lambda]$ is a basis of the vector space $M[\geq \lambda]$ and*

(b) $B \cap M[> \lambda]$ *is a basis of the vector space* $M[> \lambda]$.

First note that (b) follows from (a). Indeed, the vector space $M[> \lambda]$ is a sum of subspaces of form $M[\geq \lambda']$ for various $\lambda' > \lambda$. To prove (a), we argue by induction on $\dim M$. If $\dim M = 0$, there is nothing to prove. Therefore we may assume that $\dim M \geq 1$.

For fixed M, we argue by descending induction on λ. To begin the induction we note that if $\sum_i \langle i, \lambda \rangle$ is sufficiently large, then $M[\geq \lambda] = 0$ and there is nothing to prove. Assume that λ is given. If $M[\lambda] = 0$, then $M[\geq \lambda]$ is a sum of subspaces $M[\geq \lambda']$ with $\lambda' > \lambda$; hence the desired result holds by the induction hypothesis (on λ). Thus we may assume that $M[\lambda] \neq 0$. Then clearly $M^\lambda \neq 0$. We can find $\lambda_1 \in X^+$ such that $\lambda_1 \geq \lambda$, $M^{\lambda_1} \neq 0$ and λ_1 is maximal with these properties.

Let $M' = M[\lambda_1]$ and let $B' = B \cap M'$. Then $(M', B') \in \tilde{\mathcal{C}}$ by 27.1.7. Hence, by 27.1.4, $M'' = M/M'$, together with the image B'' of $B - B'$, is an object of $\tilde{\mathcal{C}}$. Since $M' \neq 0$, we have $\dim M'' < \dim M$; hence the induction hypothesis (on M) is applicable to M''. We see that $B'' \cap M''[\geq \lambda]$ is a basis of $M''[\geq \lambda]$. Since $M' = M'[\lambda_1]$ and $\lambda_1 \geq \lambda$, we see that $M[\geq \lambda]$ is just the inverse image of $M''[\geq \lambda]$ under the canonical map $M \to M''$; moreover, a basis for this inverse image is given by the inverse image of $B'' \cap M''[\geq \lambda]$ under the canonical map $B \to B''$. The proposition is proved.

27.2. The Subsets $B[\lambda]$

27.2.1. Let (M, B) be a based module. Let $b \in B$. We can find $\lambda \in X^+$ such that $b \in M[\geq \lambda]$ and λ is maximal with this property. Actually, λ is unique. Indeed, assume that we also have $b \in M[\geq \lambda']$ and λ' is maximal with this property. We note that $M[\geq \lambda] \cap M[\geq \lambda']$ is a sum of subspaces $M[\geq \lambda'']$ for various λ'' such that $\lambda \leq \lambda''$ and $\lambda' \leq \lambda''$ and from 27.1.8 it follows that $b \in M[\geq \lambda'']$ for some such λ''.

If $\lambda \neq \lambda'$, then λ'' satisfies $\lambda < \lambda''$ and $\lambda' < \lambda''$, and we find a contradiction with the definition of λ. Thus the uniqueness of λ is proved.

Let $B[\lambda]$ be the set of all $b \in B$ which give rise to $\lambda \in X^+$ as above. These sets clearly form a partition of B. From 27.1.8, we see that, for any $\lambda \in X^+$, the set $\sqcup_{\lambda' \in X^+; \lambda' \geq \lambda} B[\lambda']$ is a basis of $M[\geq \lambda]$ and the set $\sqcup_{\lambda' \in X^+; \lambda' > \lambda} B[\lambda']$ is a basis of $M[> \lambda]$.

Proposition 27.2.2. *Let f be a morphism in $\tilde{\mathcal{C}}$ from the based module (M, B) to the based module (M', B') (see 27.1.3). For any $\lambda \in X^+$, we have $f(B[\lambda]) \subset B'[\lambda] \cup \{0\}$.*

From the definitions, we see that $f(M[\geq \lambda]) \subset M'[\geq \lambda]$ and $f(M[> \lambda]) \subset M'[> \lambda]$. Hence if $b \in B[\lambda]$, then either $f(b) \in B'[\lambda']$ for some $\lambda' \geq \lambda$ or $f(b) = 0$. Assume that $f(b) \notin B'[\lambda]$. Then $f(b) \in M'[> \lambda]$. Using the obvious inclusion $f(M) \cap M'[> \lambda] \subset f(M[> \lambda])$, we deduce that $b \in M[> \lambda] + \ker f$. Since both $M[> \lambda]$ and $\ker f$ are generated by their intersection with B, it follows that either $b \in M[> \lambda]$ or $b \in \ker f$. The first alternative contradicts $b \in B[\lambda]$; hence the second alternative holds and we have $f(b) = 0$. The proposition follows.

27.2.3. Let (M, B) be a based module. Let $\lambda \in X^+$. We define $B[\lambda]^{hi}$ to be the set of all $b \in B$ such that $b \in M^{\lambda}$ and $\tilde{E}_i b \in v^{-1}L(M)$ for all $i \in I$. We define $B[\lambda]^{lo}$ to be the set of all $b \in B$ such that $b \in M^{w_0(\lambda)}$ and $\tilde{F}_i b \in v^{-1}L(M)$ for all $i \in I$.

Proposition 27.2.4. (a) *We have* $B[\lambda]^{hi} \subset B[\lambda]$ *and* $B[\lambda]^{lo} \subset B[\lambda]$.

(b) *Let* $p : M[\geq \lambda] \to M[\geq \lambda]/M[> \lambda] = \tilde{M}$ *be the canonical map. Note that* p *defines a bijection of* $B[\lambda]$ *with a basis* $\tilde{B}$ *of* $\tilde{M}$ *and that* $(\tilde{M}, \tilde{B})$ *belongs to* $\tilde{\mathcal{C}}$ *so that* $\tilde{B}[\lambda]^{hi}$ *and* $\tilde{B}[\lambda]^{lo}$ *are defined. Then* p *restricts to bijections* $B[\lambda]^{hi} \to \tilde{B}[\lambda]^{hi}$ *and* $B[\lambda]^{lo} \to \tilde{B}[\lambda]^{lo}$.

We prove (a). Let $b \in B[\lambda]^{hi}$. There is a unique $\lambda' \in X^+$ such that $b \in B[\lambda']$. We must prove that $\lambda = \lambda'$. We have $b \in M[\geq \lambda']$. Replacing M with $M[\geq \lambda']$, we may assume that $M = M[\geq \lambda']$. Let π be the canonical map of M onto $M'' = M/M[> \lambda']$. Then $B[\lambda']$ is mapped by π bijectively onto a basis B'' of M'' and we have $\pi(b) \in B''$. Moreover, $\pi(b)$ belongs to $B''[\lambda]^{hi}$ and we are therefore reduced to the case where $M = M''$. Thus we may assume that $M = M[\lambda']$. Now 27.1.7 reduces us further to the case where (M, B) is $\Lambda_{\lambda'}$ with its canonical basis. In this case, there are two possibilities for b: either b is in the λ'-weight space or there exist i and $b' \in B$ such that $\tilde{F}_i b' - b \in v^{-1}L(M)$. In the first case we have $b \in M^{\lambda'}$; in the second case we have $\tilde{E}_i b - b' \in v^{-1}L(M)$; hence $\tilde{E}_i b \notin v^{-1}L(M)$, in contradiction with our assumption on b. Thus we have $b \in M^{\lambda'}$, hence $\lambda = \lambda'$, as required. We have proved that $B[\lambda]^{hi} \subset B[\lambda]$. The proof of the inclusion $B[\lambda]^{lo} \subset B[\lambda]$ is entirely similar.

We prove (b). We assume that $M = M[\geq \lambda]$. It is clear that $p(B[\lambda]^{hi}) \subset \tilde{B}[\lambda]^{hi}$ and $p(B[\lambda]^{lo}) \subset \tilde{B}[\lambda]^{lo}$. Assume that $b \in B[\lambda]$ satisfies $b \notin B[\lambda]^{hi}$. We show that $p(b) \notin \tilde{B}[\lambda]^{hi}$. By our assumption, we have that either $b \in M^{\lambda'}$ with $\lambda' \neq \lambda$ or that $\tilde{E}_i b \notin v^{-1}L(M)$ for some i.

If $b \in M^{\lambda'}$ with $\lambda' \neq \lambda$, then $p(b) \in \tilde{M}^{\lambda'}$ with $\lambda' \neq \lambda$; hence $p(b) \notin \tilde{B}[\lambda]^{hi}$, as required. If $\tilde{E}_i b \notin v^{-1}L(M)$ for some $i \in I$, then there exists $b' \in B$ such that $\tilde{E}_i b - b' \in v^{-1}L(M)$ and therefore $\tilde{F}_i b' - b \in v^{-1}L(M)$.

We consider two cases according to whether or not $b' \in M[> \lambda]$. In the first case ($b' \in M[> \lambda]$), we have $\tilde{F}_i b' \in M[> \lambda]$ (since $M[> \lambda]$ is a subobject of M) hence $b \in M[> \lambda] + v^{-1}L(M)$; this implies that $b \in M[> \lambda]$ (using that $B \cap M[> \lambda]$ is a basis of $M[> \lambda]$). Then we have $p(b) = 0$ and, in particular, $p(b) \notin \tilde{B}[\lambda]^{hi}$, as required. In the second case ($b' \notin M[> \lambda]$), we have $b' \in B[\lambda]$; hence $\pi(b') \in \tilde{B}$.

Let $L(\tilde{M})$ be the **A**-submodule of $\tilde{M}$ generated by $\tilde{B}$. From $\tilde{E}_i b - b' \in v^{-1}L(M)$, we deduce $\tilde{E}_i(\pi(b)) - \pi(b') \in v^{-1}L(\tilde{M})$. In particular, we have $\tilde{E}_i(\pi(b)) \notin v^{-1}L(\tilde{M})$; hence $p(b) \notin \tilde{B}[\lambda]^{hi}$, as required. Thus we have proved the equality $p(B[\lambda]^{hi}) = \tilde{B}[\lambda]^{hi}$. The proof of the equality $p(B[\lambda]^{lo}) = \tilde{B}[\lambda]^{lo}$ is entirely similar.

27.2.5. Coinvariants. Let $(M, B) \in \tilde{\mathcal{C}}$. Let $M[\neq 0] = \oplus_{\lambda \neq 0} M[\lambda]$. The space of *coinvariants* of M is by definition the vector space $M_* = M/M[\neq 0]$. Clearly, $M[\neq 0]$ is equal to the sum of the subspaces $M[\geq \lambda']$ for various $\lambda' \in X^+ - \{0\}$; hence, from 27.2.8, it follows that $\cup_{\lambda' \neq 0} B[\lambda']$ is a basis of $M[\neq 0]$. We deduce that under the canonical map $\pi : M \to M_*$ the subset $B[0]$ of B is mapped bijectively onto a basis B_* of M_*.

Note that π is a morphism in $\mathcal{C}$ if we regard M_* with the **U**-module structure such that $M_* = M_*[0]$. We see that

(a) (M_*, B_*) is a based module with trivial action of **U**.

Proposition 27.2.6. *We have $B[0] = B[0]^{hi} = B[0]^{lo}$. This set is mapped bijectively by $\pi : M \to M_*$ onto B_*.*

To prove the first statement, we are reduced by 27.2.4(a),(b) to the case where $M = M[0]$, where it is obvious. The second statement has already been noted.

27.3. Tensor Product of Based Modules

27.3.1. Let $(M, B), (M', B')$ be two based modules with associated involutions $^- : M \to M, ^- : M' \to M'$. We will show that *the* **U***-module $M \otimes M'$ is in a natural way a based module.*

The obvious basis $B \otimes B'$ does not make $M \otimes M'$ into a based module, since the involution $^- : M \otimes M' \to M \otimes M'$ given by $\overline{m \otimes m'} = \bar{m} \otimes \bar{m}'$ is not, in general, compatible with the **U**-module structure.

We will define a new involution $\Psi : M \otimes M' \to M \otimes M'$ by $\Psi(x) = \Theta(\bar{x})$ for all $x \in M \otimes M'$; here $\Theta : M \otimes M' \to M \otimes M'$ is as in 24.1.1. Eventually, Ψ will be the associated involution of our based module.

Let $\mathcal{L}$ (resp. ${}_{\mathcal{A}}\mathcal{L}$) be the $\mathbf{Z}[v^{-1}]$-submodule (resp. $\mathcal{A}$-submodule) of $M \otimes M'$ generated by the basis $B \otimes B'$. From 24.1.6, we see that Θ leaves ${}_{\mathcal{A}}\mathcal{L}$ stable and clearly $^{-} : M \otimes M' \to M \otimes M'$ leaves ${}_{\mathcal{A}}\mathcal{L}$ stable; it follows that we have $\Psi({}_{\mathcal{A}}\mathcal{L}) \subset {}_{\mathcal{A}}\mathcal{L}$. From 24.1.2 and 4.1.3, it follows that $\Psi^2 = 1$ and $\Psi(ux) = \bar{u}\Psi(x)$ for all $u \in \mathbf{U}$ and all $x \in M \otimes M'$. We shall regard $B \times B'$ as a partially ordered set with $(b_1, b_1') \geq (b_2, b_2')$ if and only if $b_1 \in M^{\lambda_1}, b_1' \in M'^{\lambda_1'}, b_2 \in M^{\lambda_2}, b_2' \in M'^{\lambda_2'}$ where $\lambda_1 \geq \lambda_2, \lambda_1' \leq \lambda_2', \lambda_1 + \lambda_1' = \lambda_2 + \lambda_2'$.

From the definition we have, for all $b_1 \in B, b_1' \in B'$,

$$\Psi(b_1 \otimes b_1') = \sum_{b_2 \in B, b_2' \in B'} \rho_{b_1,b_1';b_2,b_2'} b_2 \otimes b_2'$$

where $\rho_{b_1,b_1';b_2,b_2'} \in \mathcal{A}$ and $\rho_{b_1,b_1';b_2,b_2'} = 0$ unless $(b_1, b_1') \geq (b_2, b_2')$. Note also that

$$\rho_{b_1,b_1';b_1,b_1'} = 1$$

and

$$\sum_{b_2 \in B, b_2' \in B'} \bar{\rho}_{b_1,b_1';b_2,b_2'} \rho_{b_2,b_2';b_3,b_3'} = \delta_{b_1,b_3} \delta_{b_1',b_3'}$$

for any $b_1, b_3 \in B$ and $b_1', b_3' \in B'$; the last condition follows from $\Psi^2 = 1$. Applying 24.2.1 to the partially ordered set $H = B \times B'$, we see that there is a unique family of elements $\pi_{b_1,b_1';b_2,b_2'} \in \mathbf{Z}[v^{-1}]$ defined for $b_1, b_2 \in B$ and $b_1', b_2' \in B'$, such that

$\pi_{b_1,b_1';b_1,b_1'} = 1$;

$\pi_{b_1,b_1';b_2,b_2'} \in v^{-1}\mathbf{Z}[v^{-1}]$ if $(b_1, b_1') \neq (b_2, b_2')$;

$\pi_{b_1,b_1';b_2,b_2'} = 0$ unless $(b_1, b_1') \geq (b_2, b_2')$;

$\pi_{b_1,b_1';b_2,b_2'} = \sum_{b_3,b_3'} \bar{\pi}_{b_1,b_1';b_3,b_3'} \rho_{b_3,b_3';b_2,b_2'}$

for all $(b_1, b_1') \geq (b_2, b_2')$.

We have the following result.

Theorem 27.3.2. (a) *For any $(b_1, b_1') \in B \times B'$, there is a unique element $b_1 \diamondsuit b_1' \in \mathcal{L}$ such that $\Psi(b_1 \diamondsuit b_1') = b_1 \diamondsuit b_1'$ and $(b_1 \diamondsuit b_1') - b_1 \otimes b_1' \in v^{-1}\mathcal{L}$.*

(b) *The element $b_1 \diamondsuit b_1'$ in (a) is equal to $b_1 \otimes b_1'$ plus a linear combination of elements $b_2 \otimes b_2'$ with $(b_2, b_2') \in B \times B'$, $(b_2, b_2') < (b_1, b_1')$ and with coefficients in $v^{-1}\mathbf{Z}[v^{-1}]$.*

(c) *The elements $b_1 \diamondsuit b_1'$ with b_1, b_1' as above, form a $\mathbf{Q}(v)$-basis $B_\diamondsuit$ of $M \otimes M'$, an $\mathcal{A}$-basis of ${}_{\mathcal{A}}\mathcal{L}$ and a $\mathbf{Z}[v^{-1}]$-basis of $\mathcal{L}$.*

$b_1 \diamondsuit b'_1$ just defined satisfy the requirements of (b),(c) and that (d) holds. It remains to show the uniqueness in (a). It is enough to show that an element $x \in v^{-1}\mathcal{L}$ such that $\bar{x} = x$ is necessarily 0. But this follows from (d).

27.3.3. The previous result, together with the known behaviour of bases at ∞ under tensor product, (see 20.2.2) shows that $(M \otimes M', B_\diamondsuit)$ is a based module with associated involution Ψ. This is by definition the tensor product of the objects $(M, B), (M', B')$.

27.3.4. Let $\lambda, \lambda' \in X^+$. Applying the previous construction to $M = {}^\omega\Lambda_\lambda$ and $M' = \Lambda_{\lambda'}$ regarded as based modules (with respect to the canonical bases), we obtain a basis of ${}^\omega\Lambda_\lambda \otimes \Lambda_{\lambda'}$, which clearly is the same as that constructed in 24.3.3. Thus, ${}^\omega\Lambda_\lambda \otimes \Lambda_{\lambda'}$, together with its canonical basis in 24.3.3, is a based module.

Proposition 27.3.5. *Let $\lambda, \lambda', \lambda'' \in X^+$.*

(a) *The* **U**-*modules $M = {}^\omega\Lambda_{\lambda+\lambda'} \otimes \Lambda_{\lambda'+\lambda''}$ and $M' = {}^\omega\Lambda_\lambda \otimes \Lambda_{\lambda'}$ with their canonical bases B, B' constructed in 24.3.3, are in $\tilde{\mathcal{C}}$; moreover, $t : M \to M'$ (see 25.1.5) is a morphism in $\tilde{\mathcal{C}}$.*

(b) *For any $\lambda_1 \in X^+$, we have $t(B[\lambda_1]) \subset B'[\lambda_1] \cup \{0\}$.*

The fact that $(M, B), (M', B')$ are objects of $\tilde{\mathcal{C}}$ has been pointed out in 27.3.4. The second assertion of (a) follows from Proposition 25.1.10. Now (b) follows from (a) and 27.2.2.

27.3.6. Associativity of tensor product. Let $(M, B), (M', B')$, and (M'', B'') be three based modules. On the **U**-module $M \otimes M' \otimes M''$, we can introduce two structures of based module: one by applying the construction in 27.3.2 first to $M \otimes M'$ and then to $(M \otimes M') \otimes M''$; the second one by applying the construction in 27.3.2 first to $M' \otimes M''$ and then to $M \otimes (M' \otimes M'')$. Let B_1, B_2 be the bases of $M \otimes M' \otimes M''$ obtained by these two constructions.

We show that $B_1 = B_2$. By definition, the associated involutions to these two structures are given by

$$\sum_{\nu',\nu''} (\Delta \otimes 1)(\Theta_{\nu'})\Theta^{12}_{\nu''}(^- \otimes ^- \otimes ^-)$$

and

$$\sum_{\nu',\nu''} (1 \otimes \Delta)(\Theta_{\nu'})\Theta^{23}_{\nu''}(^- \otimes ^- \otimes ^-)$$

respectively. These coincide by 4.2.4.

Next from the definitions, we see that the $\mathbf{Z}[v^{-1}]$-submodules of $M \otimes M' \otimes M''$ generated by B_1 or B_2 coincide; they both coincide with the $\mathbf{Z}[v^{-1}]$-submodule $\mathcal{L}$ of $M \otimes M' \otimes M''$ generated by $B \otimes B' \otimes B''$; moreover, if $\pi : \mathcal{L} \to \mathcal{L}/v^{-1}\mathcal{L}$ is the canonical projection, then $\pi(B_1) = \pi(B_2) = \pi(B \otimes B' \otimes B'')$.

To show that $B_1 = B_2$, it suffices to show that $(b\diamondsuit b')\diamondsuit b'' = b\diamondsuit(b'\diamondsuit b'')$ for any $b \in B, b' \in B', b'' \in B''$. Let $b_1 = (b\diamondsuit b')\diamondsuit b'' \in B_1$ and $b_2 = b\diamondsuit(b'\diamondsuit b'') \in B_2$. From the definitions, we have that $\pi(b_1) = \pi(b \otimes b' \otimes b'')$ and $\pi(b_2) = \pi(b \otimes b' \otimes b'')$. Hence $\pi(b_1) = \pi(b_2)$. Then $b_1 - b_2 \in v^{-1}\mathcal{L}$ and $b_1 - b_2$ is fixed by the associated involution. This forces $b_1 = b_2$, as required. Thus we may omit brackets and write $b\diamondsuit b'\diamondsuit b''$ instead of $(b\diamondsuit b')\diamondsuit b''$ or $b\diamondsuit(b'\diamondsuit b'')$. This implies automatically that the analogous associativity result is also true for more than three factors.

27.3.7. Coinvariants in a tensor product. Let $(M, B), (M', B')$ be two based modules. We form their tensor product $(M \otimes M', B_\diamondsuit)$. The following result describes the subset $B_\diamondsuit[0]$ of $B_\diamondsuit$.

Proposition 27.3.8. *Let $b \in B, b' \in B'$. We have*

$$B_\diamondsuit[0] = \cup_{\lambda' \in X^+}\{b\diamondsuit b' | b \in B[-w_0(\lambda')]^{lo}, b' \in B'[\lambda']^{hi}\}.$$

Let $b \in B, b' \in B'$ be two elements such that $b \in M^\lambda, b' \in M'^{\lambda'}$. According to 27.2.6, the condition that $b\diamondsuit b'$ belongs to $B_\diamondsuit[0]$ is that $\lambda + \lambda' = 0$ and $\tilde{F}_i(b\diamondsuit b') \in v^{-1}L(M \otimes M')$ for all i; the last condition is clearly equivalent to the condition that $\tilde{F}_i(b \otimes b') \in v^{-1}L(M \otimes M')$. By 20.2.4, our condition is equivalent to the following one: $\lambda + \lambda' = 0$, $\tilde{F}_i(b) \in v^{-1}L(M)$ and $\tilde{E}_i(b') \in v^{-1}L(M')$ for all $i \in I$. The proposition follows.

27.3.9. We consider a sequence $\lambda_1, \lambda_2, \dots, \lambda_n$ of elements of X^+. According to 27.3.6, the tensor product $\Lambda_{\lambda_1} \otimes \Lambda_{\lambda_2} \cdots \otimes \Lambda_{\lambda_n}$ is in a natural way a based module (hence has a distinguished basis) and according to 27.2.5, the space of coinvariants $(\Lambda_{\lambda_1} \otimes \Lambda_{\lambda_2} \cdots \otimes \Lambda_{\lambda_n})_*$ inherits a natural based module structure (hence has a distinguished basis).

27.3.10. Let us assume, for example, that the root datum is simply connected of type D_m, that $n = 2n'$ and that $\lambda_1 = \lambda_2 = \cdots = \lambda_n = \lambda$ is such that Λ_λ is the standard $(2m)$-dimensional module. Then we may identify the space of coinvariants $(\Lambda_{\lambda_1} \otimes \Lambda_{\lambda_2} \cdots \otimes \Lambda_{\lambda_n})_*$ naturally with the dual space of $\mathrm{End}_{\mathbf{U}}(\Lambda_\lambda^{\otimes n'})$. Hence, from 27.3.9, we obtain a distinguished basis for the algebra $\mathrm{End}_{\mathbf{U}}(\Lambda_\lambda^{\otimes n'})$, the quantum analogue of the *Brauer centralizer algebra*. This basis is of the same nature as the basis of the Hecke algebra of type A defined in [3].

CHAPTER 28

Bases for Coinvariants and Cyclic Permutations

28.1. Monomials

28.1.1. In this chapter we assume that $(I, \cdot)$ is of finite type. Let $\lambda \in X^+$. For any sequence $\mathbf{i} = (i_1, i_2, \dots, i_N)$ in I such that $s_{i_1}s_{i_2}\cdots s_{i_N}$ is a reduced expression of an element $w \in W$, we consider the element $\theta(\mathbf{i}, \lambda) = \theta_{i_1}^{(a_1)}\theta_{i_2}^{(a_2)}\cdots\theta_{i_N}^{(a_N)} \in \mathbf{f}$ where

$$a_1 = \langle s_{i_N}\cdots s_{i_2}(i_1), \lambda\rangle, \dots, a_{N-1} = \langle s_{i_N}(i_{N-1}), \lambda\rangle, a_N = \langle i_N, \lambda\rangle;$$

note that $a_1, a_2, \dots, a_N \in \mathbf{N}$, by 2.2.7.

Proposition 28.1.2. *The element $\theta(\mathbf{i}, \lambda)$ depends only on w and not on $\mathbf{i}$.*

Assume first that w is the longest element in the subgroup of W generated by two distinct elements i, j of I. In that case the assertion of the lemma is the quantum analogue of an identity of Verma, whose proof will be given in 39.3. We shall assume that this special case is known.

We now consider the general case. Let $\mathbf{i}' = (j_1, j_2, \dots, j_N)$ be another sequence like $\mathbf{i}$ (for the same w). To prove that $\theta(\mathbf{i}, \lambda) = \theta(\mathbf{i}', \lambda)$, we may assume, by 2.1.2, that $\mathbf{i}'$ is obtained from $\mathbf{i}$ by replacing a subsequence $i, j, i, j, \dots$ (m consecutive terms) of $\mathbf{i}$ by $j, i, j, i, \dots,$ (m consecutive terms), where i, j are as above and m is the order of s_is_j. But this follows immediately from the special case considered above.

28.1.3. By Proposition 28.1.2, we may use the notation $\theta(w, \lambda)$ instead of $\theta(\mathbf{i}, \lambda)$ for $w, \mathbf{i}$ as above.

Proposition 28.1.4. *The element $\theta(w, \lambda)^-\eta_\lambda \in \Lambda_\lambda$ is the unique element of the canonical basis of Λ_λ which lies in the $w(\lambda)$-weight space.*

We prove this by induction on N, the length of w. If $N = 0$, there is nothing to prove. Assume now that $N \geq 1$. Let $(i_1, i_2, \dots, i_N)$ and $(a_1, a_2, \dots, a_N)$ be as in 28.1.1. Thus $w = s_{i_1}s_{i_2}\cdots s_{i_N}$. Let $w' =$

G. Lusztig, *Introduction to Quantum Groups*, Modern Birkhäuser Classics,
DOI 10.1007/978-0-8176-4717-9_28, © Springer Science+Business Media, LLC 2010

$s_{i_2}s_{i_3}\cdots s_{i_N}$ so that $w = s_{i_1}w'$. Let b' (resp. b) be the unique element in the canonical basis of Λ_λ which lies in the $w'(\lambda)$-weight space (resp. the $w(\lambda)$-weight space). Using the induction hypothesis, we see that it is enough to prove that $b = F_{i_1}^{(a_1)}b'$. We have $w(\lambda) = s_{i_1}w'(\lambda) = w'(\lambda) - \langle i_1, w'(\lambda)\rangle i_1' = w'(\lambda) - a_1 i_1'$ so that $F_{i_1}^{(a_1)}b'$ is a non-zero vector in the same weight space as b; thus, $F_{i_1}^{(a_1)}b' = fb$ for some $f \in \mathcal{A} - \{0\}$.

Next we note that $E_{i_1}b' = 0$, since the $(w'(\lambda) + i_1')$-weight space is zero. Otherwise, the $w'^{-1}(w'(\lambda) + i_1')$-weight space would be non-zero, hence the $(\lambda + w'^{-1}(i_1'))$-weight space would be non-zero, contradicting the fact that λ is the highest weight, since $w'^{-1}(i_1') > 0$. From the definition of $\tilde{F}_{i_1}$, it then follows that $F_{i_1}^{(a_1)}b' = \tilde{F}_{i_1}^{a_1}b'$. By the properties of the basis at ∞ of Λ_λ, the previous equality implies that $f = c \mod v^{-1}\mathbf{A}$ where c is 0 or 1. Hence we have $f = c \mod v^{-1}\mathbf{Z}[v^{-1}]$ where c is as above and $f \neq 0$.

The involution $^- : \Lambda_\lambda \to \Lambda_\lambda$ keeps b, b' fixed and we have $\overline{F_{i_1}^{(a_1)})b'} = \overline{F_{i_1}^{(a_1)}}\bar{b}' = F_{i_1}^{(a_1)}b'$; hence $fb = \overline{fb} = \bar{f}b$. It follows that $\bar{f} = f$; hence $f = c$. Since $f \neq 0$ and c is 0 or 1, it follows that $f = 1$. The proposition is proved.

Proposition 28.1.5. *We have $\sigma(\theta(w_0, \lambda)) = \theta(w_0, -w_0(\lambda))$.*

Let $\mathbf{i} = (i_1, i_2, \ldots, i_N)$ be a sequence in W such that $s_{i_1}s_{i_2}\cdots s_{i_N}$ is a reduced expression of w_0. Define $a_1, a_2, \ldots, a_N$ as in 28.1.1. Then $\mathbf{i}' = (i_N, i_{N-1}, \ldots, i_1)$ is such that $s_{i_N}s_{i_{N-1}}\cdots s_{i_1}$ is a reduced expression of w_0. Let $b_1, b_2, \ldots, b_N$ be defined by

$$b_1 = \langle s_{i_1}\cdots s_{i_{N-1}}(i_N), -w_0(\lambda)\rangle, \cdots, b_{N-1} = \langle s_{i_1}(i_2), -w_0(\lambda)\rangle,$$
$$b_N = \langle i_1, -w_0(\lambda)\rangle;$$

then $b_1 = a_N, b_2 = a_{N-1}, \ldots, b_N = a_1$. Using 28.1.2, we have

$$\begin{aligned}\sigma(\theta(w_0, \lambda)) &= \sigma(\theta(\mathbf{i}, \lambda)) = \sigma(\theta_{i_1}^{(a_1)}\theta_{i_2}^{(a_2)}\cdots\theta_{i_N}^{(a_N)})\\ &= \theta_{i_N}^{(a_N)}\theta_{i_{N-1}}^{(a_{N-1})}\cdots\theta_{i_1}^{(a_1)}\\ &= \theta_{i_N}^{(b_1)}\theta_{i_{N-1}}^{(b_2)}\cdots\theta_{i_1}^{(b_N)} = \theta(\mathbf{i}', -w_0(\lambda)) = \theta(w_0, -w_0(\lambda)).\end{aligned}$$

The proposition is proved.

28.1.6. We have

(a) $$S'(\theta(w_0, \lambda)^-) = (-1)^{\langle 2\rho, \lambda\rangle}v^{-\mathbf{n}(\lambda)+c_1}\tilde{K}_\nu\sigma(\theta(w_0, \lambda))^-$$

where 2ρ and $\mathbf{n}$ are as in 2.3.1,

$$\nu = \sum_i \langle \mu(i), \lambda \rangle i$$

with $\mu(i) \in Y$ as in 2.3.2, and

$$c_1 = \sum_{i \in I} \langle \mu(i), \lambda \rangle \langle i, \lambda \rangle i \cdot i/2.$$

This follows from 2.3.2(a) and 3.3.1(d), applied to $x = \theta(w_0, \lambda)$. Note that $x \in \mathbf{f}_\nu$, where ν is as above and $\operatorname{tr} \nu = \sum_{p=1}^N \langle s_{i_N} \cdots s_{i_{p+1}}(i_p), \lambda \rangle = \langle 2\rho, \lambda \rangle$.

28.2. The Isomorphism P

28.2.1. Coinvariants and antipode. Let M, M' be two objects of $\mathcal{C}$ and let $u \in \mathbf{U}$. For $x \in M, x' \in M'$, we have

(a) $ux \otimes x' = x \otimes S(u)x'$ in the coinvariants $(M \otimes M')_*$.

We may assume that $x \in M^\lambda, x' \in M'^{\lambda'}$. First note that $x \otimes x' = 0$ (in the coinvariants) unless $\lambda + \lambda' = 0$.

We show (a) for $u = E_i$. Both sides are zero unless $\lambda + \lambda' + i' = 0$, when we have

$$E_i x \otimes x' = -v_i^{\langle i, -\lambda' - i' \rangle} x \otimes E_i x' = -x \otimes \tilde{K}_{-i} E_i x' = x \otimes S(E_i) x'$$

(in the coinvariants).

We show (a) for $u = F_i$. Both sides are zero unless $\lambda + \lambda' - i' = 0$, when we have

$$F_i x \otimes x' = -v_i^{\langle i, \lambda' \rangle} x \otimes F_i x' = -x \otimes F_i \tilde{K}_i x' = x \otimes S(F_i) x'$$

(in the coinvariants).

We show (a) for $u = K_\mu$. We have

$$K_\mu x \otimes x' = v^{\langle \mu, \lambda \rangle} x \otimes x', x \otimes S(K_\mu) x' = v^{-\langle \mu, \lambda' \rangle} x \otimes x'.$$

But we can assume that $v^{\langle \mu, \lambda \rangle} = v^{-\langle \mu, \lambda' \rangle}$.

Now if (a) holds for u, u', then it also holds for linear combinations of u, u' and for uu'. Indeed $uu'x \otimes x' = u'x \otimes S(u)x' = x \otimes S(u')S(u)x' = x \otimes S(uu')x'$ (in the coinvariants). Thus, (a) is proved.

An equivalent form of (a) is:

(b) $S'(u)x \otimes x' = x \otimes ux'$ in the coinvariants $(M \otimes M')_*$.

28.2.2. Let $\Lambda = \Lambda_\lambda$ where $\lambda \in X^+$, and let (M, B) be a based module. Let $\eta = \eta_\lambda$ and let ξ be the unique element in the canonical basis of Λ in the $w_0(\lambda)$-weight space. We define an isomorphism of vector spaces $P : \Lambda \otimes M \to M \otimes \Lambda$ by

$$P(x \otimes y) = (-1)^{\langle 2\rho, \zeta\rangle} v^{-\mathbf{n}(\zeta)} y \otimes x$$

for $x \in \Lambda^\zeta$ and $y \in M$. Here $2\rho \in Y, \mathbf{n} : X \to \mathbf{Z}$ are as in 2.3.1.

We show that P maps $E_i(\Lambda \otimes M)^{-i'}$ into $E_i(M \otimes \Lambda)$; $F_i(\Lambda \otimes M)^{i'}$ into $F_i(M \otimes \Lambda)$; and $(\Lambda \otimes M)^\zeta$ into $(M \otimes \Lambda)^\zeta$ for any $\zeta \in X$. Indeed, if $x \in \Lambda^\zeta, y \in M^{\zeta'}$, and $\zeta + \zeta' + i' = 0$, then

$$\begin{aligned} &P(E_i(x \otimes y)) = P(E_i x \otimes y + v^{\langle i, \zeta\rangle} x \otimes E_i y) \\ &= (-1)^{\langle 2\rho, \zeta\rangle} v^{-\mathbf{n}(\zeta + i')} y \otimes E_i x + (-1)^{\langle 2\rho, \zeta} v^{i \cdot i \langle i, \zeta\rangle/2 - \mathbf{n}(\zeta)} E_i y \otimes x \\ &= (-1)^{\langle 2\rho, \zeta\rangle} v^{i \cdot i \langle i, \zeta\rangle/2 - \mathbf{n}(\zeta)} (E_i y \otimes x + v^{i \cdot i \langle i, \zeta'\rangle/2} y \otimes E_i x) \\ &= (-1)^{\langle 2\rho, \zeta\rangle} v^{i \cdot i \langle i, \zeta\rangle/2 - \mathbf{n}(\zeta)} E_i(y \otimes x). \end{aligned}$$

If $x \in \Lambda^\zeta, y \in M^{\zeta'}$, and $\zeta + \zeta' - i' = 0$, then

$$\begin{aligned} &P(F_i(x \otimes y)) = P(x \otimes F_i y + v^{-i \cdot i \langle i, \zeta'\rangle/2} F_i x \otimes y) \\ &= (-1)^{\langle 2\rho, \zeta\rangle} v^{-\mathbf{n}(\zeta)} F_i y \otimes x + (-1)^{2\rho, \zeta} v^{-i \cdot i \langle i, \zeta'\rangle/2 - \mathbf{n}(\zeta - i')} y \otimes F_i x \\ &= (-1)^{\langle 2\rho, \zeta\rangle} v^{-i \cdot i \langle i, \zeta'\rangle/2 - \mathbf{n}(\zeta - i')} (y \otimes F_i x + v^{-i \cdot i \langle i, \zeta\rangle/2} F_i y \otimes x) \\ &= (-1)^{\langle 2\rho, \zeta\rangle} v^{-i \cdot i \langle i, \zeta'\rangle/2 - \mathbf{n}(\zeta - i')} F_i(y \otimes x). \end{aligned}$$

It follows that P induces an isomorphism of vector spaces $P : (\Lambda \otimes M)_* \to (M \otimes \Lambda)_*$.

28.2.3. Let $^- : M \to M$ be the associated involution of the based module (M, B). Recall that on Λ we also have an involution $^-$ associated with its natural structure of based module. Then $\Lambda \otimes M$ and $M \otimes \Lambda$ are naturally based modules with associated involution Θ^- (see 27.3.3) and the spaces of coinvariants $(\Lambda \otimes M)_*$ and $(M \otimes \Lambda)_*$ inherit from them structures of based modules (see 27.2.5) with trivial action of $\mathbf{U}$.

Proposition 28.2.4. *$P : (\Lambda \otimes M)_* \cong (M \otimes \Lambda)_*$ is an isomorphism of based modules.*

The proof will be given in 28.2.8. It will be based on a number of lemmas.

Lemma 28.2.5. *For any $y \in M^{-w_0(\lambda)}$, we have*

$$P(\xi \otimes y) = \theta(w_0, -w_0(\lambda))^- y \otimes \eta$$

(equality in $(M \otimes \Lambda)_*$).

Using 28.1.4, 28.2.1(b), 28.1.6(a), 28.1.5, we have

$$\begin{aligned}
P(\xi \otimes y) &= (-1)^{\langle 2\rho,\lambda\rangle} v^{-\mathbf{n}(w_0(\lambda))} y \otimes \xi = (-1)^{\langle 2\rho,\lambda\rangle} v^{-\mathbf{n}(w_0(\lambda))} y \otimes \theta(w_0,\lambda)^- \eta \\
&= (-1)^{\langle 2\rho,\lambda\rangle} v^{-\mathbf{n}(w_0(\lambda)} S'(\theta(w_0,\lambda)^-) y \otimes \eta \\
&= v^{-\mathbf{n}(w_0(\lambda))} v^{-\mathbf{n}(\lambda)+c_1} \tilde{K}_\nu \theta(w_0, -w_0(\lambda))^- y \otimes \eta \\
&= v^{-\mathbf{n}(w_0(\lambda))-\mathbf{n}(\lambda)+c_1+c_2} \theta(w_0, -w_0(\lambda))^- y \otimes \eta
\end{aligned}$$

(equalities in coinvariants) where $c_2 = -\sum_i \nu_i i \cdot i\langle i, \lambda\rangle/2 = -c_1$. Note also that $-\mathbf{n}(w_0(\lambda) - \mathbf{n}(\lambda) = 0$ by 2.3.1(b). The lemma is proved.

Lemma 28.2.6. *Let $b \in B$. Then*

$$\xi \otimes b = \xi \diamond b \in \Lambda \otimes M$$

and

$$b \otimes \eta = b \diamondsuit \eta \in M \otimes \Lambda.$$

From the definitions we see that $\xi \otimes b$ (resp. $b \otimes \eta$) is fixed by the involution Θ^- of $\Lambda \otimes M$ (resp. $M \otimes \Lambda$). Hence the result follows from the definition of $\xi \diamondsuit b$ and $b \diamondsuit \eta$.

In the following result, $B[\lambda]^{hi}, B[\lambda]^{lo}$ are defined in terms of (M, B) as in 27.2.3.

Lemma 28.2.7. *There is a unique bijection $B[\lambda]^{hi} \leftrightarrow B[\lambda]^{lo}$ such that the following two conditions for $b \in B[\lambda]^{hi}, b' \in B[\lambda]^{lo}$ are equivalent: $b \leftrightarrow b'$; $\theta(w_0, \lambda)^- b - b' \in M[> \lambda]$.*

Replacing M by $M[\geq \lambda]$, we are reduced to the case where $M = M[\geq \lambda]$ (see 27.2.4(a)). Then replacing M by $M/M[> \lambda]$, we are reduced to the case where $M = M[\lambda]$ (see 27.2.4(b)). Using 27.1.7, we are reduced to the case where (M, B) is Λ with its canonical basis. In this case, we have $B[\lambda]^{hi} = \{\eta\}$ and $B[\lambda]^{lo} = \{\xi\}$ and the result follows from 28.1.4.

28.2.8. Proof of Proposition 28.2.4. Let $B'_{\diamondsuit}$ (resp. $B''_{\diamondsuit}$) be the basis of $\Lambda \otimes M$ (resp. $M \otimes \Lambda$)) defined as in 27.3.3, in terms of the based modules (Λ, B_1) and (M, B). Here B_1 is the canonical basis of Λ. Let $\pi : \Lambda \otimes M \to (\Lambda \otimes M)_*$ and $\pi' : M \otimes \Lambda \to (M \otimes \Lambda)_*$ be the canonical maps. We know that π defines a bijection of $B'_{\diamondsuit}[0]$ onto a basis $(B'_{\diamondsuit})_*$ of $(\Lambda \otimes M)_*$ and π' defines a bijection of $B''_{\diamondsuit}[0]$ onto a basis $(B''_{\diamondsuit})_*$ of $(M \otimes \Lambda)_*$.

Let $b \in (B'_{\diamondsuit})_*$. Let $\tilde{b}$ be the unique element of $B'_{\diamondsuit}[0]$ such that $\pi(\tilde{b}) = b$. By 27.3.8, there exists $\lambda' \in X^+$ and elements $b_1 \in B_1[-w_0(\lambda')]^{lo}, b_2 \in B[\lambda']^{hi}$ such that $\tilde{b} = b_1 \diamondsuit b_2$. In Λ, we have that $B_1[-w_0(\lambda')]$ is empty unless $-w_0(\lambda') = \lambda$ and $B_1[\lambda]^{lo} = \{\xi\}$. Thus we have $\tilde{b} = \xi \diamondsuit b_2$ where $b_2 \in B[-w_0(\lambda)]^{hi}$ and by Lemma 28.2.6, we have $\tilde{b} = \xi \otimes b_2$. By Lemma 28.2.5, we then have $P(\tilde{b}) = \theta(w_0, -w_0(\lambda))^- b_2 \otimes \eta$ modulo the kernel of π'.

By Lemma 28.2.7, we can find an element $b'_2 \in B[-w_0(\lambda)]^{lo}$ such that

$$\theta(w_0, -w_0(\lambda))^- b_2 - b'_2 \in M[> -w_0(\lambda)].$$

Then we have $P(\tilde{b}) = b'_2 \otimes \eta$ modulo the kernel of π'. Note that

$$M[> -w_0(\lambda)] \otimes \Lambda$$

is contained in the kernel of π'.

By Lemma 28.2.6, we have $b'_2 \otimes \eta = b'_2 \diamondsuit \eta$. By 27.3.8, we have that $b'_2 \diamondsuit \eta \in B''_{\diamondsuit}[0]$. It follows that $\pi'(P(\tilde{b}))$ belongs to $\pi(B''_{\diamondsuit}[0]) = (B''_{\diamondsuit})_*$. We have therefore proved that P maps $(B'_{\diamondsuit})_*$ into $(B''_{\diamondsuit})_*$. The proposition is proved.

28.2.9. Let $\lambda_1, \lambda_2, \ldots, \lambda_n$ be a sequence of elements of X^+. As in 27.3.9, the space of coinvariants $(\Lambda_{\lambda_1} \otimes \Lambda_{\lambda_2} \cdots \otimes \Lambda_{\lambda_n})_*$ has a natural based module structure (hence has a distinguished basis).

This last based module has the following property of invariance by a cyclic permutation: there is a natural isomorphism

$$(\Lambda_{\lambda_1} \otimes \Lambda_{\lambda_2} \cdots \otimes \Lambda_{\lambda_n})_* \cong (\Lambda_{\lambda_2} \otimes \Lambda_{\lambda_3} \cdots \otimes \Lambda_{\lambda_n} \otimes \Lambda_{\lambda_1})_*$$

induced by the map

$$x_1 \otimes x_2 \cdots \otimes x_n \mapsto (-1)^{\langle 2\rho, \zeta_1 \rangle} v^{-\mathbf{n}(\zeta_1)} x_2 \otimes x_3 \cdots \otimes x_n \otimes x_1$$

where $x_p \in \Lambda_{\lambda_p}^{\zeta_p}$. This isomorphism maps the distinguished basis onto the distinguished basis (see 28.2.4). If we compose the n iterates of this isomorphism, we get the identity map of $(\Lambda_{\lambda_1} \otimes \Lambda_{\lambda_2} \cdots \otimes \Lambda_{\lambda_n})_*$, since we may assume that $\zeta_1 + \zeta_2 + \cdots + \zeta_n = 0$.

CHAPTER 29

A Refinement of the Peter-Weyl Theorem

29.1. The Subsets $\dot{\mathbf{B}}[\lambda]$ of $\dot{\mathbf{B}}$

29.1.1. In this chapter we assume that $(I, \cdot)$ is of finite type.

Let β be an element in the canonical basis $\dot{\mathbf{B}}$ of $\dot{\mathbf{U}}$. We associate to β an element $\lambda_1 \in X^+$ as follows. We have $\beta \in \dot{\mathbf{U}}1_\zeta$ for a unique $\zeta \in X$. Choose $\lambda, \lambda'' \in X^+$ such that $\lambda'' - \lambda = \zeta$ and such that $\langle i, \lambda \rangle$ is large enough for all i. Then $\beta(\xi_{-\lambda} \otimes \eta_{\lambda''})$ is in the canonical basis B of ${}^\omega\Lambda_\lambda \otimes \Lambda_{\lambda''}$, and by 27.2.1, it belongs to $B[\lambda_1]$ for a unique $\lambda_1 \in X^+$. We want to show that λ_1 depends only on β, and not on the choice of λ, λ''. It is enough to show that, if λ, λ'' are replaced by $\lambda + \lambda', \lambda' + \lambda''$, then the procedure above leads again to λ_1. This follows from 27.3.5. Thus we have a well-defined map $\dot{\mathbf{B}} \to X^+ \quad (\beta \mapsto \lambda_1)$. We shall write $\dot{\mathbf{B}}[\lambda_1]$ for the fibre of this map at λ_1. Thus we have a partition $\dot{\mathbf{B}} = \sqcup_{\lambda_1 \in X^+} \dot{\mathbf{B}}[\lambda_1]$.

29.1.2. For any $\lambda_1 \in X^+$, we denote by $\dot{\mathbf{U}}[\geq \lambda_1]$ (resp. $\dot{\mathbf{U}}[> \lambda_1]$) the $\mathbf{Q}(v)$-subspace of $\dot{\mathbf{U}}$ spanned by $\sqcup_{\lambda_2; \lambda_2 \geq \lambda_1} \dot{\mathbf{B}}[\lambda_2]$ (resp. by $\sqcup_{\lambda_2; \lambda_2 > \lambda_1} \dot{\mathbf{B}}[\lambda_2]$).

Lemma 29.1.3. *The following conditions for an element $u \in \dot{\mathbf{U}}$ are equivalent:*

(a) $u \in \dot{\mathbf{U}}[\geq \lambda_1]$;

(b) *for any $\lambda, \lambda'' \in X^+$ we have $u(\xi_{-\lambda} \otimes \eta_{\lambda''}) \in ({}^\omega\Lambda_\lambda \otimes \Lambda_{\lambda''})[\geq \lambda_1]$;*

(c) *for any object $M \in \mathcal{C}$ of finite dimension over $\mathbf{Q}(v)$ and any vector $m \in M$, we have $um \in M[\geq \lambda_1]$;*

(d) *if $\lambda_2 \in X^+$ and u acts on Λ_{λ_2} by a non-zero linear map, then $\lambda_2 \geq \lambda_1$.*

The equivalence of (a) and (b) is clear from the definition. The equivalence of (c) and (d) follows by expressing M in (c) as a direct sum of simple objects. Clearly, if u satisfies (c), then it satisfies (b). Conversely, assume that u satisfies (b); we show that it satisfies (c). We may assume that m is in a weight space of M. By 23.3.10, we can find $\lambda, \lambda'' \in X^+$ and a morphism $f : {}^\omega\Lambda_\lambda \otimes \Lambda_{\lambda''} \to M$ (in $\mathcal{C}$) such that $f(\xi_{-\lambda} \otimes \eta_{\lambda''}) = m$. We obviously have $f({}^\omega\Lambda_\lambda \otimes \Lambda_{\lambda''})[\geq \lambda_1] \subset M[\geq \lambda_1]$. Since (b) holds for u, it follows that $um = uf(\xi_{-\lambda} \otimes \eta_{\lambda''}) = f(u(\xi_{-\lambda} \otimes \eta_{\lambda''})) \in M[\geq \lambda_1]$. Thus the equivalence of (b),(c) is established. The lemma is proved.

G. Lusztig, *Introduction to Quantum Groups*, Modern Birkhäuser Classics,
DOI 10.1007/978-0-8176-4717-9_29, © Springer Science+Business Media, LLC 2010

Lemma 29.1.4. *The following conditions for an element $u \in \dot{\mathbf{U}}$ are equivalent:*

(a) $u \in \dot{\mathbf{U}}[> \lambda_1]$;

(b) *for any $\lambda, \lambda'' \in X^+$ we have*

$$u(\xi_{-\lambda} \otimes \eta_{\lambda''}) \in ({}^{\omega}\Lambda_{\lambda} \otimes \Lambda_{\lambda''})[> \lambda_1].$$

(c) *for any object $M \in \mathcal{C}$ of finite dimension over $\mathbf{Q}(v)$ and any vector $m \in M$, we have $um \in M[> \lambda_1]$;*

(d) *if $\lambda_2 \in X^+$ and u acts on Λ_{λ_2} by a non-zero linear map, then $\lambda_2 > \lambda_1$.*

This follows from the previous lemma or can be proved in the same way.

Lemma 29.1.5. *Let $\lambda_1 \in X^+$. The subspaces $\dot{\mathbf{U}}[\geq \lambda_1]$ and $\dot{\mathbf{U}}[> \lambda_1]$ of $\dot{\mathbf{U}}$ are two-sided ideals. Hence $\dot{\mathbf{U}}[\geq \lambda_1]/\dot{\mathbf{U}}[> \lambda_1]$ is naturally a $\dot{\mathbf{U}}$-bimodule.*

This follows from the descriptions 29.1.3(c), 29.1.4(c) of $\dot{\mathbf{U}}[\geq \lambda_1]$ and $\dot{\mathbf{U}}[> \lambda_1]$.

29.1.6. The $\dot{\mathbf{U}}$-module structure on Λ_{λ_1} gives us a homomorphism of algebras $\dot{\mathbf{U}} \to \mathrm{End}(\Lambda_{\lambda_1})$. This restricts to a homomorphism of algebras (without 1)

$$\dot{\mathbf{U}}[\geq \lambda_1] \to \mathrm{End}(\Lambda_{\lambda_1})$$

whose kernel is, by Lemma 29.1.4, exactly $\dot{\mathbf{U}}[> \lambda_1]$; hence we have an induced homomorphism of algebras

(a) $\dot{\mathbf{U}}[\geq \lambda_1]/\dot{\mathbf{U}}[> \lambda_1] \to \mathrm{End}(\Lambda_{\lambda_1})$ which is injective.

In particular, we have

(b) $\dim(\dot{\mathbf{U}}[\geq \lambda_1]/\dot{\mathbf{U}}[> \lambda_1]) < \infty$,

or equivalently,

(c) $\dot{\mathbf{B}}[\lambda_1]$ is a finite set for any $\lambda_1 \in X^+$.

29.2. The Finite Dimensional Algebras $\dot{\mathbf{U}}/\dot{\mathbf{U}}[P]$

29.2.1. Let P be a subset of X^+ with the following two properties:

(a) if $\lambda \in P$ and $\lambda' \in X^+$ satisfies $\lambda' \geq \lambda$, then $\lambda' \in P$;

(b) the complement of P in X^+ is finite.

Note that such P exist in abundance. We denote by $\dot{\mathbf{U}}[P]$ the subspace of $\dot{\mathbf{U}}$ generated by $\sqcup_{\lambda \in P} \dot{\mathbf{B}}[\lambda]$. From Lemma 29.1.5, we see that $\dot{\mathbf{U}}[P]$ is a two-sided ideal of $\dot{\mathbf{U}}$, and from 29.1.6(c), we see that the algebra $\dot{\mathbf{U}}/\dot{\mathbf{U}}[P]$ is finite dimensional. Note that this algebra has a unit element (unlike $\dot{\mathbf{U}}$). Indeed, since $1_\zeta \in \dot{\mathbf{B}}$ for all $\zeta \in X$, we have that $1_\zeta \in \dot{\mathbf{U}}[P]$ for all but finitely many ζ. Then $\sum_{\zeta \in X} 1_\zeta$, which is not meaningful in $\dot{\mathbf{U}}$, is meaningful in $\dot{\mathbf{U}}/\dot{\mathbf{U}}[P]$ and is the unit element there. Let $\lambda \in X^+ - P$. We show that $\dot{\mathbf{U}}[P]$ acts as zero on the $\dot{\mathbf{U}}$-module Λ_λ. Indeed, let β be an element of $\dot{\mathbf{B}} \cap P$ (these elements span P.) We have $\beta \in \dot{\mathbf{B}}[\lambda']$ for some $\lambda' \in P$. If the action of β on Λ_λ were non-zero, then from Lemma 29.1.3, it would follow that $\lambda \geq \lambda'$; using the definition of P, it would follow that $\lambda \in P$, a contradiction. We have proved that $\dot{\mathbf{U}}[P]$ acts as zero on Λ_λ, hence Λ_λ may be regarded as a $\dot{\mathbf{U}}/\dot{\mathbf{U}}[P]$-module. This module is simple. Indeed, even as a $\dot{\mathbf{U}}$-module it has no proper submodules. It is clear that for $\lambda \neq \lambda'$ in $X^+ - P$, the $\dot{\mathbf{U}}/\dot{\mathbf{U}}[P]$-modules $\Lambda_\lambda, \Lambda_{\lambda'}$ are not isomorphic (they are not isomorphic as $\dot{\mathbf{U}}$-modules).

By the standard theory of finite dimensional algebras, it follows that

$$\dim(\dot{\mathbf{U}}/\dot{\mathbf{U}}[P]) \geq \sum_\lambda (\dim \Lambda_\lambda)^2$$

(sum over all $\lambda \in X^+ - P$). On the other hand, by 29.1.6(a), we have

$$\dim(\dot{\mathbf{U}}/\dot{\mathbf{U}}[P]) = \sum_\lambda \dot{\mathbf{U}}[\geq \lambda]/\dot{\mathbf{U}}[> \lambda] \leq \sum_\lambda (\dim \Lambda_\lambda)^2$$

(both sums over all $\lambda \in X^+ - P$).

Comparing with the previous inequality, we see that

$$\dim(\dot{\mathbf{U}}/\dot{\mathbf{U}}[P]) = \sum_\lambda (\dim \Lambda_\lambda)^2$$

and

$$\dim \dot{\mathbf{U}}[\geq \lambda]/\dot{\mathbf{U}}[> \lambda] = (\dim \Lambda_\lambda)^2$$

for any $\lambda \in X^+ - P$. This implies the following result.

Proposition 29.2.2. (a) *The algebra (with 1) $\dot{\mathbf{U}}/\dot{\mathbf{U}}[P]$ is semisimple and a complete set of simple modules for it is given by Λ_λ with $\lambda \in X^+ - P$.*

(b) *For any $\lambda \in X^+$, the homomorphism $\dot{\mathbf{U}}[\geq \lambda]/\dot{\mathbf{U}}[> \lambda] \to End(\Lambda_\lambda)$ (see 29.1.6(a)) is an isomorphism.*

Actually, we get (b) for $\lambda \in X^+ - P$; but for any $\lambda \in X^+$ we can find P as above, not containing λ.

29.2.3. From the definition, we see that the finite dimensional semisimple algebra $\dot{\mathbf{U}}/\dot{\mathbf{U}}[P]$ inherits from $\dot{\mathbf{U}}$ a canonical basis, formed by the non-zero elements in the image of $\dot{\mathbf{B}}$.

29.3. The Refined Peter-Weyl Theorem

Lemma 29.3.1. *Let $\lambda \in X^+$.*

(a) *The anti-automorphisms $S, S', \sigma : \dot{\mathbf{U}} \to \dot{\mathbf{U}}$ carry $\dot{\mathbf{U}}[\geq \lambda]$ onto $\dot{\mathbf{U}}[\geq -w_0(\lambda)]$.*

(b) *The automorphism $\omega : \dot{\mathbf{U}} \to \dot{\mathbf{U}}$ carries $\dot{\mathbf{U}}[\geq \lambda]$ onto $\dot{\mathbf{U}}[\geq -w_0(\lambda)]$.*

Let $u \in \dot{\mathbf{U}}[\geq \lambda]$. Assume that $\lambda' \in X^+$ and $m \in \Lambda_{\lambda'}$ are such that $S(u)m \neq 0$. The map $\delta_{\lambda'} : {}^\omega\Lambda_{\lambda'} \otimes \Lambda_{\lambda'} \to \mathbf{Q}(v)$ (see 25.1.4) may be considered as a non-degenerate pairing; hence there exists $m' \in {}^\omega\Lambda_{\lambda'}$ such that $\delta_{\lambda'}(m' \otimes S(u)m) \neq 0$. Using 28.2.1, we see that $\delta_{\lambda'}(m' \otimes S(u)m) = \delta_{\lambda'}(um' \otimes m)$; hence $um' \neq 0$. Since ${}^\omega\Lambda_{\lambda'} \cong \Lambda_{-w_0(\lambda')}$, we see using 29.1.3, that $-w_0(\lambda') \geq \lambda$, or equivalently, that $\lambda' \geq -w_0(\lambda)$. Using again 29.1.3, we deduce that $S(u) \in \dot{\mathbf{U}}[\geq -w_0(\lambda)]$. An entirely similar proof shows that $S'(u) \in \dot{\mathbf{U}}[\geq -w_0(\lambda)]$. Thus $S(\dot{\mathbf{U}}[\geq \lambda]) \subset \dot{\mathbf{U}}[\geq -w_0(\lambda)]$ for all λ and $S'(\dot{\mathbf{U}}[\geq -w_0(\lambda)]) \subset \dot{\mathbf{U}}[\geq \lambda]$ for all λ. Since $SS' = S'S = 1$, the assertions about S and S' in (a) are proved. The assertion about σ follows from the assertion for S, using 23.1.7 and the fact that $\dot{\mathbf{U}}[\geq \lambda]$ is generated by elements in $\dot{\mathbf{B}}$ which are contained in the summands of the decomposition 23.1.2 of $\dot{\mathbf{U}}$.

We prove (b). Let $u \in \dot{\mathbf{U}}[\geq \lambda]$. Assume that $\lambda' \in X^+$ and $m \in \Lambda_{\lambda'}$ are such that $\omega(u)m \neq 0$. Then $um \neq 0$ in ${}^\omega\Lambda_{\lambda'}$ which is isomorphic to $\Lambda_{-w_0(\lambda')}$; hence, by 29.1.3, we have $-w_0(\lambda') \geq \lambda$ or equivalently, $\lambda' \geq -w_0(\lambda)$. Using again 29.1.3, we deduce that $\omega(u) \in \dot{\mathbf{U}}[\geq -w_0\lambda]$. Thus, $\omega(\dot{\mathbf{U}}[\geq \lambda]) \subset \dot{\mathbf{U}}[\geq -w_0\lambda]$. Similarly, $\omega(\dot{\mathbf{U}}[\geq -w_0(\lambda)]) \subset \dot{\mathbf{U}}[\geq \lambda]$. The lemma follows.

29.3.2. We shall use the following terminology: an element $\beta \in \dot{\mathbf{B}}$ is said to be *involutive* if $\sigma\omega(\beta) = \pm\beta$. (Recall that $\sigma\omega = \omega\sigma$ maps $\dot{\mathbf{B}}$ to $\pm\dot{\mathbf{B}}$.)

The following theorem is, in part, a summary of the results above.

Theorem 29.3.3. *Given $\lambda \in X^+$, we define $\dot{\mathbf{U}}[\geq \lambda]$ (resp. $\dot{\mathbf{U}}[> \lambda]$) as the set of all $u \in \dot{\mathbf{U}}$ with the following property: if $\lambda' \in X^+$ and u acts on $\Lambda_{\lambda'}$ by a non-zero linear map, then $\lambda' \geq \lambda$ (resp. $\lambda' > \lambda$).*

(a) *$\dot{\mathbf{U}}[\geq \lambda]$ and $\dot{\mathbf{U}}[> \lambda]$ are two-sided ideals of $\dot{\mathbf{U}}$, which are generated as vector spaces by their intersections with $\dot{\mathbf{B}}$. The quotient algebra*

$\dot{\mathbf{U}}[\geq \lambda]/\dot{\mathbf{U}}[> \lambda]$ *is isomorphic (via the action of* $\dot{\mathbf{U}}$ *on* Λ_λ*) to the algebra End(*Λ_λ*); in particular, it is finite dimensional and has a unit element, denoted by* $\mathbf{1}_\lambda$. *Let* $\pi : \dot{\mathbf{U}}[\geq \lambda] \to \dot{\mathbf{U}}[\geq \lambda]/\dot{\mathbf{U}}[> \lambda]$ *be the natural projection.*

(b) *There is a unique direct sum decomposition of* $\dot{\mathbf{U}}[\geq \lambda]/\dot{\mathbf{U}}[> \lambda]$ *into a direct sum of simple left* $\dot{\mathbf{U}}$*-modules such that each summand is generated by its intersection with the basis* $\pi(\dot{\mathbf{B}}[\lambda])$ *of* $\dot{\mathbf{U}}[\geq \lambda]/\dot{\mathbf{U}}[> \lambda]$.

(c) *There is a unique direct sum decomposition of* $\dot{\mathbf{U}}[\geq \lambda]/\dot{\mathbf{U}}[> \lambda]$ *into a direct sum of simple right* $\dot{\mathbf{U}}$*-modules such that each summand is generated by its intersection with the basis* $\pi(\dot{\mathbf{B}}[\lambda])$ *of* $\dot{\mathbf{U}}[\geq \lambda]/\dot{\mathbf{U}}[> \lambda]$.

(d) *Any summand in the decomposition (b) and any summand in the decomposition (c) have an intersection equal to a line consisting of all multiples of some element in the basis* $\pi(\dot{\mathbf{B}}[\lambda])$. *This gives a map from the set of all pairs consisting of a summand in the decomposition (b) and one in the decomposition (c), to the set* $\pi(\dot{\mathbf{B}}[\lambda])$. *This map is a bijection.*

(e) *Each summand in the decomposition (b) and each summand in the decomposition (c) contains a unique element of the form* $\pi(\beta)$ *where* $\beta \in \dot{\mathbf{B}}[\lambda]$ *is involutive.*

(f) *Let* $b, b' \in \dot{\mathbf{B}}[\lambda]$. *There exists* $b'' \in \dot{\mathbf{B}}[\lambda]$ *and* $c_{b,b',b''} \in \mathcal{A}$ *such that* $bb' = c_{b,b',b''}b'' \mod \dot{\mathbf{U}}[> \lambda]$.

(a) has already been proved. (b) follows from the definitions, using 27.1.7, 27.1.8 with $M = ({}^\omega\Lambda_{\lambda'} \otimes \Lambda_{\lambda''})[\geq \lambda]/({}^\omega\Lambda_{\lambda'} \otimes \Lambda_{\lambda''})[> \lambda]$ and with $\lambda_1 = \lambda$ (for various $\lambda', \lambda'' \in X^+$).

(c) follows from (b) using the anti-automorphism $\sigma\omega = \omega\sigma$ of $\dot{\mathbf{U}}$ which maps $\dot{\mathbf{B}}$ into itself, up to signs, (see 26.3.2) and maps $\dot{\mathbf{U}}[\geq \lambda]$ and $\dot{\mathbf{U}}[> \lambda]$ into themselves (see 29.3.1).

We prove (d). The two subspaces considered in the first sentence of (d) are a minimal left ideal and a minimal right ideal in the algebra $\dot{\mathbf{U}}[\geq \lambda]/\dot{\mathbf{U}}[> \lambda]$ which has 1, and is finite dimensional and simple (by (a)). Their intersection is therefore a line. Since both these subspaces are spanned by a subset of the basis $\pi(\dot{\mathbf{B}}[\lambda])$, the same is true about their intersection, and the first assertion of (d) follows. The map in the second sentence of (d) is obviously surjective. It is a map between two finite sets of the same cardinality $(\dim \Lambda_\lambda)^2$ (see 29.2.2); hence it is a bijection.

We prove (e). Let G be a summand in the decomposition (b). The map $\sigma\omega : \dot{\mathbf{U}} \to \dot{\mathbf{U}}$ induces an involution ι of the vector space $\dot{\mathbf{U}}[\geq \lambda]/\dot{\mathbf{U}}[> \lambda]$. The image of G under ι is a summand in the decomposition (c), which by (d) intersects G in a line spanned by a vector in $\pi(\dot{\mathbf{B}}[\lambda])$. This line is necessarily stable under ι (since ι is an involution); hence our vector in this line is preserved up to a sign by ι. This proves (e) as far as G is concerned. The same proof applies to summands in the decomposition (c).

We prove (f). The product $\pi(b)\pi(b')$ is in the intersection of the left ideal of $\dot{\mathbf{U}}[\geq \lambda]/\dot{\mathbf{U}}[> \lambda]$ generated by $\pi(b')$ with the right ideal generated by $\pi(b)$, hence, by (d), is of the form $c_{b,b',b''}\pi(b'')$ for some $b'' \in \dot{\mathbf{B}}[\lambda]$ and some $c_{b,b',b''} \in \mathbf{Q}(v)$, which is necessarily in $\mathcal{A}$ since the structure constants of the algebra $\dot{\mathbf{U}}$ with respect to $\dot{\mathbf{B}}$ are in $\mathcal{A}$.

The theorem is proved.

29.4. Cells

29.4.1. The subsets $\dot{\mathbf{B}}[\lambda]$ (for various $\lambda \in X^+$) are called *two-sided cells*; they form a partition of $\dot{\mathbf{B}}$. For each λ, the two-sided cell $\dot{\mathbf{B}}[\lambda]$ is further partitioned into subsets corresponding to the bases of the various summands in the decomposition 29.3.3(b) (these are called *left cells*) and it is also partitioned into subsets corresponding to the bases of the various summands in the decomposition 29.3.3(c) (these are called *right cells*). Then 29.3.3(d) asserts that any left cell in $\dot{\mathbf{B}}[\lambda]$ and any right cell in $\dot{\mathbf{B}}[\lambda]$ have exactly one element in common; 29.3.3(e) asserts that any left cell and any right cell contain exactly one involutive element. Since the number of left cells (or right cells) in $\dot{\mathbf{B}}[\lambda]$ is $\dim \Lambda_\lambda$, it follows that the number of involutive elements in $\dot{\mathbf{B}}[\lambda]$ is also $\dim \Lambda_\lambda$.

29.4.2. Let A be an associative algebra over a field K with a given basis B as a K-vector space. We do not assume that A has 1. The structure constants $c_{b,b',b''} \in K$ of A (where $b, b', b'' \in B$) are defined by $bb' = \sum_{b''} c_{b,b',b''}b''$.

Generalizing the definition of cells in Weyl groups (which goes back to A. Joseph), we will define certain preorders on B as follows. If $b, b' \in B$, we say that $b' \leq_L b$ (resp. $b' \leq_R b$) if there is a sequence $b = b_1, b_2, \dots, b_n = b'$ in B and a sequence $\beta_1, \beta_2, \dots, \beta_{n-1}$ in B such that $c_{\beta_s,b_s,b_{s+1}} \neq 0$ (resp. $c_{b_s,\beta_s,b_{s+1}} \neq 0$) for $s = 1, 2, \dots, n-1$. We say that $b' \leq_{LR} b$ if there is a sequence $b = b_1, b_2, \dots, b_n = b'$ in B and a sequence $\beta_1, \beta_2, \dots, \beta_{n-1}$ in B such that for any $s \in [1, n-1]$ we have either $c_{\beta_s,b_s,b_{s+1}} \neq 0$ or $c_{b_s,\beta_s,b_{s+1}} \neq 0$. Then $\leq_L, \leq_R, \leq_{LR}$ are preorders on B. We say that $b \sim_L b'$ if $b \leq_L b'$ and $b' \leq_L b$. This is an equivalence relation on B; the equivalence classes are called *left cells*. Similarly, $\leq_R$ (resp. $\leq_{LR}$) give rise to equivalence relations $\sim_R$ (resp. $\sim_{LR}$); the equivalence classes are called right cells (resp. two-sided cells).

In the case where $A = \dot{\mathbf{U}}$ and $B = \dot{\mathbf{B}}$, the definition of cells just given coincides with that given 29.4.1. (This can be easily checked.) The involutive elements in Theorem 29.3.3 are the analogues of the Duflo involutions from the theory of cells in Weyl groups.

29.4.3. We will describe explicitly the two-sided cells of $\dot{\mathbf{B}}$ in the simplest case where $I = \{i\}$ and $X = Y = \mathbf{Z}$ with $i = 1 \in Y, i' = 2 \in X$. (See 25.3.)

For each $n \geq 0$, we consider the subset

$$\mathfrak{S}(n) = \{E_i^{(a)} 1_{-n} F_i^{(b)}; n \geq a+b\} \cup \{F_i^{(b)} 1_n E_i^{(a)}; n \geq a+b\}$$

of the canonical basis $\dot{\mathbf{B}}$ (with the identification 25.3.1(c)). Note that $\mathfrak{S}(n)$ consists of $(n+1)^2$ elements. The product of two elements of $\mathfrak{S}(n)$ is given by the following equalities (modulo a linear combination of elements in $\mathfrak{S}(n+1) \cup \mathfrak{S}(n+2) \cup \cdots$):

$$E_i^{(a)} 1_{-n} F_i^{(b)} E_i^{(c)} 1_{-n} F_i^{(d)} = \begin{cases} \begin{bmatrix} n \\ b \end{bmatrix} E_i^{(a)} 1_{-n} F_i^{(d)} \text{ if } b = c, n \geq a+d \\ \begin{bmatrix} n \\ b \end{bmatrix} F_i^{(n-a)} 1_n E_i^{(n-d)} \text{ if } b = c, n \leq a+d \\ 0 \text{ if } b \neq c \end{cases}$$

$$E_i^{(a)} 1_{-n} F_i^{(b)} F_i^{(c)} 1_n E_i^{(d)} = \begin{cases} \begin{bmatrix} n \\ b \end{bmatrix} E_i^{(a)} 1_{-n} F_i^{(n-d)} \text{ if } b+c = n, d \geq a \\ \begin{bmatrix} n \\ b \end{bmatrix} F_i^{(n-a)} 1_n E_i^{(d)} \text{ if } b+c = n, d \leq a \\ 0 \text{ if } b+c \neq n \end{cases}$$

$$F_i^{(d)} 1_n E_i^{(c)} F_i^{(b)} 1_n E_i^{(a)} = \begin{cases} \begin{bmatrix} n \\ b \end{bmatrix} F_i^{(d)} 1_n E_i^{(a)} \text{ if } b = c, n \geq a+d \\ \begin{bmatrix} n \\ b \end{bmatrix} E_i^{(n-d)} 1_{-n} F_i^{(n-a)} \text{ if } b = c, n \leq a+d \\ 0 \text{ if } b \neq c \end{cases}$$

$$F_i^{(a)} 1_n E_i^{(b)} E_i^{(c)} 1_{-n} F_i^{(d)} = \begin{cases} \begin{bmatrix} n \\ b \end{bmatrix} F_i^{(a)} 1_n E_i^{(n-d)} \text{ if } b+c = n, d \geq a \\ \begin{bmatrix} n \\ b \end{bmatrix} E_i^{(n-a)} 1_{-n} F_i^{(d)} \text{ if } b+c = n, d \leq a \\ 0 \text{ if } b+c \neq n \end{cases}.$$

Hence $\mathfrak{S}(n)$ are the *two-sided cells.* The involutive elements in $\mathfrak{S}(n)$ are $E_i^{(a)} 1_{-n} F_i^{(a)}$ with $a \geq 0, 2a \leq n$ and $F_i^{(a)} 1_n E_i^{(a)}$ with $a \geq 0, 2a \leq n$, with the identification $E_i^{(a)} 1_{-n} F_i^{(a)} = F_i^{(a)} 1_n E_i^{(a)}$ if $2a = n$.

29.5. The Quantum Coordinate Algebra

29.5.1. Let $\mathbf{O}$ be the vector space of all $\mathbf{Q}(v)$-linear forms $f : \dot{\mathbf{U}} \to \mathbf{Q}(v)$ with the following property: f vanishes on $\dot{\mathbf{U}}[\geq \lambda]$ for some $\lambda \in X^+$. If $a \in \dot{\mathbf{B}}$, then the linear form $\tilde{a} : \dot{\mathbf{U}} \to \mathbf{Q}(v)$ given by $\tilde{a}(a') = \delta_{a,a'}$, for all $a \in \dot{\mathbf{B}}$, belongs to $\mathbf{O}$ and $\{\tilde{a} | a \in \dot{\mathbf{B}}\}$ is a basis of $\mathbf{O}$. This follows from 29.3.3. We define an algebra structure on $\mathbf{O}$ by the rule $\tilde{a}\tilde{b} = \sum_c \hat{m}_c^{ab} \tilde{c}$, where c runs over $\dot{\mathbf{B}}$ and $\hat{m}_c^{ab}$ are as in 25.4.1. The previous sum is well-defined: all but finitely many terms are zero. This product is associative by 25.4.1(b).

The linear map $\Delta : \mathbf{O} \to \mathbf{O} \otimes \mathbf{O}$ given by $\Delta(\tilde{c}) = \sum_{a,b} m_{ab}^c \tilde{a} \otimes \tilde{b}$ (with m_{ab}^c as in 25.4.1) is well-defined. All but finitely many terms in the sum are zero. This map is called comultiplication. It is coassociative by 25.4.1(a) and it is an algebra homomorphism by 25.4.1(c). The element $\tilde{1}_0$ is a unit element for this algebra. Consider the linear function $\mathbf{O} \to \mathbf{Q}(v)$ which takes $\tilde{a}$ to 1 if $a = 1_\lambda$ for some $\lambda \in X$, and otherwise, to zero. This is an algebra homomorphism. Thus $\mathbf{O}$ becomes a Hopf algebra called *the quantum coordinate algebra*. It is easy to see that this definition is the same as the usual one.

We call $\{\tilde{a} | a \in \dot{\mathbf{B}}\}$ the *canonical basis* of $\mathbf{O}$.

29.5.2. Let ${}_\mathcal{A}\mathbf{O}$ be the $\mathcal{A}$-submodule of $\mathbf{O}$ generated by the basis $(\tilde{a})$. Since the structure constants $m_{ab}^c, \hat{m}_c^{ab}$ are in $\mathcal{A}$, it follows that ${}_\mathcal{A}\mathbf{O}$ inherits from $\mathbf{O}$ a structure of Hopf algebra over $\mathcal{A}$. If now R is any commutative $\mathcal{A}$-algebra with 1, we can define a Hopf algebra over R by ${}_R\mathbf{O} = R \otimes_\mathcal{A} ({}_\mathcal{A}\mathbf{O})$.

CHAPTER 30

The Canonical Topological Basis of $(\mathbf{U}^- \otimes \mathbf{U}^+)\hat{}$

30.1. THE DEFINITION OF THE CANONICAL TOPOLOGICAL BASIS

30.1.1. In this chapter we assume that $(I, \cdot)$ is of finite type.

We denote by $(\mathbf{U}^- \otimes \mathbf{U}^+)\hat{}$ the closure of $\mathbf{U}^- \otimes \mathbf{U}^+$ in $(\mathbf{U} \otimes \mathbf{U})\hat{}$ (see 4.1.1). The elements of $(\mathbf{U}^- \otimes \mathbf{U}^+)\hat{}$ are possibly infinite sums of the form $\sum_{b,b' \in \mathbf{B}} c_{b,b'} b^- \otimes b'^+$ with $c_{b,b'} \in \mathbf{Q}(v)$. In this chapter we shall construct a canonical topological basis of $(\mathbf{U}^- \otimes \mathbf{U}^+)\hat{}$ which gives rise simultaneously to the canonical bases of all tensor products of type $\Lambda_\lambda \otimes {}^\omega\Lambda_{\lambda'}$.

Let $^- : (\mathbf{U}^- \otimes \mathbf{U}^+)\hat{} \to (\mathbf{U}^- \otimes \mathbf{U}^+)\hat{}$ be the ring involution defined as the unique continuous extension of $^- \otimes ^- : \mathbf{U}^- \otimes \mathbf{U}^+ \to \mathbf{U}^- \otimes \mathbf{U}^+$. Note that we have $\Theta \in (\mathbf{U}^- \otimes \mathbf{U}^+)\hat{}$ (see 4.1.2). By 24.1.6, we can write uniquely

$$\Theta = \sum_{b,b' \in \mathbf{B}; |b| = |b'|} a_{b,b'} b^- \otimes b'^+$$

where $a_{b,b'} \in \mathcal{A}$.

30.1.2. From 4.1.2, 4.1.3, it follows that the $\mathbf{Q}$-linear map

$$\Psi : (\mathbf{U}^- \otimes \mathbf{U}^+)\hat{} \to (\mathbf{U}^- \otimes \mathbf{U}^+)\hat{}$$

given by $\Psi(x) = \Theta\bar{x}$ (product in $(\mathbf{U}^- \otimes \mathbf{U}^+)\hat{}$) satisfies $\Psi^2 = 1$. We clearly have $\Psi(fx) = \bar{f}\Psi(x)$ for all $f \in \mathcal{A}$ and all x. Hence if we set

$$\Psi(b_1^- \otimes b_1'^+) = \sum_{b,b' \in \mathbf{B}; |b| = |b'|} a_{b,b'} b^- b_1^- \otimes b'^+ b_1'^+ = \sum_{b_2, b_2' \in \mathbf{B}} r_{b_1,b_1';b_2,b_2'} b_2^- \otimes b_2'^+$$

for all $b_1, b_1' \in \mathbf{B}$, then we have

$r_{b_1,b_1';b_2,b_2'} \in \mathcal{A}$;

$r_{b_1,b_1';b_2,b_2'} = 0$ unless $(b_1, b_1') \le (b_2, b_2')$ ($\le$ as in 24.3.1);

$r_{b_1,b_1';b_1,b_1'} = 1$;

$\sum_{b_2,b_2' \in \mathbf{B}} \bar{r}_{b_1,b_1';b_2,b_2'} r_{b_2,b_2';b_3,b_3'} = \delta_{b_1,b_3}\delta_{b_1',b_3'}$ for any $b_1, b_1', b_3, b_3' \in \mathbf{B}$.

G. Lusztig, *Introduction to Quantum Groups*, Modern Birkhäuser Classics,
DOI 10.1007/978-0-8176-4717-9_30, © Springer Science+Business Media, LLC 2010

The last sum is finite by the previous statements. Applying 24.2.1 to the set $H = \mathbf{B} \times \mathbf{B}$ we see that there is a unique family of elements $p_{b_1,b'_1;b_2,b'_2} \in \mathbf{Z}[v^{-1}]$ defined for $b_1, b'_1, b_2, b'_2 \in \mathbf{B}$ such that

$p_{b_1,b'_1;b_1,b'_1} = 1$;

$p_{b_1,b'_1;b_2,b'_2} \in v^{-1}\mathbf{Z}[v^{-1}]$ if $(b_1, b'_1) \neq (b_2, b'_2)$;

$p_{b_1,b'_1;b_2,b'_2} = 0$ unless $(b_1, b'_1) \leq (b_2, b'_2)$;

$p_{b_1,b'_1;b_2,b'_2} = \sum_{b_3,b'_3} \bar{p}_{b_1,b'_1;b_3,b'_3} r_{b_3,b'_3;b_2,b'_2}$

for all $(b_1, b'_1) \leq (b_2, b'_2)$. Thus we have the following result.

Proposition 30.1.3. *For any $(b_1, b'_1) \in \mathbf{B} \times \mathbf{B}$, there is a unique element $\beta_{b_1,b'_1} \in (\mathbf{U}^- \otimes \mathbf{U}^+)\hat{}$ such that $\Theta\overline{\beta_{b_1,b'_1}} = \beta_{b_1,b'_1}$ and such that $\beta_{b_1,b'_1} - b_1^- \otimes b_1'^+$ is an (infinite) linear combination of elements $b_2^- \otimes b_2'^+$ with $(b_2, b'_2) > (b_1, b'_1)$ and with coefficients in $v^{-1}\mathbf{Z}[v^{-1}]$.*

We have $\beta_{b_1,b'_1} = \sum_{b_2,b'_2} p_{b_1,b'_1;b_2,b'_2} b_2^- \otimes b_2'^+$.

30.1.4. The elements $\beta_{b_1,b'_1} \in (\mathbf{U}^- \otimes \mathbf{U}^+)\hat{}$, for various $(b_1, b'_1) \in \mathbf{B} \times \mathbf{B}$, are said to form the *canonical topological basis* of $(\mathbf{U}^- \otimes \mathbf{U}^+)\hat{}$. This is not a basis in the strict sense.

Taking $b_1 = b'_1 = 1$, we obtain an element $\Upsilon = \sum_\nu \Upsilon_\nu = \beta_{1,1}$ where $\Upsilon_\nu \in \mathbf{U}_\nu^- \otimes \mathbf{U}_\nu^+$ for all ν and

(a) $$\Upsilon_0 = 1 \otimes 1.$$

Hence Υ is an invertible element of $(\mathbf{U}^- \otimes \mathbf{U}^+)\hat{}$.

By definition, we have $\Theta\bar{\Upsilon} = \Upsilon$; hence

(b) $$\Theta = \Upsilon\bar{\Upsilon}^{-1}.$$

Note also, that if $\nu \neq 0$, then Υ_ν is a linear combination of elements $b^- \otimes b'^+$ $(b, b' \in \mathbf{B}_\nu)$ with coefficients in $v^{-1}\mathbf{Z}[v^{-1}]$. This property, together with (a),(b), characterizes Υ.

30.1.5. Let $\lambda, \lambda' \in X^+$. By the general construction in 27.3.2, the $\mathbf{U}$-module $\Lambda_\lambda \otimes {}^\omega\Lambda_{\lambda'}$ has a canonical basis $B_\diamondsuit$. It consists of elements $(b^-\eta_\lambda)\diamondsuit(b'^+\xi_{-\lambda'})$ for various $b \in \mathbf{B}(\lambda)$ and $b' \in \mathbf{B}(\lambda')$.

Note that um is a well-defined element of $\Lambda_\lambda \otimes {}^\omega\Lambda_{\lambda'}$, for any $u \in (\mathbf{U}^- \otimes \mathbf{U}^+)\hat{}$ and any $m \in \Lambda_\lambda \otimes {}^\omega\Lambda_{\lambda'}$, by regarding the last space as a $\mathbf{U} \otimes \mathbf{U}$-module. In particular, $\beta_{b_1,b'_1}(\eta_\lambda \otimes \xi_{-\lambda'})$ is well-defined.

Proposition 30.1.6. *Let $b, b' \in \mathbf{B}$.*

(a) *If $b \in \mathbf{B}(\lambda)$ and $b' \in \mathbf{B}(\lambda')$, then $\beta_{b,b'}(\eta_\lambda \otimes \xi_{-\lambda'}) = (b^- \eta_\lambda) \diamondsuit (b'^+ \xi_{-\lambda'})$.*

(b) *If either $b \notin \mathbf{B}(\lambda)$ or $b' \notin \mathbf{B}(\lambda')$, then $\beta_{b,b'}(\eta_\lambda \otimes \xi_{-\lambda'}) = 0$.*

This follows immediately from the definitions and from 27.3.2.

30.1.7. Example.. Assume $I = \{i\}$ and $X = Y = \mathbf{Z}$ with $i = 1 \in Y, i' = 2 \in X$. The canonical topological basis of $(\mathbf{U}^- \otimes \mathbf{U}^+)\hat{}$ consists of the elements

$$x_{c,d} = \sum_{s \geq 0} v_i^{-s(s+c)} \begin{bmatrix} s+d \\ s \end{bmatrix}_i F_i^{(s+c)} \otimes E_i^{(s+d)} \quad (c \geq d \geq 0)$$

and

$$y_{c,d} = \sum_{s \geq 0} v_i^{-s(s+d)} \begin{bmatrix} s+c \\ s \end{bmatrix}_i F_i^{(s+c)} \otimes E_i^{(s+d)} \quad (d \geq c \geq 0)$$

with the identification $x_{c,d} = y_{c,d}$ for $c = d$.

30.2. On the Coefficients $p_{b_1,b'_1;b_2,b'_2}$

30.2.1. The canonical topological basis in the previous section is completely determined by the set of coefficients $p_{b_1,b'_1;b_2,b'_2} \in \mathbf{Z}[v^{-1}]$ defined for all b_1, b'_1, b_2, b'_2 in $\mathbf{B}$. In this section we make a proposal for a possible topological interpretation of these coefficients, assuming that the Cartan datum is simply laced (of finite type). We shall assume that $(b_1, b'_1) \leq (b_2, b'_2)$; otherwise the coefficient is zero.

30.2.2. Let $(\mathbf{I}, H, \ldots)$ be the graph of $(I, \cdot)$ (see 14.1.3); note that $\mathbf{I} = I$. Assume that we have chosen an orientation for this graph. According to 14.5.1, to give b_1, b'_1, b_2, b'_2 in $\mathbf{B}$ is the same as to give four objects $\mathbf{V}_1, \mathbf{V}'_1, \mathbf{V}_2, \mathbf{V}'_2$ of $\mathcal{V}$ and orbits O_1, O'_1, O_2, O'_2 of $G_{\mathbf{V}_1}, G_{\mathbf{V}'_1}, G_{\mathbf{V}_2}, G_{\mathbf{V}'_2}$ on $\mathbf{E}_{\mathbf{V}_1}, \mathbf{E}_{\mathbf{V}'_1}, \mathbf{E}_{\mathbf{V}_2}, \mathbf{E}_{\mathbf{V}'_2}$, respectively (notation of 9.1.2). Hence we may write $p_{O_1,O'_1;O_2,O'_2}$ instead of $p_{b_1,b'_1;b_2,b'_2}$.

Let $\mathbf{V} = \mathbf{V}_2 \oplus \mathbf{V}'_2 \in \mathcal{V}$ and let $x \in \mathbf{E}_\mathbf{V}$ be an element such that $\mathbf{V}_2$ and $\mathbf{V}'_2$ are x-stable and the restriction of x to $\mathbf{V}_2$ (resp. $\mathbf{V}'_2$) is in O_2 (resp. in O'_2). Let J be the stabilizer of x in $G_\mathbf{V}$ and let Z be the J-orbit of $\mathbf{V}_2$ in the variety of all I-graded subspaces of $\mathbf{V}$. Note that $\mathbf{V}_2$ is a point of Z and that any $\mathbf{W} \in Z$ is x-stable. Let Z' be the subvariety of Z consisting of all subspaces $\mathbf{W} \in Z$ such that

(a) $\mathbf{W} \cap \mathbf{V}_2 \cong \mathbf{V}_1$ (in $\mathcal{V}$);

(b) $\mathbf{V}/(\mathbf{W} + \mathbf{V}_2) \cong \mathbf{V}'_1$ (in $\mathcal{V}$);

(c) the element of $\mathbf{E}_{\mathbf{W}\cap\mathbf{V}_2}$ defined by x corresponds under an isomorphism as in (a) to an element of $O_1 \subset \mathbf{E}_{\mathbf{V}_1}$;

(d) the element of $\mathbf{E}_{\mathbf{V}/(\mathbf{W}+\mathbf{V}_2)}$ defined by x corresponds under an isomorphism as in (b) to an element of $O'_1 \subset \mathbf{E}_{\mathbf{V}'_1}$.

One can hope that the coefficients of the various powers of v in $p_{O_1,O'_1;O_2,O'_2}$ are equal to the dimensions of the stalks of the intersection cohomology complex of the closure of Z' in Z at the point $\mathbf{V}_2 \in Z$, and that they are zero if $\mathbf{V}_2$ is not in that closure.

Notes on Part IV

1. The algebra $\dot{\mathbf{U}}$ has appeared in [1], in a geometric setting (in type A_n), but its definition in the general case is the same as that in type A_n. One of the main results of [1] was a topological definition of a canonical basis of $\dot{\mathbf{U}}$ (in type A_n), generalizing the author's definition of the canonical basis of $\mathbf{f}$. The method of [1] works in almost the same way for affine type A_n, but the extension to other types remains to be done.
2. In [2], Kashiwara conjectured the existence of a canonical basis of $\dot{\mathbf{U}}$, for Cartan data of finite type, and constructed a basis of the quantum coordinate algebra $\mathbf{O}$ in which the structure constants were in $\mathbf{Q}[v,v^{-1}]$; this is presumably the same as the basis in 29.5.1, in which the structure constants are in $\mathbf{Z}[v,v^{-1}]$.
3. The definition of the canonical basis $\dot{\mathbf{B}}$ of $\dot{\mathbf{U}}$, in the general case was given in [6]. Most results in Chapters 24 and 25 appeared in [6].
4. Something close to Lemma 24.2.1 has been used in [3] to attach a polynomial to two elements of a Coxeter group.
5. Expressions like those in 25.3.1 appeared in [2], and are implicit in [1].
6. I do not know what is the relation, if any, between the form (,) on $\dot{\mathbf{U}}$, in 26.1.2, and the form on $\mathbf{U}$ defined in [7].
7. Propositions 27.1.7, 27.1.8 (and also results similar to 27.2.4) appear in [2].
8. Theorem 27.3.2 is similar to results in [6]; the analogous result for more than two factors (see 27.3.6) is new.
9. The existence of a canonical basis on the space of coinvariants $(\Lambda_{\lambda_1} \otimes \Lambda_{\lambda_2} \cdots \otimes \Lambda_{\lambda_n})_*$ (see 27.3.9) is new; it was known earlier for $n = 3$, see [5]. It implies that the corresponding space of coinvariants over $\mathcal{A}$ is a free $\mathcal{A}$-module; this answers a question of D. Kazhdan. (There is a somewhat analogous result about the space of "coinvariants in the tensor product" (see [4]) of several Weyl modules with the same negative central charge over an affine Lie algebra: this space has dimension independent of the central charge. There are other analogies between the two theories, for example the invariance property under cyclic permutations (28.2.9) has a counterpart in the theory over affine Lie algebras.)
10. The results in Chapters 28, 29 and 30 are new.
11. The positivity conjecture in 25.4.2 is made plausible by the results in [1].

REFERENCES

1. A. A. Beilinson, G. Lusztig and R. MacPherson, *A geometric setting for the quantum deformation of GL_n*, Duke Math. J. **61** (1990), 655–677.
2. M. Kashiwara, *Global crystal bases of quantum groups*, Duke Math. J. **69** (1993), 455–487.
3. D. Kazhdan and G. Lusztig, *Representations of Coxeter groups and Hecke algebras*, Invent. Math. **53** (1979), 165–184.
4. ———, *Tensor structures arising from affine Lie algebras, parts I and II*, J. Amer. Math. Soc. **6** (1993), 905-947, 949–1011.
   ———, *Tensor structures arising from affine Lie algebras, parts III and IV*, J. Amer. Math. Soc. **7** (1994).
5. G. Lusztig, *Canonical bases arising from quantized enveloping algebras, II*, Common trends in mathematics and quantum field theories, (T. Eguchi et. al., eds.), Progr. Theor. Phys. Suppl., vol. 102, 1990, pp. 175–201.
6. ———, *Canonical bases in tensor products*, Proc. Nat. Acad. Sci. **89** (1992), 8177–8179.
7. M. Rosso, *Analogues de la forme de Killing et du théorème d'Harish-Chandra pour les groupes quantiques*, Ann. Sci. E. N. S. **23** (1990), 445–467.

Part V

CHANGE OF RINGS

Let R be a commutative $\mathcal{A}$-algebra with 1. The main topic of Part V is the R-algebra ${}_R\dot{\mathbf{U}}$, obtained from ${}_{\mathcal{A}}\dot{\mathbf{U}}$ by tensoring with R over $\mathcal{A}$, and its modules.

Chapter 31 contains a general discussion of ${}_R\dot{\mathbf{U}}$ and its module category.

In Chapter 32, assuming that the Cartan datum is of finite type, and that a certain root of v is given in R, we show that the integrable modules of ${}_R\dot{\mathbf{U}}$ form a braided tensor category.

In Chapter 33 we consider the specialization $v = 1$ and we establish the connection with Kac-Moody Lie algebras.

Chapters 34, 35, and 36 are concerned with the case where v is a root of unity in R. In Chapter 34 we establish various properties of Gaussian binomial coefficients at roots of 1. In Chapter 35 we construct a quantum analogue of the Frobenius homomorphism (under some rather mild assumptions). This includes as a special case the classical Frobenius homomorphism over fields of positive characteristic and also the exceptional isogenies (in small characteristic) defined by Chevalley [1]. In Chapter 36 we study the Hopf algebra ${}_R\mathfrak{u}$, which in some sense, is the kernel of the Frobenius homomorphism. This algebra is finite dimensional if R is a field and the Cartan datum is of finite type.

CHAPTER 31

The Algebra ${}_R\dot{\mathbf{U}}$

31.1. Definition of ${}_R\dot{\mathbf{U}}$

31.1.1. From now on, R will be a fixed commutative ring with 1, with a given invertible element $\mathbf{v}$. We shall regard R as an $\mathcal{A}$-algebra via the ring homomorphism $\phi : \mathcal{A} \to R$ such that $\phi(v^n) = \mathbf{v}^n$ for all $n \in \mathbf{Z}$.

We consider the R-algebras

$$ {}_R\mathbf{f} = R \otimes_{\mathcal{A}} ({}_{\mathcal{A}}\mathbf{f}) \text{ and } {}_R\dot{\mathbf{U}} = R \otimes_{\mathcal{A}} ({}_{\mathcal{A}}\dot{\mathbf{U}}). $$

We have a direct sum decomposition ${}_R\mathbf{f} = \oplus_\nu({}_R\mathbf{f}_\nu)$ where ν runs over $\mathbf{N}[I]$ and ${}_R\mathbf{f}_\nu = R \otimes_{\mathcal{A}} ({}_{\mathcal{A}}\mathbf{f}_\nu)$. The canonical bases $\mathbf{B}$ and $\dot{\mathbf{B}}$ of ${}_{\mathcal{A}}\mathbf{f}, {}_{\mathcal{A}}\dot{\mathbf{U}}$ give rise to R-bases of ${}_R\mathbf{f}, {}_R\dot{\mathbf{U}}$ consisting of elements $1 \otimes b$ where b is in $\mathbf{B}$ or $\dot{\mathbf{B}}$; we write b instead of $1 \otimes b$. In particular the elements $1_\lambda \in {}_R\dot{\mathbf{U}}$ are well-defined for all $\lambda \in X$. They satisfy as in $\dot{\mathbf{U}}$, $1_\lambda 1_{\lambda'} = \delta_{\lambda,\lambda'}1_\lambda$.

The structure constants $m_{ab}^c, \hat{m}_c^{ab}$ of $\dot{\mathbf{U}}$ (see 25.4.1) can be regarded as elements of R via the ring homomorphism $\phi : \mathcal{A} \to R$. The identities 25.4.1(a)–(d) are clearly satisfied in R.

The comultiplication of ${}_R\dot{\mathbf{U}}$ (a collection of maps as in 23.1.5) is defined by the same formulas as in 25.4.1.

31.1.2. The ${}_{\mathcal{A}}\mathbf{f} \otimes_{\mathcal{A}} ({}_{\mathcal{A}}\mathbf{f}^{opp})$-module structure $(x \otimes x') : u \mapsto x^+ux'^-$ on ${}_{\mathcal{A}}\dot{\mathbf{U}}$, by change of scalars, induces a ${}_R\mathbf{f} \otimes_R ({}_R\mathbf{f}^{opp})$-module structure on ${}_R\dot{\mathbf{U}}$ denoted in the same way. Similarly, the ${}_{\mathcal{A}}\mathbf{f} \otimes_{\mathcal{A}} ({}_{\mathcal{A}}\mathbf{f}^{opp})$-module structure $(x \otimes x') : u \mapsto x^-ux'^+$ on ${}_{\mathcal{A}}\dot{\mathbf{U}}$, by change of scalars, induces a ${}_R\mathbf{f} \otimes_R ({}_R\mathbf{f}^{opp})$-module structure on ${}_R\dot{\mathbf{U}}$ denoted in the same way.

From 23.2.2 we deduce that

(a) the elements $b^+1_\lambda b'^-$ $(b, b' \in \mathbf{B}, \lambda \in X)$ form a basis of the R-module ${}_R\dot{\mathbf{U}}$;

(b) the elements $b^-1_\lambda b'^+$ $(b, b' \in \mathbf{B}, \lambda \in X)$ form a basis of the R-module ${}_R\dot{\mathbf{U}}$;

(c) the R-algebra ${}_R\dot{\mathbf{U}}$ is generated by the elements $E_i^{(n)}1_\lambda, F_i^{(n)}1_\lambda$ for various $i \in I$, $n \geq 0$ and $\lambda \in X$.

G. Lusztig, *Introduction to Quantum Groups*, Modern Birkhäuser Classics,
DOI 10.1007/978-0-8176-4717-9_31, © Springer Science+Business Media, LLC 2010

31.1.3. We will give an alternative construction of ${}_R\dot{\mathbf{U}}$ in terms of ${}_R\mathbf{f}$.

Let ${}_RA$ be the algebra generated by the symbols $x^+1_\zeta x'^-, x^-1_\zeta x'^+$ with $x \in {}_R\mathbf{f}_\nu, x' \in {}_R\mathbf{f}_{\nu'}$, for various ν, ν', and $\zeta \in X$; these symbols are subject to the following relations:

$(\theta_i^{(a)})^+1_\zeta(\theta_j^{(b)})^- = (\theta_j^{(b)})^-1_{\zeta+ai'+bj'}(\theta_i^{(a)})^+$ if $i \neq j$;

$$(\theta_i^{(a)})^+1_{-\zeta}(\theta_i^{(b)})^- = \sum_{t\geq 0}\phi(\begin{bmatrix} a+b-\langle i,\zeta\rangle \\ t\end{bmatrix}_i)(\theta_i^{(b-t)})^-1_{-\zeta+(a+b-t)i'}(\theta_i^{(a-t)})^+;$$

$$(\theta_i^{(b)})^-1_\zeta(\theta_i^{(a)})^+ = \sum_{t\geq 0}\phi(\begin{bmatrix} a+b-\langle i,\zeta\rangle \\ t\end{bmatrix}_i)(\theta_i^{(a-t)})^+1_{\zeta-(a+b-t)i'}(\theta_i^{(b-t)})^-;$$

$x^+1_\zeta = 1_{\zeta+\nu}x^+, x^-1_\zeta = 1_{\zeta-\nu}x^-$ for $x \in \mathbf{f}_\nu$;

$(x^+1_\zeta)(1_{\zeta'}x'^-) = \delta_{\zeta,\zeta'}x^+1_\zeta x'^-$, $(x^-1_\zeta)(1_{\zeta'}x'^+) = \delta_{\zeta,\zeta'}x^-1_\zeta x'^+$;

$(x^+1_\zeta)(1_{\zeta'}x'^+) = \delta_{\zeta,\zeta'}1_{\zeta+\nu}(xx')^+$, $(x^-1_\zeta)(1_{\zeta'}x'^-) = \delta_{\zeta,\zeta'}1_{\zeta-\nu}(xx')^-$ if $x \in {}_R\mathbf{f}_\nu$;

$(rx+r'x')^+1_\zeta = rx^+1_\zeta + r'x'^+1_\zeta$, $(rx+r'x')^-1_\zeta = rx^-1_\zeta + r'x'^-1_\zeta$ if $x, x' \in {}_R\mathbf{f}_\nu$ and $r, r' \in R$.

If x or x' in $x^+1_\zeta x'^-$ or $x^-1_\zeta x'^+$ is 1, we omit writing it.

We have an obvious surjective R-algebra homomorphism ${}_RA \to {}_R\dot{\mathbf{U}}$. Using the relations of ${}_RA$, we easily see that the symbols $x^+1_\zeta x'^-$ generate ${}_RA$ as an R-module. In other words, the elements $b^+1_\zeta b'^-$, with $b, b' \in \mathbf{B}$ and $\zeta \in X$, generate ${}_RA$ as an R-module. Since they form an R-basis of ${}_R\dot{\mathbf{U}}$, they must also form an R-basis of ${}_RA$ and we deduce that:

(a) *the natural algebra homomorphism ${}_RA \to {}_R\dot{\mathbf{U}}$ is an isomorphism.*

31.1.4. There is a natural R-linear involution $\sigma : {}_R\mathbf{f} \to {}_R\mathbf{f}$; it is given by a change of rings from the analogous involution for $R = \mathcal{A}$, which is the restriction of $\sigma : \mathbf{f} \to \mathbf{f}$.

The automorphism $\omega : \dot{\mathbf{U}} \to \dot{\mathbf{U}}$ restricts to an automorphism $\omega : {}_\mathcal{A}\dot{\mathbf{U}} \to {}_\mathcal{A}\dot{\mathbf{U}}$; tensoring with R, we obtain an R-algebra automorphism $\omega : {}_R\dot{\mathbf{U}} \to {}_R\dot{\mathbf{U}}$.

31.1.5. As in 23.1.4, we say that a ${}_R\dot{\mathbf{U}}$-module M is *unital* if

(a) for any $m \in M$ we have $1_\lambda m = 0$ for all but finitely many $\lambda \in X$;

(b) for any $m \in M$ we have $\sum_{\lambda\in X} 1_\lambda m = m$.

We then have a direct sum decomposition (as an abelian group) $M = \oplus_{\lambda\in X} M^\lambda$ where $M^\lambda = 1_\lambda M$; we can regard M as an R-module by $rm = \sum_\lambda (r1_\lambda)(m)$ for $r \in R, m \in M$. Then the decomposition above is as an R-module. The unital ${}_R\dot{\mathbf{U}}$-modules are the objects of an abelian category ${}_R\mathcal{C}$ with the morphisms being homomorphisms of ${}_R\dot{\mathbf{U}}$-modules.

31.1.6. Let $M \in {}_R\mathcal{C}$, let $i \in I$ and let $n \in \mathbf{Z}$. We define R-linear maps $E_i^{(n)} : M \to M, F_i^{(n)} : M \to M$ by $E_i^{(n)} m = \sum_\lambda (E_i^{(n)} 1_\lambda) m$ and $F_i^{(n)} m = \sum_\lambda (F_i^{(n)} 1_\lambda) m$ for all $m \in M$. (Recall that $E_i^{(n)} 1_\lambda$ and $F_i^{(n)} 1_\lambda$ are elements of $\dot{\mathbf{B}} \subset \dot{\mathbf{U}}$, hence are well-defined in ${}_R\dot{\mathbf{U}}$.) It follows immediately from the definitions that $\theta_i^{(n)} \mapsto (E_i^{(n)} : M \to M)$ and $\theta_i^{(n)} \mapsto (F_i^{(n)} : M \to M)$ define two ${}_R\mathbf{f}$-module structures on M, denoted by $x, m \mapsto x^+ m$ and $x, m \mapsto x^- m$ respectively. We have

(a) $E_i^{(n)} M^\lambda \subset M^{\lambda + ni'}$, $F_i^{(n)} M^\lambda \subset M^{\lambda - ni'}$ for any $i \in I, n \in \mathbf{Z}$ and $\lambda \in X$.

Moreover, for any $\zeta \in X$ and any $m \in M^\zeta$, we have

(b) $E_i^{(a)} F_j^{(b)} m = F_j^{(b)} E_i^{(a)} m$ if $i \neq j$;

(c) $E_i^{(a)} F_i^{(b)} m = \sum_{t\ge 0} \phi(\left[\begin{smallmatrix} a-b+\langle i,\zeta\rangle \\ t \end{smallmatrix}\right]_i) F_i^{(b-t)} E_i^{(a-t)} m$;

(d) $F_i^{(b)} E_i^{(a)} m = \sum_{t\ge 0} \phi(\left[\begin{smallmatrix} -a+b-\langle i,\zeta\rangle \\ t \end{smallmatrix}\right]_i) E_i^{(a-t)} F_i^{(b-t)} m$.

31.1.7. Conversely, let M be an R-module with a given direct sum decomposition $M = \oplus_{\zeta\in X} M^\zeta$ and given R-linear maps $E_i^{(n)}, F_i^{(n)} : M \to M$ (for $i \in I, n \in \mathbf{Z}$) satisfying 31.1.6(a)–(d) and such that $E_i^{(n)} = F_i^{(n)} = 0$ for $n < 0$. Assume that $\theta_i^{(n)} \mapsto (E_i^{(n)} : M \to M)$ and $\theta_i^{(n)} \mapsto (F_i^{(n)} : M \to M)$ define two ${}_R\mathbf{f}$-module structures on M, denoted by $x, m \mapsto x^+ m$ and $x, m \mapsto x^- m$, respectively. Then this structure comes from a well-defined structure of unital ${}_R\dot{\mathbf{U}}$-module on M. Indeed, it is clear that this structure gives an ${}_R A$-module structure on M hence a ${}_R\dot{\mathbf{U}}$-module structure (see 31.1.3).

31.1.8. Let $M, M' \in {}_R\mathcal{C}$. The tensor product $M \otimes_R M'$ (as R-modules) will be regarded as a ${}_R\dot{\mathbf{U}}$-module by the rule $c(x \otimes x') = \sum_{a,b} \phi(\hat{m}_c^{ab}) ax \otimes bx'$. (All but finitely many terms in the last sum are zero.) The fact that the rule above defines an ${}_R\dot{\mathbf{U}}$-module structure follows from the identity 25.4.1(c). This ${}_R\dot{\mathbf{U}}$-module is unital, by the identity 25.4.1(d). Thus $M \otimes_R M'$ is naturally an object of ${}_R\mathcal{C}$.

Now let M, M', M'' be three objects of ${}_R\mathcal{C}$. By the previous construction, the R-module $M \otimes_R M' \otimes_R M''$ can be regarded as an object of ${}_R\mathcal{C}$ in two ways, $(M \otimes_R M') \otimes_R M''$ and $M \otimes_R (M' \otimes_R M'')$. In fact these two ways coincide; this follows from the identity 25.4.1(b).

31.1.9. From the definitions it is clear that, in the case where $R = \mathbf{Q}(v), \mathbf{v} = v$, we have ${}_R\mathcal{C} = \mathcal{C}$ and the tensor product just defined coincides with the one introduced earlier for $\mathcal{C}$.

31.1.10. To any object M of ${}_R\mathcal{C}$, we associate (as in 3.4.4) a new object ${}^{\omega}M$ of ${}_R\mathcal{C}$ as follows. ${}^{\omega}M$ has the same underlying R-module as M. By definition, for any $u \in {}_R\dot{\mathbf{U}}$, the operator u on ${}^{\omega}M$ coincides with the operator $\omega(u)$ on M.

31.1.11. If $R' \to R$ is a homomorphism of commutative $\mathcal{A}$-algebras with 1, we have ${}_R\dot{\mathbf{U}} = R \otimes_{R'} ({}_{R'}\dot{\mathbf{U}})$ and for any object $M \in {}_{R'}\mathcal{C}$, we may regard $R \otimes_{R'} M$ naturally as an object in ${}_R\mathcal{C}$ with the induced ${}_R\dot{\mathbf{U}}$-module structure. This gives a functor ${}_{R'}\mathcal{C} \to {}_R\mathcal{C}$ called *change of rings*, or *change of scalars*. It commutes with tensor products (as in 31.1.8) and with the operation ω in 31.1.10.

31.1.12. Let $(Y', X', \dots)$ be another root datum of type $(I, \cdot)$ and let $f : Y' \to Y, g : X \to X'$ be a morphism of root data. This induces a homomorphism $\phi : \mathbf{U}' \to \mathbf{U}$ between the corresponding Drinfeld-Jimbo algebras (see 3.1.2). For each $\zeta' \in X'$ and $\zeta \in X$ such that $g(\zeta) = \zeta'$ let ${}_{\mathcal{A}}\dot{\phi} : {}_{\mathcal{A}}\dot{\mathbf{U}}'1_{\zeta'} \cong {}_{\mathcal{A}}\dot{\mathbf{U}}1_{\zeta}$ be as in 23.2.5. By tensoring with R this gives rise to ${}_R\dot{\phi} : {}_R\dot{\mathbf{U}}'1_{\zeta'} \cong {}_R\dot{\mathbf{U}}1_{\zeta}$. Let M be a unital ${}_R\dot{\mathbf{U}}$-module. We can regard M as a unital ${}_R\dot{\mathbf{U}}'$-module by the following rule: if $m \in M^{\zeta}$ and $u \in {}_R\dot{\mathbf{U}}'1_{\zeta'}$ then um is defined to be $({}_R\dot{\phi}(u))m$ if $\zeta' = g(\zeta)$, and 0, otherwise. This gives a functor from unital ${}_R\dot{\mathbf{U}}$-modules to unital ${}_R\dot{\mathbf{U}}'$-modules.

31.1.13. Let $\lambda \in X$. The $\mathcal{A}$-submodule ${}_{\mathcal{A}}M_{\lambda}$ of the Verma module M_{λ} is a unital ${}_{\mathcal{A}}\dot{\mathbf{U}}$-submodule (see 23.3.2); by change of scalars, it gives rise to an object ${}_RM_{\lambda}$ of ${}_R\mathcal{C}$, called an R-Verma module. We have an exact sequence in ${}_R\mathcal{C}$:

$$\oplus_{i,n>0}({}_R\dot{\mathbf{U}}1_{\lambda+ni'}) \to {}_R\dot{\mathbf{U}}1_{\lambda} \to M_{\lambda} \to 0,$$

where the first map has components given by right multiplication by $1_{\lambda+ni'}E_i^{(n)}$ and the second map is given by $u \mapsto u1$ (1 is the canonical generator of M_{λ}). This is deduced by tensoring with R from the analogous exact sequence over $\mathcal{A}$.

Let $M \in {}_R\mathcal{C}$ and let $m \in M^\lambda$ be such that $E_i^{(n)} m = 0$ for all $i \in I$ and all $n > 0$. From the previous exact sequence, we see that there is a unique morphism $t : M_\lambda \to M$ such that $t(1) = m$.

31.2. Integrable ${}_R\dot{\mathbf{U}}$-modules

31.2.1. In this section we assume that the root datum is Y-regular (except in 31.2.4). Let $\lambda, \lambda' \in X^+$. The $\mathcal{A}$-submodule ${}_\mathcal{A}\Lambda_{\lambda'}$ of $\Lambda_{\lambda'}$ is a unital ${}_\mathcal{A}\dot{\mathbf{U}}$-submodule (see 23.3.7); by change of scalars, it gives rise to an object ${}_R\Lambda_{\lambda'} = R \otimes_\mathcal{A} ({}_\mathcal{A}\Lambda_{\lambda'})$ of ${}_R\mathcal{C}$.

Similarly, the $\mathcal{A}$-submodule ${}^\omega_\mathcal{A}\Lambda_\lambda \otimes_\mathcal{A} ({}_\mathcal{A}\Lambda_{\lambda'})$ of ${}^\omega\Lambda_\lambda \otimes \Lambda_{\lambda'}$ is a unital ${}_\mathcal{A}\dot{\mathbf{U}}$-submodule (see 23.3.9), in fact a tensor product in ${}_\mathcal{A}\mathcal{C}$; by change of scalars, it gives rise to the object ${}^\omega_R\Lambda_\lambda \otimes_R ({}_R\Lambda_{\lambda'})$ of ${}_R\mathcal{C}$.

Let $\zeta = \lambda' - \lambda \in X$. Consider the following morphisms of ${}_R\dot{\mathbf{U}}$-modules

$$\begin{array}{c}
(\oplus_{i,n>\langle i,\lambda'\rangle}({}_R\dot{\mathbf{U}}1_{\zeta-ni'})) \oplus (\oplus_{i,n>\langle i,\lambda\rangle}({}_R\dot{\mathbf{U}}1_{\zeta+ni'})) \\
\downarrow f \\
{}_R\dot{\mathbf{U}}1_\zeta \\
\downarrow \pi \\
{}^\omega_R\Lambda_\lambda \otimes_R ({}_R\Lambda_{\lambda'}) \\
\downarrow \\
0
\end{array}$$

where f has components given by right multiplication by $1_{\lambda-ni'}F_i^{(n)}$ (in the first group of summands), $1_{\lambda+ni'}E_i^{(n)}$ (in the second group of summands) and $\pi(u) = u(\xi_{-\lambda} \otimes \eta_{\lambda'})$. We write $\xi_{-\lambda}$ instead of $1 \otimes \xi_{-\lambda}$ and similarly for $\eta_{\lambda'}$.

Proposition 31.2.2. *The sequence above is exact.*

When R is $\mathcal{A}$ and $\mathbf{v} = v$, this is a restatement of 23.3.8. The general case follows from this by taking the tensor product with R, by the right exactness of tensor products.

We can state the previous proposition in the following equivalent form.

Corollary 31.2.3. *π is surjective and its kernel is the left ideal*

$$\sum_{i,n>\langle i,\lambda'\rangle} {}_R\dot{\mathbf{U}}F_i^{(n)}1_\zeta + \sum_{i,n>\langle i,\lambda\rangle} {}_R\dot{\mathbf{U}}E_i^{(n)}1_\zeta \qquad \textit{of} \;\; {}_R\dot{\mathbf{U}}.$$

31.2.4. In this subsection, the root datum is arbitrary. An object $M \in {}_R\mathcal{C}$ is said to be *integrable* if for any $m \in M$ and any $i \in I$ there exists $n_0 \geq 1$ such that $E_i^{(n)}m = F_i^{(n)}m = 0$ for all $n \geq n_0$. In the case where $R = \mathbf{Q}(v), \mathrm{v} = v$, this coincides with the earlier definition of an integrable object of $\mathcal{C}$.

From the definitions we see immediately that:

(a) if $M, M' \in {}_R\mathcal{C}$ are integrable, then $M \otimes_R M' \in {}_R\mathcal{C}$ is integrable;

(b) if $R' \to R$ is as in 31.1.11, and if $M' \in {}_{R'}\mathcal{C}$ is integrable, then $R \otimes_{R'} M' \in {}_R\mathcal{C}$ is integrable.

Let ${}_R\mathcal{C}'$ be the the category of integrable unital ${}_R\dot{\mathbf{U}}$-modules, regarded as a full subcategory of ${}_R\mathcal{C}$.

31.2.5. Returning to the assumptions of 31.2.1, we note that ${}_R\Lambda_\lambda$ and ${}^\omega_R\Lambda_\lambda \otimes_R ({}_R\Lambda_{\lambda'})$ are integrable. Indeed, this is already known over $\mathbf{Q}(v)$; from this, the result over $\mathcal{A}$ follows, since our objects over $\mathcal{A}$ are imbedded in the corresponding objects over $\mathbf{Q}(v)$ and finally, this implies the general case, by 31.2.4(b) with $R' = \mathcal{A}$.

Proposition 31.2.6. *Let $M \in {}_R\mathcal{C}$; let $\lambda, \lambda' \in X^+$. Let $\tilde{M}$ be the R-submodule of $M^{\lambda'-\lambda}$ consisting of all m such that $E_i^{(n)}m = 0$ for all i and all $n > \langle i, \lambda\rangle$ and such that $F_i^{(n)}m = 0$ for all i and all $n > \langle i, \lambda'\rangle$. Then the map $Hom_{{}_R\dot{\mathbf{U}}}({}^\omega_R\Lambda_\lambda \otimes_R ({}_R\Lambda_{\lambda'}), M) \to \tilde{M}$ given by $f \mapsto f(\xi_{-\lambda} \otimes \eta_{\lambda'})$ is an isomorphism.*

This follows immediately from Corollary 31.2.3.

Proposition 31.2.7. *Let $M \in {}_R\mathcal{C}$. Then M is integrable if and only if it satisfies the following condition:*

(a) M is a sum of subobjects each isomorphic to a quotient object of some ${}^\omega_R\Lambda_\lambda \otimes_R ({}_R\Lambda_{\lambda'})$ with $\lambda, \lambda' \in X^+$.

We know already that any object of the form ${}^\omega_R\Lambda_\lambda \otimes_R ({}_R\Lambda_{\lambda'})$ with $\lambda, \lambda' \in X^+$ is integrable. It follows immediately that, if M is as in (a), then M is integrable. We now prove the converse.

Assume that M is integrable and that $m \in M^\zeta$ where $\zeta \in X$. We can find integers $a_i, a'_i \in \mathbf{N}$ such that $E_i^{(a)}m = 0$ for all i and all $a > a_i$ and $F_i^{(a')}m = 0$ for all i and all $a' > a'_i$. Since the root datum is Y-regular, we can find $\lambda \in X$ such that $\langle i, \lambda\rangle \geq a_i$ and $\langle i, \lambda + \zeta\rangle \geq a'_i$ for all i. Let $\lambda' = \lambda + \zeta$. Then $\langle i, \lambda'\rangle \geq a'_i$ for all i. By the previous proposition, there exists a morphism $f : {}^\omega_R\Lambda_\lambda \otimes_R ({}_R\Lambda_{\lambda'}) \to M$ in ${}_R\mathcal{C}$ such that $f(\xi_{-\lambda} \otimes \eta_{\lambda'}) = m$. The image of f is a quotient of ${}^{\omega}_{R}\Lambda_\lambda \otimes_R ({}_R\Lambda_{\lambda'})$ and it contains m. Hence M satisfies (a).

31.3. Highest Weight Modules

31.3.1. In this section, we assume that the root datum is X-regular.

Let M be an object of ${}_R\mathcal{C}$. We say that M is a *highest weight module*, with highest weight $\lambda \in X$, if there exists a vector $m \in M^\lambda$ such that

(a) $E_i^{(n)} m = 0$ for all i and all $n > 0$;

(b) $M = \{x^- m | x \in {}_R\mathbf{f}\}$; and

(c) M^λ is a free R-module of rank one with generator m.

In this case, we have $M = \sum_{\lambda' \leq \lambda} M^{\lambda'}$.

Proposition 31.3.2. *Assume that R is a field.*

(a) *For any $\lambda \in X$, there exists a simple object (unique up to isomorphism) ${}_R L_\lambda$ of ${}_R\mathcal{C}$ which is a highest weight module with highest weight λ.*

(b) *If $\lambda \neq \lambda'$ then ${}_R L_\lambda$ is not isomorphic to ${}_R L_{\lambda'}$.*

(c) *If M is a highest weight module in ${}_R\mathcal{C}$ with highest weight λ, then M has a unique maximal subobject; the corresponding quotient object is isomorphic to ${}_R L_\lambda$.*

Let M be as in (c). A subobject M' of M is distinct from M if and only if $M' \subset \sum_{\lambda' < \lambda} M^{\lambda'}$. This shows that the sum of all subobjects of M distinct from M is a subobject distinct from M. Thus, M has a unique maximal subobject, hence a unique simple quotient object, which is clearly a highest weight module with highest weight λ. Applying this to the Verma module ${}_R M_\lambda$, which is a highest weight module with highest weight λ, we obtain a simple quotient ${}_R L_\lambda$ of this Verma module; this proves the existence part of (a). If L' is a simple object of ${}_R\mathcal{C}$ which is a highest weight module with highest weight λ, then, by 31.1.13, we can find a non-zero morphism ${}_R M_\lambda \to L'$. This is necessarily surjective. Since ${}_R M_\lambda$ has a unique simple quotient, we must have that L' is isomorphic to ${}_R L_\lambda$. Thus (a) and (c) are proved. (b) is now obvious.

CHAPTER 32

Commutativity Isomorphism

32.1. The Isomorphism ${}_f\mathcal{R}_{M,M'}$

32.1.1. In this chapter we assume that the Cartan datum is of finite type.

Proposition 32.1.2. *Let $M \in {}_R\mathcal{C}$.*

(a) *M is integrable if and only if it is a sum of subobjects which are finitely generated as R-modules.*

(b) *If M is integrable, then for any $m \in M$ there exists a number $N \geq 0$ such that $x^+ m = 0$ for all $x \in {}_R f_\nu$ with $tr\ \nu \geq N$.*

Assume first that M is integrable. By 31.2.7, M is a sum of subobjects which are quotients of objects of the form ${}^{\omega}_{R}\Lambda_\lambda \otimes_R ({}_R\Lambda_{\lambda'})$ with $\lambda, \lambda' \in X^+$; these objects are finitely generated (free) R-modules. It remains to show that an object $M \in {}_R\mathcal{C}$ which is finitely generated as an R-module, is integrable and satisfies (b). This follows from the fact that there are only finitely many $\lambda \in X$ such that $M^\lambda \neq 0$, together with the fact that the root datum is X-regular. If $x \in {}_R\mathbf{f}_\nu$ and $m \in M^\lambda$, then $x^+ m \in M^{\lambda+\nu}$ and $x^- m \in M^{\lambda-\nu}$.

32.1.3. Let $f : X \times X \to \mathbf{Q}$ be a function such that

$$\begin{aligned} & f(\zeta+\nu, \zeta'+\nu') - f(\zeta, \zeta') \\ \text{(a)} \qquad & = -\sum_i \nu_i \langle i, \zeta' \rangle (i \cdot i/2) - \sum_i \nu'_i \langle i, \zeta \rangle (i \cdot i/2) - \nu \cdot \nu' \end{aligned}$$

for all $\zeta, \zeta' \in X$ and all $\nu, \nu' \in \mathbf{Z}[I]$.

Such f exists: for example, we can choose a set of representatives H for the cosets $X/\mathbf{Z}[I]$ and an arbitrary function $c : H \times H \to \mathbf{Q}$, and set for any $h, h' \in H$ and $\nu, \nu' \in \mathbf{Z}[I]$:

$$f(h+\nu, h'+\nu') = c(h, h') - \sum_i \nu_i \langle i, h' \rangle (i \cdot i/2) - \sum_i \nu'_i \langle i, h \rangle (i \cdot i/2) - \nu \cdot \nu'.$$

This function satisfies (a) and conversely, any function satisfying (a), is of this form for a unique function c for fixed H. A function f satisfying (a) clearly satisfies the following identities:

G. Lusztig, *Introduction to Quantum Groups*, Modern Birkhäuser Classics,
DOI 10.1007/978-0-8176-4717-9_32, © Springer Science+Business Media, LLC 2010

(b) $f(\zeta,\zeta'+i')-f(\zeta,\zeta')=-\langle i,\zeta\rangle i\cdot i/2$,
$f(\zeta+i',\zeta')-f(\zeta,\zeta')=-\langle i,\zeta'\rangle i\cdot i/2$,
$f(\zeta-i',\zeta')-f(\zeta,\zeta')=\langle i,\zeta'\rangle i\cdot i/2$,
$f(\zeta,\zeta'-i')-f(\zeta,\zeta')=\langle i,\zeta\rangle i\cdot i/2$,
for all $\zeta,\zeta'\in X$ and $i\in I$.

32.1.4. We fix an integer $d\ge 1$ and a function $f:X\times X\to\mathbf{Q}$ as in 32.1.3, such that the values of f are contained in $\frac{1}{d}\mathbf{Z}$. Such f exists even with integer values: it suffices to take the function c in 32.1.3 with integer values. Assume that we are given an element $\tilde{\mathbf{v}}\in R$ such that $\tilde{\mathbf{v}}^d=\mathbf{v}$. For any rational number $q\in\frac{1}{d}\mathbf{Z}$, we will write $\mathbf{v}^q$ instead of $\tilde{\mathbf{v}}^{dq}$. This is the usual power of $\mathbf{v}$, when q is an integer.

Given two objects M,M' in ${}_R\mathcal{C}'$, we define an (invertible) linear operator $\Pi_f:M\otimes M'\to M\otimes M'$ by $\Pi_f(m\otimes m')=\mathbf{v}^{f(\lambda,\lambda')}m\otimes m'$ for $m\in M^\lambda, m'\in M'^{\lambda'}$. Let $\mathbf{s}:M'\otimes M\to M\otimes M'$ be the isomorphism of R-modules given by $\mathbf{s}(m'\otimes m)=m\otimes m'$. We define the R-linear map $\Theta:M\otimes_R M'\to M\otimes_R M'$ by

$$\Theta(m\otimes m')=\sum_\nu\sum_{b,b'\in\mathbf{B}_\nu}\phi(p_{b,b'})b^-m\otimes b'^+m'$$

where $\Theta=\sum_\nu\sum_{b,b'\in\mathbf{B}_\nu}p_{b,b'}b^-\otimes b'^+$ is as in 4.1.2, 24.1.6 (with $p_{b,b'}\in\mathcal{A}$) and $\phi:\mathcal{A}\to R$ is as in 31.1.1. By 32.1.2 applied to M', only finitely many terms in the sum are non-zero for any given m,m'.

Similarly, the R-linear map $\bar{\Theta}:M\otimes_R M'\to M\otimes_R M'$ given by

$$\bar{\Theta}(m\otimes m')=\sum_\nu\sum_{b,b'\in\mathbf{B}_\nu}\phi(\overline{p_{b,b'}})b^-m\otimes b'^+m'$$

for any $m\in M, m'\in M'$, is well-defined. From 4.1.3, we see that $\Theta,\bar{\Theta}:M\otimes_R M'\to M\otimes_R M'$ are inverse to each other.

Theorem 32.1.5. *Let ${}_f\mathcal{R}_{M,M'}=\Theta\Pi_f\mathbf{s}:M'\otimes M\to M\otimes M'$. Then ${}_f\mathcal{R}_{M,M'}$ is an isomorphism in ${}_R\mathcal{C}$.*

Let (f',d') be another pair like (f,d), but with $d'=1$; thus f' has values in $\mathbf{Z}$. Assume that the theorem holds for f replaced by f'; we show that it holds for f. Since $f(\lambda,\lambda')-f'(\lambda,\lambda')$ is constant when λ,λ' run through fixed cosets of $\mathbf{Z}[I]$ in X, the operator $\mathbf{s}\Pi_{f'}^{-1}\Pi_f\mathbf{s}:M'\otimes M\to M'\otimes M$ is an isomorphism in ${}_R\mathcal{C}$. Since ${}_f\mathcal{R}_{M,M'}={}_{f'}\mathcal{R}_{M,M'}\mathbf{s}\Pi_{f'}^{-1}\Pi_f\mathbf{s}$, our claim follows. Thus, in the rest of the proof we shall assume that $d=1$ so that f takes values in $\mathbf{Z}$; we then have $\tilde{\mathbf{v}}=\mathbf{v}$.

From the remark preceding the theorem, we see that ${}_f\mathcal{R}_{M,M'}$ is an isomorphism of R-modules; its inverse is $s^{-1}\Pi_f^{-1}\bar{\Theta} : M \otimes M' \to M' \otimes M$. We must show that $u\Theta\Pi_f(m \otimes m') = \Theta\Pi_f u(m' \otimes m)$ for all homogeneous m, m' and all $u \in {}_R\dot{\mathbf{U}}$. Using the characterization 31.2.7 of integrable objects, we are reduced to the case where both M, M' are of the form ${}^{\omega}_{R}\Lambda^{\lambda} \otimes_R ({}_R\Lambda_{\lambda'})$; since such objects are obtained by change of rings from the analogous objects over $\mathcal{A}$, we may assume that $R = \mathcal{A}$. This can obviously be reduced to the case where $R = \mathbf{Q}(v)$. We may assume therefore that $R = \mathbf{Q}(v)$. It suffices to show that

$$\Delta(u)\Theta\Pi_f(m \otimes m') = \Theta(\Pi_f{}^t\Delta(u)(m \otimes m'))$$

for all $u \in \mathbf{U}$, $m \in M, m' \in M'$. (Here ${}^t\Delta(u)$ is as in 3.3.4.) Let $\alpha : \mathbf{U} \otimes \mathbf{U} \to \mathbf{U} \otimes \mathbf{U}$ be the algebra automorphism given on the generators by

$$\alpha(E_i \otimes 1) = E_i \otimes \tilde{K}_{-i},$$
$$\alpha(F_i \otimes 1) = F_i \otimes \tilde{K}_i,$$
$$\alpha(1 \otimes E_i) = \tilde{K}_{-i} \otimes E_i,$$
$$\alpha(1 \otimes F_i) = \tilde{K}_i \otimes F_i,$$
$$\alpha(K_\mu \otimes K_{\mu'}) = K_\mu \otimes K_{\mu'}.$$

We have the identity $\bar{\Delta}(u) = \alpha({}^t\Delta(u))$ for all $u \in \mathbf{U}$. Indeed, both sides can be regarded as algebra homomorphisms $\mathbf{U} \to \mathbf{U} \otimes \mathbf{U}$; hence it suffices to check that they agree on the generators E_i, F_i, K_μ, which is immediate. Therefore, the identity in 24.1.2(a) can be rewritten as follows:

$$\Delta(u)\Theta(m \otimes m') = \Theta(\alpha({}^t\Delta(u))(m \otimes m')).$$

We will show that

(a) $$\alpha({}^t\Delta(u)) = \Pi_f{}^t\Delta(u)\Pi_f^{-1} : M \otimes M' \to M \otimes M',$$

for all $u \in \mathbf{U}$. Therefore we obtain

$$\Delta(u)\Theta(m \otimes m') = \Theta(\Pi_f{}^t\Delta(u)\Pi_f^{-1}(m \otimes m')),$$

for all m, m'. This implies

$$\Delta(u)\Theta\Pi_f(m \otimes m') = \Theta(\Pi_f{}^t\Delta(u)(m \otimes m')),$$

for all m, m', as required.

It remains to show (a). Clearly, if (a) holds for u, u' then it holds for uu', and for linear combinations of u, u'. Hence it suffices to verify (a) in the case where u is one of the generators E_i, F_i, K_μ. The verification for K_μ is trivial. We apply both sides of (a) (with $u = E_i$, resp. F_i) to $m \otimes m'$ where $m \in M^\zeta, m' \in M'^{\zeta'}$. The left hand side is

$$E_i m \otimes m' + \tilde{K}_{-i} m \otimes E_i m'$$

(resp. $m \otimes F_i m' + F_i m \otimes \tilde{K}_i m'$). The right hand side is

$$v^{-f(\zeta,\zeta')}(v^{f(\zeta,\zeta'+i')} m \otimes E_i m' + v^{f(\zeta+i',\zeta')} E_i m \otimes \tilde{K}_i m')$$

(resp. $v^{-f(\zeta,\zeta')}(v^{f(\zeta-i',\zeta')} F_i m \otimes m' + v^{f(\zeta,\zeta'-i')} \tilde{K}_{-i} m \otimes F_i m')$). It remains to use the identities 32.1.3(b) for f. The theorem is proved.

In the following corollary, we do not assume the existence of roots of $\mathbf{v}$.

Corollary 32.1.6. *If M, M' are in ${}_R\mathcal{C}'$, then $M \otimes M'$ and $M' \otimes M$ are isomorphic objects of ${}_R\mathcal{C}'$.*

Indeed, f can be chosen with integer values.

32.2. The Hexagon Property

32.2.1. Let M, M', M'' be three objects of ${}_R\mathcal{C}'$. We define a linear isomorphism ${}_f\Pi' : M \otimes M' \otimes M'' \to M \otimes M' \otimes M''$ by

$${}_f\Pi'(m \otimes m' \otimes m'') = \mathbf{v}^{f(\lambda'',\lambda+\lambda')-f(\lambda'',\lambda')-f(\lambda'',\lambda)} m \otimes m' \otimes m''$$

for all $m \in M^\lambda, m' \in M^{\lambda'}, m'' \in M^{\lambda''}$.

We define a linear isomorphism ${}_f\Pi'' : M'' \otimes M \otimes M' \to M'' \otimes M \otimes M'$ by

$${}_f\Pi''(m'' \otimes m \otimes m') = \mathbf{v}^{f(\lambda+\lambda',\lambda'')-f(\lambda,\lambda'')-f(\lambda',\lambda'')} m'' \otimes m \otimes m'$$

for all $m \in M^\lambda, m' \in M^{\lambda'}, m'' \in M^{\lambda''}$.

Proposition 32.2.2. (a) *The map*

$${}_f\mathcal{R}_{M'',M\otimes M'}({}_f\Pi')^{-1} : M \otimes M' \otimes M'' \to M'' \otimes M \otimes M'$$

coincides with the composition

$$M \otimes M' \otimes M'' \xrightarrow{1_M \otimes {}_f\mathcal{R}_{M'',M'}} M \otimes M'' \otimes M' \xrightarrow{{}_f\mathcal{R}_{M'',M} \otimes 1_{M'}} M'' \otimes M \otimes M'.$$

(b) *The map*

$${}_f\mathcal{R}_{M\otimes M',M''}({}_f\Pi'')^{-1} : M'' \otimes M \otimes M' \to M \otimes M' \otimes M''$$

coincides with the composition

$$M'' \otimes M \otimes M' \xrightarrow{{}_f\mathcal{R}_{M,M''} \otimes 1_{M'}} M \otimes M'' \otimes M' \xrightarrow{1_M \otimes {}_f\mathcal{R}_{M',M''}} M \otimes M' \otimes M''.$$

Let (f', d') be another pair like (f, d), but with $d' = 1$; thus f' has values in $\mathbf{Z}$. Assume that the proposition holds for f replaced by f'; as in the proof of Theorem 32.1.5, we see that it also holds for f'. Thus, in the rest of the proof, we shall assume that $d = 1$ so that f takes values in $\mathbf{Z}$; we then have $\tilde{\mathbf{v}} = \mathbf{v}$.

Using the characterization of integrable objects given in 31.2.7, we are reduced to the case where each of M, M', M'' is of the form ${}^{\omega}_{R}\Lambda^{\lambda} \otimes_R ({}_R\Lambda_{\lambda'})$; since such objects are obtained by change of rings from the analogous objects over $\mathcal{A}$, we may assume that $R = \mathcal{A}$; this case can be obviously reduced to the case where $R = \mathbf{Q}(v)$. We may assume therefore that $R = \mathbf{Q}(v)$.

Let $m \in M^{\lambda}, m' \in M'^{\lambda'}, m'' \in M''^{\lambda''}$. Using the definitions and 4.2.2(b), we have

$$\begin{aligned}
&{}_f\mathcal{R}_{M'',M\otimes M'}(m \otimes m' \otimes m'') = v^{f(\lambda'',\lambda+\lambda')} \sum_{\nu} \Theta_\nu(m'' \otimes (m \otimes m')) \\
&= v^{f(\lambda'',\lambda+\lambda')} \sum_{\nu',\nu''} v^{-f(\nu'',\lambda)+f(0,\lambda)} \Theta^{12}_{\nu'} \Theta^{13}_{\nu''}(m'' \otimes m \otimes m')
\end{aligned}$$

and

$$\begin{aligned}
&({}_f\mathcal{R}_{M'',M} \otimes 1_{M'})(1_M \otimes {}_f\mathcal{R}_{M'',M'})(m \otimes m' \otimes m'') \\
&= v^{f(\lambda'',\lambda')} \sum_{\nu} ({}_f\mathcal{R}_{M'',M} \otimes 1_{M'}) \Theta^{23}_{\nu}(m \otimes m'' \otimes m') \\
&= \sum_{\nu',\nu''} v^{f(\lambda'',\lambda')} v^{f(\lambda''-\nu'',\lambda)} \Theta^{12}_{\nu'} \Theta^{13}_{\nu''}(m'' \otimes m \otimes m').
\end{aligned}$$

On the other hand, using the definitions and 4.2.2(a), we have

$$\begin{aligned}
&{}_f\mathcal{R}_{M\otimes M',M''}(m'' \otimes m \otimes m') = v^{f(\lambda+\lambda',\lambda'')} \sum_{\nu} \Theta_\nu((m \otimes m') \otimes m'') \\
&= v^{f(\lambda+\lambda',\lambda'')} \sum_{\nu',\nu''} v^{f(\lambda',\nu'')-f(\lambda',0)} \Theta^{23}_{\nu'} \Theta^{13}_{\nu''}(m \otimes m' \otimes m'')
\end{aligned}$$

and

$$(1_M \otimes {}_f\mathcal{R}_{M',M''})({}_f\mathcal{R}_{M,M''} \otimes 1_{M'})(m'' \otimes m \otimes m')$$
$$= v^{f(\lambda,\lambda'')}(1_M \otimes {}_f\mathcal{R}_{M',M''}) \sum_\nu \Theta^{12}_\nu (m \otimes m'' \otimes m')$$
$$= v^{f(\lambda,\lambda'')} v^{f(\lambda',\lambda''+\nu'')} \sum_{\nu',\nu''} \Theta^{23}_{\nu'} \Theta^{13}_{\nu''} (m \otimes m' \otimes m'').$$

The proposition follows since $f(\lambda'' - \nu'', \lambda) - f(\lambda'', \lambda) = f(0, \lambda) - f(\nu'', \lambda)$ and $f(\lambda', \lambda'' + \nu'') - f(\lambda', \lambda'') = f(\lambda', \nu'') - f(\lambda', 0)$ for $\nu'' \in \mathbf{Z}[I]$.

Lemma 32.2.3. *Let $d \geq 1$ be the order of the torsion subgroup of $X/\mathbf{Z}[I]$. There exists a symmetric $\mathbf{Z}$-bilinear pairing $f : X \times X \to \frac{1}{d}\mathbf{Z}$ which satisfies 32.1.3(a).*

We can find a direct sum decomposition $X = X_1 \oplus X_2$ such that $\mathbf{Z}[I]$ is contained in X_1 as a subgroup of index d. There is a unique symmetric bilinear pairing $f_1 : X_1 \times X_1 \to \frac{1}{d}\mathbf{Z}$ such that $f_1(i', j') = -i \cdot j$ for all $i, j \in I$. For $x_1, x_1' \in X_1$ and $x_2, x_2' \in X_2$, we set $f(x_1 + x_2, x_1' + x_2') = f_1(x_1, x_1')$. This has the required properties.

Proposition 32.2.4 (Hexagon property). *Let $M, M', M'' \in {}_R\mathcal{C}$. Let (f, d) be as in the previous lemma. Assume that we are given an element $\tilde{\mathbf{v}} \in R$ such that $\tilde{\mathbf{v}}^d = \mathbf{v}$. Then the map ${}_f\mathcal{R}_{M'',M \otimes M'} : M \otimes M' \otimes M'' \to M'' \otimes M \otimes M'$ coincides with the composition*

$$M \otimes M' \otimes M'' \xrightarrow{1_M \otimes {}_f\mathcal{R}_{M'',M'}} M \otimes M'' \otimes M' \xrightarrow{{}_f\mathcal{R}_{M'',M} \otimes 1_{M'}} M'' \otimes M \otimes M'$$

and the map

$${}_f\mathcal{R}_{M \otimes M',M''} : M'' \otimes M \otimes M' \to M \otimes M' \otimes M''$$

coincides with the composition

$$M'' \otimes M \otimes M' \xrightarrow{{}_f\mathcal{R}_{M,M''} \otimes 1_{M'}} M \otimes M'' \otimes M' \xrightarrow{1_M \otimes {}_f\mathcal{R}_{M',M''}} M \otimes M' \otimes M''.$$

This follows from Proposition 32.2.2, since in our case, ${}_f\Pi', {}_f\Pi''$ are the identity maps.

CHAPTER 33

Relation with Kac-Moody Lie Algebras

33.1. The Specialization $v = 1$

33.1.1. Let ${}_R{}'\mathbf{f}$ be the free associative algebra over R with generators θ_i $(i \in I)$. As for $'\mathbf{f}$, which corresponds to the case $R = \mathbf{Q}(v)$, we have a natural direct sum decomposition ${}_R{}'\mathbf{f} = \oplus_\nu({}_R{}'\mathbf{f}_\nu)$ where ν runs over $\mathbf{N}[I]$; each $'\mathbf{f}_\nu$ is a free R-module of finite rank.

Let ${}_R\tilde{\mathbf{f}}$ be the quotient of the algebra ${}_R{}'\mathbf{f}$ by the two-sided ideal of ${}_R{}'\mathbf{f}$ generated by the elements

$$\Phi_{i,j} = \sum_{p+p'=1-\langle i,j'\rangle} (-1)^{p'} \phi(\begin{bmatrix} p+p' \\ p \end{bmatrix}_i) \theta_i^p \theta_j \theta_i^{p'}$$

for various $i \neq j$ in I. Recall that $\phi : \mathcal{A} \to R$ is given.

Let ${}_R\tilde{\mathbf{f}}_\nu$ be the image of ${}_R{}'\mathbf{f}_\nu$ under the natural map ${}_R{}'\mathbf{f} \to {}_R\tilde{\mathbf{f}}$. It is clear that we have a direct sum decomposition ${}_R\tilde{\mathbf{f}} = \oplus_\nu({}_R\tilde{\mathbf{f}}_\nu)$. From the definition, we have, for any ν, an exact sequence of R-modules

$$\oplus_{\nu',\nu'';i\neq j}({}_R{}'\mathbf{f}_{\nu'}) \otimes_R ({}_R{}'\mathbf{f}_{\nu''}) \to {}_R{}'\mathbf{f}_\nu \to {}_R\tilde{\mathbf{f}}_\nu \to 0$$

where the indices satisfy $\nu', \nu'' \in \mathbf{N}[I]$ and $\nu' + \nu'' + (1 - \langle i, j'\rangle)i + j = \nu$; the first map has components $x, x' \mapsto x\Phi_{i,j}x'$. If we take this exact sequence for $R = \mathbf{Q}[v, v^{-1}]$ and we tensor it over $\mathbf{Q}[v, v^{-1}]$ with $\mathbf{Q}(v)$ or with R_0 (a field of characteristic zero, regarded as an $\mathcal{A}$-algebra or $\mathbf{Q}[v, v^{-1}]$-algebra via $v \mapsto 1$), we obtain again exact sequences, by the right exactness of tensor product. We deduce that

$${}_{R_0}\tilde{\mathbf{f}}_\nu = R_0 \otimes ({}_{\mathbf{Q}[v,v^{-1}]}\tilde{\mathbf{f}}_\nu)$$

and

$${}_{\mathbf{Q}(v)}\tilde{\mathbf{f}}_\nu = \mathbf{Q}(v) \otimes ({}_{\mathbf{Q}[v,v^{-1}]}\tilde{\mathbf{f}}_\nu).$$

Since ${}_{\mathbf{Q}[v,v^{-1}]}\tilde{\mathbf{f}}_\nu$ is a finitely generated $\mathbf{Q}[v, v^{-1}]$-module and $\mathbf{Q}(v)$ is the quotient field of $\mathbf{Q}[v, v^{-1}]$, we deduce that

(a) $\dim_{\mathbf{Q}(v)}({}_{\mathbf{Q}(v)}\tilde{\mathbf{f}}_\nu) \leq \dim_{R_0}({}_{R_0}\tilde{\mathbf{f}}_\nu)$.

G. Lusztig, *Introduction to Quantum Groups*, Modern Birkhäuser Classics,
DOI 10.1007/978-0-8176-4717-9_33,

By 1.4.3, there is a unique (surjective) algebra homomorphism ${}_{\mathbf{Q}(v)}\tilde{\mathbf{f}} \to \mathbf{f}$ which takes θ_i to θ_i for all i (and preserves 1). It is clear that this homomorphism maps ${}_{\mathbf{Q}(v)}\tilde{\mathbf{f}}_\nu$ onto $\mathbf{f}_\nu$, hence

(b) $\dim_{\mathbf{Q}(v)} \mathbf{f}_\nu \le \dim_{\mathbf{Q}(v)}({}_{\mathbf{Q}(v)}\tilde{\mathbf{f}}_\nu)$ for any $\nu \in \mathbf{N}[I]$.

Note that ${}_{R_0}\tilde{\mathbf{f}}$ is the R_0-algebra defined by the generators $\theta_i \quad (i \in I)$ and the Serre relations

$$\sum_{p+p'=1-\langle i,j'\rangle} (-1)^{p'}(\theta_i^p/p!)\theta_j(\theta_i^{p'}/p'!) = 0$$

for various $i \neq j$ in I. Thus it is the enveloping algebra of (the upper triangular part of) the corresponding Kac-Moody Lie algebra over R_0.

33.1.2. Assume now that the root datum is Y-regular and X-regular. Let $\lambda \in X^+$ and let $M = {}_{R_0}\Lambda_\lambda \in {}_{R_0}\mathcal{C}'$. The linear maps $E_i, F_i : M \to M$ satisfy in our case:

(a) $E_i M^\zeta \subset M^{\zeta+i'}, \quad F_i M^\zeta \subset M^{\zeta-i'}$ for any $i \in I$ and $\zeta \in X$;

(b) $(E_iF_j - F_jE_i)m = \delta_{i,j}\langle i,\zeta\rangle m$ for any $i,j \in I$ and $m \in M^\zeta$;

(c) $\sum_{p+p'=1-\langle i,j'\rangle}(-1)^{p'}(E_i^p/p!)E_j(E_i^{p'}/p'!) = 0 : M \to M$ for any $i \neq j$ in I;

(d) $\sum_{p+p'=1-\langle i,j'\rangle}(-1)^{p'}(F_i^p/p!)F_j(F_i^{p'}/p'!) = 0 : M \to M$ for any $i \neq j$ in I.

This shows that M is an integrable highest weight module of the Kac-Moody Lie algebra attached to the root datum. By results in [3], namely, the complete reducibility theorem of Weyl-Kac and the Gabber-Kac theorem, M is simple as a module of that Lie algebra and the R_0-linear map

$${}_{R_0}\tilde{\mathbf{f}}/\sum_i {}_{R_0}\tilde{\mathbf{f}}\theta_i^{\langle i,\lambda\rangle+1} \to M$$

given by

$$\theta_{i_1}\theta_{i_2}\cdots\theta_{i_p} \mapsto F_{i_1}F_{i_2}\cdots F_{i_p}\eta_\lambda$$

is an isomorphism. It follows that

(e) ${}_{R_0}\Lambda_\lambda$ is a simple object of ${}_{R_0}\mathcal{C}$ and

(f) for given $\nu \in \mathbf{N}[I]$, we have

$$\dim_{R_0}({}_{R_0}\tilde{\mathbf{f}}_\nu) = \dim_{R_0}({}_{R_0}\Lambda_\lambda^{\lambda-\nu})$$

provided that $\langle i,\lambda\rangle$ are large enough for all i.

From the definition of Λ_λ and ${}_{R_0}\Lambda_\lambda$, it is clear that

(g) $\dim_{R_0}({}_{R_0}\Lambda_\lambda^{\lambda-\nu}) = \dim_{\mathbf{Q}(v)} \Lambda_\lambda^{\lambda-\nu}$ for any λ, ν and

(h) for given $\nu \in \mathbf{N}[I]$, we have $\dim_{\mathbf{Q}(v)}(\mathbf{f}_\nu) = \dim_{\mathbf{Q}(v)} \Lambda_\lambda^{\lambda-\nu}$ provided that $\langle i, \lambda\rangle$ are large enough for all i.

From (f),(g),(h) we deduce that

$$\dim_{\mathbf{Q}(v)}(\mathbf{f}_\nu) = \dim_{R_0}({}_{R_0}\tilde{\mathbf{f}}_\nu)$$

for all ν. Combining this with the inequalities 33.1.1(a),(b), we see that those inequalities are in fact equalities. In particular, the natural surjective homomorphism ${}_{\mathbf{Q}(v)}\tilde{\mathbf{f}} \to \mathbf{f}$ must be an isomorphism. Similarly, the natural surjective homomorphism ${}_{R_0}\tilde{\mathbf{f}} \to {}_{R_0}\mathbf{f}$ is an isomorphism since

$$\dim_{R_0}({}_{R_0}\tilde{\mathbf{f}}_\nu) \geq \dim_{R_0}(\mathbf{f}_\nu) = \dim_{\mathbf{Q}(v)}(\mathbf{f}_\nu) = \dim_{R_0}({}_{R_0}\tilde{\mathbf{f}}_\nu)$$

for all ν.

Thus we have the following result.

Theorem 33.1.3. (a) *The natural algebra homomorphism* ${}_{\mathbf{Q}(v)}\tilde{\mathbf{f}} \to \mathbf{f}$ *is an isomorphism.*

(b) *We have* $\dim_{\mathbf{Q}(v)} \mathbf{f}_\nu = \dim_{R_0}({}_{R_0}\tilde{\mathbf{f}}_\nu)$ *for any* ν.

(c) *The natural algebra homomorphism* ${}_{R_0}\tilde{\mathbf{f}} \to {}_{R_0}\mathbf{f}$ *is an isomorphism.*

(d) *If* $\lambda \in X^+$, *then the dimension of the weight spaces of* Λ_λ *are the same as those of the simple integrable highest weight representation of the corresponding Kac-Moody Lie algebra.*

33.1.4. Remark. Parts (a), (b) and (c) of the theorem hold for arbitrary root data, since only the Cartan datum is used in their statement.

Corollary 33.1.5. *The algebra* $\mathbf{U}$ *can be defined by the generators* E_i $(i \in I)$, F_i $(i \in I)$, K_μ $(\mu \in Y)$ *and the relations 3.1.1(a)–(d), together with the quantum Serre relations for the* E_i*'s and for the* F_i*'s.*

33.2. The Quasi-Classical Case

33.2.1. In this section we assume that the $\mathcal{A}$-algebra R is a field of characteristic zero and $\phi(v_i) = \pm 1$ in R for all $i \in I$. We then say that we are in the *quasi-classical* case; this is justified by the results in this section. We also assume that $(I, \cdot)$ is without odd cycles (see 2.1.3). Then, by 2.1.3, we can find a function

(a) $i \mapsto a_i$ from I to $\{0,1\}$ such that $a_i + a_j = 1$ whenever $\langle i, j'\rangle < 0$.

Let R_0 be the ring R with a new $\mathcal{A}$-algebra structure in which $v \in \mathcal{A}$ is mapped to $1 \in R$. We want to relate the algebras ${}_R\tilde{\mathbf{f}}$ and ${}_{R_0}\tilde{\mathbf{f}}$. We cannot do this directly, but must enlarge them first as follows. Let A be the R-algebra defined by the generators $\theta_i, \tilde{K}_i \quad (i \in I)$ subject to the following relations: the θ_i satisfy the relations of ${}_R\tilde{\mathbf{f}}$, the $\tilde{K}_i$ commute among themselves and $\tilde{K}_i\theta_j = \phi(v_i)^{\langle i,j'\rangle}\theta_j\tilde{K}_i$, for all $i, j \in I$.

Let A_0 be the R-algebra defined by the generators $\theta_i, \tilde{K}_i \quad (i \in I)$ subject to the following relations: the θ_i satisfy the relations of ${}_{R_0}\tilde{\mathbf{f}}$, the $\tilde{K}_i$ commute among themselves and $\tilde{K}_i\theta_j = \phi(v_i)^{\langle i,j'\rangle}\theta_j\tilde{K}_i$, for all $i, j \in I$.

It is clear that, as an R-vector space, A (resp. A_0) is the tensor product of ${}_R\tilde{\mathbf{f}}$ (resp. ${}_{R_0}\tilde{\mathbf{f}}$) with the group algebra of $\mathbf{Z}^I$ over R, with basis given by the monomials in $\tilde{K}_i$.

Proposition 33.2.2. (a) *The assignment $E_i \mapsto E_i' = E_i\tilde{K}_i^{a_i}$ and $\tilde{K}_i \mapsto \tilde{K}_i$ for all i, defines an isomorphism of R-algebras $A_0 \to A$.*

(b) $\dim_R({}_R\tilde{\mathbf{f}}_\nu) = \dim_R({}_{R_0}\tilde{\mathbf{f}}_\nu)$ *for all ν.*

(c) *The natural algebra homomorphism ${}_R\tilde{\mathbf{f}} \to {}_R\mathbf{f}$ is an isomorphism.*

Let $i, j \in I$ be distinct. A simple computation shows that we have in A:

$$E_i'^p E_j' E_i'^{p'} = \phi(v_i)^{a_i(p+p')(p+p'-1)}\phi(v_i)^{\langle i,j'\rangle(pa_i+p'a_j)}E_i^pE_jE_i^{p'}\tilde{K}_i^{(p+p')a_i}\tilde{K}_j^{a_j}.$$

The factor $\phi(v_i)^{a_i(p+p')(p+p'-1)}$ is 1 since $\phi(v_i) = \pm 1$ and the exponent is even. By the definition of a_i, we have

$$\phi(v_i)^{\langle i,j'\rangle(pa_i+p'a_j)} = \phi(v_i)^{\langle i,j'\rangle(p+p')a_j}\phi(v_i)^{\langle i,j'\rangle p}$$

and this equals $\phi(v_i)^{\langle i,j'\rangle p}$, if $p + p' = 1 - \langle i, j'\rangle$. Note also that $\phi([n]_i^!) = \phi(v_i)^{n(n-1)/2}n!$, since $\phi(v_i) = \pm 1$. Hence, if $p + p' = 1 - \langle i, j'\rangle$, then

$$\begin{aligned}\phi([p]_i^!)\phi([p']_i^!) &= \phi(v_i)^{(p(p-1)+p'(p'-1))/2}p!p'!\\ &= \phi(v_i)^{(p+p')(p+p'-1)/2}\phi(v_i)^{-pp'}\phi(v_i)^{p(1-p)}p!p'!\\ &= \phi(v_i)^{(p+p')(p+p'-1)/2}\phi(v_i)^{\langle i,j'\rangle p}p!p'!.\end{aligned}$$

It follows that

$$\sum_{p+p'=1-\langle i,j'\rangle}(-1)^{p'}(E_i'^p/p!)E_j'(E_i'^{p'}/p'!) = \phi(v_i)^{\langle i,j'\rangle(1-\langle i,j'\rangle)/2}$$

$$\sum_{p+p'=1-\langle i,j'\rangle}(-1)^{p'}(E_i{}^p/\phi([p]_i^!))E_j(E_i{}^{p'}/\phi([p']_i^!))\tilde{K}_i^{(1-\langle i,j'\rangle)a_i}\tilde{K}_j^{a_j}.$$

This shows that the assignment in (a) preserves relations, hence defines an algebra homomorphism $A_0 \to A$. The same proof shows that the assignment $E_i \mapsto E_i\tilde{K}_i^{-a_i}$ and $\tilde{K}_i \mapsto \tilde{K}_i$ defines an algebra homomorphism $A \to A_0$; it is clear that this is the inverse of the previous homomorphism. This proves (a).

For any ν, the isomorphism in (a) maps the subspace ${}_{R_0}\tilde{\mathbf{f}}_\nu$ isomorphically onto the subspace $({}_R\tilde{\mathbf{f}}_\nu)\tilde{K}$ where $\tilde{K}$ is a monomial in the $\tilde{K}_i$ depending only on ν. This proves (b).

The homomorphism in (c) maps ${}_R\tilde{\mathbf{f}}_\nu$ onto ${}_R\mathbf{f}_\nu$ for any ν. Using (b), it follows that this restriction ${}_R\tilde{\mathbf{f}}_\nu \to {}_R\mathbf{f}_\nu$ is an isomorphism. This implies (c). The proposition is proved.

The next result compares the algebras ${}_{R_0}\dot{\mathbf{U}}$ and ${}_R\dot{\mathbf{U}}$.

Proposition 33.2.3. *There is a unique isomorphism of R-algebras* $f: {}_{R_0}\dot{\mathbf{U}} \to {}_R\dot{\mathbf{U}}$ *such that*

$$f(E_i 1_\zeta) = \phi(v_i)^{a_i\langle i,\zeta\rangle} E_i 1_\zeta, f(F_i 1_\zeta) = \phi(v_i)^{(1-a_i)\langle i,\zeta\rangle+1} F_i 1_\zeta, f(1_\zeta) = 1_\zeta$$

for all $i \in I, \zeta \in X$. *Here* a_i *is as in 33.2.1(a).*

To construct f, we take advantage of the fact that for the algebra ${}_{R_0}\dot{\mathbf{U}}$, we know a simple presentation by generators and relations, while for ${}_R\dot{\mathbf{U}}$ we do not. It will be easier to first construct a functor from ${}_R\mathcal{C}$ to ${}_{R_0}\mathcal{C}$ and then to show that it comes from an algebra homomorphism.

Let M be an object of ${}_R\mathcal{C}$. The linear maps $E_i, F_i : M \to M \quad (i \in I)$ satisfy

(a) $E_i M^\lambda \subset M^{\lambda+i'}, \quad F_i M^\lambda \subset M^{\lambda-i'}$ for any $i \in I$ and $\lambda \in X$;

(b) $(E_iF_j - F_jE_i)m = \delta_{i,j}\phi(v_i)^{\langle i,\lambda\rangle-1}\langle i,\lambda\rangle m$ for any $i,j \in I$ and $m \in M^\lambda$;

(c) $\sum_{p+p'=1-\langle i,j'\rangle}(-1)^{p'}\phi(v_i)^{p\langle i,j'\rangle}(E_i^p/p!)E_j(E_i^{p'}/p'!) = 0 : M \to M$ for any $i \neq j$ in I;

(d) $\sum_{p+p'=1-\langle i,j'\rangle}(-1)^{p'}\phi(v_i)^{p\langle i,j'\rangle}(F_i^p/p!)F_j(F_i^{p'}/p'!) = 0 : M \to M$ for any $i \neq j$ in I.

For any $a \in \mathbf{Z}$, we define linear maps $P_{i,a} : M \to M$ by $P_{i,a}(m) = \phi(v_i)^{a\langle i,\lambda\rangle}m$ for $m \in M^\lambda$. We define linear maps $E_i', F_i' : M \to M$ in terms of the function 33.2.1(a) by $E_i' = E_iP_{i,a_i}, F_i' = \phi(v_i)F_iP_{i,1-a_i}$. As in the proof of the previous proposition, we can check that:

(a1) $E_i'M^\lambda \subset M^{\lambda+i'}, \quad F_i'M^\lambda \subset M^{\lambda-i'}$ for any $i \in I$ and $\lambda \in X$;

(b1) $(E_i'F_j' - F_j'E_i')m = \delta_{i,j}\langle i,\lambda\rangle m$ for any $i,j \in I$ and $m \in M^\lambda$;

(c1) $\sum_{p+p'=1-\langle i,j'\rangle}(-1)^{p'}(E_i'^p/p!)E_j'(E_i'^{p'}/p'!) = 0 : M \to M$ for any $i \neq j$ in I;

(d1) $\sum_{p+p'=1-\langle i,j'\rangle}(-1)^{p'}(F_i'^p/p!)F_j'(F_i'^{p'}/p'!) = 0 : M \to M$ for any $i \neq j$ in I. Since ${}_{R_0}\mathbf{f} = {}_{R_0}\tilde{\mathbf{f}}$ (see 33.1.3) has the presentation given by the Serre relations, it follows that M with its weight space decomposition and with the linear maps E_i', F_i' is an object of ${}_{R_0}\mathcal{C}$.

Thus, we have defined a functor $\Gamma : {}_R\mathcal{C} \to {}_{R_0}\mathcal{C}$. This functor has the following obvious property: it associates to M an object $\Gamma(M)$ with the same underlying R-module as M and any endomorphism of M in ${}_R\mathcal{C}$ is at the same time an endomorphism of $\Gamma(M)$ in ${}_{R_0}\mathcal{C}$. Applying the functor Γ to ${}_R\dot{\mathbf{U}}$, regarded as a left module over itself, we obtain a structure of a unital left ${}_{R_0}\dot{\mathbf{U}}$-module on ${}_R\dot{\mathbf{U}}$. Thus, we have an R-bilinear pairing ${}_{R_0}\dot{\mathbf{U}} \times {}_R\dot{\mathbf{U}} \to {}_R\dot{\mathbf{U}}$, denoted by $a, b \mapsto a * b$; this has the properties $(aa') * b = a * (a' * b)$ and $(a * b)b' = a * (bb')$. The last property holds since right multiplication by b is an endomorphism of ${}_R\dot{\mathbf{U}}$ in ${}_R\mathcal{C}$, hence also in ${}_{R_0}\mathcal{C}$. We define a map $f : {}_{R_0}\dot{\mathbf{U}} \to {}_R\dot{\mathbf{U}}$ by $f(a) = \sum_{\zeta \in X} a * 1_\zeta$; only finitely many terms in the sum are non-zero. We have

$$
\begin{aligned}
f(a)f(a') &= \sum_{\zeta,\zeta'}(a * 1_\zeta)(a' * 1_{\zeta'}) \\
&= \sum_{\zeta'} a * \sum_{\zeta} 1_\zeta(a' * 1_{\zeta'}) \\
&= \sum_{\zeta'} a * (a' * 1_{\zeta'}) = \sum_{\zeta'} aa' * 1_{\zeta'} = f(aa').
\end{aligned}
$$

It is clear that f has the specified values on the algebra generators $E_i 1_\zeta, F_i 1_\zeta, 1_\zeta$ of ${}_{R_0}\dot{\mathbf{U}}$.

We show that f is an isomorphism. The elements $E_i 1_\zeta, F_i 1_\zeta, 1_\zeta$ are algebra generators of ${}_R\dot{\mathbf{U}}$, since $\phi([n]_i^!) = \pm n!$ is invertible in R for any $i \in I$ and any $n \geq 0$. It follows that f is surjective. As in 31.1.2, ${}_R\dot{\mathbf{U}}$ (resp. ${}_{R_0}\dot{\mathbf{U}}$) is a free ${}_R\mathbf{f} \otimes {}_R\mathbf{f}^{opp}$-module (resp. ${}_{R_0}\mathbf{f} \otimes {}_{R_0}\mathbf{f}^{opp}$-module) with generators 1_ζ (under $(x \otimes x') : u \mapsto x^+ u x'^-$), and f carries the subspace of ${}_{R_0}\dot{\mathbf{U}}$ spanned by $x^+ 1_\zeta x'^-$ with

$$
x \in \oplus_{\text{ tr } \nu \leq N}({}_{R_0}\mathbf{f}_\nu), x' \in \oplus_{\text{ tr } \nu \leq N'}({}_{R_0}\mathbf{f}_\nu)
$$

and fixed ζ onto the analogous subspace of ${}_R\dot{\mathbf{U}}$. Since these subspaces are both of the same (finite) dimension over R, the restriction of f must be an isomorphism between them. This implies that f is an isomorphism. The uniqueness of f is obvious. The proposition is proved.

Corollary 33.2.4. *Assume that the root datum is Y-regular and X-regular and that $\lambda \in X^+$. Then ${}_R\Lambda_\lambda$ is a simple object of ${}_R\mathcal{C}$.*

By the proof of 33.2.3, this is equivalent to the statement that ${}_{R_0}\Lambda_\lambda$ is a simple object of ${}_{R_0}\mathcal{C}$. (See 33.1.2(e).)

CHAPTER 34

Gaussian Binomial Coefficients at Roots of 1

34.1.1. Let l be an integer ≥ 1. In this chapter we assume that the $\mathcal{A}$-algebra R, with $\phi : \mathcal{A} \to R$ and with $\mathbf{v} = \phi(v)$, is such that R is an integral domain and the following hold

$$\mathbf{v}^{2l} = 1 \text{ and } \mathbf{v}^{2t} \neq 1 \text{ for all } 0 < t < l.$$

All the identities proved in 1.3 for Gaussian binomial coefficients imply, after applying ϕ, corresponding identities in R. However, certain identities will be satisfied only in R. We shall now give some examples of such identities.

Lemma 34.1.2. (a) *If $t \geq 1$ is not divisible by l, and $a \in \mathbf{Z}$ is divisible by l, then $\phi(\left[\begin{smallmatrix} a \\ t \end{smallmatrix}\right]) = 0$.*

(b) *If $a_1 \in \mathbf{Z}$ and $t_1 \in \mathbf{N}$, then we have*

$$\phi\left(\begin{bmatrix} la_1 \\ lt_1 \end{bmatrix}\right) = \mathbf{v}^{l^2(a_1+1)t_1} \binom{a_1}{t_1}.$$

(c) *Let $a \in \mathbf{Z}$ and $t \in \mathbf{N}$. Write $a = a_0 + la_1$ with $a_0, a_1 \in \mathbf{Z}$ such that $0 \leq a_0 \leq l-1$ and $t = t_0 + lt_1$ with $t_0, t_1 \in \mathbf{N}$ such that $0 \leq t_0 \leq l-1$. We have*

$$\phi\left(\begin{bmatrix} a \\ t \end{bmatrix}\right) = \mathbf{v}^{(a_0 t_1 - a_1 t_0)l + (a_1+1)t_1 l^2} \phi\left(\begin{bmatrix} a_0 \\ t_0 \end{bmatrix}\right) \binom{a_1}{t_1}.$$

Here $\binom{a_1}{t_1} \in \mathbf{Z}$ is an ordinary binomial coefficient. We prove (a) for $a \geq 0$ by induction on a. If $a = 0$, we have trivially $\left[\begin{smallmatrix} a \\ t \end{smallmatrix}\right] = 0$; if $a = l$, the equality $\phi(\left[\begin{smallmatrix} a \\ t \end{smallmatrix}\right]) = 0$ follows directly from the definitions.

Assume now that $a \geq 2l$ and that (a) holds for $a - l$ instead of a. By 1.3.1(e), we have

$$\phi\left(\begin{bmatrix} a \\ t \end{bmatrix}\right) = \sum_{t'+t''=t} \mathbf{v}^{(al-l)t''-lt'} \phi\left(\begin{bmatrix} al-l \\ t' \end{bmatrix}\right) \phi\left(\begin{bmatrix} l \\ t'' \end{bmatrix}\right).$$

G. Lusztig, *Introduction to Quantum Groups*, Modern Birkhäuser Classics,
DOI 10.1007/978-0-8176-4717-9_34, © Springer Science+Business Media, LLC 2010

For each term in the sum, we have that either t' or t'' is not divisible by l; hence the sum is zero by the induction hypothesis. This proves (a) for $a \geq 0$. We now prove (a), assuming that $a < 0$. Write $t = t_0 + lt_1$ with $0 < t_0 < l$. We have

$$\phi\left(\begin{bmatrix} a \\ t \end{bmatrix}\right) = (-1)^t \phi\left(\begin{bmatrix} -a+t-1 \\ t \end{bmatrix}\right)$$
$$= \sum_{t'+t''=t} \mathbf{v}^{(-a+lt_1)t''-(t_0-1)t'} \phi\left(\begin{bmatrix} -a+lt_1 \\ t' \end{bmatrix}\right) \phi\left(\begin{bmatrix} t_0-1 \\ t'' \end{bmatrix}\right).$$

Consider a term in the sum corresponding to (t', t''). Since $-a+lt_1 \geq 0$ is divisible by l, we see from the earlier part of the proof that $\phi(\left[\begin{smallmatrix} -a+lt_1 \\ t' \end{smallmatrix}\right]) = 0$ unless t' is divisible by l. But then t'' is congruent to t modulo l. Hence t'' is congruent to t_0 modulo l. It follows that $t'' \geq t_0$, hence $\left[\begin{smallmatrix} t_0-1 \\ t'' \end{smallmatrix}\right] = 0$. Hence our sum is zero and (a) is proved.

We prove (c), assuming (b). In the setup of (c) we have

$$\phi\left(\begin{bmatrix} a \\ t \end{bmatrix}\right) = \sum_{t'+t''=t} \mathbf{v}^{a_0 t'' - la_1 t'} \phi\left(\begin{bmatrix} a_0 \\ t' \end{bmatrix}\right) \phi\left(\begin{bmatrix} la_1 \\ t'' \end{bmatrix}\right).$$

By (a), the sum may be restricted to indices such that $t'' = lt_1''$ for some $t_1'' \in \mathbf{N}$ and such that $t' \leq a_0$. Then t' is congruent to t modulo l, hence t' is congruent to t_0 modulo l. Since both t', t_0 are in $[0, l-1]$, we must have $t' = t_0$ and therefore $t'' = lt_1$. Thus,

$$\text{(d)} \qquad \phi\left(\begin{bmatrix} a \\ t \end{bmatrix}\right) = \mathbf{v}^{la_0 t_1 - la_1 t_0} \phi\left(\begin{bmatrix} a_0 \\ t_0 \end{bmatrix}\right) \phi\left(\begin{bmatrix} la_1 \\ lt_1 \end{bmatrix}\right).$$

This shows that (c) is a consequence of (b).

We prove (b) for $a_1 \geq 0$ by induction on a_1. The case where $a_1 = 0$ or 1 is trivial. Assume now that $a_1 > 0$. By 1.3.1(e), we have

$$\phi\left(\begin{bmatrix} la_1 \\ lt_1 \end{bmatrix}\right) = \sum_{t'+t''=lt_1} \mathbf{v}^{l(a_1-1)t''-lt'} \phi\left(\begin{bmatrix} la_1-l \\ t' \end{bmatrix}\right) \phi\left(\begin{bmatrix} l \\ t'' \end{bmatrix}\right).$$

By (a), we may assume that the sum is restricted to indices divisible by l, namely $t' = lt_1', t'' = lt_1''$ with $t_1' + t_2' = t_1$. Using the induction hypothesis, we get

$$\phi\left(\begin{bmatrix} la_1 \\ lt_1 \end{bmatrix}\right) = \sum_{t_1'+t_1''=t_1} \mathbf{v}^{l^2(a_1 t_1 - t_1' + t_1'')} \binom{a_1-1}{t_1'} \binom{1}{t_1''} = \mathbf{v}^{l^2(a_1+1)t_1} \binom{a_1}{t_1}.$$

We have used the identity 1.3.1(e), specialized for $v = 1$. This proves (b) for $a_1 \geq 0$.

We now prove (b) assuming that $a_1 < 0$. We have

$$\begin{aligned}\phi\left(\begin{bmatrix} la_1 \\ lt_1 \end{bmatrix}\right) &= (-1)^{lt_1}\phi\left(\begin{bmatrix} -la_1 + lt_1 - 1 \\ lt_1 \end{bmatrix}\right) \\ &= (-1)^{lt_1}\phi\left(\begin{bmatrix} (l-1) + l(-a_1 + t_1 - 1) \\ lt_1 \end{bmatrix}\right) \\ &= (-1)^{lt_1}\mathbf{v}^{l(l-1)t_1}\phi\left(\begin{bmatrix} l(-a_1 + t_1 - 1) \\ lt_1 \end{bmatrix}\right).\end{aligned}$$

The last equality follows from (d). By the part of (b) that is already proved, we have

$$\begin{aligned}\phi\left(\begin{bmatrix} l(-a_1 + t_1 - 1) \\ lt_1 \end{bmatrix}\right) &= \mathbf{v}^{l^2(-a_1+t_1)t_1}\binom{-a_1 + t_1 - 1}{t_1} \\ &= \mathbf{v}^{l^2(-a_1+t_1)t_1}(-1)^{t_1}\binom{a_1}{t_1}.\end{aligned}$$

It follows that

$$\begin{aligned}\phi\left(\begin{bmatrix} la_1 \\ lt_1 \end{bmatrix}\right) &= (-1)^{lt_1+t_1}\mathbf{v}^{l(l-1)t_1+l^2(-a_1+t_1)t_1}\binom{a_1}{t_1} \\ &= (-1)^{(l+1)t_1}\mathbf{v}^{lt_1+l^2t_1}\mathbf{v}^{l^2(a_1+1)t_1}\binom{a_1}{t_1}.\end{aligned}$$

It remains to observe that

(e) $\mathbf{v}^{l^2+l} = (-1)^{l+1}$.

Indeed, if l is even, we have $\mathbf{v}^l = -1$ and both sides of (e) are -1; if l is odd, then $\mathbf{v}^l = \pm 1$ and both sides of (e) are 1. The lemma is proved.

34.1.3. Let $p \geq 0$. We have

(a) $\phi([lp]^!/([l]!)^p) = p!\mathbf{v}^{l^2p(p-1)/2}$.

We prove (a) by induction on p. If $p = 0$ or 1, then (a) is trivial. Assume that $p \geq 2$. We have

$$\phi([lp]^!/([l]!)^p) = \phi([l(p-1)]^!/([l]!)^{p-1})\phi\left(\begin{bmatrix} lp \\ l \end{bmatrix}\right).$$

Using 34.1.2 and the induction hypothesis, we see that

$$\phi([lp]^!/([l]!)^p) = (p-1)!p\mathbf{v}^{l^2(p-1)(p-2)/2+l^2(p+1)} = p!\mathbf{v}^{l^2p(p-1)}.$$

This proves (a).

Lemma 34.1.4. *Assume that* $0 \le r \le a < l$. *We have*

(a) $\sum_{q=0}^{l-a-1}(-1)^{l-r+1+q}\mathbf{v}^{-(l-r)(a-l+1+q)+q}\phi(\left[\begin{smallmatrix}l-r\\ q\end{smallmatrix}\right]) = \mathbf{v}^{l(a-r)}\phi(\left[\begin{smallmatrix}a\\ r\end{smallmatrix}\right])$.

In the left hand side of (a) we may replace $\mathbf{v}^{l^2-l}$ by $(-1)^{l+1}$. Note also that $l - r \ge 1$; hence

$$\sum_{q=0}^{l-r}(-1)^q v^{q(1-l+r)}\begin{bmatrix}l-r\\ q\end{bmatrix} = 0$$

(see 1.3.4). Hence the left hand side of (a) equals

$$(-1)^{r+1}\mathbf{v}^{r(a+1-l)-la}\sum_{q=l-a}^{l-r}(-1)^q\mathbf{v}^{q(1-l+r)}\phi\left(\begin{bmatrix}l-r\\ q\end{bmatrix}\right)$$

$$= (-1)^{r+1}\mathbf{v}^{r(a+1-l)-la}\sum_{q'=0}^{a-r}(-1)^{l-r-q'}\mathbf{v}^{(l-r-q')(1-l+r)}\phi\left(\begin{bmatrix}l-r\\ q'\end{bmatrix}\right).$$

For any q' in the last sum, we have by 34.1.2(c), $\phi(\left[\begin{smallmatrix}l-r\\ q'\end{smallmatrix}\right]) = \mathbf{v}^{lq'}\phi(\left[\begin{smallmatrix}-r\\ q'\end{smallmatrix}\right])$. Thus the left hand side of (a) is

$$(-1)^{a+l-r+1}\mathbf{v}^{r(a+1-l)-la}\sum_{q'=0}^{a-r}(-1)^{a-r-q'}\mathbf{v}^{lq'+(l-r-q')(1-l+r)}\phi\left(\begin{bmatrix}-r\\ q'\end{bmatrix}\right)$$

$$= (-1)^{a+l-r+1}\mathbf{v}^{l-(l-r)l-la}\sum_{q'=0}^{a-r}\mathbf{v}^{-q'+r(a-r-q')}\phi(\begin{bmatrix}-1\\ a-r-q'\end{bmatrix})\phi\left(\begin{bmatrix}-r\\ q'\end{bmatrix}\right)$$

$$= (-1)^{a+l-r+1}\mathbf{v}^{l-(l-r)l-la}\phi\left(\begin{bmatrix}-r-1\\ a-r\end{bmatrix}\right)$$

$$= (-1)^{l+1}\mathbf{v}^{l-(l-r)l-la}\phi\left(\begin{bmatrix}a\\ a-r\end{bmatrix}\right) = \mathbf{v}^{l(a-r)}\phi\left(\begin{bmatrix}a\\ r\end{bmatrix}\right).$$

We have used $\mathbf{v}^l = \mathbf{v}^{-l}$. The lemma is proved.

CHAPTER 35

The Quantum Frobenius Homomorphism

35.1. Statements of results

35.1.1. In this chapter we fix an integer $l \geq 1$. Then the integers $l_i \geq 1$, the new Cartan datum $(I, \circ)$ and the new root datum $(Y^*, X^*, \dots)$ of type $(I, \circ)$ are defined in terms of $l, (I, \cdot), (Y, X, \dots)$ as in 2.2.4, 2.2.5.

35.1.2. The assumptions (a),(b) below will be in force in this chapter.

(a) for any $i \neq j$ in I such that $l_j \geq 2$, we have $l_i \geq -\langle i, j' \rangle + 1$;

(b) $(I, \cdot)$ is without odd cycles (see 2.1.3).

Note that (a) is automatically satisfied in the simply laced case: in that case, we have $l_i = l$ for all i; in the general case, the assumption (a) can be violated only by finitely many l. Note also that (b) is automatically satisfied if $(I, \cdot)$ is of finite type.

35.1.3. Let l' be one of the integers $l, 2l$, if l is odd, and let l' be equal to $2l$, if l is even. Let $\mathcal{A}'$ be the quotient of $\mathcal{A}$ by the two-sided ideal generated by the l'-th cyclotomic polynomial $f_{l'} \in \mathcal{A}$. Thus, $(f_1, f_2, f_3, \dots) = (v - 1, v + 1, v^2 + v + 1, \dots)$.

In this chapter, we assume that the given ring homomorphism $\phi : \mathcal{A} \to R$ factors through a ring homomorphism $\mathcal{A}' \to R$, or that R is an $\mathcal{A}'$-algebra or, equivalently, that $f_{l'}(\mathbf{v}) = 0$ in R, where $\mathbf{v} = \phi(v)$.

35.1.4. When $(I, \cdot)$ is replaced by $(I, \circ)$, the element $v_i \in \mathcal{A}$, whose definition depends on the Cartan datum, becomes $v_i^* = v^{i \circ i/2} = v_i^{l_i^2}$.

For any $P \in \mathcal{A}$, we denote by P_i^* the element obtained from P by substituting v by v_i^*. For each $i \in I$, we set $\mathbf{v}_i = \phi(v_i)$ and $\mathbf{v}_i^* = \phi(v_i^*) = \mathbf{v}_i^{l_i^2}$.

Lemma 35.1.5. (a) *Let $a \in l_i\mathbf{Z}$ and $t \in l_i\mathbf{N}$. We have $\phi(\left[\begin{smallmatrix} a \\ t \end{smallmatrix}\right]_i) = \phi(\left[\begin{smallmatrix} a/l_i \\ t/l_i \end{smallmatrix}\right]_i^*)$ (equality in R).*

G. Lusztig, *Introduction to Quantum Groups*, Modern Birkhäuser Classics,
DOI 10.1007/978-0-8176-4717-9_35, © Springer Science+Business Media, LLC 2010

(b) *Let $a \in l_i\mathbf{Z}$ and let $t \in \mathbf{N}$ be non-divisible by l_i. We have $\phi(\left[\begin{smallmatrix} a \\ t \end{smallmatrix}\right]_i) = 0$ (equality in R).*

It suffices to prove this for $R = \mathcal{A}'$. This is an integral domain in which $\mathbf{v}_i^{2l_i} = 1$ and $\mathbf{v}_i^{2t} \neq 1$ for all $0 < t < l_i$. We prove (a). Applying 34.1.2(b) to v_i and l_i instead of v and l, we see that $\phi(\left[\begin{smallmatrix} a \\ t \end{smallmatrix}\right]_i) = \mathbf{v}_i^{(a+l_i)t} \binom{a/l_i}{t/l_i}$. Applying the same result to v_i^* and 1 instead of v and l, we see that $\phi(\left[\begin{smallmatrix} a/l_i \\ t/l_i \end{smallmatrix}\right]_i^*) = \mathbf{v}_i^{*(a+l_i)t/l_i^2} \binom{a/l_i}{t/l_i}$. It remains to use the equality $\mathbf{v}_i^* = \mathbf{v}_i^{l_i^2}$. The proof of (b) is entirely similar; it uses 34.1.2(a).

35.1.6. Let $\mathbf{f}^*$ (resp. $\dot{\mathbf{U}}^*$) be the $\mathbf{Q}(v)$-algebra defined like $\mathbf{f}$ (resp. like $\dot{\mathbf{U}}$), in terms of the Cartan matrix $(I, \circ)$ (resp. in terms of $(Y^*, X^*, \dots)$). Then the R-algebras ${}_R\mathbf{f}, {}_R\mathbf{f}^*, {}_R\dot{\mathbf{U}}, {}_R\dot{\mathbf{U}}^*$ are well-defined.

We state the main results of this chapter.

Theorem 35.1.7. *Recall that R is an $\mathcal{A}'$-algebra. There is a unique R-algebra homomorphism $Fr : {}_R\mathbf{f} \to {}_R\mathbf{f}^*$ such that for all $i \in I$ and $n \in \mathbf{Z}$, $Fr(\theta_i^{(n)})$ equals $\theta_i^{(n/l_i)}$ if $n \in l_i\mathbf{Z}$, and equals 0, otherwise.*

Theorem 35.1.8. *There is a unique R-algebra homomorphism $Fr' : {}_R\mathbf{f}^* \to {}_R\mathbf{f}$ such that $Fr'(\theta_i^{(n)}) = \theta_i^{(nl_i)}$ for all $i \in I$ and all $n \in \mathbf{Z}$.*

Theorem 35.1.9. *There is a unique R-algebra homomorphism $Fr : {}_R\dot{\mathbf{U}} \to {}_R\dot{\mathbf{U}}^*$ such that for all $i \in I, n \in \mathbf{Z}$ and $\zeta \in X$, we have:*

$Fr(E_i^{(n)}1_\zeta)$ equals $E_i^{(n/l_i)}1_\zeta$ if $n \in l_i\mathbf{Z}$ and $\zeta \in X^*$, and equals 0, otherwise;

$Fr(F_i^{(n)}1_\zeta)$ equals $F_i^{(n/l_i)}1_\zeta$ if $n \in l_i\mathbf{Z}$ and $\zeta \in X^*$, and equals 0, otherwise.

We give a proof of the last theorem, assuming that theorem 35.1.7 is known. Using the presentation of the algebra ${}_R\dot{\mathbf{U}}$ in terms of ${}_R\mathbf{f}$ given in 31.1.3, and the analogous presentation of the algebra ${}_R\dot{\mathbf{U}}^*$ in terms of ${}_R\mathbf{f}^*$, we see that it is enough to prove that the assignment

$$x^+1_\zeta x^- \mapsto Fr(x^+)1_\zeta Fr(x^-), x^-1_\zeta x^+ \mapsto Fr(x^-)1_\zeta Fr(x^+)$$

for $x, x' \in {}_R\mathbf{f}$, $\zeta \in X^*$,

$$x^+1_\zeta x^- \mapsto 0, x^-1_\zeta x^+ \mapsto 0$$

for $x, x' \in {}_R\mathbf{f}$, $\zeta \in X - X^*$ respects the relations described in 31.1.3. (Here, Fr is the homomorphism given by Theorem 35.1.7.) This is immediate, using Lemma 35.1.5.

35.1.10. Remark. The homomorphism Fr constructed in Theorem 35.1.9 is called the *quantum Frobenius homomorphism.* It is compatible with the comultiplication on ${}_R\dot{\mathbf{U}}$ and ${}_R\dot{\mathbf{U}}^*$ (proof by verification on the generators $E_i^{(n)}1_\zeta$ and $F_i^{(n)}1_\zeta$).

35.1.11. The uniqueness part of Theorems 35.1.7 and 35.1.8 is clear. To prove the existence part of these theorems, we note that the general case follows by change of scalars from the case where $R = \mathcal{A}'$. Since $\mathcal{A}'$ is an integral domain, it is contained in its quotient field K and the algebras ${}_{\mathcal{A}'}\mathbf{f}, {}_{\mathcal{A}'}\mathbf{f}^*$ are naturally imbedded in the corresponding algebras over K. Thus, if the theorems are known over K then, by restriction, we see that they hold over $\mathcal{A}'$. We are thus reduced to proving the theorems assuming that R is the quotient field of $\mathcal{A}'$. The proof in this case will be given in 35.2, 35.5.

35.2 Proof of Theorem 35.1.8

35.2.1. In the rest of this chapter (except in 35.5.2, 35.5.3), we assume that R is the quotient field of $\mathcal{A}'$. Note that R is a field of characteristic zero and that the order of $\mathbf{v}^2 = \phi(v^2)$ in the multiplicative group of R is l. Thus, we have $\mathbf{v}^{2l} = 1$ and $\mathbf{v}^{2t} \neq 1$ for all $0 < t < l$. By the definition of l_i, we have $\mathbf{v}_i^{2l_i} = 1$ and $\mathbf{v}_i^{2t} \neq 1$ for all $0 < t < l_i$. In particular, $\phi([n]_i^!)$ is invertible in R, if $0 \le n < l_i$. For any i we have $\mathbf{v}_i^* = \pm 1$ since $\mathbf{v}_i^{2l_i^2} = 1$; hence, when dealing with the algebras ${}_R\mathbf{f}^*, {}_R\dot{\mathbf{U}}^*$, we are in the *quasi-classical case* (see 33.2).

Lemma 35.2.2. *The R-algebra ${}_R\mathbf{f}$ is generated by the elements $\theta_i^{(l_i)}$ $(i \in I)$ and by the elements θ_i for $i \in I$ such that $l_i \ge 2$.*

Recall that the R-algebra ${}_R\mathbf{f}$ is generated by the elements $\theta_i^{(n)}$ for various i and $n \ge 0$. Writing $n = a + l_i b$ with $0 \le a < l_i$ and $b \in \mathbf{N}$, we have $\theta_i^{(n)} = \mathbf{v}_i^{abl_i}\theta_i^{(a)}\theta_i^{(l_ib)}$ (using 34.1.2). On the other hand,

(a) $\theta_i^{(a)} = \phi([a]_i^!)^{-1}\theta_i^a$ (see 35.2.1) and

(b) $\theta_i^{(l_ib)} = (b!)^{-1}\mathbf{v}_i^{l_i^2b(b-1)/2}(\theta_i^{(l_i)})^b$.

The equality (b) follows from the equalities

$$(\theta_i^{(l_i)})^b = [l_ib]_i^!/([l_i]_i^!)^b\theta_i^{(l_ib)} \quad \text{in } \mathbf{f}$$

and

$$\phi([l_ib]_i^!/([l_i]_i^!)^b) = b!\phi(v_i^{l_i^2b(b-1)/2})$$

(see 34.1.3). The lemma is proved.

35.2.3. We now give the proof of Theorem 35.1.8. As observed in 35.1.11, it is enough to prove the existence statement in 35.1.8, assuming that R is as in 35.2.1.

We will show that there exists an algebra homomorphism ${}_R\mathbf{f}^* \to {}_R\mathbf{f}$ such that $\theta_i \mapsto \theta_i^{(l_i)}$ for all i. Since the assumption 35.1.2(b) is in force, the algebra ${}_R\mathbf{f}^*$ has a presentation given by the generators θ_i and the Serre-type relations; this follows from 33.2.2(c) . Since $\phi(([n]^!)_i^*) = \mathbf{v}_i^{l_i^2 n(n-1)/2} n!$ (see the proof of Lemma 35.2.2), we see that it suffices to prove that, for any $i \neq j$, we have the following identity in ${}_R\mathbf{f}$:

(a) $\sum_{p+p'=1-\langle i,j'\rangle l_j/l_i} (-1)^{p'} v_i^{l_i^2 p(p-1)/2} v_i^{l_i^2 p'(p'-1)/2} \frac{(\theta_i^{(l_i)})^p}{p!} \theta_j^{(l_j)} \frac{(\theta_i^{(l_i)})^{p'}}{p'!} = 0.$

Using $(\theta_i^{(l_i)})^b = b! \mathbf{v}_i^{l_i^2 b(b-1)/2} \theta_i^{(l_i b)}$ (equality in ${}_R\mathbf{f}$, see 34.1.3), we can rewrite (a) in the following equivalent form:

(b) $\sum_{p+p'=1-\langle i,j'\rangle l_j/l_i} (-1)^{p'} \theta_i^{(l_i p)} \theta_j^{(l_j)} \theta_i^{(l_i p')} = 0.$

It remains to prove (b). Let $\alpha = -\langle i, j'\rangle$. For any $q \in [0, l_i - 1]$ we set

$$g_q = \sum_{r+s=\alpha l_j + l_i - q} (-1)^r v_i^{r(l_i-1-q)} \theta_i^{(r)} \theta_j^{(l_j)} \theta_i^{(s)} \in {}_\mathcal{A}\mathbf{f}.$$

This is $f_{i,j;l_j,\alpha l_j + l_i - q;-1}$ in the notation of 7.1.1.

Let $g = \sum_{q=0}^{l_i-1} (-1)^q v_i^{\alpha l_j q + l_i q - q} g_q \theta_i^{(q)}$. By the higher order quantum Serre relations (see 7.1.5(b)), we have $g_q = 0$ for all $q \in [0, l_i - 1]$; hence $g = 0$. On the other hand, setting $s' = s + q$, we have $g = \sum_{r,s'; r+s'=\alpha l_j + l_i} c_{r,s'} \theta_i^{(r)} \theta_j^{(l_j)} \theta_i^{(s')}$, where

$$c_{r,s'} = \sum_{q=0}^{l_i-1} (-1)^{r+q} v_i^{r(l_i-1-q)+\alpha l_j q + l_i q - q} \begin{bmatrix} s' \\ q \end{bmatrix}_i.$$

Taking the image of $g = 0$ under the obvious map ${}_\mathcal{A}\mathbf{f} \to {}_R\mathbf{f}$, we obtain

$$\sum_{r,s'; r+s'=\alpha l_j + l_i} \phi(c_{r,s'}) \theta_i^{(r)} \theta_j^{(l_j)} \theta_i^{(s')} = 0 \quad \text{in } {}_R\mathbf{f}.$$

For fixed s', we write $s' = a + l_i b$ where $0 \le a \le l_i - 1$. We have $\phi(\begin{bmatrix} s' \\ q \end{bmatrix}_i) = \mathbf{v}_i^{-bql_i} \phi(\begin{bmatrix} a \\ q \end{bmatrix}_i)$ (see 34.1.2); hence

$$\begin{aligned}
\phi(c_{r,s'}) &= (-1)^r \mathbf{v}_i^{r(l_i-1)} \sum_{q=0}^{l_i-1} (-1)^q \mathbf{v}_i^{q(s'-1)-bql_i} \phi\left(\begin{bmatrix} a \\ q \end{bmatrix}_i\right) \\
&= (-1)^r \mathbf{v}_i^{r(l_i-1)} \sum_{q=0}^{a} (-1)^q \mathbf{v}_i^{q(a-1)} \phi\left(\begin{bmatrix} a \\ q \end{bmatrix}_i\right) \\
&= \delta_{0,a} (-1)^{\alpha l_j + l_i - l_i b} \mathbf{v}_i^{(l_i-1)(\alpha l_j + l_i - l_i b)} \\
&= \delta_{0,a} (-1)^{1-b+\alpha l_j / l_i}.
\end{aligned}$$

We have used the identity $\mathbf{v}_i^{l_i(l_i-1)} = (-1)^{l_i+1}$, see 34.1.2(e). (b) follows. Thus, we have an algebra homomorphism $Fr' : {}_R\mathbf{f}^* \to {}_R\mathbf{f}$ such that $Fr'(\theta_i) = \theta_i^{(l_i)}$ for all i.

To complete the proof we must compute $Fr'(\theta_i^{(n)})$ for $n \geq 0$. We have

$$Fr'(\theta_i^{(n)}) = \phi(([n]^!)_i^*)^{-1} Fr'(\theta_i)^n = \mathbf{v}_i^{-l_i^2 n(n-1)/2}(n!)^{-1}(\theta_i^{(l_i)})^n = \theta_i^{(nl_i)}.$$

Theorem 35.1.8 is proved.

35.3. Structure of Certain Highest Weight Modules of ${}_R\dot{\mathbf{U}}$

Proposition 35.3.1. *Assume that $i \neq j$ in I satisfy $l_i \geq -\langle i, j'\rangle + 1$. The following identity holds in ${}_R\mathbf{f}$:*

$$\theta_i^{(l_i)}\theta_j = \sum_{r=0}^{-\langle i,j'\rangle} \mathbf{v}_i^{l_i(-\langle i,j'\rangle - r)} \phi\left(\begin{bmatrix} -\langle i,j'\rangle \\ r \end{bmatrix}_i\right) \theta_i^{(r)}\theta_j\theta_i^{(l_i-r)}.$$

We set $\alpha = -\langle i, j'\rangle$. Using Corollary 7.1.7 with $m = l_i, n = 1$, we see that we are reduced to checking the identity

$$\sum_{q=0}^{l_i-\alpha-1} (-1)^{l_i-r+1+q}\mathbf{v}_i^{-(l_i-r)(\alpha-l_i+1+q)+q}\phi\left(\begin{bmatrix} l_i - r \\ q \end{bmatrix}_i\right) = \mathbf{v}_i^{l_i(\alpha-r)}\phi\left(\begin{bmatrix} \alpha \\ r \end{bmatrix}_i\right)$$

for any $r \in [0, \alpha]$. This follows from Lemma 34.1.4.

Proposition 35.3.2. *Assume that the root datum is X-regular. Let $\lambda \in X$ be such that $\langle i, \lambda\rangle \in l_i\mathbf{Z}$ for all $i \in I$. Let M be a simple highest weight module with highest weight λ in ${}_R\mathcal{C}$ and let η be a generator of the R-vector space M^λ.*

(a) *If $\zeta \in X$ satisfies $M^\zeta \neq 0$, then $\zeta = \lambda - \sum_i l_i n_i i'$, where $n_i \in \mathbf{N}$. In particular, $\langle i, \zeta\rangle \in l_i\mathbf{Z}$ for all $i \in I$.*

(b) *If $i \in I$ is such that $l_i \geq 2$, then E_i, F_i act as zero on M.*

(c) *For any $r \geq 0$, let M'_r be the subspace of M spanned by the vectors $F_{i_1}^{(l_{i_1})}F_{i_2}^{(l_{i_2})}\cdots F_{i_r}^{(l_{i_r})}\eta$ for various sequences $i_1, i_2, \ldots, i_r$ in I. Let $M' = \sum_r M'_r$. Then $M' = M$.*

Clearly, M'_r is spanned by vectors in M^ζ where ζ is of the form $\zeta = \lambda - \sum_i l_i n_i i'$, with $n_i \in \mathbf{N}$. Such ζ satisfies $\langle i, \zeta\rangle \in l_i\mathbf{Z}$ for all $i \in I$. We use the fact that, for $j \in I$, the integer $l_j\langle i, j'\rangle$ is divisible by l_i.

We show by induction on $r \geq 0$, that

(d) $E_i M'_r = 0, F_i M'_r = 0$ for any $i \in I$ such that $l_i \geq 2$.

Assume first that $r = 0$. Then $E_i M'_0 = 0$ is obvious. Assume that for some $i \in I$ such that $l_i \geq 2$, we have $x = F_i \eta \neq 0$. For any $j \in I$, we have

$$E_j x = E_j F_i \eta = F_i E_j \eta + \delta_{i,j} \phi\left([\langle i, \lambda \rangle]_i\right) \eta = \delta_{i,j} \phi\left(\begin{bmatrix} \langle i, \lambda \rangle \\ 1 \end{bmatrix}_i\right) \eta.$$

Since $\langle i, \lambda \rangle \in l_i \mathbf{Z}$, and $l_i \geq 2$, we have $\phi(\left[\begin{smallmatrix} \langle i,\lambda \rangle \\ 1 \end{smallmatrix}\right]_i) = 0$; thus, $E_j x = 0$. If $n \geq 2$, then $E_j^{(n)} x = E_j^{(n)} F_i \eta$ is an R-linear combination of $F_i E_j^{(n)} \eta$ and of $E_j^{(n-1)} \eta$, hence is again zero. Thus, $E_j^{(n)} x = 0$ for all $j \in I$ and all $n > 0$; since $x \in M^{\lambda - i'}$, there exists a unique morphism in ${}_R\mathcal{C}$ from the Verma module $M_{\lambda - i'}$ into M which takes the canonical generator to x; its image is a subobject of M containing x but not η. Since M is simple, we must have $x = 0$. Thus (d) holds for $r = 0$.

Assume now that $r \geq 1$ and that (d) holds for $r - 1$. To show that it holds for r, it suffices to show that $E_i F_j^{(l_j)} m = 0, F_i F_j^{(l_j)} m = 0$ for any i, j in I such that $l_i \geq 2$ and any $m \in M'_{r-1}{}^{\zeta}$. If $l_j \geq 2$, then $E_i F_j^{(l_j)} m$ is an R-linear combination of $F_j^{(l_j)} E_i m$ and of $F_j^{l_j - 1} m$, hence is zero since $E_i m = 0, F_j m = 0$, by the induction hypothesis. If $l_j = 1$, then $E_i F_j m = F_j E_i m + \delta_{i,j} \phi(\left[\begin{smallmatrix} \langle i,\zeta \rangle \\ 1 \end{smallmatrix}\right]_i) m$ where $\langle i, \zeta \rangle \in l_i \mathbf{Z}$, hence $\phi(\left[\begin{smallmatrix} \langle i,\zeta \rangle \\ 1 \end{smallmatrix}\right]_i) = 0$, as above; since $E_i m = 0$, by the induction hypothesis, we have again $E_i F_j m = 0$.

If $i \neq j$, then from the identity in 35.3.1, we deduce by interchanging i, j and applying σ, that $F_i F_j^{(l_j)} m$ is an R-linear combination of $F_j^{(l_j - r)} F_i F_j^{(r)} m$ for various r with $0 \leq r \leq -\langle j, i' \rangle < l_j$. For such r we have $F_i F_j^{(r)} m = 0$. (Indeed, if $l_j \geq 2$, then $F_i F_j^{(r)} m = 0$, by the induction hypothesis; if $l_j = 1$, then $r = 0$ and $F_i F_j^{(r)} m = F_i m = 0$, again by the induction hypothesis.) Thus, we have $F_i F_j^{(l_j)} m = 0$. If $i = j$, then $F_i F_j^{(l_j)} m = F_j^{(l_j)} F_i m = 0$, by the induction hypothesis. This completes the inductive proof of (d).

Next we show by induction on $r \geq 0$ that

(e) $E_i^{(l_i)} M'_r \subset M'_{r-1}$ for any $i \in I$,

where, by convention, $M'_{-1} = 0$. This is clear for $r = 0$. Assume now that $r \geq 1$. We must show that $E_i^{(l_i)} F_j^{(l_j)} m' \in M'_{r-1}$ for any j and any $m \in M'_{r-1}$. Now $E_i^{(l_i)} F_j^{(l_j)} m'$ is an R-linear combination of $F_j^{(l_j)} E_i^{(l_i)} m'$ (which is in M'_{r-1} by the induction hypothesis) and of elements $F_j^{(l_j - t)} E_i^{(l_i - t)} m'$ with $t > 0$ such that $t \leq l_i, t \leq l_j$ (which are zero if $t < l_i$ or if $t = l_i, t < l_j$, by (d), and are in M'_{r-1} if $t = l_i = l_j$). Thus, (e) is proved.

From (d), (e), 35.2.2, and the obvious inclusion $F_i^{(l_i)}M'_r \subset M'_{r+1}$, we see that $\sum_r M'_r$ is an ${}_R\dot{\mathbf{U}}$-submodule of M. (It is certainly equal to the sum of its intersections with the weight spaces of M since it is spanned by homogeneous elements.) Since M is simple, we must have $M = \sum_r M'_r$. Thus (c) is proved. Now (b) follows from (d) and (c).

We prove (a). Let $\zeta \in X$ be such that $M^\zeta \neq 0$. By (c), we have $M'^\zeta \neq 0$. Then, as we have seen at the beginning of the proof, ζ is of the required form. The proposition is proved.

Corollary 35.3.3. *There is a unique unital ${}_R\dot{\mathbf{U}}^*$-module structure on M in which the ζ-weight space is the same as that in the ${}_R\dot{\mathbf{U}}$-module M, for any $\zeta \in X^* \subset X$, and such that $E_i, F_i \in {}_R\mathbf{f}^*$ act as $E_i^{(l_i)}, F_i^{(l_i)} \in {}_R\mathbf{f}$. Moreover, this is a simple highest weight module for ${}_R\dot{\mathbf{U}}^*$ with highest weight $\lambda \in X^*$.*

We define operators $e_i, f_i : M \to M$ for $i \in I$ by $e_i = E_i^{(l_i)}, f_i = F_i^{(l_i)}$. Using Theorem 35.1.8, we see that the e_i satisfy the Serre-type relations of ${}_R\mathbf{f}^*$ and that the f_i satisfy the Serre-type relations of ${}_R\mathbf{f}^*$.

If $\zeta \in X - X^*$ we have $M^\zeta = 0$, by 35.3.2(a). If $\zeta \in X^*$ and $m \in M^\zeta$, then, by 31.1.6(c), we have that $(e_i f_j - f_j e_i)(m)$ is equal to $\delta_{i,j}\phi(\left[\begin{smallmatrix}\langle i,\zeta\rangle \\ l_i\end{smallmatrix}\right]_i)m$ plus an R-linear combination of elements of the form $F_i^{l_i-t}E_i^{l_i-t}(m)$ with $0 < t < l_i$ which are zero by 35.3.2(b). Since $\langle i,\zeta\rangle \in l_i\mathbf{Z}$, we see from 35.1.5 that

$$\phi\left(\begin{bmatrix}\langle i,\zeta\rangle \\ l_i\end{bmatrix}_i\right) = \phi\left(\begin{bmatrix}\langle i,\zeta\rangle/l_i \\ 1\end{bmatrix}_i^*\right).$$

Therefore, $(e_i f_j - f_j e_i)(m) = \delta_{i,j}\phi(\left[\begin{smallmatrix}\langle i,\zeta\rangle/l_i \\ 1\end{smallmatrix}\right]_i^*)m$. It is clear that $e_i(M^\zeta) \subset M^{\zeta + l_i i'}$ and $f_i(M^\zeta) \subset M^{\zeta - l_i i'}$.

Thus, we have a unital ${}_R\dot{\mathbf{U}}^*$-module structure on M. By 35.3.2(c), this is a highest weight module of ${}_R\dot{\mathbf{U}}^*$ with highest weight λ. This ${}_R\dot{\mathbf{U}}^*$-module is simple. Indeed, assume that M'' is a non-zero ${}_R\dot{\mathbf{U}}^*$-submodule of M. Then M'' is the sum of its intersections with the various M^ζ (with $\zeta \in X$) and is stable under all $E_i^{(l_i)}, F_i^{(l_i)} : M \to M$ (in the ${}_R\dot{\mathbf{U}}$-module structure). Now M'' is automatically stable under E_i^a, F_i^a (in the ${}_R\dot{\mathbf{U}}$-module structure) for any i and a such that $0 < a < l_i$, since these act as zero on M. Using now lemma 35.2.2, we see that M'' is stable under $E_i^{(n)}, F_i^{(n)}$ for any i and any $n \in \mathbf{N}$. Thus, M'' is a (non-zero) ${}_R\dot{\mathbf{U}}$-submodule of M; hence $M'' = M$. The corollary is proved.

Corollary 35.3.4. *Assume, in addition, that the root datum is Y-regular and that $\lambda \in X^+$. Then the simple highest weight module for ${}_R\dot{\mathbf{U}}^*$ defined in Corollary 35.3.3 is ${}_R\Lambda_\lambda$.*

Indeed, the module ${}_R\Lambda_\lambda$ of ${}_R\dot{\mathbf{U}}^*$ is simple. By the assumption 35.1.2(b), we may apply 33.2.4.

35.4. A Tensor Product Decomposition of ${}_R\mathbf{f}$

35.4.1. Definition. Let $\mathfrak{f}$ be the R-subalgebra of ${}_R\mathbf{f}$ generated by the elements θ_i for various i such that $l_i \geq 2$. (Note that without the assumption 35.1.2(a), the definition of $\mathfrak{f}$ should be more complicated.) We have $\mathfrak{f} = \oplus_\nu \mathfrak{f}_\nu$ where $\mathfrak{f}_\nu = {}_R\mathbf{f}_\nu \cap \mathfrak{f}$.

Theorem 35.4.2. (a) *If $i \in I$ and $y \in \mathfrak{f}_\nu$, the difference $\theta_i^{(l_i)}y - \mathbf{v}_i^{-l_i\langle i,\nu\rangle}y\theta_i^{(l_i)}$ belongs to $\mathfrak{f}$.*

(b) *The R-linear map $\chi : {}_R\mathbf{f}^* \otimes_R \mathfrak{f} \to {}_R\mathbf{f}$ given by $x \otimes y \mapsto Fr'(x)y$ is an isomorphism of vector spaces.*

We prove (a). If (a) holds for y and y', then it also holds for yy'. Hence it suffices to prove (a) when y is one of the algebra generators of $\mathfrak{f}$. Thus, we may assume that $y = \theta_j$ where j satisfies $l_j \geq 2$. By our assumption, we then have $l_i \geq -\langle i, j'\rangle + 1$. Therefore, we may use the identity in Proposition 35.3.1, and we see that $\theta_i^{(l_i)}\theta_j - \mathbf{v}_i^{-l_i\langle i,j'\rangle}\theta_j\theta_i^{(l_i)}$ is an R-linear combination of products $\theta_i^{(r)}\theta_j\theta_i^{(l_i-r)}$ with $0 < r \leq -\langle i, j'\rangle < l_i$; these products are contained in $\mathfrak{f}$, by the definition of $\mathfrak{f}$. This proves (a).

We prove (b). We first show that χ is surjective. Using Lemma 35.2.2, we see that ${}_R\mathbf{f}$ is spanned as an R-vector space by products $x_1x_2\cdots x_p$ where the factors are either in $\mathfrak{f}_\nu$ for some ν (factors of the first kind) or of the form $\theta_i^{(l_i)}$ (factors of the second kind).

By (a), any product x_sx_{s+1} with x_s (resp. x_{s+1}) a factor of the first kind (resp. of the second kind) is equal to $\mathbf{v}_i^n x_{s+1}x_s$ plus an element of $\mathfrak{f}_{\nu'}$ for some n and some ν'. Applying this fact repeatedly, we see that $x_1x_2\cdots x_p$ is a linear combination of analogous words in which any factor of the second kind appears to the left of any factor of the first kind. It follows that χ is surjective.

It remains to show that χ is injective. Recall that the elements of $\mathbf{B}$ may be regarded as an R-basis of ${}_R\mathbf{f}^*$. Assume that for each $b \in \mathbf{B}$, we are given an element $y_b \in \mathfrak{f}$ such that $y_b = 0$ for all but finitely many b and such that we have a relation $\sum_b Fr'(b)y_b = 0$ in ${}_R\mathbf{f}$. We must prove that $y_b = 0$ for all b. We may assume that each y_b belongs to $\mathfrak{f}_\nu$ for some ν. Assume that $y_b \neq 0$ for some b. Then we may consider the largest integer N such that there exists b with $y_b \neq 0$ and $\operatorname{tr}|b| = N$.

In this proof we shall assume, as we may, that $(Y, X, \dots)$ is both Y-regular and X-regular. Let $\lambda \in X^+$ and $\lambda' \in X$; assume that $\langle i, \lambda \rangle \in l_i\mathbf{Z}$ for all i (i.e., that $\lambda \in X^*$). We consider the objects $M = {}_RL_\lambda, M' = {}_RM_{\lambda'}$ of ${}_R\mathcal{C}$ (see 31.3.2, 31.1.13); let η, η' be generators of the R-vector spaces $M^\lambda, M'^{\lambda'}$. Then $M' \otimes M \in {}_R\mathcal{C}$.

In $M' \otimes M$ we have $\sum_b Fr'(b)^- y_b^-(\eta' \otimes \eta) = 0$. We have $y_b^-(\eta' \otimes \eta) = v^{n(b)} y_b^-(\eta') \otimes \eta$, for some integer $n(b)$, since any element of $\mathfrak{f}_\nu$ with $\nu \neq 0$ annihilates η (see 35.3.2(b)). Hence we have

(c) $\sum_b v^{n(b)} Fr'(b)^-(y_b^-(\eta') \otimes \eta) = 0$ in $M' \otimes M$.

Let $M_1 = \oplus M^{\lambda_1} \subset M$ where the sum is taken over all $\lambda_1 \in X$ of the form $\lambda - \sum_i l_i p_i i'$ with $\sum_i p_i = N$. Let $\pi : M \to M_1$ be the obvious projection. We apply $1 \otimes \pi : M' \otimes M \to M' \otimes M_1$ to the equality (c). We obtain

(d) $\sum_b v^{n(b)} y_b^-(\eta') \otimes Fr'(b)^-(\eta) = 0$

where the sum is taken over b subject to $\operatorname{tr}|b| = N$.

By 35.3.3, we may regard M as a ${}_R\dot{\mathbf{U}}^*$-module; this is a simple highest weight module of ${}_R\dot{\mathbf{U}}^*$ which is just ${}_R\Lambda_\lambda$ (see 35.3.4). Note also that $Fr'(b)^-\eta$ in the ${}_R\dot{\mathbf{U}}$-module structure is the same as $b^-\eta$ in the ${}_R\dot{\mathbf{U}}^*$-module structure.

We shall assume, as we may, that $\langle i, \lambda \rangle$ are not only divisible by l_i, but are also large for all i, so that the vectors $b^-\eta \in M$ are linearly independent when b is subject to $\operatorname{tr}|b| = N$ (a finite set of b's). Here we use that $M = {}_R\Lambda_\lambda$ as a ${}_R\dot{\mathbf{U}}^*$-module. Then from (d) we deduce that $y_b^-(\eta') = 0$, hence $y_b = 0$ for all b such that $\operatorname{tr}|b| = N$. (We use the fact that M' is a Verma module.) This is a contradiction. The theorem is proved.

35.4.3. We assume that the root datum is simply connected. Then there is a unique $\lambda \in X^+$ such that $\langle i, \lambda \rangle = l_i - 1$ for all i. Let η be the canonical generator of ${}_R\Lambda_\lambda$.

Proposition 35.4.4. *The map $x \mapsto x^-\eta$ is an R-linear isomorphism $\mathfrak{f} \to {}_R\Lambda_\lambda$.*

Let $J = \sum_{i, n \geq l_i} ({}_R\mathbf{f}\theta_i^{(n)})$. It suffices to show that $J \oplus \mathfrak{f} = {}_R\mathbf{f}$. An equivalent statement is that $\sigma(J) \oplus \mathfrak{f} = {}_R\mathbf{f}$ since $\mathfrak{f}$ is σ-stable. We have $\sigma(J) = \sum_{i, n \geq l_i} \theta_i^{(n)} {}_R\mathbf{f}$. If i, n are such that $n \geq l_i$, then we can write $n = a + l_i b$ with $0 \leq a < l_i$ and $b \geq 1$ and we use the formulas in the proof of Lemma 35.2.2. We see that $\theta_i^{(n)} \subset \theta_i^{(l_i)} {}_R\mathbf{f}$. It follows that $\sigma(J) = \sum_i \theta_i^{(l_i)} {}_R\mathbf{f}$. The fact that $\sum_i \theta_i^{(l_i)} {}_R\mathbf{f}$ and $\mathfrak{f}$ are complementary subspaces of ${}_R\mathbf{f}$ follows easily from Theorem 35.4.2(b). The proposition follows.

35.5. Proof of Theorem 35.1.7

35.5.1. As we have seen in 35.1.11, we only have to prove existence in 35.1.7, assuming that R is as in 35.2.1.

By Theorem 35.4.2, there exists a unique R-linear map $P : {}_R\mathbf{f} \to {}_R\mathbf{f}^*$ such that $P(\theta_{i_1}^{(l_{i_1})} \cdots \theta_{i_n}^{(l_{i_n})} \theta_{j_1} \cdots \theta_{j_r})$ is equal to $\theta_{i_1} \cdots \theta_{i_n}$ if $r = 0$ and to 0 if $r > 0$. (Here $i_1, \dots, i_n$ is any sequence in I and $j_1, \dots, j_r$ is any sequence in I such that $l_{j_1} \geq 2, \dots, l_{j_r} \geq 2$.)

We show that P is an algebra homomorphism. It suffices to show that

(a) $P(x\theta_i) = P(x)P(\theta_i)$ for any $x \in {}_R\mathbf{f}$ and any i such that $l_i \geq 2$ and

(b) $P(x\theta_i^{(l_i)}) = P(x)P(\theta_i^{(l_i)})$ for any $x \in {}_R\mathbf{f}$ and any i.

(a) is obvious. We prove (b) for $x = \theta_{i_1}^{(l_{i_1})} \cdots \theta_{i_n}^{(l_{i_n})} \theta_{j_1} \cdots \theta_{j_r}$ by induction on $r \geq 0$. The case where $r = 0$ is trivial. Assume that $r \geq 1$. We have $x = x'\theta_j$ where $j = j_r$ and $x' = \theta_{i_1}^{(l_{i_1})} \cdots \theta_{i_n}^{(l_{i_n})} \theta_{j_1} \cdots \theta_{j_{r-1}}$.

If $i \neq j$ then, using the identity in 35.3.1 (after applying σ to it), we see that $x'\theta_j\theta_i^{(l_i)}$ is equal to a multiple of $x'\theta_i^{(l_i)}\theta_j$ plus a linear combination of terms of the form $x'\theta_i^{(r)}\theta_j\theta_i^{(l_i-r)}$ where $0 < r < l_i$.

By (a) and the induction hypothesis, we have

$$P(x'\theta_i^{(l_i)}\theta_j) = P(x'\theta_i^{(l_i)})P(\theta_j) = 0$$

and

$$P(x'\theta_i^{(r)}\theta_j\theta_i^{(l_i-r)}) = P(x'\theta_i^{(r)}\theta_j)P(\theta_i^{(l_i-r)}) = 0.$$

It follows that $P(x'\theta_j\theta_i^{(l_i)}) = 0$. If $i = j$, then $x'\theta_j\theta_i^{(l_i)} = x'\theta_i^{(l_i)}\theta_j$ and the same proof as the one above shows that $P(x'\theta_j\theta_i^{(l_i)}) = 0$. On the other hand, we have from the definition $P(x) = P(x'\theta_j) = 0$. Hence (b) holds in this case; both sides are zero.

To complete the proof we must compute $P(\theta_i^{(n)})$ for $n \geq 0$. Writing $n = a + l_i b$ with $0 \leq a < l_i$ and $b \in \mathbf{N}$, we have $\theta_i^{(n)} = \mathbf{v}_i^{abl_i}\theta_i^{(a)}\theta_i^{(l_i b)}$ as in Lemma 35.2.2, hence $P(\theta_i^{(n)}) = \mathbf{v}_i^{abl_i}P(\theta_i^{(a)})P(\theta_i^{(l_i b)})$. This is zero if $a > 0$, i.e., if n is not divisible by l_i. If $a = 0$, then

$$\begin{aligned} P(\theta_i^{(n)}) &= P(\theta_i^{(l_i b)}) = (b!)^{-1}\mathbf{v}_i^{l_i^2 b(b-1)/2}P(\theta_i^{(l_i)})^b \\ &= (b!)^{-1}(\mathbf{v}_i^*)^{b(b-1)/2}\theta_i^b = \theta_i^{(b)}. \end{aligned}$$

The theorem is proved.

35.5.2. We now discuss to what extent the assumptions 35.1.2(a),(b) are necessary. Theorem 35.1.8 depends only on the assumption 35.1.2(b). This assumption can be replaced by the assumption that l is odd; then essentially the same proof will work (using the results in 33.1, instead of those in 33.2). If the Cartan datum is of finite type, then as pointed out in 35.1.2, the assumption 35.1.2(b) is automatically satisfied; hence 35.1.8 holds in this case.

We now discuss Theorem 35.1.7. Here we may again substitute the assumption 35.1.2(b) by the assumption that l is odd. If the Cartan datum is irreducible, of finite type, 35.1.2(b) is automatically satisfied, but 35.1.2(a) can fail; more precisely, if 35.1.2(a) is not satisfied, then we may assume that

(a) we have $\langle i, j' \rangle = -2$ for some $i, j \in I$ and $l = 2$ or

(b) we have $I = \{i, j\}$ with $\langle i, j' \rangle = -3, \langle j, i' \rangle = -1$ and $l = 2$ or 3.

In case (a), the algebra ${}_R\mathbf{f}$ is known in terms of explicit generators and relations (see [6]) and the statement of 35.1.7 can be verified by checking that these relations are satisfied in ${}_R\mathbf{f}^*$.

In case (b), the explicit presentation of the algebra ${}_R\mathbf{f}$ is not known for general l; however, for small l (for example $l = 2$ or 3), it is possible to again write generators and relations, using the formulas in [6], and with their help to verify 35.1.7. We omit further details.

We see that 35.1.8, 35.1.7 (hence also 35.1.9) hold unconditionally in the case where the Cartan datum is of finite type. It is likely that they hold without any restriction whatsoever.

35.5.3. Frobenius homomorphism in the classical case. We now assume that l is a prime number and that the $\mathcal{A}$-algebra R is such that $v = 1$ and $l = 0$ in R. (For example, R could be the finite field with l elements.) Let $l' = l$ if l is odd and let $l' = 4$ if $l = 2$. Then the value of the l'-th cyclotomic polynomial at $v = 1$ is divisible by l; hence if we define $\mathcal{A}'$ as in 35.1.3 (with the present choice of l'), we have that R is an $\mathcal{A}'$-algebra. Hence Theorems 35.1.7, 35.1.8, 35.1.9 hold in this case. (For Cartan data of finite type, the assumptions in 35.1.2 are not needed; for infinite types, 35.1.2(a) is needed, and 35.1.2(b) is needed only if $l = 2$.) For Cartan data of finite type, we thus obtain the transpose of the classical Frobenius map or of an exceptional isogeny in the sense of Chevalley.

CHAPTER 36

The Algebras ${}_R\mathfrak{f}$, ${}_R\mathfrak{u}$

36.1. The Algebra ${}_R\mathfrak{f}$

36.1.1. In this chapter we assume that the Cartan datum is simply laced. As in the previous chapter, we fix an integer $l \geq 1$. We preserve the assumptions of 35.1.1– 35.1.3. Note that in this case, 35.1.2(a) is automatically satisfied. Note also that in this case, we have $l_i = l$ and $\mathbf{v}_i = \mathbf{v}$ for all i.

36.1.2. We define an R-algebra ${}_R\mathfrak{f}$ as follows. If $R = \mathcal{A}'$, then ${}_R\mathfrak{f}$ is the R-subalgebra of ${}_R\mathbf{f}$ generated by the elements $\theta_i^{(n)}$ for various i, n such that $0 \leq n < l$. In the general case, we define ${}_R\mathfrak{f} = R \otimes_{\mathcal{A}'} ({}_{\mathcal{A}'}\mathfrak{f})$. We have a direct sum decomposition ${}_R\mathfrak{f} = \oplus_\nu ({}_R\mathfrak{f}_\nu)$ indexed by $\nu \in \mathbf{N}[I]$; for $R = \mathcal{A}'$, it is induced by the analogous decomposition of ${}_R\mathbf{f}$ and, in general, is obtained by extension of scalars from the special case $R = \mathcal{A}'$.

From the definitions we see that in the case where R is the quotient field of $\mathcal{A}'$, ${}_R\mathfrak{f}$ is the R-subalgebra of ${}_R\mathbf{f}$ generated by the elements $\theta_i^{(n)}$ for various i, n such that $0 \leq n < l$, or equivalently, by the elements θ_i for various i (if $l \geq 2$) and by 1. (We use the fact that $\phi([n]^!)$ is non-zero in this field for $0 \leq n < l$.) It follows that in this case, ${}_R\mathfrak{f}$ is the same as the algebra $\mathfrak{f}$ defined in 35.4.1.

We shall need the following result.

Lemma 36.1.3. *Let $i, j \in I$ be such that $\langle i, j' \rangle \neq 0$. Let $m, n \in \mathbf{N}$ be such that $m \in l\mathbf{N}$ and $n < l$.*

(a) $\theta_i^{(m)}\theta_j^{(n)} \in {}_{\mathcal{A}'}\mathbf{f}$ is an $\mathcal{A}'$-linear combination of elements $u_s\theta_i^{(s)}$ where $s \in [0, m]$ is divisible by l and $u_s \in {}_{\mathcal{A}'}\mathfrak{f}$.

(b) $\theta_i^{(n)}\theta_j^{(m)} \in {}_{\mathcal{A}'}\mathbf{f}$ is an $\mathcal{A}'$-linear combination of elements $\theta_i^{(s)}u'_s$ where $s \in [0, m]$ is divisible by l and $u'_s \in {}_{\mathcal{A}'}\mathfrak{f}$.

We have $\langle i, j' \rangle = -1$, since the Cartan datum is simply laced. We prove (a). We may assume that $m > 0$. Then $m \geq l$, hence $m \geq n + 1$ and 7.1.7 is applicable. Thus we can express $\theta_i^{(m)}\theta_j^{(n)}$ as an $\mathcal{A}'$-linear combination of terms $\theta_i^{(r)}\theta_j^{(n)}\theta_i^{(s')}$ where $r, s' \in \mathbf{N}, r + s' = m, m - n \leq s' \leq m$. For such a term, we have $r \leq n < l$, hence $\theta_i^{(r)}\theta_i^{(n)} \in {}_{\mathcal{A}'}\mathfrak{f}$ and $\theta_i^{(s')}$ is either $\theta_i^{(m)}$

G. Lusztig, *Introduction to Quantum Groups*, Modern Birkhäuser Classics,
DOI 10.1007/978-0-8176-4717-9_36, © Springer Science+Business Media, LLC 2010

or, if $s' < m$, a power of $\mathbf{v}$ times $\theta_i^{(s'-m+l)}\theta_i^{(m-l)}$ (see 34.1.2). In the last expression we have $m - l \in l\mathbf{Z}$ and $0 < s' - m + l < l$; (a) follows. Now (b) follows from (a) by using the involution σ.

Theorem 36.1.4. *The R-module ${}_R\mathfrak{f}_\nu$ is free for any $\nu \in \mathbf{N}[I]$.*

Lemma 36.1.5. . *Assume that the root datum is simply connected. Let $\lambda \in X^+$ be defined by $\langle i, \lambda \rangle = l_i - 1$ for all i. Let η be the canonical generator of ${}_R\Lambda_\lambda$. The map $x \mapsto x^-\eta$ is an isomorphism ${}_R\mathfrak{f} \to {}_R\Lambda_\lambda$.*

It suffices to prove this in the case where $R = \mathcal{A}'$; the general case follows by change of rings. The fact that this map is injective follows from Proposition 35.4.4 (over the quotient field). We prove surjectivity. The argument is similar to the one in the proof of 35.4.2. Note that ${}_R\mathbf{f}$ is spanned as an R-module by products $x_1x_2\cdots x_p$ where the factors are either of the form $\theta_i^{(n)}$ with $0 \le n < l$ (factors of the first kind) or of the form $\theta_i^{(m)}$ with $m \in l\mathbf{N}$ (factors of the second kind).

By Lemma 36.1.3, any product x_tx_{t+1} with x_t (resp. x_{t+1}) a factor of the second kind (resp. of the first kind) is an $\mathcal{A}'$-linear combination of products of the form $x'_1x'_2\cdots x'_{r-1}x'_r$ where $x'_1, x'_2, \ldots, x'_{r-1}$ are factors of the first kind and x'_r is a factor of the second kind. Applying this fact repeatedly, we see that $x_1x_2\cdots x_p$ is a linear combination of analogous words in which any factor of the first kind appears to the left of any factor of the second kind.

Since the R-module ${}_R\Lambda_\lambda$ is generated by elements $x^-\eta$ with $x \in {}_R\mathbf{f}$, we see from the previous argument that ${}_R\Lambda_\lambda$ is generated by elements $x'^-x_1^-x_2^-\cdots x_p^-\eta$ where $x' \in {}_R\mathfrak{f}$ and $x_1, x_2, \ldots, x_p \in {}_R\mathbf{f}$ are factors of the second kind.

Since $(\theta_i^{(m)})^-\eta = 0$, for any m such that $m > \langle i, \lambda \rangle = l_i - 1$, we have $(\theta_i^{(m)})^-\eta = 0$ for any $m \in l\mathbf{N}$ such that $m \neq 0$. It follows that the R-module ${}_R\Lambda_\lambda$ is generated by elements $x'^-\eta$ with $x' \in {}_R\mathfrak{f}$. The lemma follows.

36.1.6. Proof of Theorem 36.1.4. We may assume that the root datum is simply connected. Hence Lemma 36.1.5 is applicable. The isomorphism in that lemma is compatible with the direct sum decompositions according to ν; it remains to observe that the canonical basis of ${}_R\Lambda_\lambda$ provides a basis for the summand corresponding to ν.

The following result is an integral version of Theorem 35.4.2(b); here R is not assumed to be a field.

Theorem 36.1.7. *The R-linear map $\chi : {}_R\mathbf{f}^* \otimes_R ({}_R\mathfrak{f}) \to {}_R\mathbf{f}$ given by $x \otimes y \mapsto Fr'(x)y$ is an isomorphism of R-modules.*

It is enough to prove this in the case where $R = \mathcal{A}'$; the general case follows by change of rings. The fact that χ is surjective has already been proved in the course of proving Lemma 36.1.5 (actually the products in that proof are in the opposite order of what we need now, so we must apply σ to them). Next we note that χ is a homomorphism between two free $\mathcal{A}'$-modules (the freeness of ${}_R\mathbf{f}^*$ and of ${}_R\mathbf{f}$ is already known; the freeness of ${}_R\mathfrak{f}$ follows from 36.1.4). Hence to prove that χ is injective over $\mathcal{A}'$, it is enough to prove the corresponding statement for the quotient field of $\mathcal{A}'$. That statement is already known (see 35.4.2(b)). The theorem is proved.

36.1.8. Let ${}_R\mathfrak{f} \to {}_R\mathbf{f}$ be the R-algebra homomorphism induced by change of scalars from the analogous homomorphism for $R = \mathcal{A}'$, which is the obvious imbedding.

Corollary 36.1.9. *The natural algebra homomorphism ${}_R\mathfrak{f} \to {}_R\mathbf{f}$ is an imbedding; its image is the R-subalgebra of ${}_R\mathbf{f}$ generated by the elements $\theta_i^{(n)}$ for various i, n such that $0 \le n < l$.*

We shall identify ${}_R\mathfrak{f}$ with a subalgebra of ${}_R\mathbf{f}$, as above.

36.2. The Algebras ${}_R\dot{\mathfrak{u}}$, ${}_R\mathfrak{u}$

36.2.1. Let ${}_R\dot{\mathfrak{u}}$ be the R-subalgebra of ${}_R\dot{\mathbf{U}}$ generated by the elements $E_i^{(n)}1_\zeta, F_i^{(n)}1_\zeta$ for various i, n such that $0 \le n < l$ and various $\zeta \in X$. Note that ${}_R\dot{\mathfrak{u}}$ is the free ${}_R\mathfrak{f} \otimes_R ({}_R\mathfrak{f}^{opp})$-submodule of ${}_R\dot{\mathbf{U}}$ with basis (1_ζ) (the module structure being $(x \otimes x') : u \mapsto x^+ u x'^-$); the same statement holds for the module structure $(x \otimes x') : u \mapsto x^- u x'^+$.

Lemma 36.2.2. *${}_R\dot{\mathfrak{u}}$ is closed under comultiplication.*

This is easily proved by checking on the algebra generators of ${}_R\dot{\mathfrak{u}}$.

36.2.3. In the rest of this chapter we assume that $l = l'$ is odd. Then $\mathbf{v}^l = 1$. We introduce a certain completion ${}_R\hat{\mathfrak{u}}$ of ${}_R\dot{\mathfrak{u}}$ as follows. Note that any element ${}_R\dot{\mathfrak{u}}$ can be written uniquely as a sum

(a) $\sum_{\zeta,\zeta' \in X} x_{\zeta,\zeta'}$

where $x_{\zeta,\zeta'} \in 1_\zeta({}_R\dot{\mathfrak{u}})1_{\zeta'}$ are zero except for finitely many pairs (ζ, ζ').

We now relax the last condition and we consider infinite formal sums (a) in which the only requirement is that there exists a finite subset $F \subset X$

such that $x_{\zeta,\zeta'} \in 1_\zeta({}_R\dot{\mathbf{u}})1_{\zeta'}$ are zero unless $\zeta - \zeta' \in F$. The set of all such formal sums is denoted by ${}_R\hat{\mathbf{u}}$. (Note that the set F varies from element to element of ${}_R\hat{\mathbf{u}}$.) The R-algebra structure of ${}_R\dot{\mathbf{u}}$ extends in an obvious way to an R-algebra structure on ${}_R\hat{\mathbf{u}}$; this algebra has a unit element $\sum_\zeta 1_\zeta$. Note that that the two ${}_R\mathfrak{f} \otimes_R ({}_R\mathfrak{f}^{opp})$-module structures on ${}_R\dot{\mathbf{u}}$ extend in an obvious way to two ${}_R\mathfrak{f} \otimes_R ({}_R\mathfrak{f}^{opp})$-module structures on ${}_R\hat{\mathbf{u}}$.

For any X^*-coset c in X, we define $1_c = \sum_{\zeta \in c} 1_c \in {}_R\hat{\mathbf{u}}$. Let J (resp. J') be the R-submodule of ${}_R\hat{\mathbf{u}}$ generated by the elements $x^+1_c x'^-$ (resp. $x^-1_c x'^+$) for various $c \in X/X^*$ and $x, x' \in {}_R\mathfrak{f}$.

Lemma 36.2.4. (a) *$F_i^{(b)}u \subset J$ for any $u \in J$ and any i, b such that $0 \le b < l$.*

(b) *J is an R-subalgebra of ${}_R\hat{\mathbf{u}}$ and $J = J'$.*

To prove (a), we may assume that $u = E_{i_1}^{(a_1)} \cdots E_{i_p}^{(a_p)} 1_c x'^-$ where $a_1, \ldots, a_p \in [0, l-1], c \in X/X^*$ and $x' \in {}_R\mathfrak{f}$. We argue by induction on p. If $p = 0$, the result is trivial. Assume that $p \ge 1$. Let $x_1 = \theta_{i_2}^{(a_2)} \cdots \theta_{i_p}^{(a_p)}$. We have

$$u = 1_{c'} E_{i_1}^{(a_1)} x_1^+ x'^-$$

for some $c' \in X/X^*$. If $i \neq i_1$, the desired result follows immediately. Assume that $i = i_1$. We have

$$F_i^{(b)}u =$$

$$\sum_{\zeta \in c'} \sum_{t \ge 0; t \le a_1; t \le b} \phi\left(\begin{bmatrix} a_1 + b - \langle i, \zeta \rangle \\ t \end{bmatrix}\right) E_i^{(a_1-t)} 1_{\zeta-(a_1+b-t)i'} F_i^{(b-t)} x_1^+ x'^-.$$

For each t, ζ in the sum we have $0 \le t < l$ and $a_1 + b - \langle i, \zeta \rangle = a_1 + b - \langle i, \zeta_0 \rangle$ mod $l\mathbf{Z}$, for some fixed element ζ_0 of c'. We have

$$\phi\left(\begin{bmatrix} a_1 + b - \langle i, \zeta \rangle \\ t \end{bmatrix}\right) = \phi\left(\begin{bmatrix} a_1 + b - \langle i, \zeta_0 \rangle \\ t \end{bmatrix}\right)$$

(see 34.1.2); here we use the hypothesis that l is odd. Hence we have

$$F_i^{(b)}u =$$

$$\sum_{t \ge 0; t \le a_1; t \le b} \phi\left(\begin{bmatrix} a_1 + b - \langle i, \zeta_0 \rangle \\ t \end{bmatrix}\right) E_i^{(a_1-t)} F_i^{(b-t)} (\sum_{\zeta \in c'} 1_{\zeta - a_1 i'}) x_1^+ x'^-.$$

Note that $\sum_{\zeta \in c'} 1_{\zeta - a_1 i'} = 1_{c''}$ for some $c'' \in X/X^*$. Using now the induction hypothesis, we see that $F_i^{(b)}u \in J$; (a) is proved. Using repeatedly (a) and the identities $1_c 1_{c'} = \delta_{c,c'} 1_c$, for $c, c' \in X/X^*$, we see that J is a subalgebra of ${}_R\hat{\mathbf{u}}$. Again, using (a) repeatedly, starting with $1_c x'^+ \in J$ for $c \in X/X^*, x' \in {}_R\mathbf{f}$, we see that $J' \subset J$. By symmetry, we have $J \subset J'$, hence $J = J'$. The lemma is proved.

36.2.5. Definition. ${}_R\mathbf{u}$ is the R-subalgebra $J = J'$ of ${}_R\hat{\mathbf{u}}$.

Note that the algebra ${}_R\mathbf{u}$ has a unit element $\sum_c 1_c$; here c runs through the set X/X^*. The set X/X^* is finite, since by the definition of X^*, the map $\zeta \to \langle i, \zeta\rangle \mod l$ defines an injective map $X/X^* \to (\mathbf{Z}/l\mathbf{Z})^I$.

From the definition, we see that ${}_R\mathbf{u}$ is the free ${}_R\mathfrak{f} \otimes_R ({}_R\mathfrak{f}^{opp})$-submodule of ${}_R\hat{\mathbf{u}}$ with basis $\{1_c | c \in X/X^*\}$ (the module structure being $(x \otimes x') : u \mapsto x^+ u x'^-$); the same statement holds for the module structure $(x \otimes x') : u \mapsto x^- u x'^+$.

Notes on Part V

1. The results in Chapter 32 are due to to Drinfeld, for $R = \mathbf{Q}(v)$. The extension to the case where R is a field and v is a root of 1 in R is new; it answers a question that Drinfeld asked me in January 1990.
2. The fact that the simple integrable modules of a Kac-Moody Lie algebra admit a quantum deformation (Chapter 33) was proved in [4]; for Cartan data of finite type this was also stated in [8], but the proof there has a serious gap. (It appears [2] that, for Cartan data of finite type, this result was known to Drinfeld.) The results in 33.2 are new.
3. The results in Chapter 34 have appeared (for l odd) in [5].
4. The quantum Frobenius homomorphism, for Cartan data of finite type and with some restrictions on l , was implicit in [5] and explicit in [7]; its generalization given in Chapter 35 is new.
5. In the case where R is a field of characteristic zero, v is a root of 1 in R, and the Cartan datum is of finite type, ${}_R\mathfrak{u}$ is the finite dimensional Hopf algebra defined in [6], [7] (with some restrictions on the order of v). The extension to infinite types is new.

References

1. C. Chevalley, *Séminaire sur la classification des groupes de Lie algébriques*, Ecole Norm. Sup. Paris, 1956–58.
2. V. G. Drinfeld, *On almost cocommutative Hopf algebras*, (Russian), Algebra i Analiz **1** (1989), 30–46.
3. V. G. Kac, *Infinite dimensional Lie algebras*, Birkhäuser, Boston, 1983.
4. G. Lusztig, *Quantum deformations of certain simple modules over enveloping algebras*, Adv. Math. **70** (1988), 237–249.
5. ———, *Modular representations and quantum groups*, Contemp. Math. **82** (1989), 59–77, Amer. Math. Soc., Providence, R. I..
6. ———, *Finite dimensional Hopf algebras arising from quantized universal enveloping algebras*, J. Amer. Math. Soc. **3** (1990), 257–296.
7. ———, *Quantum groups at roots of 1*, Geom. Dedicata **35** (1990), 89–114.
8. M. Rosso, *Finite dimensional representations of the quantum analog of a complex semisimple Lie algebra*, Comm. Math. Phys. **117** (1988), 581–593.

Part VI

BRAID GROUP ACTION

In the classical theory of semisimple Lie algebras, the Weyl group plays an important role. Now the Weyl group is not quite a subgroup of the corresponding simply connected Lie group; only a finite covering of it is. As Tits has shown [9], one can choose such a covering which is naturally a quotient of the braid group. In particular, there is a small obstruction to making the Weyl group act on a simple integrable module for the Lie algebra; what acts naturally is a quotient of the braid group, which is a finite covering of the Weyl group. Since in this case, the obstruction involves only signs, it is almost invisible. In the quantum case, the obstruction becomes quite serious, and in this case, not even a finite covering of the Weyl group can be made to act; the braid group still acts, but in general not through a finite quotient.

In Part VI we explain how the braid group acts on integrable **U**-modules and on **U** itself. (In fact there are several braid group actions, but they are related to each other in a simple way.)

The symmetries $T'_{i,e}, T''_{i,e}$ of an integrable **U**-module have already been introduced in Chapter 5. In Chapter 39 it is shown that these symmetries satisfy the braid group relations, hence they define braid group actions. These symmetries are studied simultaneously with the analogous symmetries of **U** (see Chapter 37) which also satisfy the braid group relations.

In Chapters 38 and 40 we study the connection between the symmetries of **U** and the inner product $(,)$ on $\mathbf{f}$. In Chapter 41 we define a braid group action on ${}_R\dot{\mathbf{U}}$ and on its integrable modules for any R.

In Chapter 42 we assume that the Cartan datum is simply laced and of finite type and we use the braid group actions to give a purely combinatorial parametrization of the canonical basis **B** in terms of reduced expressions for the longest element of W.

CHAPTER 37

The Symmetries $T'_{i,e}$, $T''_{i,e}$ of U

37.1. Definition of the Symmetries

37.1.1. In this section we fix $i \in I$ and $e = \pm 1$. Recall that in 5.2.1 we have defined some symmetries $T'_{i,e}, T''_{i,e} : M \to M$ for any integrable module M of $\mathbf{U}$. In the following proposition we define analogous symmetries $T'_{i,e}, T''_{i,e} : \mathbf{U} \to \mathbf{U}$.

Proposition 37.1.2. (a) *For any $u \in \mathbf{U}$, there is a unique element $u'' \in \mathbf{U}$ such that for any integrable $\mathbf{U}$-module M and any $z \in M$, we have $T''_{i,-e}(u''z) = uT''_{i,-e}(z)$.*

(b) *The map $u \mapsto u''$ is an automorphism of the algebra $\mathbf{U}$, denoted by $T'_{i,e}$.*

(c) *For any $u \in \mathbf{U}$, there is a unique element $u' \in \mathbf{U}$ such that for any integrable $\mathbf{U}$-module M and any $z \in M$, we have $T'_{i,e}(u'z) = uT'_{i,e}(z)$.*

(d) *The map $u \mapsto u'$ is an automorphism of the algebra $\mathbf{U}$, denoted by $T''_{i,-e}$. It is the inverse of $T'_{i,e} : \mathbf{U} \to \mathbf{U}$.*

Thus, for any integrable $\mathbf{U}$-module M, any $z \in M$, and any $u \in \mathbf{U}$, we have

$$T'_{i,e}(uz) = T'_{i,e}(u)T'_{i,e}(z)$$

and

$$T''_{i,-e}(uz) = T''_{i,-e}(u)T''_{i,-e}(z).$$

37.1.3. The proof will be given in 37.2.3. The proof will give at the same time the following formulas for the values of the automorphisms $T''_{i,-e}, T'_{i,e} : \mathbf{U} \to \mathbf{U}$ on the generators of the algebra $\mathbf{U}$.

$T'_{i,e}(E_i) = -\tilde{K}_{ei}F_i$, $T'_{i,e}(F_i) = -E_i\tilde{K}_{-ei}$;

$T'_{i,e}(E_j) = \sum_{r+s=-\langle i,j'\rangle}(-1)^r v_i^{er} E_i^{(r)} E_j E_i^{(s)}$ for $j \neq i$;

$T'_{i,e}(F_j) = \sum_{r+s=-\langle i,j'\rangle}(-1)^r v_i^{-er} F_i^{(s)} F_j F_i^{(r)}$ for $j \neq i$;

$T'_{i,e}(K_\mu) = K_{\mu-\langle \mu,i'\rangle i}$;

G. Lusztig, *Introduction to Quantum Groups*, Modern Birkhäuser Classics,
DOI 10.1007/978-0-8176-4717-9_37, © Springer Science+Business Media, LLC 2010

$T''_{i,-e}(E_i) = -F_i\tilde{K}_{-ei}, \quad T''_{i,-e}(F_i) = -\tilde{K}_{ei}E_i;$
$T''_{i,-e}(E_j) = \sum_{r+s=-\langle i,j'\rangle}(-1)^r v_i^{er} E_i^{(s)} E_j E_i^{(r)}$ for $j \neq i$;
$T''_{i,-e}(F_j) = \sum_{r+s=-\langle i,j'\rangle}(-1)^r v_i^{-er} F_i^{(r)} F_j F_i^{(s)}$ for $j \neq i$;
$T''_{i,-e}(K_\mu) = K_{\mu-\langle \mu,i'\rangle i}.$

More generally, we have, for any $n \geq 0$:

$T'_{i,e}(E_i^{(n)}) = (-1)^n v_i^{en(n-1)} \tilde{K}_{eni} F_i^{(n)};$
$T'_{i,e}(F_i^{(n)}) = (-1)^n v_i^{-en(n-1)} E_i^{(n)} \tilde{K}_{-eni};$
$T'_{i,e}(E_j^{(n)}) = \sum_{r+s=-\langle i,j'\rangle n}(-1)^r v_i^{er} E_i^{(r)} E_j^{(n)} E_i^{(s)}$ for $j \neq i$;
$T'_{i,e}(F_j^{(n)}) = \sum_{r+s=-\langle i,j'\rangle n}(-1)^r v_i^{-er} F_i^{(s)} F_j^{(n)} F_i^{(r)}$ for $j \neq i$;
$T''_{i,-e}(E_i^{(n)}) = (-1)^n v_i^{en(n-1)} F_i^{(n)} \tilde{K}_{-eni};$
$T''_{i,-e}(F_i^{(n)}) = (-1)^n v_i^{-en(n-1)} \tilde{K}_{eni} E_i^{(n)};$
$T''_{i,-e}(E_j^{(n)}) = \sum_{r+s=-\langle i,j'\rangle n}(-1)^r v_i^{er} E_i^{(s)} E_j^{(n)} E_i^{(r)}$ for $j \neq i$;
$T''_{i,-e}(F_j^{(n)}) = \sum_{r+s=-\langle i,j'\rangle n}(-1)^r v_i^{-er} F_i^{(r)} F_j^{(n)} F_i^{(s)}$ for $j \neq i$.

37.2. Calculations in Rank 2

37.2.1. Given $i \neq j$ in I, we set $\alpha = -\langle i,j'\rangle \in \mathbf{N}$. For any $m,n \in \mathbf{Z}$, we set (compare 7.1.1)

$$x_{i,j;n,m;e} = x_{n,m;e} = f^+_{n,m;e} = \sum_{r+s=m}(-1)^r v_i^{er(\alpha n-m+1)} E_i^{(r)} E_j^{(n)} E_i^{(s)} \in \mathbf{U}^+;$$

$$x'_{i,j;n,m;e} = x'_{n,m;e} = \sum_{r+s=m}(-1)^r v_i^{er(\alpha n-m+1)} E_i^{(s)} E_j^{(n)} E_i^{(r)} \in \mathbf{U}^+;$$

$$y_{i,j;n,m;e} = y_{n,m;e} = \sum_{r+s=m}(-1)^r v_i^{-er(\alpha n-m+1)} F_i^{(s)} F_j^{(n)} F_i^{(r)} \in \mathbf{U}^-;$$

$$\begin{aligned} y'_{i,j;n,m;e} &= y'_{n,m;e} = f^-_{n,m;-e} \\ &= \sum_{r+s=m}(-1)^r v_i^{-er(\alpha n-m+1)} F_i^{(r)} F_j^{(n)} F_i^{(s)} \in \mathbf{U}^-. \end{aligned}$$

With this notation, we have the following result.

Lemma 37.2.2. *Let M be an object of $\mathcal{C}'$ and let $z \in M$. We have*

(a) $T''_{i,-e}(x_{n,\alpha n;e}z) = E_j^{(n)} T''_{i,-e}(z)$;

(b) $T'_{i,e}(x'_{n,\alpha n;e}z) = E_j^{(n)} T'_{i,e}(z)$;

(c) $T''_{i,-e}(y_{n,\alpha n;e}z) = F_j^{(n)} T''_{i,-e}(z)$;

(d) $T'_{i,e}(y'_{n,\alpha n;e}z) = F_j^{(n)} T'_{i,e}(z)$.

We may assume that $z \in M^\lambda$. We have $x_{n,\alpha n;e}z \in M^{\lambda+\alpha ni'+nj'}$. Let

$$g = \langle i, \lambda\rangle, g' = \langle i, \lambda + \alpha ni' + nj'\rangle = g + \alpha n.$$

We have

$$T''_{i,-e}(x_{n,\alpha n;e}z) = \sum_{a,b,c\in\mathbf{N};-a+b-c=g'} (-1)^b v_i^{-e(-ac+b)} E_i^{(a)} F_i^{(b)} E_i^{(c)} x_{n,\alpha n;e}z.$$

By 7.1.3(a) and 7.1.5(a), we have

$$E_i^{(c)} x_{n,\alpha n;e} = v_i^{e\alpha cn} x_{n,\alpha n;e} E_i^{(c)}.$$

Hence

$$T''_{i,-e}(x_{n,\alpha n;e}z)$$
$$= \sum_{a,b,c\in\mathbf{N};-a+b-c=g'} (-1)^b v_i^{-e(-ac+b)+e\alpha cn} E_i^{(a)} F_i^{(b)} x_{n,\alpha n;e} E_i^{(c)} z.$$

By 7.1.3(b) and (a), this can be written:

$$\sum_{a,b,c\in\mathbf{N};-a+b-c=g'} \sum_{b'=0}^{b} (-1)^{b+b'} v_i^{-e(-ac+b)+e\alpha cn-e(bb'-b')}$$
$$\times E_i^{(a)} \tilde{K}_{-eb'i} x_{n,\alpha n-b';e} F_i^{(b-b')} E_i^{(c)} z$$
$$= \sum_{a,b,c\in\mathbf{N};-a+b-c=g'} \sum_{b'=0}^{b} (-1)^{b+b'} v_i^{-e(-ac+b)}$$
$$\times v_i^{e\alpha cn-e(bb'-b')-eb'\langle i,\lambda+ci'-bi'+\alpha ni'+nj'\rangle} E_i^{(a)} x_{n,\alpha n-b';e} F_i^{(b-b')} E_i^{(c)} z$$
$$= \sum_{a,b,c\in\mathbf{N};-a+b-c=g'} \sum_{b'=0}^{b} \sum_{a'=0}^{a} (-1)^{b+b'+a'} v_i^t \begin{bmatrix} \alpha n - b' + a' \\ a' \end{bmatrix}_i$$
$$\times x_{n,\alpha n-b'+a';e} E_i^{(a-a')} F_i^{(b-b')} E_i^{(c)} z$$

where

(e) $t = e(ac - b + \alpha cn + bb' + b' - b'g - 2b'c - \alpha b'n + \alpha an + aa' - a' - 2ab')$.

We make a change of variable: $a'' = a - a', b'' = b - b'$. The exponent (e) then becomes

$$t = e(a'(-1 + b'' - a'' - c - g) + (a''c - b'') + (a'' + c)(-a'' + b'' - c - g))$$

We use the condition $-a' - a'' + b' + b'' - c = g + \alpha n$ which is equivalent to $-a + b - c = g'$.

We have $x_{n,\alpha n - b' + a';e} = 0$ unless $0 \le \alpha n - b' + a' \le \alpha n$ (see 7.1.5), or equivalently, $0 \le -a'' + b'' - c - g \le \alpha n$; hence we may add the condition $0 \le -a'' + b'' - c - g \le \alpha n$ in the summation without changing the sum.

In the presence of the equation $-a' - a'' + b' + b'' - c = g + \alpha n$, the inequality $a' \ge 0$ implies $b' \ge 0$ since $a'' - b'' + c + g + \alpha n \ge 0$. The sum becomes

$$\sum_{a'',b'',c \in \mathbf{N}; 0 \le -a''+b''-c-g \le \alpha n} (-1)^{b''} v_i^{e((a''c-b'')+(a''+c)(-a''+b''-c-g))}$$
$$\times \sum_{a' \in \mathbf{N}} (-1)^{a'} v_i^{ea'(-1-a''+b''-c-g)} \begin{bmatrix} -a''+b''-c-g \\ a' \end{bmatrix}_i x_{n,-a''+b''-c-g;e}$$
$$\times E_i^{(a'')} F_i^{(b'')} E_i^{(c)} z \, .$$

The sum over a' is zero unless $-a'' + b'' - c - g = 0$ (see 1.3.4). Hence the sum becomes

$$\sum_{a'',b'',c \in \mathbf{N}; -a''+b''-c=g} (-1)^{b''} v_i^{e(a''c-b'')} x_{n,0;e} E_i^{(a'')} F_i^{(b'')} E_i^{(c)} z$$

which equals $E_j^{(n)} T''_{i,-e}(z)$. This proves (a).

Using 5.2.3(b), we have

$$T''_{i,-e} z = (-1)^{\langle i,\lambda \rangle} v_i^{-e\langle i,\lambda \rangle} T'_{i,-e} z$$

and

$$T''_{i,-e}(x_{n,\alpha n;e} z) = (-1)^{\langle i,\lambda \rangle + \alpha n} v_i^{-e\langle i,\lambda \rangle - e\alpha n} T'_{i,-e}(x_{n,\alpha n;e} z).$$

Hence (a) implies

$$T'_{i,-e}(x_{n,\alpha n;e} z) = (-1)^{\alpha n} v_i^{e\alpha n} E_j^{(n)} T'_{i,-e}(z).$$

Substituting here

$$x'_{n,an;-e} = (-1)^{an} v_i^{-ean} x_{n,an;e},$$

we obtain

$$T'_{i,-e}(x'_{n,an;-e}z) = E_j^{(n)} T'_{i,-e}(z).$$

Replacing e by $-e$, we obtain (b).

We now apply (a),(b) to z and ${}^\omega M$ instead of M. The action of $T''_{i,-e}$ (resp. $T'_{i,e}$) in ${}^\omega M$ is the same as the action of $T'_{i,-e}$ (resp. $T''_{i,e}$) in M; thus, (a) and (b) for ${}^\omega M$ imply

$$T'_{i,-e}(\omega(x_{n,an;e})z) = F_j^{(n)} T'_{i,-e}(z)$$

and

$$T''_{i,e}(\omega(x'_{n,an;e})z) = F_j^{(n)} T''_{i,e}(z) \qquad \text{for} \quad M.$$

We have $\omega(x_{n,an;e}) = y'_{n,an;-e}$ and $\omega(x'_{n,an;e}) = y_{n,an;-e}$. Hence

$$T'_{i,-e}(y'_{n,an;-e}z) = F_j^{(n)} T'_{i,-e}z$$

and

$$T''_{i,e}(y_{n,an;-e}z) = F_j^{(n)} T''_{i,e}z.$$

Replacing e by $-e$, we obtain (d) and (c).

37.2.3. Proof of Proposition 37.1.2. We show the uniqueness of u'' in 37.1.2(a). It suffices to show that, if $\tilde{u} \in \mathbf{U}$ satisfies $T''_{i,-e}(\tilde{u}z) = 0$ for any integrable $\mathbf{U}$-module M and any $z \in M$, then $\tilde{u} = 0$. Since $T''_{i,-e}$ is invertible on M, we have that $\tilde{u}$ annihilates any integrable $\mathbf{U}$-module. But this implies $\tilde{u} = 0$ (see 3.5.4). Thus the uniqueness in 37.1.2(a) is proved.

To prove existence, we observe that, if (a) holds for two elements u_1, u_2, then it also holds for u_1u_2 and $au_1 + bu_2$ for any $a, b \in \mathbf{Q}(v)$; indeed, it is clear that $(u_1u_2)'' = u_1''u_2''$ and $(au_1 + bu_2)'' = au_1'' + bu_2''$.

Hence it suffices to prove the existence of u'' in the case where u is one of the generators E_j, F_j, K_μ of $\mathbf{U}$. For K_μ, this follows from 5.2.6. For E_i, F_i this follows from 5.2.4. For E_j, F_j with $j \neq i$, this follows from 37.2.2.

This proves (a). The argument above shows that $u \mapsto u''$ is an algebra homomorphism. In an entirely similar way we see that (c) holds and that $u \mapsto u'$ is an algebra homomorphism.

It remains to show that $(u'')' = u$ and $(u')'' = u$ for all u. For any M, z as above, we have, using 5.2.3(a):

$$(u')''z = T'_{i,e}T''_{i,-e}(u')''z = T'_{i,e}(u'T''_{i,e}z) = uT'_{i,e}T''_{i,e}z = uz.$$

Thus $(u')'' - u$ annihilates any integrable $\mathbf{U}$-module, hence $(u')'' = u$. The identity $(u'')' = u$ is proved in the same way. The proposition is proved.

37.2.4. We have

$$\omega T'_{i,e}\omega = T''_{i,e} : \mathbf{U} \to \mathbf{U} \text{ and } \sigma T'_{i,e}\sigma = T''_{i,-e} : \mathbf{U} \to \mathbf{U}.$$

The symmetries $T'_{i,e}, T''_{i,e}$ are related to the involution $^- : \mathbf{U} \to \mathbf{U}$ as follows:

$$\overline{T'_{i,e}(u)} = T'_{i,-e}(\bar{u}), \quad \overline{T''_{i,e}(u)} = T''_{i,-e}(\bar{u}) \text{ for all } u \in \mathbf{U}.$$

Let $u \in \mathbf{U}$ be such that $\tilde{K}_i u \tilde{K}_{-i} = v_i^n u$ We have

$$T''_{i,e}(u) = (-1)^n v_i^{en} T'_{i,e}(u),$$

or equivalently,

$$T''_{i,e}(u) = (-1)^n T'_{i,e}(\tilde{K}_{ei} u \tilde{K}_{-ei}).$$

These formulas are proved by checking on generators.

Proposition 37.2.5. *Let $i \neq j$ in I. For any $m \in \mathbf{Z}$, we have*

(a) $T'_{i,e}(x'_{i,j;1,m;e}) = x_{i,j;1,-\langle i,j'\rangle-m;e}$;

(b) $T''_{i,-e}(x_{i,j;1,m;e}) = x'_{i,j;1,-\langle i,j'\rangle-m;e}$.

The two formulas are equivalent, by 37.1.2(d), so it suffices to prove (a). For $m < 0$ or $m > -\langle i,j'\rangle$, both sides of (a) are zero. So we may assume that $0 \le m \le -\langle i,j'\rangle$. We argue by descending induction on m. For $m = -\langle i,j'\rangle$, (a) follows from 37.2.2(b). Assume now that (a) is known for some m such that $1 \le m \le -\langle i,j'\rangle$. We show that it is then also true for $m-1$.

Recall from 7.1.2(b) the identity

$$-F_i x_{i,j;1,m;e} + x_{i,j;1,m;e} F_i = [-\langle i,j'\rangle - m + 1]_i \tilde{K}_{-ei} x_{i,j;1,m-1;e}.$$

Applying to it the algebra isomorphism $\sigma : \mathbf{U} \to \mathbf{U}^{opp}$, which interchanges $x_{i,j;1,m;e}$ and $x'_{i,j;1,m;e}$, we deduce that

$$-x'_{i,j;1,m;e} F_i + F_i x'_{i,j;1,m;e} = [-\langle i,j'\rangle - m + 1]_i x'_{i,j;1,m-1;e} \tilde{K}_{ei}.$$

We apply $T'_{i,e}$ to this; using the induction hypothesis, we have

$$\begin{aligned}
&[-\langle i,j'\rangle - m + 1]_i T'_{i,e}(x'_{i,j;1,m-1;e}) \tilde{K}_{-ei} \\
&= T'_{i,e}(x'_{i,j;1,m;e}) E_i \tilde{K}_{-ei} - E_i \tilde{K}_{-ei} T'_{i,e}(x'_{i,j;1,m;e}) \\
&= x_{i,j;1,-\langle i,j'\rangle-m;e} E_i \tilde{K}_{-ei} - E_i \tilde{K}_{-ei} x_{i,j;1,-\langle i,j'\rangle-m;e} \\
&= (x_{i,j;1,-\langle i,j'\rangle-m;e} E_i - v_i^{e(\langle i,j'\rangle+2m)} E_i x_{i,j;1,-\langle i,j'\rangle-m;e}) \tilde{K}_{-ei} \\
&= [-\langle i,j'\rangle - m + 1]_i x_{i,j;1,-\langle i,j'\rangle-m+1;e} \tilde{K}_{-ei}.
\end{aligned}$$

The last equality follows from 7.1.2(a). Since $[-\langle i,j'\rangle - m+1]_i \neq 0$, we deduce that $T'_{i,e}(x'_{i,j;1,m-1;e}) = x_{i,j;1,-\langle i,j'\rangle -m+1;e}$. This proves the induction step. The proposition is proved.

37.3. Relation of the Symmetries with Comultiplication

37.3.1. We define two elements

$$\mathbf{L}''_i = \sum_n (-1)^n v_i^{-n(n-1)/2}\{n\}_i F_i^{(n)} \otimes E_i^{(n)},$$

$$\mathbf{L}'_i = \sum_n v_i^{n(n-1)/2}\{n\}_i F_i^{(n)} \otimes E_i^{(n)}$$

of $(\mathbf{U}\otimes\mathbf{U})^{\widehat{}}$ (see 4.1.1 and 4.1.2).

Proposition 37.3.2. *The following equality holds in* $(\mathbf{U}\otimes\mathbf{U})^{\widehat{}}$, *for any* $u \in \mathbf{U}$ *:*

(a) $$(T'_{i,-1}\otimes T'_{i,-1})\Delta(T''_{i,1}u) = \mathbf{L}'_i\Delta(u)\mathbf{L}''_i.$$

It suffices to show that for any integrable $\mathbf{U}$-modules M, N, the two sides of (a) act in the same way on $M\otimes N$. Let $x \in M, y \in N$. Using 5.3.4 twice, we have

$$\begin{aligned}
&\mathbf{L}'_i\Delta(u)\mathbf{L}''_i(x\otimes y) = L''^{-1}_i(uL''_i(x\otimes y))\\
&= ((T''_{i,1})^{-1}\otimes(T''_{i,1})^{-1})T''_{i,1}(u(T''_{i,1})^{-1}(T''_{i,1}x\otimes T''_{i,1}y))\\
&= ((T''_{i,1})^{-1}\otimes(T''_{i,1})^{-1})T''_{i,1}(u)(T''_{i,1}x\otimes T''_{i,1}y))\\
&= ((T'_{i,-1}\otimes T'_{i,-1})\Delta(T''_{i,1}u))(x\otimes y).
\end{aligned}$$

The proposition is proved.

CHAPTER 38

Symmetries and Inner product on f

38.1. The Algebras $\mathbf{f}[i], {}^\sigma\mathbf{f}[i]$

38.1.1. In this section we fix $i \in I$. For any $j \in I$ distinct from i and for any $m \in \mathbf{Z}$, we set

$$f(i,j;m) = \sum_{r+s=m} (-1)^r v_i^{-r(-\langle i,j'\rangle - m+1)} \theta_i^{(r)} \theta_j \theta_i^{(s)} \in \mathbf{f},$$

$$f'(i,j;m) = \sum_{r+s=m} (-1)^r v_i^{-r(-\langle i,j'\rangle - m+1)} \theta_i^{(s)} \theta_j \theta_i^{(r)} \in \mathbf{f}.$$

($f(i,j;m)$ is the same as the element $f_{i,j;1,m;-1}$ in 7.1.1.)

Let $\mathbf{f}[i]$ (resp. ${}^\sigma\mathbf{f}[i]$) be the subalgebra of $\mathbf{f}$ generated by the elements $f(i,j;m)$ (resp. $f'(i,j;m)$) for various $j \in I$, distinct from i, and various $m \in \mathbf{Z}$. Since $\sigma : \mathbf{f} \to \mathbf{f}$ interchanges $f(i,j;m)$ and $f'(i,j;m)$, we have $\sigma(\mathbf{f}[i]) = {}^\sigma\mathbf{f}[i]$.

Lemma 38.1.2. (a) $\mathbf{f} = \sum_{t \geq 0} \theta_i^t \mathbf{f}[i]$;

(b) $\mathbf{f} = \sum_{t \geq 0} {}^\sigma\mathbf{f}[i]\theta_i^t$.

Note that (b) follows from (a) by applying σ. We prove (a). We consider a product $y_1 y_2 \cdots y_n$ of elements in $\mathbf{f}$ in which each factor is either θ_i or is of the form $f(i,j;m)$ for some j different from i and some $m \in \mathbf{Z}$. Assume that there are two consecutive factors $y_a = f(i,j;m)$ and $y_{a+1} = \theta_i$. Using the formula

$$-v_i^{-(-\langle i,j'\rangle - 2m)} \theta_i f(i,j;m) + f(i,j;m)\theta_i = [m+1]_i f(i,j;m+1)$$

which follows from 7.1.2(a), we see that we may replace $y_a y_{a+1}$ by a scalar multiple of $y_{a+1} y_a$ plus a scalar multiple of $f(i,j;m+1)$. Using this procedure repeatedly, we see that $y_1 y_2 \cdots y_n$ is equal to a linear combination of products $y'_1 y'_2 \cdots y'_{n'}$ which do not have consecutive factors of the form described above, hence have the property that for some $s \geq 0$, we have $y'_1 = y'_2 = \cdots = y'_s = \theta_i$ and $y'_{s+1}, \ldots, y'_{n'}$ are of the form $f(i,j;m)$ for various j, m. Thus, $y_1 y_2 \cdots y_n$ belongs to $\sum_{t \geq 0} \theta_i^t \mathbf{f}[i]$. Now any word in the generators $\theta_j \quad (j \in I)$ is of the form $y_1 y_2 \cdots y_n$ with y_a as above, since $\theta_j = f(i,j;0)$ for $j \neq i$. The lemma follows.

G. Lusztig, *Introduction to Quantum Groups*, Modern Birkhäuser Classics,
DOI 10.1007/978-0-8176-4717-9_38, © Springer Science+Business Media, LLC 2010

Lemma 38.1.3. *There is a unique algebra isomorphism $x \mapsto g(x)$ of $\mathbf{f}[i]$ onto ${}^\sigma\mathbf{f}[i]$ such that $T''_{i,1}(x^+) = g(x)^+$ for any $x \in \mathbf{f}[i]$. We have $T'_{i,-1}(x'^+) = g^{-1}(x')^+$ for any $x' \in {}^\sigma\mathbf{f}[i]$.*

Since $T''_{i,1}, T'_{i,-1}$ are inverse algebra isomorphisms, it suffices to note that

$$T''_{i,1}(f(i,j;m)^+) = f'(i,j;-\langle i,j'\rangle - m)^+$$

and

$$T'_{i,-1}(f'(i,j;m)^+) = f(i,j;-\langle i,j'\rangle - m)^+,$$

for any $j \neq i$ and any $m \in \mathbf{Z}$. (See 37.2.5.)

Lemma 38.1.4. *Assume that $x \in \mathbf{f}$ satisfies $T''_{i,1}(x^+) \in \mathbf{U}^+$. Then ${}_ir(x) =$ 0.*

We may assume that x is homogeneous. By 3.1.6, we have

$$\text{(a)} \qquad x^+F_i - F_ix^+ = \frac{r_i(x)^+\tilde{K}_i - \tilde{K}_{-i}({}_ir(x)^+)}{v_i - v_i^{-1}}.$$

By 38.1.2, we can write

$$r_i(x)/(v_i - v_i^{-1}) = \sum_{t\geq 0} \theta_i^{(t)} y_t$$

and

$${}_ir(x)/(v_i - v_i^{-1}) = \sum_{t\geq 0} \theta_i^{(t)} z_t$$

where y_t, z_t belong to $\mathbf{f}[i]$ and are homogeneous. By 38.1.3, we have $T''_{i,1}(y_t^+) \in \mathbf{U}^+$ and $T''_{i,1}(z_t^+) \in \mathbf{U}^+$ for all $t \geq 0$. Applying $T''_{i,1}$ to both sides of (a), we obtain

$$\begin{aligned} &- T''_{i,1}(x^+)\tilde{K}_{-i}E_i + \tilde{K}_{-i}E_iT''_{i,1}(x^+) \\ \text{(b)} \qquad &= \sum_{t\geq 0}(-1)^t v_i^{-t(t-1)} F_i^{(t)} \tilde{K}_{ti}(T''_{i,1}(y_t^+)\tilde{K}_{-i} - \tilde{K}_iT''_{i,1}(z_t^+)). \end{aligned}$$

By our assumption, the left hand side of (b) is in $\tilde{K}_{-i}\mathbf{U}^+$, hence so is the right hand side. Using the triangular decomposition in $\mathbf{U}$, we deduce from (b) that $T''_{i,1}(y_t^+) = 0$ for all $t > 0$ and $T''_{i,1}(z_t^+) = 0$ for all $t \geq 0$. Since $T''_{i,1}$ is bijective, this implies $z_t^+ = 0$ for all t, hence $z_t = 0$ for all t and ${}_ir(x) = 0$.

Lemma 38.1.5. *Let x_t $(t \geq 0)$ be elements in the kernel of ${}_ir$, which are zero for all but finitely many t. Assume that $\sum_t \theta_i^{(t)} x_t = 0$. Then $x_t = 0$ for all t.*

This follows from 16.1.2(c) in the setup of 17.3.1.

Proposition 38.1.6. (a) *The following three subspaces of* $\mathbf{f}$ *coincide:*

$\mathbf{f}[i]$; $\{x \in \mathbf{f} | T''_{i,1}(x^+) \in \mathbf{U}^+\}$; *and* $\{x \in \mathbf{f} | {}_ir(x) = 0\}$.

(b) *The following three subspaces of* $\mathbf{f}$ *coincide:*

${}^\sigma\mathbf{f}[i]$; $\{x \in \mathbf{f} | T'_{i,-1}(x^+) \in \mathbf{U}^+\}$; *and* $\{x \in \mathbf{f} | r_i(x) = 0\}$.

Note that (b) follows from (a) by applying σ. We prove (a). By 38.1.3 and 38.1.4, the first space in (a) is contained in the second and the second in the third. Let $x \in \mathbf{f}$ be such that ${}_ir(x) = 0$. It remains to prove that $x \in \mathbf{f}[i]$.

By 38.1.2, we can write $x = \sum_{t \geq 0} \theta_i^{(t)} x_t$ where $x_t \in \mathbf{f}[i]$ for all t. By 38.1.3 and 38.1.4, we have ${}_ir(x_t) = 0$ for all t. We then have $0 = (x_0 - x) + \sum_{t \geq 1} \theta_i^{(t)} x_t$ where $x_0 - x$ and x_t for $t > 0$ are in the kernel of ${}_ir$. Using 38.1.5, we deduce that $x_0 - x = 0$ and $x_t = 0$ for $t > 0$. In particular, we have $x = x_0 \in \mathbf{f}[i]$.

The proposition is proved.

Lemma 38.1.7.

$$r(f(i,j;m)) = 1 \otimes f(i,j;m) + \sum_{t=0}^{m} \prod_{h=0}^{m-t-1} (1 - v_i^{-2h+2m-2\alpha-2}) v_i^{t(m-t)} f(i,j;t) \otimes \theta_i^{(m-t)},$$

$$r(f'(i,j;m)) = f'(i,j;m) \otimes 1 + \sum_{t=0}^{m} \prod_{h=0}^{m-t-1} (1 - v_i^{-2h+2m-2\alpha-2}) v_i^{t(m-t)} \theta_i^{(m-t)} \otimes f'(i,j;t).$$

We set $\alpha = -\langle i, j' \rangle$. Using 1.4.2 and the fact that r is an algebra homomorphism (see 1.2.2), we compute

$$r(f(i,j;m)) = \sum_{r'+r''+s'+s''=m} (-1)^{r'+r''} v_i^{-(r'+r'')(\alpha-m+1)+r'r''+s's''}$$

$$\times (\theta_i^{(r')} \otimes \theta_i^{(r'')})(\theta_j \otimes 1 + 1 \otimes \theta_j)(\theta_i^{(s')} \otimes \theta_i^{(s'')})$$

$$= \sum_{r'+r''+s'+s''=m} (-1)^{r'+r''} v_i^{-(r'+r'')(\alpha-m+1)+r'r''+s's''-r''\alpha+2r''s'}$$

$$\times \begin{bmatrix} r''+s'' \\ r'' \end{bmatrix}_i \theta_i^{(r')}\theta_j\theta_i^{(s')} \otimes \theta_i^{(r''+s'')}$$

$$+ \sum_{r'+r''+s'+s''=m} (-1)^{r'+r''} v_i^{-(r'+r'')(\alpha-m+1)+r'r''+s's''+2r''s'-s'\alpha}$$

$$\times \begin{bmatrix} r'+s' \\ r' \end{bmatrix}_i \theta_i^{(r'+s')} \otimes \theta_i^{(r'')}\theta_j\theta_i^{(s'')}$$

$$= \sum_{r',s'} (-1)^{r'} v_i^{s'(m-r'-s')-r'(\alpha-m+1)} \sum_{r'',s'';r''+s''=m-r'-s'} (-1)^{r''}$$

$$\times v_i^{r''(r'+s'-\alpha-(\alpha-m+1))} \begin{bmatrix} r''+s'' \\ r'' \end{bmatrix}_i \theta_i^{(r')}\theta_j\theta_i^{(r'')} \otimes \theta_i^{(r''+s'')}$$

$$+ \sum_{r'',s''} (-1)^{r''} v_i^{r''(m-r''-s''-(\alpha-m+1))+(m-r''-s'')(s''+r''-\alpha)}$$

$$\times \sum_{r',s';r'+s'=m-r''-s''} (-1)^{r'} v_i^{r'(-1+s'+r')} \begin{bmatrix} r'+s' \\ r' \end{bmatrix}_i \theta_i^{(r'+s')} \otimes \theta_i^{(r'')}\theta_j\theta_i^{(s'')}$$

$$= \sum_{r',s'} (-1)^{r'} v_i^{s'(m-r'-s')-r'(\alpha-m+1)} \prod_{h=0}^{m-r'-s'-1} (1 - v_i^{-2h+2m-2\alpha-2})$$

$$\times \theta_i^{(r')}\theta_j\theta_i^{(s')} \otimes \theta_i^{(m-r'-s')} + \sum_{r''+s''=m} (-1)^{r''} v_i^{-r''(\alpha-m+1)} 1 \otimes \theta_i^{(r'')}\theta_j\theta_i^{(s'')}$$

$$= 1 \otimes f(i,j;m) + \sum_{t=0}^{m} \prod_{h=0}^{m-t-1} (1 - v_i^{-2h+2m-2\alpha-2}) v_i^{t(m-t)} f(i,j;t) \otimes \theta_i^{(m-t)}.$$

This proves the first formula of the lemma. The second formula follows from the first, using the formula $r(\sigma(x)) = (\sigma \otimes \sigma)^t r(x)$ (see 1.2.8(a)).

Lemma 38.1.8. *Let $x \in \mathbf{f}[i]$ and let $x' = g(x)$ be the corresponding element in ${}^{\sigma}\mathbf{f}[i]$ (see 38.1.3).*

(a) *We have $r(x) \in \mathbf{f}[i] \otimes \mathbf{f}$ and $r(x') \in \mathbf{f} \otimes {}^{\sigma}\mathbf{f}[i]$.*

(b) *Let ${}'r(x) \in \mathbf{f}[i] \otimes \mathbf{f}[i]$ be defined by $r(x) - {}'r(x) \in \mathbf{f}[i] \otimes \theta_i\mathbf{f}$. We use the direct sum decomposition $\mathbf{f} = \mathbf{f}[i] \oplus \theta_i\mathbf{f}$ (see 38.1.5).*

Let $''r(x') \in {}^\sigma\mathbf{f}[i] \otimes {}^\sigma\mathbf{f}[i]$ *be defined by* $r(x) - ''r(x') \in \mathbf{f}\theta_i \otimes {}^\sigma\mathbf{f}[i]$. *We use the direct sum decomposition* $\mathbf{f} = {}^\sigma\mathbf{f}[i] \oplus \mathbf{f}\theta_i$. *We then have* $(g \otimes g)('r(x)) = ''r(x')$, *where* g *is as in 38.1.3.*

We prove (a). If $x_1, x_2 \in \mathbf{f}$ satisfy $r(x_1) \in \mathbf{f}[i] \otimes \mathbf{f}$ and $r(x_2) \in \mathbf{f}[i] \otimes \mathbf{f}$, then $r(x_1x_2) \in \mathbf{f}[i] \otimes \mathbf{f}$ since r respects products and $\mathbf{f}[i]$ is closed under multiplication. Hence to check that $r(x) \in \mathbf{f}[i] \otimes \mathbf{f}$ for $x \in \mathbf{f}[i]$, it is enough to check this in the special case where $x = f(i,j;m)$. But this follows from the previous lemma. This proves the first assertion of (a). The second assertion of (a) is proved in the same way.

We now prove (b). Let $(z_h)_{h\in H}$ be a basis of the vector space $\mathbf{f}[i]$, consisting of homogeneous elements; then $(g(z_h)_{h\in H})$ is a basis of the vector space ${}^\sigma\mathbf{f}[i]$ (see 38.1.3). Let $f(h) = |z_h|, f'(h) = |g(z_h)|$ be the corresponding elements of $\mathbf{N}[I]$.

Using (a), we can write uniquely

$$r(x) = \sum_{n\geq 0; h,h'\in H} c_{n;h,h'} z_h \otimes \theta_i^{(n)} z_{h'},$$

$$r(x') = \sum_{n\geq 0; h,h'\in H} d_{n;h,h'} g(z_h)\theta_i^{(n)} \otimes g(z_{h'}),$$

where $c(n;h,h'), d(n;h,h') \in \mathbf{Q}(v)$ are zero for all but finitely many indices.

By 3.1.5, we have

(c) $\Delta(x^+) = \sum_{n\geq 0;h,h'\in H} c_{n;h,h'} z_h^+ \tilde{K}_{f(h')+ni} \otimes E_i^{(n)} z_{h'}^+$;

(d) $\Delta(x'^+) = \sum_{n\geq 0;h,h'\in H} d_{n;h,h'} g(z_h)^+ E_i^{(n)} \tilde{K}_{f'(h')} \otimes g(z_{h'})^+$.

By definition, we have $\Delta(x'^+) = \Delta(T''_{i,1}(x^+))$. Applying $T'_{i,-1} \otimes T'_{i,-1}$ to (d) we obtain

$$(T'_{i,-1} \otimes T'_{i,-1})\Delta(T''_{i,1}(x^+))$$
$$= \sum_{n\geq 0;h,h'\in H} d_{n;h,h'} (-1)^n v_i^{-n(n-1)} z_h^+ \tilde{K}_{-ni} F_i^{(n)} \tilde{K}_{f(h')} \otimes z_{h'}^+.$$

By 37.3.2, we can replace $(T'_{i,-1} \otimes T'_{i,-1})\Delta(T''_{i,1}(x^+))$ by $\mathbf{L}'_i \Delta(x^+) \mathbf{L}''_i$, which by (c) equals

$$\mathbf{L}'_i \Big(\sum_{n\geq 0;h,h'\in H} c_{n;h,h'} z_h^+ \tilde{K}_{f(h')+ni} \otimes E_i^{(n)} z_{h'}^+ \Big) \mathbf{L}''_i.$$

Thus, we have

$$\sum_{n\geq 0;h,h'\in H} d_{n;h,h'} (-1)^n v_i^{-n(n-1)} z_h^+ \tilde{K}_{-ni} F_i^{(n)} \tilde{K}_{f(h')} \otimes z_{h'}^+$$
$$= \mathbf{L}'_i \Big(\sum_{n\geq 0;h,h'\in H} c_{n;h,h'} z_h^+ \tilde{K}_{f(h')+ni} \otimes E_i^{(n)} z_{h'}^+ \Big) \mathbf{L}''_i$$

or equivalently,

$$\Big(\sum_{n\geq 0;h,h'\in H} d_{n;h,h'}(-1)^n v_i^{-n(n-1)} z_h^+ \tilde{K}_{-ni} F_i^{(n)} \tilde{K}_{f(h')} \otimes z_{h'}^+\Big)\mathbf{L}'_i$$
$$= \mathbf{L}'_i\Big(\sum_{n\geq 0;h,h'\in H} c_{n;h,h'} z_h^+ \tilde{K}_{f(h')+ni} \otimes E_i^{(n)} z_{h'}^+\Big)$$

(equality in $(\mathbf{U}\otimes\mathbf{U})\hat{}$).

We consider the $\mathbf{U}\otimes\mathbf{U}$-module ${}^\omega M_0 \otimes {}^\omega M_0$, where M_0 is a Verma module. Both sides of the previous equality act naturally on this $\mathbf{U}\otimes\mathbf{U}$-module. Applying them to the vector $\xi\otimes\xi$, where $\xi \in {}^\omega M_0$ is the canonical generator, we obtain the equality

$$\sum_{n,t\geq 0;h,h'\in H} v_i^{t(t-1)/2}\{t\}_i(-1)^n v_i^{-n(n-1)} d_{n;h,h'}$$
$$\times\, z_h^+ \tilde{K}_{-ni} F_i^{(n)} \tilde{K}_{f(h')} F_i^{(t)}\xi \otimes z_{h'}^+ E_i^{(t)}\xi$$

(e)

$$= \sum_{n,t\geq 0;h,h'\in H} v_i^{t(t-1)/2}\{t\}_i c_{n;h,h'}(F_i^{(t)} z_h^+ \tilde{K}_{f(h')+ni}\xi \otimes E_i^{(t)} E_i^{(n)}) z_{h'}^+\xi,$$

(equality in ${}^\omega M_0 \otimes {}^\omega M_0$). Since $F_i\xi = 0$, only the summands with $n = t = 0$ contribute to the left hand side of (e). Let $\pi : {}^\omega M_0 / E_i {}^\omega M_0$ be the canonical map. After applying $1\otimes\pi$ to (e), the summands with $(n,t)\neq(0,0)$ (in the right hand side) go to zero; hence (e) implies the equality

$$\sum_{h,h'\in H} d_{0;h,h'} z_h^+\xi \otimes \pi(z_{h'}^+\xi) = \sum_{h,h'\in H} c_{0;h,h'} z_h^+\xi \otimes \pi(z_{h'}^+\xi)$$

in ${}^\omega M_0 \otimes ({}^\omega M_0 / E_i {}^\omega M_0)$.

The vectors $z_h^+\xi$ are linearly independent in ${}^\omega M_0$; hence we deduce that

$$\sum_{h'\in H} (d_{0;h,h'} - c_{0;h,h'})\pi(z_{h'}^+\xi) = 0$$

in ${}^\omega M_0 / E_i {}^\omega M_0$), for all h. Hence

$$\sum_{h'\in H} (d_{0;h,h'} - c_{0;h,h'}) z_{h'} \in \theta_i\mathbf{f},$$

for all h. Using the fact that $\mathbf{f}[i] \cap (\theta_i\mathbf{f}) = 0$ (see 38.1.5), we deduce that $d_{0;h,h'} - c_{0;h,h'} = 0$ for all h, h'.

By definition, we have ${}'r(x) = \sum_{h,h'\in H} c_{0;h,h'} z_h \otimes z_{h'}$, and ${}''r(x') = \sum_{h,h'\in H} d_{0;h,h'} g(z_h)\otimes g(z_{h'})$ and the equalities $d_{0;h,h'} = c_{0;h,h'}$ imply that $(g\otimes g)({}'r(x)) = {}''r(x')$. The lemma is proved.

38.2. A Computation of Inner Products

Proposition 38.2.1. *For all $x, y \in \mathbf{f}[i]$, we have*

(a) $(g(x), g(y)) = (x, y)$.

Assume that we are given two elements $z', z'' \in \mathbf{f}[i]$ such that

$$(g(z'), g(y')) = (z', y')$$

and

$$(g(z''), g(y'')) = (z'', y'')$$

for all $y', y'' \in \mathbf{f}[i]$. We show that we then have

$$(g(z'z''), g(y)) = (z'z'', y),$$

for any $y \in \mathbf{f}[i]$. By definition, we have

$$(z'z'', y) = (z' \otimes z'', r(y))$$

and

$$(g(z'z''), g(y)) = (g(z') \otimes g(z''), r(g(y))).$$

We have $(z'', \theta_i \mathbf{f}) = 0$ since ${}_ir(z'') = 0$ (see 1.2.13(a)); hence

$$(z' \otimes z'', r(y)) = (z' \otimes z'', {}' r(y))$$

(with the notations of lemma 38.1.8). Similarly, $(g(z'), \mathbf{f}\theta_i) = 0$, hence

$$\begin{aligned}(g(z') \otimes g(z''), r(g(y))) &= (g(z') \otimes g(z''), {}'' r(g(y))) \\ &= ((g \otimes g)(z' \otimes z''), (g \otimes g)' r(y)).\end{aligned}$$

The last equality comes from lemma 38.1.8. By our hypothesis, we have $((g \otimes g)(z' \otimes z''), (g \otimes g)' r(y)) = (z' \otimes z'', {}' r(y))$. Combining the equalities above, we obtain $(g(z'z''), g(y)) = (z'z'', y)$, as claimed.

Since the algebra $\mathbf{f}[i]$ is generated by the elements $f(i, j; m)$, we see that it is enough to prove (a) under the assumption that $x = f(i, j; m)$. We can assume also that y is homogeneous of the same degree as $f(i, j; m)$. Since $y \in \mathbf{f}[i]$, this forces y to be a scalar multiple of $f(i, j; m)$. Thus, we are reduced to verifying the identity

$$(f(i, j; m), f(i, j; m)) = (f'(i, j; m'), f'(i, j; m'))$$

for any m, m' such that $m + m' = \alpha = -\langle i, j' \rangle$. (See the proof of 38.1.3.)

We have $f(i,j;m) = \theta_j \theta_i^{(m)}$ mod $\theta_i \mathbf{f}$; since $(f(i,j;m), \theta_i \mathbf{f}) = 0$, as above, we have $(f(i,j;m), f(i,j;m)) = (f(i,j;m), \theta_j \theta_i^{(m)})$. This is equal, by definition, to $(r(f(i,j;m)), \theta_j \otimes \theta_i^{(m)})$ and, by 38.1.7, this equals

(b) $\prod_{h=0}^{m-1}(1 - v_i^{-2h+2m-2\alpha-2})(\theta_j, \theta_j)(\theta_i^{(m)}, \theta_i^{(m)})$.

Similarly, we have $f'(i,j;m') = \theta_i^{(m')}\theta_j$ mod $\mathbf{f}\theta_i$; since

$$(f'(i,j;m'), \mathbf{f}\theta_i) = 0,$$

we have

$$(f'(i,j;m'), f'(i,j;m')) = (f'(i,j;m'), \theta_i^{(m')}\theta_j).$$

This is equal, by definition, to $(r(f'(i,j;m')), \theta_i^{(m')} \otimes \theta_j)$ and, by 38.1.7, this equals

(c) $\prod_{h=0}^{m'-1}(1 - v_i^{-2h+2m'-2\alpha-2})(\theta_j, \theta_j)(\theta_i^{(m')}, \theta_i^{(m')})$.

We substitute $(\theta_i^{(m)}, \theta_i^{(m)})$ and $(\theta_i^{(m')}, \theta_i^{(m')})$ in (b) and (c) by the expressions given by 1.4.4; we see that the expressions (b),(c) are equal. The proposition is proved.

38.2.2. We say that a sequence $\mathbf{h} = (i_1, i_2, \ldots, i_n)$ in I is *admissible* if for any a, b such that $1 \leq a \leq b \leq n$, we have

(a) $T'_{i_a,-1}T'_{i_{a+1},-1} \cdots T'_{i_{b-1},-1}(E_{i_b}) \in \mathbf{U}^+$ and

(b) $T''_{i_b,1}T''_{i_{b-1},1} \cdots T''_{i_{a+1},1}(E_{i_a}) \in \mathbf{U}^+$.

Assume that an admissible sequence $\mathbf{h}$ as above is fixed, and that we are given $0 \leq p \leq n$. An element $x \in \mathbf{f}$ is said to be *adapted* to $(\mathbf{h}, p)$, if for any a such that $1 \leq a \leq p$ we have

(c) $T'_{i_a,-1}T'_{i_{a+1},-1} \cdots T'_{i_p,-1}(x^+) \in \mathbf{U}^+$,

and for any b such that $p + 1 \leq b \leq n$, we have

(d) $T''_{i_b,1}T''_{i_{b-1},1} \cdots T''_{i_{p+1},1}(x^+) \in \mathbf{U}^+$.

Given such x and given a sequence $\mathbf{c} = (c_1, c_2, \ldots, c_n) \in \mathbf{N}^n$, we define $L(\mathbf{h}, \mathbf{c}, p, x) \in \mathbf{f}$ by

$$\begin{aligned} &L(\mathbf{h}, \mathbf{c}, p, x)^+ \\ &= E_{i_{p+1}}^{(c_{p+1})} T'_{i_{p+1},-1}(E_{i_{p+2}}^{(c_{p+2})}) \cdots T'_{i_{p+1},-1}T'_{i_{p+2},-1} \cdots T'_{i_{n-1},-1}(E_{i_n}^{(c_n)}) \\ &\quad \times x^+ T''_{i_p,1}T''_{i_{p-1},1} \cdots T''_{i_2,1}(E_{i_1}^{(c_1)}) \cdots T''_{i_p,1}(E_{i_{p-1}}^{(c_{p-1})})E_{i_p}^{(c_p)}. \end{aligned}$$

The definition is correct, since the factors in the right hand side are in $\mathbf{U}^+$, by (a),(b).

Proposition 38.2.3. *Let* $\mathbf{c} = (c_1, c_2, \dots, c_n) \in \mathbf{N}^n, \mathbf{c}' = (c'_1, c'_2, \dots, c'_n) \in \mathbf{N}^n$ *and let* $x, x' \in \mathbf{f}$ *be adapted to* $(\mathbf{h}, p)$. *We have the equality of inner products* $(L(\mathbf{h}, \mathbf{c}, p, x), L(\mathbf{h}, \mathbf{c}', p, x')) = (x, x') \prod_{s=1}^{n} (\theta_{i_s}^{(c_s)}, \theta_{i_s}^{(c'_s)})$.

For any $i \in I$, $t, t' \in \mathbf{N}$ and $y, y' \in \mathbf{f}[i] = \ker({}_ir)$, we have

(a) $(\theta_i^{(t)} y, \theta_i^{(t')} y') = (\theta_i^{(t)}, \theta_i^{(t')})(y, y')$.

Indeed, we may assume that y, y' are homogeneous. We can write $r(y') = \sum y'_1 \otimes y'_2$ with y'_1, y'_2 homogeneous, and $|y'_1| \notin \{i, 2i, 3i, \dots\}$ (see 38.1.8(a)) and our assertion follows easily from the definitions. Similarly, if $y, y' \in {}^\sigma\mathbf{f}[i]$, we have

(b) $(y\theta_i^{(t)}, y'\theta_i^{(t')}) = (\theta_i^{(t)}, \theta_i^{(t')})(y, y')$.

Assume first that $p < n$ and that the proposition is true when p is replaced by $p + 1$. Let $\tilde{\mathbf{c}}, \tilde{\mathbf{c}}'$ be the sequences defined by $\tilde{c}_{p+1} = \tilde{c}'_{p+1} = 0$ and $\tilde{c}_s = c_s, \tilde{c}'_s = c'_s$ for $s \neq p + 1$. Let $\tilde{x} = g(x) \in \mathbf{f}, \tilde{x}' = g(x') \in \mathbf{f}$, where g is as in 38.1.3 (with $i = i_{p+1}$). Let $\tilde{p} = p + 1$. It is clear that $\tilde{x}, \tilde{x}'$ are well-defined and that they are adapted to $(\mathbf{h}, p + 1)$. We have from the definitions

$$L(\mathbf{h}, \mathbf{c}, p, x)^+ = E_{i_{p+1}}^{(c_{p+1})} T'_{i_{p+1},-1}(L(\mathbf{h}, \tilde{\mathbf{c}}, \tilde{p}, \tilde{x})^+).$$

By our assumptions, we have

$$T'_{i_{p+1},-1}(L(\mathbf{h}, \tilde{\mathbf{c}}, \tilde{p}, \tilde{x})^+) \in \mathbf{U}^+;$$

hence

$$L(\mathbf{h}, \tilde{\mathbf{c}}, \tilde{p}, \tilde{x}) \in {}^\sigma\mathbf{f}[i_{p+1}]$$

and

$$T'_{i_{p+1},-1}(L(\mathbf{h}, \tilde{\mathbf{c}}, \tilde{p}, \tilde{x})^+) = (g^{-1}(L(\mathbf{h}, \tilde{\mathbf{c}}, \tilde{p}, \tilde{x})))^+$$

where g is as above. Similarly

$$L(\mathbf{h}, \mathbf{c}', p, x')^+ = E_{i_{p+1}}^{(c'_{p+1})} T'_{i_{p+1},-1}(L(\mathbf{h}, \tilde{\mathbf{c}}', \tilde{p}, \tilde{x}')^+)$$

and

$$L(\mathbf{h}, \tilde{\mathbf{c}}', \tilde{p}, \tilde{x}') \in {}^\sigma\mathbf{f}[i_{p+1}], T'_{i_{p+1},-1}(L(\mathbf{h}, \tilde{\mathbf{c}}', \tilde{p}, \tilde{x}')^+) = (g^{-1}(L(\mathbf{h}, \tilde{\mathbf{c}}', \tilde{p}, \tilde{x}')))^+.$$

Using (a), we have

$$\begin{aligned}
&(L(\mathbf{h}, \mathbf{c}, p, x), L(\mathbf{h}, \mathbf{c}', p, x')) \\
&= (\theta_{i_{p+1}}^{(c_{p+1})} g^{-1}(L(\mathbf{h}, \tilde{\mathbf{c}}, \tilde{p}, \tilde{x})), \theta_{i_{p+1}}^{(c'_{p+1})} g^{-1}(L(\mathbf{h}, \tilde{\mathbf{c}}', \tilde{p}, \tilde{x}'))) \\
&= (\theta_{i_{p+1}}^{(c_{p+1})}, \theta_{i_{p+1}}^{(c'_{p+1})})(g^{-1}(L(\mathbf{h}, \tilde{\mathbf{c}}, \tilde{p}, \tilde{x})), g^{-1}(L(\mathbf{h}, \tilde{\mathbf{c}}', \tilde{p}, \tilde{x}'))) \\
&= (\theta_{i_{p+1}}^{(c_{p+1})}, \theta_{i_{p+1}}^{(c'_{p+1})})(L(\mathbf{h}, \tilde{\mathbf{c}}, \tilde{p}, \tilde{x}), L(\mathbf{h}, \tilde{\mathbf{c}}', \tilde{p}, \tilde{x}')).
\end{aligned}$$

The last equality follows from 38.2.1.

Using now this hypothesis, we see that the last expression equals $(g(x),g(x'))\prod_{s=1}^{n}(\theta_{i_s}^{(c_s)},\theta_{i_s}^{(c'_s)})$. We now replace $(g(x),g(x'))$ by (x,x') and we see that the formula in the proposition holds in this case. Using this argument repeatedly, we are reduced to the case where $p=n$. In the rest of the proof we assume that $p=n$. We now have

$$L(\mathbf{h},\mathbf{c},n,x)^+ = x^+T''_{i_n,1}T''_{i_{n-1},1}\cdots T''_{i_2,1}(E_{i_1}^{(c_1)})\cdots T''_{i_n,1}(E_{i_{n-1}}^{(c_{n-1})})E_{i_n}^{(c_n)}.$$

We now assume that $n\geq 1$ and that the result is known for $n-1$ instead of n.

Let $\tilde{\mathbf{h}}$ be the sequence $(i_1,i_2,\ldots,i_{n-1})$. Let $\tilde{\mathbf{c}}=(c_1,c_2,\ldots,c_{n-1}),\tilde{\mathbf{c}}'=(c'_1,c'_2,\ldots,c'_{n-1})$. Let $\tilde{x}=g^{-1}x,\tilde{x}'=g^{-1}x'$ where g is as in 38.1.3, with $i=i_n$. It is clear that $\tilde{x},\tilde{x}'$ are well-defined and they are adapted to $(\tilde{\mathbf{h}},n-1)$. We have from the definitions

$$L(\mathbf{h},\mathbf{c},n,x)^+ = T''_{i_n,1}(L(\mathbf{h},\tilde{\mathbf{c}},n-1,\tilde{x})^+)E_{i_n}^{(c_n)},$$

and similarly

$$L(\mathbf{h},\mathbf{c}',n,x')^+ = T''_{i_n,1}(L(\mathbf{h},\tilde{\mathbf{c}}',n-1,\tilde{x}')^+)E_{i_n}^{(c'_n)}.$$

By arguments almost identical to the ones above (using (b) instead of (a)) we see that

$$\begin{aligned}
&(L(\mathbf{h},\mathbf{c},n,x),L(\mathbf{h},\mathbf{c}',n,x'))\\
&=(g(L(\mathbf{h},\tilde{\mathbf{c}},n-1,\tilde{x}))\theta_{i_n}^{(c_n)},g(L(\mathbf{h},\tilde{\mathbf{c}}',n-1,\tilde{x}'))\theta_{i_n}^{(c'_n)})\\
&=(g(L(\mathbf{h},\tilde{\mathbf{c}},n-1,\tilde{x})),g(L(\mathbf{h},\tilde{\mathbf{c}}',n-1,\tilde{x}')))(\theta_{i_n}^{(c_n)},\theta_{i_n}^{(c'_n)})\\
&=(L(\mathbf{h},\tilde{\mathbf{c}},n-1,\tilde{x}),L(\mathbf{h},\tilde{\mathbf{c}}',n-1,\tilde{x}'))(\theta_{i_n}^{(c_n)},\theta_{i_n}^{(c'_n)})\\
&=(g^{-1}(x),g^{-1}(x'))\prod_{s=1}^{n}(\theta_{i_s}^{(c_s)},\theta_{i_s}^{(c'_s)})\\
&=(x,x')\prod_{s=1}^{n}(\theta_{i_s}^{(c_s)},\theta_{i_s}^{(c'_s)}).
\end{aligned}$$

Using this argument repeatedly, we are reduced to the case where $n=0$. We now have $L(\mathbf{h},\mathbf{c},n,x)=x$ and the result is obvious. The proposition is proved.

CHAPTER 39

Braid Group Relations

39.1. Preparatory Results

39.1.1. Study of the symmetries $T'_{i,e}$ on Λ_λ. In this and the next subsection, we assume that the root datum is X-regular and Y-regular. Let $\lambda \in X^+$. Let $\eta \in \Lambda_\lambda$ be as in 3.5.7.

Lemma 39.1.2. *Let $\mathbf{i} = (i_1, i_2, \dots, i_N)$ be a sequence in I such that $s_{i_1} s_{i_2} \cdots s_{i_N}$ is a reduced expression of an element of W. We have*

(a) $T'_{i_1,e} T'_{i_2,e} \cdots T'_{i_N,e} \eta = F_{i_1}^{(a_1)} F_{i_2}^{(a_2)} \cdots F_{i_N}^{(a_N)} \eta$, *where*

$$a_1 = \langle s_{i_N} \cdots s_{i_2}(i_1), \lambda \rangle, \dots, a_{N-1} = \langle s_{i_N}(i_{N-1}), \lambda \rangle, a_N = \langle i_N, \lambda \rangle;$$

note that $a_1, a_2, \dots, a_N \in \mathbf{N}$ (see 2.2.7).

We argue by induction on $N \geq 0$. For $N = 0$, there is nothing to prove. Assume that $N \geq 1$. Let $\eta(\mathbf{i})$ be the left hand side of (a). Let $\mathbf{i}' = (i_2, i_3, \dots, i_N)$. Then the induction hypothesis is applicable to $\mathbf{i}'$ and it remains to show that

(b) $T'_{i_1,e}(\eta(\mathbf{i}')) = F_{i_1}^{(a_1)}(\eta(\mathbf{i}'))$.

Note that $\eta(\mathbf{i}')$ belongs to the $s_{i_2} s_{i_3} \cdots s_{i_N}(\lambda)$-weight space of Λ_λ, and we have, by definition, $a_1 = \langle i_1, s_{i_2} s_{i_3} \cdots s_{i_N}(\lambda) \rangle$. Using 5.2.2(a), we see that (b) would follow from the equality $E_{i_1}(\eta(\mathbf{i}')) = 0$.

Since $E_{i_1}(\eta(\mathbf{i}'))$ belongs to the $s_{i_2} s_{i_3} \cdots s_{i_N}(\lambda) + i'_1$-weight space, it is enough to prove that this weight space is zero. If this weight space were non-zero, then the $s_{i_N} \cdots s_{i_3} s_{i_2}(s_{i_2} s_{i_3} \cdots s_{i_N}(\lambda) + i'_1)$-weight space, which is the $(\lambda + s_{i_N} \cdots s_{i_3} s_{i_2}(i'_1))$-weight space, would also be non-zero. But then we would have $\lambda + s_{i_N} \cdots s_{i_3} s_{i_2}(i'_1)) \leq \lambda$, hence $s_{i_N} \cdots s_{i_3} s_{i_2}(i'_1) \leq 0$ which contradicts the assumption that $s_{i_1} s_{i_2} \cdots s_{i_N}$ is a reduced expression. The lemma is proved.

39.1.3. A lemma on monomials. Assume that $s_{i_1} s_{i_2} \cdots s_{i_n}$ is a reduced expression in W. Let $\lambda \in X$ be such that $\langle i_p, \lambda \rangle \geq 0$ for $p = 1, 2, \dots, n$. Set

$$a_n = \langle i_n, \lambda \rangle, a_{n-1} = \langle s_{i_n}(i_{n-1}), \lambda \rangle, \dots, a_1 = \langle s_{i_n} s_{i_{n-1}} \cdots s_{i_2}(i_1), \lambda \rangle.$$

Note that $a_1, a_2, \dots, \in \mathbf{N}$. Let $x = F_{i_1}^{(a_1)} F_{i_2}^{(a_2)} \cdots F_{i_n}^{(a_n)}$.

G. Lusztig, *Introduction to Quantum Groups*, Modern Birkhäuser Classics,
DOI 10.1007/978-0-8176-4717-9_39, © Springer Science+Business Media, LLC 2010

Lemma 39.1.4. *For any $i \in I$, we have*

$$E_i x = x E_i + y_i(\tilde{K}_{-i} v_i^{\langle i,\lambda\rangle -1} - \tilde{K}_i v_i^{-\langle i,\lambda\rangle +1}) \text{ for some } y_i \in \mathbf{U}^-.$$

We argue by induction on n. For $n = 0$, there is nothing to prove. Assume that $n \geq 1$. Let $x' = F_{i_2}^{(a_2)} F_{i_3}^{(a_3)} \cdots F_{i_n}^{(a_n)}$. By the induction hypothesis, we have

$$E_i x' = x' E_i + y_i'(\tilde{K}_{-i} v_i^{\langle i,\lambda\rangle -1} - \tilde{K}_i v_i^{-\langle i,\lambda\rangle +1})$$

for some $y_i' \in \mathbf{U}^-$. Hence

$$\begin{aligned} E_i x &= E_i F_{i_1}^{(a_1)} x' \\ &= F_{i_1}^{(a_1)} E_i x' - \delta_{i,i_1} F_{i_1}^{(a_1-1)} \frac{\tilde{K}_{-i} v_i^{a_1-1} - \tilde{K}_i v_i^{-a_1+1}}{v_i - v_i^{-1}} x' \\ &= F_{i_1}^{(a_1)} x' E_i + F_{i_1}^{(a_1)} y_i'(\tilde{K}_{-i} v_i^{\langle i,\lambda\rangle -1} - \tilde{K}_i v_i^{-\langle i,\lambda\rangle +1}) \\ &\quad - \delta_{i,i_1} F_{i_1}^{(a_1-1)} \frac{\tilde{K}_{-i} v_i^{a_1-1} - \tilde{K}_i v_i^{-a_1+1}}{v_i - v_i^{-1}} x'. \end{aligned}$$

It remains to note that

$$a_1 = \langle i_1, -a_2 i_2' - a_3 i_3' - \cdots - a_n i_n' + \lambda\rangle,$$

which follows easily from the definitions.

39.2. Braid Group Relations for U in Rank 2

39.2.1. In this section we assume that I consists of two elements $i \neq j$ and that the Cartan datum is of finite type or, equivalently, $\langle i, j'\rangle\langle j, i'\rangle$ is equal to $0, 1, 2$, or 3. We define accordingly $h = h(i,j)$ to be $2, 3, 4$ or 6 (see 2.1.1). In this case, the root datum is automatically X-regular and Y-regular.

By interchanging if necessary i and j, we may assume that either $\langle i, j'\rangle = \langle j, i'\rangle = 0$, or $\langle j, i'\rangle = -1$ and $\langle i, j'\rangle \in \{-1, -2, -3\}$. We shall consider each case separately.

39.2.2. Assume that $\langle i, j'\rangle = -3$, $\langle j, i'\rangle = -1$, so that $h = 6$. Note that $v_j = v_i^3$. As in 37.2.1, we set

$$x_{1,m;e} = \sum_{r+s=m} (-1)^r v_i^{er(4-m)} E_i^{(r)} E_j E_i^{(s)} \in \mathbf{U}^+,$$

$$x'_{1,m;e} = \sum_{r+s=m} (-1)^r v_i^{er(4-m)} E_i^{(s)} E_j E_i^{(r)} \in \mathbf{U}^+.$$

Recall from 7.1.2 that

$$-v_i^{e(3-2m)} E_i x_{1,m;e} + x_{1,m;e} E_i = [m+1]_i x_{1,m+1;e}$$

and

$$-F_i x_{1,m;e} + x_{1,m;e} F_i = [4-m]_i \tilde{K}_{-ei} x_{1,m-1;e}.$$

Applying σ, which interchanges $x_{1,m;e}$ and $x'_{1,m;e}$, we deduce that

$$-v_i^{e(3-2m)} x'_{1,m;e} E_i + E_i x'_{1,m;c} = [m+1]_i x'_{1,m+1;e}$$

and

$$-x'_{1,m;e} F_i + F_i x'_{1,m;e} = [4-m]_i x'_{1,m-1;e} \tilde{K}_{ei}.$$

By 37.2.5, we have $T'_{i,e}(E_j) = x_{1,3;e}$ and $T''_{i,-e}(E_j) = x'_{1,3;e}$. By 37.2.5 with the roles of i, j interchanged, we have $T'_{j,e}(E_i) = -v_j^e E_j E_i + E_i E_j = x'_{1,1;e}$ and similarly, $T''_{j,-e}(E_i) = x_{1,1;e}$. By 37.2.5, we have $T''_{i,-e}(x_{1,1;e}) = x'_{1,2;e}$ and $T'_{i,e}(x'_{1,1;e}) = x_{1,2;e}$. We have

$$\begin{aligned} T''_{j,-e}(x'_{1,2;e}) &= [2]_i^{-1} T''_{j,-e}(-v_i^e x'_{1,1;e} E_i + E_i x'_{1,1;e}) \\ &= [2]_i^{-1}(-v_i^e T''_{j,-e}(x'_{1,1;e}) T''_{j,-e}(E_i) + T''_{j,-e}(E_i) T''_{j,-e}(x'_{1,1;e})) \\ &= [2]_i^{-1}(-v_i^e E_i x_{1,1;e} + x_{1,1;e} E_i) = x_{1,2;e}. \end{aligned}$$

We set

$$\hat{x} = [3]_i^{-1}(-v_i^{-e} x_{1,2;e} x_{1,1;e} + x_{1,1;e} x_{1,2;e})$$

and

$$\hat{x}' = [3]_i^{-1}(-v_i^{-e} x'_{1,1;e} x'_{1,2;e} + x'_{1,2;e} x'_{1,1;e}).$$

We have

$$\begin{aligned} T'_{j,e}(x_{1,3;e}) &= [3]_i^{-1} T'_{j,e}(-v_i^{-e} E_i x_{1,2;e} + x_{1,2;e} E_i) \\ &= [3]_i^{-1}(-v_i^{-e} T'_{j,e}(E_i) T'_{j,e}(x_{1,2;e}) + T'_{j,e}(x_{1,2;e}) T'_{j,e}(E_i)) \\ &= [3]_i^{-1}(-v_i^{-e} x'_{1,1;e} x'_{1,2;e} + x'_{1,2;e} x'_{1,1;e}) = \hat{x}', \end{aligned}$$

$$\begin{aligned} T''_{j,-e}(x'_{1,3;e}) &= [3]_i^{-1} T''_{j,-e}(-v_i^{-e} x'_{1,2;e} E_i + E_i x'_{1,2;e}) \\ &= [3]_i^{-1}(-v_i^{-e} T''_{j,-e}(x'_{1,2;e}) T''_{j,-e}(E_i) + T''_{j,-e}(E_i) T''_{j,-e}(x'_{1,2;e})) \\ &= [3]_i^{-1}(-v_i^{-e} x_{1,2;e} x_{1,1;e} + x_{1,1;e} x_{1,2;e}) = \hat{x}, \end{aligned}$$

$$\begin{aligned}T'_{i,e}(\hat{x}') &= [3]_i^{-1}T'_{i,e}(-v_i^{-e}x'_{1,1;e}x'_{1,2;e}+x'_{1,2;e}x'_{1,1;e})\\ &= [3]_i^{-1}(-v_i^{-e}T'_{i,e}(x'_{1,1;e})T'_{i,e}(x'_{1,2;e})+T'_{i,e}(x'_{1,2;e})T'_{i,e}(x'_{1,1;e}))\\ &= [3]_i^{-1}(-v_i^{-e}x_{1,2;e}x_{1,1;e}+x_{1,1;e}x_{1,2;e})=\hat{x}.\end{aligned}$$

From the previous formulas we see that

(a) $$E_j \xrightarrow{T'_{i,e}} x_{1,3;e} \xrightarrow{T'_{j,e}} \hat{x}' \xrightarrow{T'_{i,e}} \hat{x} \xrightarrow{T'_{j,e}} x'_{1,3;e} \xrightarrow{T'_{i,e}} E_j$$

and

(b) $$E_i \xrightarrow{T'_{j,e}} x'_{1,1;e} \xrightarrow{T'_{i,e}} x_{1,2;e} \xrightarrow{T'_{j,e}} x'_{1,2;e} \xrightarrow{T'_{i,e}} x_{1,1;e} \xrightarrow{T'_{j,e}} E_i.$$

Thus, $T'_{i,e}T'_{j,e}T'_{i,e}T'_{j,e}T'_{i,e}(E_j)=E_j$ and $T'_{j,e}T'_{i,e}T'_{j,e}T'_{i,e}T'_{j,e}(E_i)=E_i$. Applying ω and using 37.2.4, we deduce that $T''_{i,e}T''_{j,e}T''_{i,e}T''_{j,e}T''_{i,e}(F_j)=F_j$ and $T''_{j,e}T''_{i,e}T''_{j,e}T''_{i,e}T''_{j,e}(F_i)=F_i$. Replacing e by $-e$ we deduce

$$T'_{i,e}T'_{j,e}T'_{i,e}T'_{j,e}T'_{i,e}(F_j)=F_j \qquad \text{and}$$

$$T'_{j,e}T'_{i,e}T'_{j,e}T'_{i,e}T'_{j,e}(F_i)=F_i.$$

From (a),(b), it follows that

$$T'_{j,e}T'_{i,e}T'_{j,e}T'_{i,e}T'_{j,e}T'_{i,e}(E_j)=T'_{j,e}E_j=-\tilde{K}_{ej}F_j,$$

$$T'_{i,e}T'_{j,e}T'_{i,e}T'_{j,e}T'_{i,e}T'_{j,e}(E_j)=T'_{i,e}T'_{j,e}T'_{i,e}T'_{j,e}T'_{i,e}(-\tilde{K}_{ej}F_j)=-\tilde{K}_{ej}F_j,$$

$$T'_{j,e}T'_{i,e}T'_{j,e}T'_{i,e}T'_{j,e}T'_{i,e}(E_i)=T'_{j,e}T'_{i,e}T'_{j,e}T'_{i,e}T'_{j,e}(-\tilde{K}_{ei}F_i)=-\tilde{K}_{ei}F_i,$$

$T'_{i,e}T'_{j,e}T'_{i,e}T'_{j,e}T'_{i,e}T'_{j,e}(E_i)=T'_{i,e}E_i=-\tilde{K}_{ei}F_i$. Thus, the automorphisms $T'_{i,e}T'_{j,e}T'_{i,e}T'_{j,e}T'_{i,e}T'_{j,e}$ and $T'_{j,e}T'_{i,e}T'_{j,e}T'_{i,e}T'_{j,e}T'_{i,e}$ coincide on the generators E_i,E_j; similarly, they coincide on the generators F_i,F_j and one checks easily that they coincide on each K_μ, hence are equal. Taking inverses, we see that the automorphisms

$$T''_{i,-e}T''_{j,-e}T''_{i,-e}T''_{j,-e}T''_{i,-e}T''_{j,-e} \qquad \text{and}$$

$$T''_{j,-e}T''_{i,-e}T''_{j,-e}T''_{i,-e}T''_{j,-e}T''_{i,-e}$$

are equal.

39.2.3. Assume that $\langle i,j'\rangle = -2, \langle j,i'\rangle = -1$, so that $h = 4$. Note that $v_j = v_i^2$. As in 37.2.1, we set

$$x_{1,m;e} = \sum_{r+s=m} (-1)^r v_i^{er(3-m)} E_i^{(r)} E_j E_i^{(s)} \in \mathbf{U}^+,$$
$$x'_{1,m;e} = \sum_{r+s=m} (-1)^r v_i^{er(3-m)} E_i^{(s)} E_j E_i^{(r)} \in \mathbf{U}^+.$$

Recall from 7.1.2 that

$$-v_i^{e(2-2m)} E_i x_{1,m;e} + x_{1,m;e} E_i = [m+1]_i x_{1,m+1;e}.$$

Applying σ, which interchanges $x_{1,m;e}$, and $x'_{1,m;e}$, we deduce that

$$-v_i^{e(2-2m)} x'_{1,m;e} E_i + E_i x'_{1,m;e} = [m+1]_i x'_{1,m+1;e}.$$

By 37.2.5, we have $T'_{i,e}E_j = x_{1,2;e}$ and $T''_{i,-e}E_j = x'_{1,2;e}$. By 37.2.5 with the roles of i,j interchanged, we have

$$T'_{j,e}(E_i) = -v_j^e E_j E_i + E_i E_j = x'_{1,1;e}$$

and similarly

$$T''_{j,-e}(E_i) = x_{1,1;e}.$$

By 37.2.5, we have $T''_{i,-e}(x_{1,1;e}) = x'_{1,1;e}$. We have

$$\begin{aligned} T''_{j,-e}(x'_{1,2;e}) &= [2]_i^{-1} T''_{j,-e}(-x'_{1,1;e}E_i + E_i x'_{1,1;e}) \\ &= [2]_i^{-1}(-T''_{j,-e}(x'_{1,1;e})T''_{j,-e}(E_i) + T''_{j,-e}(E_i)T''_{j,-e}(x'_{1,1;e})) \\ &= [2]_i^{-1}(-E_i x_{1,1;e} + x_{1,1;e}E_i) = x_{1,2;e}. \end{aligned}$$

From the previous formulas we see that

(a) $$E_j \xrightarrow{T'_{i,e}} x_{1,2;e} \xrightarrow{T'_{j,e}} x'_{1,2;e} \xrightarrow{T'_{i,e}} E_j$$

and

(b) $$E_i \xrightarrow{T'_{j,e}} x'_{1,1;e} \xrightarrow{T'_{i,e}} x_{1,1;e} \xrightarrow{T'_{j,e}} E_i.$$

Thus,

$$T'_{i,e}T'_{j,e}T'_{i,e}(E_j) = E_j \qquad \text{and}$$
$$T'_{j,e}T'_{i,e}T'_{j,e}(E_i) = E_i.$$

Applying ω and using 37.2.4, we deduce that

$$T''_{i,e}T''_{j,e}T''_{i,e}(F_j) = F_j \qquad \text{and}$$

$$T''_{j,e}T''_{i,e}T''_{j,e}(F_i) = F_i.$$

Replacing e by $-e$, we deduce that

$$T'_{i,e}T'_{j,e}T'_{i,e}(F_j) = F_j \qquad \text{and}$$

$$T'_{j,e}T'_{i,e}T'_{j,e}(F_i) = F_i.$$

From (a),(b), it follows that

$$T'_{j,e}T'_{i,e}T'_{j,e}T'_{i,e}(E_j) = T'_{j,e}E_j = -\tilde{K}_{ej}F_j,$$

$$T'_{i,e}T'_{j,e}T'_{i,e}T'_{j,e}(E_j) = T'_{i,e}T'_{j,e}T'_{i,e}(-\tilde{K}_{ej}F_j) = -\tilde{K}_{ej}F_j,$$

$$T'_{j,e}T'_{i,e}T'_{j,e}T'_{i,e}(E_i) = T'_{j,e}T'_{i,e}T'_{j,e}(-\tilde{K}_{ei}F_i) = -\tilde{K}_{ei}F_i,$$

$$T'_{i,e}T'_{j,e}T'_{i,e}T'_{j,e}(E_i) = T'_{i,e}E_i = -\tilde{K}_{ei}F_i.$$

Thus, the automorphisms $T'_{i,e}T'_{j,e}T'_{i,e}T'_{j,e}$ and $T'_{j,e}T'_{i,e}T'_{j,e}T'_{i,e}$ coincide on the generators E_i, E_j; similarly, they coincide on the generators F_i, F_j and one checks easily that they coincide on each K_μ, hence are equal. Taking inverses, we see that the automorphisms $T''_{i,-e}T''_{j,-e}T''_{i,-e}T''_{j,-e}$ and $T''_{j,-e}T''_{i,-e}T''_{j,-e}T''_{i,-e}$ are equal.

39.2.4. Assume that $\langle i,j'\rangle = \langle j,i'\rangle = -1$, so that $h = 3$. Note that $v_j = v_i$. As in 37.2.1, we set

$$x_{1,1;e} = \sum_{r+s=1} (-1)^r v_i^{er} E_i^{(r)} E_j E_i^{(s)} \in \mathbf{U}^+,$$

$$x'_{1,1;e} = \sum_{r+s=1} (-1)^r v_i^{er} E_i^{(s)} E_j E_i^{(r)} \in \mathbf{U}^+.$$

By 37.2.5, we have $T'_{i,e}(E_j) = x_{1,1;e}$ and $T''_{i,-e}(E_j) = x'_{1,1;e}$. By 37.2.5, with the roles of i,j interchanged, we have $T'_{j,e}(E_i) = x'_{1,1;e}$ and $T''_{j,-e}(E_i) = x_{1,1;e}$.

From the previous formulas we see that

(a) $$E_j \xrightarrow{T'_{i,e}} x_{1,1;e} \xrightarrow{T'_{j,e}} E_i,$$

and

(b) $$E_i \xrightarrow{T'_{j,e}} x'_{1,1;e} \xrightarrow{T'_{i,e}} E_j.$$

Thus,

$$T'_{j,e}T'_{i,e}(E_j) = E_i \qquad \text{and}$$

$$T'_{i,e}T'_{j,e}(E_i) = E_j.$$

Applying ω and using 37.2.4, we deduce that

$$T''_{j,e}T''_{i,e}(F_j) = F_i \qquad \text{and}$$

$$T''_{i,e}T''_{j,e}(F_i) = F_j.$$

Replacing e by $-e$, we deduce that

$$T'_{i,e}T'_{j,e}(F_i) = F_j \qquad \text{and}$$

$$T'_{j,e}T'_{i,e}(F_j) = F_i.$$

From (a), (b), it follows that

$$T'_{i,e}T'_{j,e}T'_{i,e}(E_j) = T'_{i,e}E_i = -\tilde{K}_{ei}F_i,$$

$$T'_{j,e}T'_{i,e}T'_{j,e}(E_j) = T'_{j,e}T'_{i,e}(-\tilde{K}_{ej}F_j) = -\tilde{K}_{ei}F_i,$$

$$T'_{i,e}T'_{j,e}T'_{i,e}(E_i) = T'_{i,e}T'_{j,e}(-\tilde{K}_{ei}F_i) = -\tilde{K}_{ej}F_j,$$

$$T'_{j,e}T'_{i,e}T'_{j,e}(E_i) = T'_{j,e}E_j = -\tilde{K}_{ej}F_j.$$

Thus, the automorphisms $T'_{j,e}T'_{i,e}T'_{j,e}$ and $T'_{i,e}T'_{j,e}T'_{i,e}$ coincide on the generators E_i, E_j; similarly, they coincide on the generators F_i, F_j and one checks easily that they coincide on each K_μ, hence are equal. Taking inverses, we see that the automorphisms $T''_{j,-e}T''_{i,-e}T''_{j,-e}$ and $T''_{i,-e}T''_{j,-e}T''_{i,-e}$ are equal.

39.2.5. Assume that $\langle i, j' \rangle = \langle j, i' \rangle = 0$, so that $h = 2$. By definition, we have

$$T'_{i,e}(E_j) = E_j, T''_{i,-e}(E_j) = E_j, T'_{j,e}(E_i) = E_i, T''_{j,-e}(E_i) = E_i,$$

and similar formulas with F_i, F_j instead of E_i, E_j. We have

$$T'_{j,e}T'_{i,e}(E_j) = T'_{j,e}(E_j) = -\tilde{K}_{ej}F_j,$$

$$T'_{i,e}T'_{j,e}(E_j) = T'_{i,e}(-\tilde{K}_{ej}F_j) = -\tilde{K}_{ej}F_j.$$

Thus, $T'_{j,e}T'_{i,e}(E_j) = T'_{i,e}T'_{j,e}(E_j)$. Similarly, $T'_{j,e}T'_{i,e}(F_j) = T'_{i,e}T'_{j,e}(F_j)$. Thus, $T'_{j,e}T'_{i,e}, T'_{i,e}T'_{j,e}$ coincide on E_j, F_j and, by symmetry, also on E_i, F_i;

they clearly coincide on K_μ; hence they are equal. Taking inverses, we see that the automorphisms $T''_{j,-e}T''_{i,-e}, T''_{i,-e}T''_{j,-e}$ are equal.

39.3. The Quantum Verma Identities

39.3.1. In this section, we preserve the assumptions of 39.2.1. From 39.2.2–39.2.5, we see that for any p such that $0 \le p \le h-1$, the following elements belong to $\mathbf{U}^+$ (each product contains p factors T' or T''):

$(\cdots T'_{i,e}T'_{j,e}T'_{i,e})(E_j)$; $(\cdots T'_{j,e}T'_{i,e}T'_{j,e})(E_i)$; $(\cdots T''_{i,-e}T''_{j,-e}T''_{i,-e})(E_j)$; $(\cdots T''_{j,-e}T''_{i,-e}T''_{j,-e})(E_i)$. It follows that the sequences $(i,j,i,...)$ and $(j,i,j,\dots)$, with h terms each, are admissible in the sense of 38.2.2.

Consider the following four sets of elements of $\mathbf{U}^+$; each element written below is a product of h elements of $\mathbf{U}^+$ and $(c_1,c_2,\dots,c_h)$ runs through $\mathbf{N}^h$:

(a) $\{E_i^{(c_1)}T'_{i,-1}(E_j^{(c_2)})T'_{i,-1}T'_{j,-1}(E_i^{(c_3)})\cdots\}$;

(b) $\{E_j^{(c_1)}T'_{j,-1}(E_i^{(c_2)})T'_{j,-1}T'_{i,-1}(E_j^{(c_3)})\cdots\}$;

(c) $\{E_i^{(c_1)}T''_{i,1}(E_j^{(c_2)})T''_{i,1}T''_{j,1}(E_i^{(c_3)})\cdots\}$;

(d) $\{E_j^{(c_1)}T''_{j,1}(E_i^{(c_2)})T''_{j,1}T''_{i,1}(E_j^{(c_3)})\cdots\}$.

Using 38.2.3 (with $x = 1$ and $p = n$ or $p = 0$) we see that each of the sets (a),(b) is an orthogonal set for an inner product on $\mathbf{U}^+$, hence each of the sets (a), (b) consists of linearly independent vectors. This implies, by the results in 37.2.4, that each of the sets (c), (d) consists of linearly independent vectors.

Lemma 39.3.2. *Each of the four sets (a)–(d) in 39.3.1 is a basis of the $\mathbf{Q}(v)$-vector space $\mathbf{U}^+$.*

These sets consist of homogeneous elements. The number $n(\nu)$ of elements in the intersection of one of these sets with $\mathbf{U}^+_\nu$ is the same for any of the four sets. By 39.3.1, it suffices to show that $\dim_{\mathbf{Q}(v)} \mathbf{U}^+_\nu \le n(\nu)$ for all ν. Using 33.1.3(b), we see that it suffices to show that

(a) $\dim_{\mathbf{Q}}({}_{\mathbf{Q}}\tilde{\mathbf{f}}_\nu) \le n(\nu)$ for all ν

(notations of 33.1.1; $\mathbf{Q}$ is regarded as an $\mathcal{A}$-algebra with $v \to 1$).

We shall examine the various cases separately. Assume first that $\langle i,j'\rangle = \langle j,i'\rangle = 0$. Then θ_i,θ_j commute; hence any word in θ_i,θ_j can be expressed in ${}_{\mathbf{Q}}\tilde{\mathbf{f}}$ as a linear combination of monomials $\theta_j^a\theta_i^b$.

Assume next that $\langle i,j'\rangle = -1, \langle j,i'\rangle = -1$ so that $h = 3$. We set $\theta_{ij} = -\theta_j\theta_i + \theta_i\theta_j \in {}_{\mathbf{Q}}\tilde{\mathbf{f}}$. Then θ_{ij} commutes with both θ_i and θ_j by the

Serre relations. Hence any word in θ_i, θ_j can be expressed in ${}_{\mathbf{Q}}\tilde{\mathbf{f}}$ as a linear combination of monomials $\theta_j^a \theta_{ij}^b \theta_i^c$.

Assume next that $\langle i, j' \rangle = -2, \langle j, i' \rangle = -1$ so that $h = 4$. We set $\theta_{ij} = -\theta_j \theta_i + \theta_i \theta_j, 2\theta_{iij} = -\theta_{ij}\theta_i + \theta_i \theta_{ij}$ (in ${}_{\mathbf{Q}}\tilde{\mathbf{f}}$). Using the Serre relations, we see that any word in θ_i, θ_j can be expressed in ${}_{\mathbf{Q}}\tilde{\mathbf{f}}$ as a linear combination of monomials $\theta_j^a \theta_{ij}^b \theta_{iij}^c \theta_i^d$. This is a special case of the Poincaré-Birkhoff-Witt theorem which can be checked directly.

Finally, assume that $\langle i, j' \rangle = -3, \langle j, i' \rangle = -1$ so that $h = 6$. We define the following elements in ${}_{\mathbf{Q}}\tilde{\mathbf{f}}$ inductively:

$$\theta_{ij} = -\theta_j \theta_i + \theta_i \theta_j, 2\theta_{iij} = -\theta_{ij}\theta_i + \theta_i \theta_{ij},$$

$$3\theta_{iiij} = -\theta_{iij}\theta_i + \theta_i \theta_{iij}, 3\theta_{iiijj} = -\theta_{ij}\theta_{iij} + \theta_{iij}\theta_{ij}.$$

Using the Serre relations, we see that any word in θ_i, θ_j can be expressed in ${}_{\mathbf{Q}}\tilde{\mathbf{f}}$ as a linear combination of monomials $\theta_j^a \theta_{ij}^b \theta_{iiijj}^c \theta_{iij}^d \theta_{iiij}^e \theta_i^f$. This is again a special case of the Poincaré-Birkhoff-Witt theorem which can be checked directly. The estimate (a) follows. The lemma is proved.

39.3.3. Remark. It is easy to see that $\mathbf{U}^+$ is a noetherian domain. Indeed, it is enough to show that ${}_{\mathbf{Q}}\tilde{\mathbf{f}}$ is a noetherian domain and for this we note that the associated graded ring for a suitable filtration is the algebra of (commutative) polynomials in finitely many variables.

39.3.4. Let J (resp. J') be the $\mathbf{Q}(v)$-subspace of $\mathbf{U}^+$ spanned by the elements in 39.3.1(b) such that $(c_2, c_3, \ldots, c_{h-1}) = (0, 0, \ldots, 0)$ (resp. such that $(c_2, c_3, \ldots, c_{h-1}) \neq (0, 0, \ldots, 0)$).

Lemma 39.3.5. *J' is a two-sided ideal of $\mathbf{U}^+$. We may identify canonically the quotient algebra $\mathbf{U}^+/J'$ with the $\mathbf{Q}(v)$-algebra defined by the generators E_i, E_j and the relation $E_i E_j = v_j^{-1} E_j E_i$.*

We identify $\mathbf{f} = \mathbf{U}^+$ via $x \mapsto x^+$. Then we may regard $(,)$ as a bilinear form on $\mathbf{U}^+$. From 38.2.3, we see that for any ν, each of the subspaces $J \cap \mathbf{U}_\nu^+$ and $J' \cap \mathbf{U}_\nu^+$ is the orthogonal of the other with respect to $(,)$. To check that J' is a left ideal, we must check that $E_i J' \subset J'$ (the inclusion $E_j J' \subset J'$ is obvious).

From the definition of ${}_i r$, we have that ${}_i r(E_j^a E_i^b)$ is a scalar multiple of $E_j^a E_i^{b-1}$. Hence ${}_i r(J) \subset J$. This implies, by taking orthogonals, that $E_i J' \subset J'$. Thus, J' is a left ideal. Similarly, J' is a right ideal. Hence the quotient algebra $\mathbf{U}^+/J'$ is well-defined. Clearly, J maps isomorphically onto it. Hence the products $E_j^a E_i^b$ with $a, b \in \mathbf{N}$ form a basis of $\mathbf{U}^+/J'$. It remains to observe that $E_i E_j = v_j^{-1} E_j E_i$ holds in $\mathbf{U}^+/J'$ since in $\mathbf{U}^+$, we have $-v_j^e E_j E_i + E_i E_j = T'_{j,e} E_i \in J'$. The lemma is proved.

39.3.6. Let $\lambda \in X^+$. We define two sequences $(a_1, a_2, \ldots, a_h)$ and $(b_1, b_2, \ldots, b_h)$ in $\mathbf{Z}^h$ by $a_1 = \langle \cdots s_j s_i s_j(i), \lambda \rangle, b_1 = \langle \cdots s_i s_j s_i(j), \lambda \rangle$; both products have $(h-1)$ factors s; $a_2 = \langle \cdots s_j s_i(j), \lambda \rangle, b_2 = \langle \cdots s_i s_j(i), \lambda \rangle$; both products have $(h-2)$ factors s, etc. Note that $a_p, b_p \in \mathbf{N}$, since $s_i s_j s_i \cdots$ and $s_j s_i s_j \cdots$ (h factors) are reduced expressions in W.

Let $x = F_i^{(a_1)} F_j^{(a_2)} F_i^{(a_3)} \ldots$ and $y = F_j^{(b_1)} F_i^{(b_2)} F_j^{(b_3)} \ldots$; both products have h factors F. Note that x, y belong to $\mathbf{U}_\nu^-$ where

$$\nu = a_1 i + a_2 j + a_3 i + \cdots = b_1 j + b_2 i + b_3 j + \cdots$$

Both sums have h terms. Then ν is given by $\langle 4i + 6j, \lambda \rangle i + \langle 2i + 4j, \lambda \rangle j$ if $h = 6$, $\langle 2i + 2j, \lambda \rangle i + \langle i + 2j, \lambda \rangle j$ if $h = 4$, $\langle i + j, \lambda \rangle i + \langle i + j, \lambda \rangle j$ if $h = 3$, $\langle i, \lambda \rangle i + \langle j, \lambda \rangle j$ if $h = 2$.

The following result is a quantum analogue of an identity of Verma [10].

Proposition 39.3.7. $x = y$.

The result is trivial when $h = 2$. We assume that $h \geq 3$.

For any $i_1 \in I$, both $E_{i_1} x$ and $E_{i_1} y$ belong to the left ideal of $\mathbf{U}$ generated by E_{i_1} and $(\tilde{K}_{-i_1} v_{i_1}^{\langle i_1, \lambda \rangle - 1} - \tilde{K}_{i_1} v_{i_1}^{-\langle i_1, \lambda \rangle + 1})$ (see 39.1.4). We may assume that Y is generated by i, j. Then there exists $\lambda' \in X$ such that $\langle i_1, \lambda' \rangle = \langle i_1, \lambda \rangle - 1$ for $i_1 \in I$. Applying x and y to the generator 1 of the Verma module $M_{\lambda'}$, we obtain vectors $x1, y1 \in M_{\lambda'}$ such that $E_{i_1}(x1) = 0, E_{i_1}(y1) = 0$ for $i_1 \in I$. Note that $x1, y1$ belong to the λ''-weight space of $M_{\lambda'}$, where $\lambda'' = \lambda' - \nu_i i' - \nu_j j'$. Hence there are well-defined morphisms of $\mathbf{U}$-modules $\beta, \gamma : M_{\lambda''} \to M_{\lambda'}$ which take the generator $1 \in M_{\lambda''}$ to $x1, y1$, respectively.

Using the explicit expression for ν_i, ν_j in 39.3.6, we see that $\langle i_1, \lambda'' \rangle = -\langle \tilde{i}_1, \lambda \rangle - 1$, where $\tilde{i}_1 = i_i$ if $h = 4$ or 6 and $\tilde{i} = j, \tilde{j} = i$ if $h = 3$. In particular, we have $\langle i_1, \lambda'' \rangle \leq -1$ for $i_1 \in I$. From 6.3.2, it then follows that the Verma module $M_{\lambda''}$ is simple.

Since x and y are non-zero (see 39.3.3), we have that $x1 \neq 0, y1 \neq 0$, hence β, γ must be isomorphisms onto their images. These images may be regarded as non-zero left principal ideals of $\mathbf{U}^-$, if we identify $M_{\lambda'}$ with $\mathbf{U}^-$ in the obvious way. By 39.3.3, $\mathbf{U}^+$, hence also $\mathbf{U}^-$, has the Ore property, hence these two ideals must have a non-zero intersection. This intersection is a non-zero submodule of the simple $\mathbf{U}$-module $\beta(M_{\lambda''})$, hence it coincides with it. Similarly, it coincides with $\gamma(M_{\lambda''})$. Thus, we have $\beta(M_{\lambda''}) = \gamma(M_{\lambda''})$. Since these two modules have the same (one-dimensional) λ''-weight space, it follows that $x1 = fy1$ for some $f \in \mathbf{Q}(v)-$

$\{0\}$. Thus, $x = fy$ in $\mathbf{U}^-$. We shall identify the algebras $\mathbf{U}^-$ and $\mathbf{U}^+$ via $F_i \to E_i, F_j \to E_j$.

We now apply to both sides of the equality $x = yf$ (in $\mathbf{U}^+$) the natural homomorphism $\mathbf{U}^+ \to \mathbf{U}^+/J'$ (see 39.3.5). Since in $\mathbf{U}^+ \to \mathbf{U}^+/J'$ we have $E_iE_j = v_j^{-1}E_jE_i$, we see that x, y are mapped to

$$v_j^s E_i^{(a_1+a_3+\cdots)} E_j^{(a_2+a_4+\cdots)}, v_j^t E_i^{(b_2+b_4+\cdots)} E_j^{(b_1+b_3+\cdots)}$$

respectively, where

$$s = \sum_{p<q} a_p a_q, t = \sum_{p'<q'} b_{p'} b_{q'}$$

and p is even, q is odd, p' is odd, q' is even. It follows that $v_j^s = f v_j^t$. From the definition of the sequences $(a_1, a_2, \dots, a_h), (b_1, b_2, \dots, b_h)$, we see that $s = t$, hence $f = 1$. The proposition is proved.

39.3.8. Remarks. (a) The use of non-commutative ring theory in the proof of the fact that the two left ideals considered above have non-zero intersection can be avoided as follows. Let $\nu' \in \mathbf{N}[I]$. The intersection of either of our left ideals with $\mathbf{U}^-_{\nu+\nu'}$ has dimension $n(\nu')$ (notation of 39.3.2); on the other hand, $\dim \mathbf{U}^-_{\nu+\nu'} = n(\nu+\nu')$. To show that the intersection is non-zero, it suffices therefore to show that $2n(\nu') > n(\nu+\nu')$ if ν' has large enough coordinates (here ν is fixed as in 39.3.6). But $n(\nu')$ is explicitly computable and the previous inequality is easily checked.

(b) In the simply laced case there is a much shorter proof of the quantum Verma identities (see 42.1.2(h)).

39.4. Proof of the Braid Group Relations

In the following lemma, we preserve the assumptions of 39.2.1.

Lemma 39.4.1. *Let M be any integrable $\mathbf{U}$-module. We have*

$$T'_{i,e}T'_{j,e}T'_{i,e}\cdots = T'_{j,e}T'_{i,e}T'_{j,e}\cdots : M \to M;$$

both products have h factors.

By the complete reducibility theorem 6.3.6, it suffices to prove the lemma in the case where $M = \Lambda_\lambda$ with $\lambda \in X^+$. Let $\eta \in \Lambda_\lambda$ be as in 3.5.7. Using 39.1.2, we see that

(a) $(T'_{i,e}T'_{j,e}T'_{i,e}\cdots)\eta = (F_i^{(a_1)}F_j^{(a_2)}F_i^{(a_3)}\cdots)\eta$ and

(b) $(T'_{j,e}T'_{i,e}T'_{j,e}\cdots)\eta = (F_j^{(b_1)}F_i^{(b_2)}F_j^{(b_3)}\cdots)\eta$
(all products have h factors) where $(a_1, a_2, \ldots, a_h), (b_1, b_2, \ldots, b_h)$ are as in 39.3.6.

By 39.3.7, the right hand sides of (a),(b) coincide. It follows that so do the left hand sides:

(c) $(T'_{i,e}T'_{j,e}T'_{i,e}\cdots)\eta = (T'_{j,e}T'_{i,e}T'_{j,e}\cdots)\eta.$

Let $u \in \mathbf{U}$. Using (c) and the equality in $\mathbf{U}$

$$(T'_{i,e}T'_{j,e}T'_{i,e}\cdots)(u) = (T'_{j,e}T'_{i,e}T'_{j,e}\cdots)(u)$$

(see 39.2) we see that

$$\begin{aligned}(T'_{i,e}T'_{j,e}T'_{i,e}\cdots)(u\eta) &= (T'_{i,e}T'_{j,e}T'_{i,e}\cdots)(u)((T'_{i,e}T'_{j,e}T'_{i,e}\cdots)\eta)\\ &= (T'_{j,e}T'_{i,e}T'_{j,e}\cdots)(u)((T'_{j,e}T'_{i,e}T'_{j,e}\cdots)\eta)\\ &= (T'_{j,e}T'_{i,e}T'_{j,e}\cdots)(u\eta).\end{aligned}$$

Since any vector in Λ_λ is of the form $u\eta$ for some $u \in \mathbf{U}$ we see that $T'_{i,e}T'_{j,e}T'_{i,e}\cdots = T'_{j,e}T'_{i,e}T'_{j,e}\cdots : M \to M$. The lemma is proved.

39.4.2. Now $(I, \cdot)$ is again an arbitrary Cartan datum.

Theorem 39.4.3. *For any $i \neq j$ in I such that $h = h(i,j) < \infty$ (see 2.1.1), we have the following equalities (all products have h factors):*

(a) $T'_{i,e}T'_{j,e}T'_{i,e}\cdots = T'_{j,e}T'_{i,e}T'_{j,e}\cdots,$

(b) $T''_{i,-e}T''_{j,-e}T''_{i,-e}\cdots = T''_{j,-e}T''_{i,-e}T''_{j,-e}\cdots$

as automorphisms of $\mathbf{U}$ *and*

(c) $T'_{i,e}T'_{j,e}T'_{i,e}\cdots = T'_{j,e}T'_{i,e}T'_{j,e}\cdots,$

(d) $T''_{i,-e}T''_{j,-e}T''_{i,-e}\cdots = T''_{j,-e}T''_{i,-e}T''_{j,-e}\cdots$

as linear maps $M \to M$ where M is any integrable $\mathbf{U}$*-module M.*

We prove (c). Let $\mathbf{U}'$ be the algebra defined like $\mathbf{U}$ by replacing I by $I' = \{i, j\}$ and keeping the other data $X, Y, \langle,\rangle, \ldots$ unchanged.

We may restrict M to an (integrable) $\mathbf{U}'$-module in an obvious way. Since (c) is true for this restriction, by 39.4.1, it is also true for the original M. Thus (c) is proved. (d) follows from (c) by taking inverses.

We prove (a). Let $u \in \mathbf{U}$. Let $u_1 = (T'_{i,e}T'_{j,e}T'_{i,e}\cdots)(u) \in \mathbf{U}$ and $u_2 = (T'_{j,e}T'_{i,e}T'_{j,e}\cdots)(u) \in \mathbf{U}$. For any integrable $\mathbf{U}$-module M, and any $m \in M$ we have (using (c) twice):

$$\begin{aligned}&u_1((T'_{j,e}T'_{i,e}T'_{j,e}\cdots)m) = u_1((T'_{i,e}T'_{j,e}T'_{i,e}\cdots)m)\\ &= (T'_{i,e}T'_{j,e}T'_{i,e}\cdots)(um) = (T'_{j,e}T'_{i,e}T'_{j,e}\cdots)(um)\\ &= u_2((T'_{j,e}T'_{i,e}T'_{j,e}\cdots)m).\end{aligned}$$

Since $T'_{j,e}T'_{i,e}T'_{j,e}\cdots : M \to M$ is an isomorphism, it follows that $u_1 - u_2$ acts as zero on M. Since M is arbitrary, it follows (see 3.5.4) that $u_1 = u_2$. This proves (a). In the same way we deduce (b) from (d). The theorem is proved.

39.4.4. Let $e = \pm 1$. Let $w \in W$. We define algebra isomorphisms $T''_{w,e} : \mathbf{U} \to \mathbf{U}$ and $T'_{w,e} : \mathbf{U} \to \mathbf{U}$ by

$$T''_{w,e} = T''_{i_1,e}T''_{i_2,e}\cdots T''_{i_N,e} \qquad \text{and}$$

$$T'_{w,e} = T'_{i_1,e}T'_{i_2,e}\cdots T'_{i_N,e},$$

where $s_{i_1}s_{i_2}\cdots s_{i_N}$ is a reduced expression of w. ($T''_{w,e}, T'_{w,e}$ are independent of the choice of reduced expression, by 2.1.2 and 39.4.3.) From the definition, we have

$$T'_{ww',e} = T'_{w,e}T'_{w',e}, T''_{ww',e} = T''_{w,e}T''_{w',e},$$

if $w, w' \in W$ satisfy $l(ww') = l(w)l(w')$. Thus we have four actions of the braid group on the algebra $\mathbf{U}$.

39.4.5. By 5.2.3 and 37.2.4, we have

$$T''_{w,e} = (T'_{w^{-1},-e})^{-1},\ \omega T'_{w,e}\omega = T''_{w,e} : \mathbf{U} \to \mathbf{U},\ \sigma T'_{w,e}\sigma = T''_{w,-e} : \mathbf{U} \to \mathbf{U}.$$

Let us define, for any linear map $P : \mathbf{U} \to \mathbf{U}$, a new linear map $\bar{P} : \mathbf{U} \to \mathbf{U}$ by $\overline{P(u)} = \bar{P}(\bar{u})$ for all $u \in \mathbf{U}$. With this notation, we have

$$\bar{T}'_{w,e} = T'_{w,-e}, \bar{T}''_{w,e} = T''_{w,-e}.$$

(see 37.2.4).

39.4.6. Let $u \in \mathbf{U}$ be such that, for any $i \in I$, we have $\tilde{K}_i u \tilde{K}_{-i} = v_i^{n_i} u$ for some integer n_i. For any $w \in W$, we have $T''_{w,e}(u) = (-1)^a v^b T'_{w,e}(u)$ where a, b are integers depending only on w and (n_i) but not on u. This follows from 37.2.4.

39.4.7. For any integrable $\mathbf{U}$-module M and any $w \in W$, we define $\mathbf{Q}(v)$-linear isomorphisms $T''_{w,e} : M \to M$ and $T'_{w,e} : M \to M$ by

$$T''_{w,e} = T''_{i_1,e}T''_{i_2,e}\cdots T''_{i_N,e}, T'_{w,e} = T'_{i_1,e}T'_{i_2,e}\cdots T'_{i_N,e},$$

where $s_{i_1}s_{i_2}\cdots s_{i_N}$ is a reduced expression of w. ($T''_{w,e}, T'_{w,e}$ are independent of the choice of reduced expression, by 2.1.2 and 39.4.3.)

From the definition, we have

$$T'_{ww',e} = T'_{w,e}T'_{w',e}, T''_{ww',e} = T''_{w,e}T''_{w',e},$$

if $w, w' \in W$ satisfy $l(ww') = l(w)l(w')$ (identities as maps $M \to M$). Thus, we have four actions of the braid group on M. We have $T''_{w,e} = (T'_{w^{-1},-e})^{-1} : M \to M$ for all w. For any $u \in \mathbf{U}, m \in M$, we have

$$T'_{w,e}(um) = T'_{w,e}(u)(T'_{w,e}m), T''_{w,e}(um) = T''_{w,e}(u)(T''_{w,e}m).$$

CHAPTER 40

Symmetries and $\mathbf{U}^+$

40.1. Preparatory Results

Lemma 40.1.1. *Assume that we are given $i \neq j$ in I. Let $h = h(i,j) \le \infty$ be as in 2.1.1. Let p be an integer such that $0 \le p < h$. We denote by $T''_{i,j;p}$ (resp. $T'_{i,j;p}$) the automorphism $\cdots T''_{i,1}T''_{j,1}T''_{i,1}$ (resp. $\cdots T'_{i,-1}T'_{j,-1}T'_{i,-1}$) of $\mathbf{U}$; both products have p factors. Then $T''_{i,j;p}(E_j)$ and $T''_{j,i;p}(E_i)$ belong to the subalgebra of $\mathbf{U}$ generated by E_i, E_j; $T'_{i,j;p}(E_j)$ and $T'_{j,i;p}(E_i)$ belong to the subalgebra of $\mathbf{U}$ generated by E_i, E_j.*

In the case where $h < \infty$, the lemma is contained in 39.2. In the rest of the proof we assume that $h = \infty$, or equivalently, that $\alpha\alpha' \ge 4$ where $\alpha = -\langle i, j' \rangle$ and $\alpha' = -\langle j, i' \rangle$. To symplify notation, we set

$x(i,j;m) = x_{i,j;1,m;-1}, \quad x(j,i;m) = x_{j,i;1,m;-1}$ and

$x'(i,j;m) = x'_{i,j;1,m;-1}, \quad x'(j,i;m) = x'_{j,i;1,m;-1}$

in $\mathbf{U}^+$ (see 37.2.1).

Let Z be the subalgebra of $\mathbf{U}^+$ generated by the two elements $x(i,j;\alpha)$ and $x(i,j;\alpha-1)$. Let Z' be the subalgebra of $\mathbf{U}^+$ generated by the two elements $x(j,i;\alpha')$ and $x(j,i;\alpha'-1)$. We prove by induction on m the following statement:

(a) if $m \ge 1$, then $T'_{j,-1}(x(i,j;m)) \in Z'$.

For $m = 1$, this follows from the computation

$$T'_{j,-1}(x(i,j;1)) = T'_{j,-1}(x'(j,i;1)) = x(j,i;\alpha'-1).$$

Assume now that $m > 1$. Applying $T'_{j,-1}$ to

$$-v_i^{-(\alpha-2m+2)}E_i x(i,j;m-1) + x(i,j;m-1)E_i = [m]_i x(i,j;m),$$

we obtain

$$\begin{aligned} &- v_i^{-(\alpha-2m+2)} x(j,i;\alpha')T'_{j,-1}(x(i,j;m-1)) \\ &+ T'_{j,-1}(x(i,j;m-1))x(j,i;\alpha') \\ &= [m]_i T'_{j,-1}(x(i,j;m)). \end{aligned}$$

G. Lusztig, *Introduction to Quantum Groups*, Modern Birkhäuser Classics,
DOI 10.1007/978-0-8176-4717-9_40, © Springer Science+Business Media, LLC 2010

This equality, together with the induction hypothesis, shows that $T'_{j,-1}(x(i,j;m)) \in Z'$; (a) is proved.

We have

(b) $T'_{j,-1}(Z) \subset Z'$, provided that $\alpha \geq 2$ and $T'_{i,-1}(Z') \subset Z$, provided that $\alpha' \geq 2$.

The first assertion of (b) follows from (a); the second one follows from the first by symmetry. We show by induction on $p \geq 1$ that

(c) if $\alpha \geq 2$ and $\alpha' \geq 2$, then $T'_{i,j;p}(E_j)$ is contained in Z' (if p is even) and in Z (if p is odd).

For $p = 1$, we have $T'_{i,j;1}(E_j) = T'_{i,-1}(E_j) = x(i,j;\alpha) \in Z$. The induction step is given by (b). This proves (c). In particular, under the assumptions of (c), $T'_{i,j;p}(E_j)$ belongs to the subalgebra of $\mathbf{U}$ generated by E_i, E_j; by symmetry, we have also that $T'_{j,i;p}(E_i)$ belongs to the subalgebra of $\mathbf{U}$ generated by E_i, E_j (for all p).

We next consider the case where one of α, α' is 1 and the other is ≥ 4. We may assume that $\alpha \geq 4, \alpha' = 1$. Let Z_1 be the subalgebra of $\mathbf{U}$ generated by the elements $x(i,j;\alpha), x(i,j;\alpha-1)$ and $x(i,j;\alpha-2)$. Let Z'_1 be the subalgebra of $\mathbf{U}$ generated by the two elements $x'(i,j;1)$ and $x'(i,j;2)$. From $T'_{i,-1}(x'(i,j;m)) = x(i,j;\alpha - m)$ we see that

(d) $T'_{i,-1}(Z'_1) \subset Z_1$.

We show by induction on m that

(e) $T'_{j,-1}(x(i,j;m)) \in Z'_1$ for any $m \geq 2$.

We have

$$T'_{j,-1}(x(i,j;1)) = T'_{j,-1}(x'(j,i;1)) = x(j,i;0) = E_i.$$

We also have

$$T'_{j,-1}(E_i) = x(j,i;1) = x'(i,j;1).$$

It follows that

$$\begin{aligned} T'_{j,-1}(x(i,j;2)) &= [2]_i^{-1} T'_{j,-1}(-v_i^{-(\alpha-2)} E_i x(i,j;1) + x(i,j;1)E_i) \\ &= [2]_i^{-1}(-v_i^{-(\alpha-2)} x'(i,j;1)E_i - E_i x'(i,j;1)) = x'(i,j;2). \end{aligned}$$

Thus $T'_{j,-1}(x(i,j;2)) \in Z'_1$ so that (e) holds for $m = 2$. We now assume that $m > 2$. Then

$$\begin{aligned} T'_{j,-1}(x(i,j;m)) &= [m]_i^{-1} T'_{j,-1}(-v_i^{-(\alpha-2m+2)} E_i x(i,j;m-1) \\ &\quad + x(i,j;m-1)E_i) \\ &= [m]_i^{-1}(-v_i^{-(\alpha-2m+2)} x'(i,j;1) T'_{j,-1}(x(i,j;m-1)) \\ &\quad + T'_{j,-1}(x(i,j;m-1)) x'(i,j;1)) \end{aligned}$$

and this is in Z_1', by the induction hypothesis. Thus (e) is proved.

We show by induction on $p \geq 1$ that

(f) $T'_{i,j;p}(E_j)$ is contained in Z_1' (if p is even) and in Z_1 (if p is odd); $T'_{j,i;p}(E_i)$ is contained in Z_1' (if p is odd) and in Z_1 (if p is even).

We have $T'_{i,j;1}(E_j) = T'_{i,-1}(E_j) = x(i,j;\alpha) \in Z_1$ and $T'_{j,i;1}(E_i) = T'_{j,-1}E_i = x(j,i;1) = x'(i,j;1) \in Z_1'$. Thus (f) holds for $p = 1$. The induction step follows from (d),(e). To apply (e), we use that $\alpha \geq 4$. In particular, $T'_{i,j;p}(E_j)$ and $T'_{j,i;p}(E_i)$ belong to the subalgebra of $\mathbf{U}$ generated by E_i, E_j. This completes the proof of the second assertion of the lemma. The proof of the first assertion is entirely similar.

Lemma 40.1.2. *Let $w \in W$ and let $i \in I$ be such that $l(ws_i) = l(w) + 1$. We have*

(a) $T''_{w,e}(E_i) \in \mathbf{U}^+$;

(b) $T'_{w,e}(E_i) \in \mathbf{U}^+$.

We prove (a) for $e = 1$ by induction on $l(w)$. When $l(w) = 1$, the result is trivial. Assume now that $l(w) > 1$. We can find $j \in I$ such that $l(ws_j) = l(w) - 1$. Note that $i \neq j$. By standard properties of Coxeter groups, there is a unique element $w' \in W$ such that $l(w's_j) = l(w')+1, l(w's_i) = l(w')+1$ and $w = w'y$ where either

(c) $y = s_i s_j s_i \cdots s_j$ ($p \geq 2$ factors) and $l(w) = l(w') + p, l(y) = p$ or

(d) $y = s_j s_i s_j \cdots s_j$ ($p \geq 1$ factors) and $l(w) = l(w') + p, l(y) = p$.

We have necessarily $p < h(i,j)$, since $l(ws_i) = l(w) + 1$. According to 40.1.1, $T''_{y,1}(E_i)$ belongs to the subalgebra of $\mathbf{U}$ generated by E_i, E_j. Then $T''_{w,1}(E_i) = T''_{w',1}T''_{y,1}(E_i)$ is in the subalgebra of $\mathbf{U}$ generated by $T''_{w',1}(E_i)$ and $T''_{w',1}(E_j)$. Since, by the induction hypothesis, we have $T''_{w',1}(E_i) \in \mathbf{U}^+$ and $T''_{w',1}(E_j) \in \mathbf{U}^+$, it follows that $T''_{w,1}(E_i) \in \mathbf{U}^+$. This proves (a) assuming that $e = 1$.

We apply $\bar{\ } : \mathbf{U} \to \mathbf{U}$ to $T''_{w,1}(E_i) \in \mathbf{U}^+$; using 39.4.5, we obtain $T''_{w,-1}(E_i) \in \mathbf{U}^+$. Thus, (a) is proved. The proof of (b) is entirely similar.

Proposition 40.1.3. *Let $\mathbf{h} = (i_1, i_2, \dots, i_n)$ be a sequence in I such that $s_{i_1} s_{i_2} \cdots s_{i_n}$ is a reduced expression for some $w \in W$.*

Then $T''_{i_1,e}T''_{i_2,e} \cdots T''_{i_{n-1},e}(E_{i_n}) \in \mathbf{U}^+$ and $T'_{i_1,e}T'_{i_2,e} \cdots T'_{i_{n-1},e}(E_{i_n}) \in \mathbf{U}^+$.

This follows immediately from the previous lemma.

40.2. The Subspace $\mathbf{U}^+(w,e)$ of $\mathbf{U}^+$

Proposition 40.2.1. *Let $w \in W$ and let $e = \pm 1$. Let $\mathbf{h} = (i_1, i_2, \ldots, i_n)$ be a sequence in I such that $s_{i_1} s_{i_2} \cdots s_{i_n}$ is a reduced expression for w.*

(a) *$\mathbf{h}$ is admissible (see 38.2.2).*

(b) *The elements $E_{i_1}^{(c_1)} T'_{i_1,e}(E_{i_2}^{(c_2)}) \cdots T'_{i_1,e} T'_{i_2,e} \cdots T'_{i_{n-1},e}(E_{i_n}^{(c_n)})$ (for various sequences $\mathbf{c} = (c_1, c_2, \ldots, c_n) \in \mathbf{N}^n$) form a basis for a subspace $\mathbf{U}^+(w,e)$ of $\mathbf{U}^+$ which does not depend on $\mathbf{h}$.*

(c) *The elements $E_{i_1}^{(c_1)} T''_{i_1,e}(E_{i_2}^{(c_2)}) \cdots T''_{i_1,e} T''_{i_2,e} \cdots T''_{i_{n-1},e}(E_{i_n}^{(c_n)})$ (for various sequences $\mathbf{c} = (c_1, c_2, \ldots, c_n) \in \mathbf{N}^n$) form a basis for the subspace $\mathbf{U}^+(w,e)$ of $\mathbf{U}^+$ defined in (b).*

(d) *Let $i \in I$ be such that $l(s_i w) = l(w) - 1$. Then $E_i \mathbf{U}^+(w,e) \subset \mathbf{U}^+(w,e)$.*

(a) follows from definitions using 40.1.3. We prove (b) assuming that $e = -1$. We shall regard $(,)$ as a pairing on $\mathbf{U}^+$, via the isomorphism $\mathbf{f} \to \mathbf{U}^+$ given by $x \mapsto x^+$. The fact that the set of vectors in (b) is linearly independent follows from the fact that it is an orthogonal set with respect to $(,)$ (we use (a) and 38.2.3, with $p = 0$ and $x = 1$). Let $\mathbf{U}^+(\mathbf{h}, e)$ be the subspace spanned by this set of vectors. To show that $\mathbf{U}^+(\mathbf{h}, e)$ depends only on w and not on $\mathbf{h}$, it suffices, by 2.1.2, to check the following statement: if $\mathbf{h}'$ is obtained from $\mathbf{h}$ by replacing h consecutive indices $i, j, i, \ldots$ in $\mathbf{h}$ by the h indices $j, i, j, \ldots$ (for some $i \neq j$ with $h = h(i,j) < \infty$), then $\mathbf{U}^+(\mathbf{h}, e) = \mathbf{U}^+(\mathbf{h}', e)$. Using the fact that the T''s are algebra automorphisms of $\mathbf{U}$ satisfying the braid relations, we see that the last equality would be a consequence of the analogous equality in the case where I is replaced by $\{i, j\}$. But in that case, the desired equality holds by 39.3.2; both sides are equal to $\mathbf{U}^+$. This proves (b) for $e = -1$. Now (b) for $e = 1$ follows from (b) for $e = -1$ by applying $^- : \mathbf{U}^+ \to \mathbf{U}^+$ (see 39.4.5). Using 39.4.6, we see that (c) follows from (b). We now prove (d). If i is as in (d), then we can find a sequence $\mathbf{h} = (i_1, i_2, \ldots, i_n)$ in I such that $i_1 = i$ and $s_{i_1} s_{i_2} \cdots s_{i_n}$ is a reduced expression for w. From the definitions, it is clear that for this $\mathbf{h}$, we have $E_i \mathbf{U}^+(\mathbf{h}, e) \subset \mathbf{U}^+(\mathbf{h}, e)$; (d) follows.

Corollary 40.2.2. *Assume that the Cartan datum is of finite type. Then $\mathbf{U}^+(w_0, e) = \mathbf{U}^+$. Hence, given $e = \pm 1$ and a sequence $\mathbf{h} = (i_1, i_2, \ldots, i_n)$ in I such that $s_{i_1} s_{i_2} \cdots s_{i_n}$ is a reduced expression for w_0, the vectors*

$$E_{i_1}^{(c_1)} T'_{i_1,e}(E_{i_2}^{(c_2)}) \cdots T'_{i_1,e} T'_{i_2,e} \cdots T'_{i_{n-1},e}(E_{i_n}^{(c_n)})$$

(for various sequences $\mathbf{c} = (c_1, c_2, \ldots, c_n) \in \mathbf{N}^n$) form a basis for the $\mathbf{Q}(v)$-vector space $\mathbf{U}^+$; moreover, the vectors

$$E_{i_1}^{(c_1)} T''_{i_1,e}(E_{i_2}^{(c_2)}) \cdots T''_{i_1,e} T''_{i_2,e} \cdots T''_{i_{n-1},e}(E_{i_n}^{(c_n)})$$

(for various sequences $\mathbf{c} = (c_1, c_2, \ldots, c_n) \in \mathbf{N}^n$) form a basis for the $\mathbf{Q}(v)$-vector space $\mathbf{U}^+$.

For any $i \in I$, we have $l(s_i w_0) = l(w_0) - 1$, hence $E_i \mathbf{U}^+(w_0, e) \subset \mathbf{U}^+(w_0, e)$ (see 40.2.1(d)). Thus, $\mathbf{U}^+(w_0, e)$ is a left ideal of $\mathbf{U}^+$. It clearly contains 1, hence it is equal to $\mathbf{U}^+$.

40.2.3. It would be interesting to find an extension of the statement of 40.2.2 to the case of an affine Cartan datum $(I, \cdot)$. (For a result in this direction, in the case of affine SL_2, see [1], [4].)

The following approach may be useful in finding such an extension. Assume that we are given a sequence $\mathbf{h} = (\ldots, i_{-2}, i_{-1}, i_0, i_1, i_2, \ldots)$ of elements in I (infinite in both directions) such that for any integers $a \le b$, the product $s_{i_a} s_{i_{a+1}} \cdots s_{i_b}$ has length $(b - a + 1)$ in W. Infinite sequences like $\mathbf{h}$ above are known to exist for Cartan data of affine type.

We also assume as given an integer p. Let P be the set of all $x \in \mathbf{f}$ such that $T''_{i_r,1} T''_{i_{r-1},1} \cdots T''_{i_{p+1},1} x^+ \in \mathbf{U}^+$ for all $r \ge p+1$ and $T'_{i_s,-1} T'_{i_{s+1},-1} \cdots T'_{i_p,-1} x^+ \in \mathbf{U}^+$ for all $s \le p$.

For any sequence $\mathbf{c} = (\ldots, c_{-2}, c_{-1}, c_0, c_1, c_2, \ldots)$ of numbers in $\mathbf{N}$ (infinite in both directions and such that $c_n = 0$ for all but finitely many n) we define $L(\mathbf{h}, \mathbf{c}, p, x) \in \mathbf{f}$ by the following formula

$$L(\mathbf{h}, \mathbf{c}, p, x)^+ = (E_{i_{p+1}}^{(c_{p+1})} T'_{i_{p+1},-1}(E_{i_{p+2}}^{(c_{p+2})}) T'_{i_{p+1},-1} T'_{i_{p+2},-1}(E_{i_{p+3}}^{(c_{p+3})}) \cdots)$$
$$\times\, x^+ (\cdots T''_{i_p,1} T''_{i_{p-1},1}(E_{i_{p-2}}^{(c_{p-2})}) T''_{i_p,1}(E_{i_{p-1}}^{(c_{p-1})}) E_{i_p}^{(c_p)}).$$

This is well-defined, since the factors on the left and on the right of x^+ belong to $\mathbf{U}^+$, by 40.1.3. Now let $\mathbf{c}' = (\ldots, c'_{-2}, c'_{-1}, c'_0, c'_1, c'_2, \ldots)$ be another sequence like $\mathbf{c}$ and let $x' \in P$.

The following result is a consequence of 38.2.3.

Proposition 40.2.4. *We have the equality of inner products*

$$(L(\mathbf{h}, \mathbf{c}, p, x), L(\mathbf{h}, \mathbf{c}', p, x')) = (x, x') \prod_{s \in \mathbf{Z}} (\theta_{i_s}^{(c_s)}, \theta_{i_s}^{(c'_s)}).$$

40.2.5. Let $\mathbf{U}^+(>)$ (resp. $\mathbf{U}^+(<)$) be the subspace of $\mathbf{U}^+$ spanned by the elements $E_{i_{p+1}}^{(c_{p+1})} T'_{i_{p+1},-1}(E_{i_{p+2}}^{(c_{p+2})}) T'_{i_{p+1},-1} T'_{i_{p+2},-1}(E_{i_{p+3}}^{(c_{p+3})}) \cdots$ (resp. by the elements $\cdots T''_{i_p,1}T''_{i_{p-1},1}(E_{i_{p-2}}^{(c_{p-2})}T''_{i_p,1}(E_{i_{p-1}}^{(c_{p-1})})E_{i_p}^{(c_p)})$ for various sequences $(c_{p+1}, c_{p+2}, \dots)$ in $\mathbf{N}$, with $c_n = 0$ for large n (resp. for various sequences $(c_p, c_{p-1}, \dots)$ in $\mathbf{N}$, with $c_{-n} = 0$ for large n.)

The previous proposition implies that the natural map

(a) $\mathbf{U}^+(>) \otimes P \otimes \mathbf{U}^+(<) \to \mathbf{U}^+$

given by multiplication is injective. It would be interesting to show that, in the affine case, the map (a) is an isomorphism and to describe explicitly the space P.

CHAPTER 41

Integrality Properties of the Symmetries

41.1 Braid Group Action on $\dot{\mathbf{U}}$

41.1.1. Let $e = \pm 1$ and let $i \in I$. The symmetry $T'_{i,e} : \mathbf{U} \to \mathbf{U}$ (resp. $T''_{i,e} : \mathbf{U} \to \mathbf{U}$) induces for each λ', λ'' a linear isomorphism ${}_{\lambda'}\mathbf{U}_{\lambda''} \to {}_{s_i(\lambda')}\mathbf{U}_{s_i(\lambda'')}$ (notation of 23.1.1; $s_i : X \to X$ is as in 2.2.6). Taking direct sums, we obtain an algebra automorphism $T'_{i,e} : \dot{\mathbf{U}} \to \dot{\mathbf{U}}$ (resp. $T''_{i,e} : \dot{\mathbf{U}} \to \dot{\mathbf{U}}$) such that $T'_{i,e}(1_\lambda) = 1_{s_i(\lambda)}$ (resp. $T''_{i,e}(1_\lambda) = 1_{s_i(\lambda)}$) for all λ and $T'_{i,e}(uxx'u') = T'_{i,e}(u)T'_{i,e}(x)T'_{i,e}(x')T'_{i,e}(u')$ (resp. $T''_{i,e}(uxx'u') = T''_{i,e}(u)T''_{i,e}(x)T''_{i,e}(x')T''_{i,e}(u')$) for all $u, u' \in \mathbf{U}$ and $x, x' \in \dot{\mathbf{U}}$. Then $T'_{i,e}$ is an automorphism of the algebra $\dot{\mathbf{U}}$ with inverse $T''_{i,-e}$. These automorphisms satisfy braid group relations just like those of $\mathbf{U}$.

41.1.2. From the formulas in 37.1.3, we deduce that

$$T'_{i,e}(E_i^{(n)}1_\lambda) = (-1)^n v_i^{-en(n+\langle i,\lambda\rangle+1)} F_i^{(n)} 1_{s_i(\lambda)};$$

$$T'_{i,e}(F_i^{(n)}1_\lambda) = (-1)^n v_i^{-en(n-\langle i,\lambda\rangle-1)} E_i^{(n)} 1_{s_i(\lambda)};$$

$$T'_{i,e}(E_j^{(n)}1_\lambda) = \sum_{r+s=-\langle i,j'\rangle n} (-1)^r v_i^{er} E_i^{(r)} E_j^{(n)} E_i^{(s)} 1_{s_i(\lambda)} \text{ for } j \neq i;$$

$$T'_{i,e}(F_j^{(n)}1_\lambda) = \sum_{r+s=-\langle i,j'\rangle n} (-1)^r v_i^{-er} F_i^{(s)} F_j^{(n)} F_i^{(r)} 1_{s_i(\lambda)} \text{ for } j \neq i;$$

$$T''_{i,-e}(E_i^{(n)}1_\lambda) = (-1)^n v_i^{en(n+\langle i,\lambda\rangle-1)} F_i^{(n)} 1_{s_i(\lambda)};$$

$$T''_{i,-e}(F_i^{(n)}1_\lambda) = (-1)^n v_i^{en(n-\langle i,\lambda\rangle+1)} E_i^{(n)} 1_{s_i(\lambda)};$$

$$T''_{i,-e}(E_j^{(n)}1_\lambda) = \sum_{r+s=-\langle i,j'\rangle n} (-1)^r v_i^{er} E_i^{(s)} E_j^{(n)} E_i^{(r)} 1_{s_i(\lambda)} \text{ for } j \neq i;$$

$$T''_{i,-e}(F_j^{(n)}1_\lambda) = \sum_{r+s=-\langle i,j'\rangle n} (-1)^r v_i^{-er} F_i^{(r)} F_j^{(n)} F_i^{(s)} 1_{s_i(\lambda)} \text{ for } j \neq i.$$

It follows that $T'_{i,e}, T''_{i,e}$ restrict to $\mathcal{A}$-algebra automorphisms ${}_{\mathcal{A}}\dot{\mathbf{U}} \to {}_{\mathcal{A}}\dot{\mathbf{U}}$. They take 1_λ to $1_{s_i(\lambda)}$ for any $\lambda \in X$.

G. Lusztig, *Introduction to Quantum Groups*, Modern Birkhäuser Classics,
DOI 10.1007/978-0-8176-4717-9_41, © Springer Science+Business Media, LLC 2010

The following result is an integral version of 40.1.3.

Proposition 41.1.3. *Let* $\mathbf{h} = (i_1, i_2, \dots, i_n)$ *be a sequence in* I *such that* $s_{i_1} s_{i_2} \cdots s_{i_n}$ *is a reduced expression for some* $w \in W$*; let* $t \in \mathbf{Z}$*. Then*

(a) $T''_{i_1,e} T''_{i_2,e} \cdots T''_{i_{n-1},e}(E_{i_n}^{(t)}) \in {}_{\mathcal{A}}\mathbf{U}^+$;

(b) $T'_{i_1,e} T'_{i_2,e} \cdots T'_{i_{n-1},e}(E_{i_n}^{(t)}) \in {}_{\mathcal{A}}\mathbf{U}^+$.

Let u be the left hand side of (a). Let $\zeta' \in X$; define $\zeta \in X$ by $\zeta = s_{i_1} s_{i_2} \cdots s_{i_{n-1}}(\zeta')$. We have $u 1_\zeta = T''_{i_1,e} T''_{i_2,e} \cdots T''_{i_{n-1},e}(E_{i_n}^{(t)} 1_{\zeta'})$. Hence, by 41.1.2, we have $u 1_\zeta \in {}_{\mathcal{A}}\dot{\mathbf{U}}$. On the other hand, by 40.1.3, we have $u \in \mathbf{U}^+$. Thus, to prove (a), it suffices to prove the following statement: if $x \in \mathbf{f}$ and $\zeta \in X$ satisfy $x^+ 1_\zeta \in {}_{\mathcal{A}}\dot{\mathbf{U}}$, then $x \in {}_{\mathcal{A}}\mathbf{f}$. This follows immediately from 23.2.2. The proof of (b) is entirely similar.

The following result is an integral version of 40.2.1.

Proposition 41.1.4. *Let* $w \in W$ *and let* $e = \pm 1$*. Let* $\mathbf{h} = (i_1, i_2, \dots, i_n)$ *be a sequence in* I *such that* $s_{i_1} s_{i_2} \cdots s_{i_n}$ *is a reduced expression for* w*. Then*

(a) *the elements* $E_{i_1}^{(c_1)} T'_{i_1,e}(E_{i_2}^{(c_2)}) \cdots T'_{i_1,e} T'_{i_2,e} \cdots T'_{i_{n-1},e}(E_{i_n}^{(c_n)})$ *(for various sequences* $\mathbf{c} = (c_1, c_2, \dots, c_n) \in \mathbf{N}^n$*) form an* $\mathcal{A}$*-basis for an* $\mathcal{A}$*-submodule* ${}_{\mathcal{A}}\mathbf{U}^+(w,e)$ *of* $\mathbf{U}^+(w,e)$ *which does not depend on* $\mathbf{h}$*;*

(b) *the elements* $E_{i_1}^{(c_1)} T''_{i_1,e}(E_{i_2}^{(c_2)}) \cdots T''_{i_1,e} T''_{i_2,e} \cdots T''_{i_{n-1},e}(E_{i_n}^{(c_n)})$ *(for various sequences* $\mathbf{c} = (c_1, c_2, \dots, c_n) \in \mathbf{N}^n$*) form an* $\mathcal{A}$*-basis for* ${}_{\mathcal{A}}\mathbf{U}^+(w,e)$ *in (a).*

(c) *Let* $i \in I$ *be such that* $l(s_i w) = l(w) - 1$ *and let* $t \in \mathbf{Z}$*. Then* $E_i^{(t)} {}_{\mathcal{A}}\mathbf{U}^+(w,e) \subset {}_{\mathcal{A}}\mathbf{U}^+(w,e)$.

Using the method of 40.2.1, we see that it suffices to prove (a) in the case where I consists of two elements i, j and $h(i,j) < \infty$. In that case, the result follows from the analysis in [7]. (If $i \cdot i = j \cdot j$, this can also be deduced from Lemma 42.1.2.)

41.1.5. With the notations of 41.1.4, let $\theta_{\mathbf{h}}^{\mathbf{c}} \in {}_{\mathcal{A}}\mathbf{f}$ be the element corresponding to $E_{i_1}^{(c_1)} T'_{i_1,-1}(E_{i_2}^{(c_2)}) \cdots T'_{i_1,-1} T'_{i_2,-1} \cdots T'_{i_{n-1},-1}(E_{i_n}^{(c_n)})$ under the isomorphism $\mathbf{f} \to \mathbf{U}^+$ given by $x \mapsto x^+$.

Proposition 41.1.6. *Let* $w, n, \mathbf{h}, \mathbf{c}$ *be as in 41.1.4. Let* $\pi : \mathcal{L}(\mathbf{f}) \to \mathcal{L}(\mathbf{f})/v^{-1}\mathcal{L}(\mathbf{f})$ *be the canonical projection. We have* $\theta_{\mathbf{h}}^{\mathbf{c}} \in \mathcal{L}(\mathbf{f})$ *and there is a unique element* b *of the canonical basis* $\mathbf{B}$ *such that* $\pi(\theta_{\mathbf{h}}^{\mathbf{c}}) = \pm\pi(b)$.

From 38.2.3, we have $(\theta_{\mathbf{h}}^{\mathbf{c}}, \theta_{\mathbf{h}}^{\mathbf{c}}) \in 1 + v^{-1}\mathbf{Z}[[v^{-1}]] \cap \mathbf{Q}(v)$. This implies the proposition by 14.2.2(a).

Proposition 41.1.7. *Assume that the Cartan datum is of finite type. Then*

(a) ${}_{\mathcal{A}}\mathbf{U}^+(w_0, e) = {}_{\mathcal{A}}\mathbf{U}^+$.

This follows from 41.1.4 in the same way as 40.2.2 follows from 40.2.1.

41.1.8. Braid group action on ${}_R\dot{\mathbf{U}}$. Let R be as in 31.1.1. Let $e = \pm 1$. For any $i \in I$, the $\mathcal{A}$-algebra automorphism $T'_{i,e} : {}_{\mathcal{A}}\dot{\mathbf{U}} \to {}_{\mathcal{A}}\dot{\mathbf{U}}$ (resp. $T''_{i,e} : {}_{\mathcal{A}}\dot{\mathbf{U}} \to {}_{\mathcal{A}}\dot{\mathbf{U}}$) induces, upon tensoring with R, an R-algebra automorphism $T'_{i,e} : {}_R\dot{\mathbf{U}} \to {}_R\dot{\mathbf{U}}$ (resp. $T''_{i,e} : {}_R\dot{\mathbf{U}} \to {}_R\dot{\mathbf{U}}$). These automorphisms satisfy the braid group relations on ${}_R\dot{\mathbf{U}}$ just like they did over $\mathbf{Q}(v)$ (this holds over for $\mathcal{A}$ by reduction to $\mathbf{Q}(v)$, since ${}_{\mathcal{A}}\dot{\mathbf{U}}$ is imbedded in $\dot{\mathbf{U}}$, and then it holds in general by change of rings from $\mathcal{A}$ to R). Similarly, we have ${T'_{i,e}}^{-1} = T''_{i,-e}$ as automorphisms of ${}_R\dot{\mathbf{U}}$.

41.1.9. Braid group action and the quantum Frobenius homomorphism. Let R be as in 35.1.3. In the setup of 35.1.9, the homomorphism $Fr : {}_R\dot{\mathbf{U}} \to {}_R\dot{\mathbf{U}}^*$ is compatible with the braid group actions on ${}_R\dot{\mathbf{U}}$ and ${}_R\dot{\mathbf{U}}^*$. The proof is by checking on generators.

41.2. Braid Group Action on Integrable ${}_R\dot{\mathbf{U}}$-Modules

41.2.1. In the following proposition we assume that the root datum is Y-regular and we consider $\lambda, \lambda' \in X^+$.

Proposition 41.2.2. *The symmetries $T'_{i,e}, T''_{i,e}$ of the $\dot{\mathbf{U}}$-module ${}^\omega\Lambda_\lambda \otimes \Lambda_{\lambda'}$ map the ${}_{\mathcal{A}}\dot{\mathbf{U}}$-submodule ${}^\omega_{\mathcal{A}}\Lambda_\lambda \otimes_{\mathcal{A}} ({}_{\mathcal{A}}\Lambda_{\lambda'})$ into itself.*

Let $m \in {}^\omega_{\mathcal{A}}\Lambda_\lambda \otimes_{\mathcal{A}} ({}_{\mathcal{A}}\Lambda_{\lambda'})$. By definition (see 5.2.1), the vector $T'_{i,e}(m)$ is given by a sum of infinitely many terms such that all but a finite number of terms (depending on m) are zero. The finitely many terms that can be non-zero are of the form um where $u \in {}_{\mathcal{A}}\dot{\mathbf{U}}$. They belong to ${}^\omega_{\mathcal{A}}\Lambda_\lambda \otimes_{\mathcal{A}} ({}_{\mathcal{A}}\Lambda_{\lambda'})$ since this is an ${}_{\mathcal{A}}\dot{\mathbf{U}}$-submodule. Thus this submodule is stable under $T'_{i,e}$. The same argument shows that it is stable under $T''_{i,e}$. The proposition is proved.

41.2.3. Let R be as in 31.1.1. Let M be an integrable object in ${}_R\mathcal{C}$. We define R-linear maps $T'_{i,e} : M \to M$ and $T''_{i,e} : M \to M$ by the formulas in 5.2.1, in which v is regarded as an element of R, by the $\mathcal{A}$-algebra structure on R. It is clear that these operators are well-defined.

Proposition 41.2.4. (a) *The operators $T'_{i,e} : M \to M$ satisfy the braid group relations. The same holds for the operators $T''_{i,e} : M \to M$.*

(b) *We have ${T'_{i,e}}^{-1} = T''_{i,-e}$ as operators $M \to M$.*

(c) *For any $u \in {}_R\dot{\mathbf{U}}$ and any $m \in M$, we have $T'_{i,e}(um) = T'_{i,e}(u)T'_{i,e}(m)$ and $T''_{i,e}(um) = T''_{i,e}(u)T''_{i,e}(m)$.*

Using the functor in 31.1.12 in the case where $(Y', X', \dots)$ is the simply connected root datum of type $(I, \cdot)$, we can reduce the general case to the case where the root datum is simply connected, hence Y-regular. In that case, using the characterization of integrable objects given in 31.2.7, we are reduced to the special case where $M = {}^{\omega}_{R}\Lambda_\lambda \otimes_R ({}_R\Lambda_{\lambda'})$ with $\lambda, \lambda' \in X^+$. Indeed, suppose that M is a sum of ${}_R\dot{\mathbf{U}}$-submodules M_α and that the proposition holds for each M_α. Then it clearly holds for M. Suppose that M is a quotient of an integrable object M' such that the proposition holds for M'; then it clearly holds for M.

In this special case, the proposition follows by change of scalars from the case $R = \mathcal{A}$ which in turn follows from the already known case where $R = \mathbf{Q}(v)$. The proposition is proved.

CHAPTER 42

The *ADE* Case

42.1. Combinatorial Description of the Left Colored Graph

42.1.1. In this chapter we assume that the Cartan datum is simply laced and of finite type.

Lemma 42.1.2. *Consider the $\mathbf{Q}(v)$-algebra with generators α, β and relations $\alpha^2\beta - (v+v^{-1})\alpha\beta\alpha + \beta\alpha^2 = 0, \quad \beta^2\alpha - (v+v^{-1})\beta\alpha\beta + \alpha\beta^2 = 0$. Set $\gamma = \alpha\beta - v^{-1}\beta\alpha$. For $x = \alpha, \beta$ or γ, and $n \ge 0$, we set $x^{(n)} = x^n/[n]^!$; for $n < 0$, we set $x^{(n)} = 0$. We have*

(a) $\alpha\gamma = v\gamma\alpha, \quad v\beta\gamma = \gamma\beta$;

(b) $\alpha^{(p)}\beta^{(q)} = \sum_n v^{-(p-n)(q-n)}\beta^{(q-n)}\gamma^{(n)}\alpha^{(p-n)}$;

(c) $\gamma^{(m)} = \sum_{j'+j''=m}(-1)^{j'}v^{-j'}\beta^{(j')}\alpha^{(m)}\beta^{(j'')}$;

(d) $\alpha^{(p)}\beta^{(q)}\alpha^{(r)} = \sum_{m,n\ge 0; m+n=p+r-q}\begin{bmatrix} m+n \\ m \end{bmatrix}\beta^{(r-m)}\alpha^{(p+r)}\beta^{(p-n)}$, *if* $p+r \ge q$;

(e) $\beta^{(p)}\alpha^{(q)}\beta^{(r)} = \sum_{m,n\ge 0; m+n=p+r-q}\begin{bmatrix} m+n \\ m \end{bmatrix}\alpha^{(r-m)}\beta^{(p+r)}\alpha^{(p-n)}$, *if* $p+r \ge q$;

(f) $\alpha^{(p)}\beta^{(q)}\alpha^{(r)} = \sum_{n; n\le p} v^{-(p-n)(q-n)}\begin{bmatrix} p-n+r \\ p-n \end{bmatrix}\beta^{(q-n)}\gamma^{(n)}\alpha^{(p-n+r)}$;

(g) $\beta^{(p)}\alpha^{(q)}\beta^{(r)} = \sum_{n; n\le r} v^{-(q-n)(r-n)}\begin{bmatrix} r-n+p \\ r-n \end{bmatrix}\beta^{(r-n+p)}\gamma^{(n)}\alpha^{(q-n)}$;

(h) $\alpha^{(p)}\beta^{(p+r)}\alpha^{(r)} = \beta^{(r)}\alpha^{(p+r)}\beta^{(p)}$.

(a) is obvious.

Now (b) is obvious when $p \le 0$ or $q \le 0$. For $q = 1$, (b) states that $\alpha^{(p)}\beta = v^{-p}\beta\alpha^{(p)} + \gamma\alpha^{(p-1)}$; this is proved by induction on $p \ge 1$, using (a). Assume now that $q \ge 2$ and that (b) is known when q is replaced by $q-1$. We write (b) for $(p, q-1)$ and multiply it on the right by β. Using the case $q = 1$, we substitute

$$\beta^{(q-1-n)}\gamma^{(n)}\alpha^{(p-n)}\beta = \beta^{(q-1-n)}\gamma^{(n)}(v^{-p+n}\beta\alpha^{(p-n)} + \gamma\alpha^{(p-n-1)}).$$

This can be rearranged using (a) and yields (b) for (p,q). Thus (b) is proved.

To prove (c), we replace $\alpha^{(m)}\beta^{(j'')}$ in the right hand side of (c) by the expression provided by (b); we perform cancellations, and we obtain (c).

G. Lusztig, *Introduction to Quantum Groups*, Modern Birkhäuser Classics,
DOI 10.1007/978-0-8176-4717-9_42, © Springer Science+Business Media, LLC 2010

To prove (d), we replace $\alpha^{(p)}\beta^{(q)}$ on the left hand side and $\alpha^{(p+r)}\beta^{(p-n)}$ on the right hand side by the expressions provided by (b); we perform cancellations, and we obtain (d). Now (e) follows from (d) by symmetry; (f) and (g) follow immediately from (b) and (h) is a special case of either (d) or (e). Note that (h) is a special case of the quantum Verma identity 39.3.7.

42.1.3. Let $\mathbf{H}$ be the set of all sequences $\mathbf{h} = (i_1, i_2, \dots, i_n)$ in I such that $s_{i_1}s_{i_2}\cdots s_{i_n}$ is a reduced expression for w_0. (Thus, $n = l(w_0)$.)

We shall regard $\mathbf{H}$ as the set of vertices of a graph in which $\mathbf{h} = (i_1, i_2, \dots, i_n)$ and $\mathbf{h}' = (j_1, j_2, \dots, j_n)$ are joined if $\mathbf{h}'$ is obtained from $\mathbf{h}$ by

(a) replacing three consecutive entries i, j, i in $\mathbf{h}$ (with $i \cdot j = -1$) by j, i, j or by

(b) replacing two consecutive entries i, j in $\mathbf{h}$ (with $i \cdot j = 0$) by j, i.

For such joined $(\mathbf{h}, \mathbf{h}')$, i.e., in case (a) (resp. (b)) we define a map $R_{\mathbf{h}}^{\mathbf{h}'} : \mathbf{N}^n \cong \mathbf{N}^n$ as follows. This map takes $\mathbf{c} = (c_1, \dots, c_n) \in \mathbf{N}^n$ to $\mathbf{c}' = (c'_1, \dots, c'_n) \in \mathbf{N}^n$ which has the same coordinates as $\mathbf{c}$ except in the three (resp. two) consecutive positions at which $(\mathbf{h}, \mathbf{h}')$ differ; if (a, b, c) (resp. (a, b)) are the coordinates of $\mathbf{c}$ at those three (resp. two) positions, the coordinates of $\mathbf{c}'$ at those positions are

$$(b + c - \min(a, c), \min(a, c), a + b - \min(a, c)) \qquad (\text{resp. } (b, a)).$$

It is easy to check that $R_{\mathbf{h}}^{\mathbf{h}'}$ is a bijection; its inverse is $R_{\mathbf{h}'}^{\mathbf{h}}$.

From 2.1.2, it follows that

(c) the graph $\mathbf{H}$ is connected.

42.1.4. Given $\mathbf{h} = (i_1, \dots, i_n) \in \mathbf{H}$ and $\mathbf{c} = (c_1, \dots, c_n) \in \mathbf{N}^n$, we define

(a) $E_{\mathbf{h}}^{\mathbf{c}} =$

$$E_{i_1}^{(c_1)} T'_{i_1,-1}(E_{i_2}^{(c_2)}) T'_{i_1,-1} T'_{i_2,-1}(E_{i_3}^{(c_3)}) \cdots T'_{i_1,-1} T'_{i_2,-1} \cdots T'_{i_{n-1},-1}(E_{i_n}^{(c_n)}).$$

According to 41.1.3, 40.2.2, the elements $E_{\mathbf{h}}^{\mathbf{c}}$ ($\mathbf{c} \in \mathbf{N}^n$) are contained in ${}_{\mathcal{A}}\mathbf{U}^+$ and form a $\mathbf{Q}(v)$-basis of $\mathbf{U}^+$. We shall denote this basis by $B_{\mathbf{h}}$. Hence, given $\mathbf{h}, \mathbf{h}' \in \mathbf{H}$ and $\mathbf{c} \in \mathbf{N}^n$, we can write uniquely

$$E_{\mathbf{h}}^{\mathbf{c}} = \sum_{\mathbf{c}' \in \mathbf{N}^n} \gamma_{\mathbf{h},\mathbf{h}'}^{\mathbf{c},\mathbf{c}'} E_{\mathbf{h}'}^{\mathbf{c}'}$$

where $\gamma_{\mathbf{h},\mathbf{h}'}^{\mathbf{c},\mathbf{c}'} \in \mathbf{Q}(v)$.

Proposition 42.1.5. (a) *Assume that* $\mathbf{h}, \mathbf{h}'$ *are joined in the graph* $\mathbf{H}$. *For* $\mathbf{c}, \mathbf{c}' \in \mathbf{N}^n$, $\gamma_{\mathbf{h},\mathbf{h}'}^{\mathbf{c},\mathbf{c}'}$ *is in* $\mathbf{Z}[v^{-1}]$. *Its constant term is 1 if* $\mathbf{c}' = R_{\mathbf{h}}^{\mathbf{h}'}(\mathbf{c})$ *and is zero otherwise.*

(b) *For* $\mathbf{h} \in \mathbf{H}$, *let* $\mathcal{L}_{\mathbf{h}}$ *be the* $\mathbf{Z}[v^{-1}]$*-submodule of* $\mathbf{U}^+$ *generated by the basis* $B_{\mathbf{h}}$. *Then* $\mathcal{L}_{\mathbf{h}}$ *is independent of* $\mathbf{h} \in \mathbf{H}$. *We denote it by* $\mathcal{L}$.

(c) *For* $\mathbf{h} \in \mathbf{H}$, *let* $\pi : \mathcal{L} \to \mathcal{L}/v^{-1}\mathcal{L}$ *be the canonical projection. Then* $\pi(B_{\mathbf{h}})$ *is a* $\mathbf{Z}$*-basis of* $\mathcal{L}/v^{-1}\mathcal{L}$, *independent of* $\mathbf{h} \in \mathbf{H}$; *we denote it by* B.

Assume that the proposition is known in the special case in which I consists of two elements i, j. Using the definitions and the fact that the $T'_{i,-1} : \mathbf{U} \to \mathbf{U}$ are algebra homomorphisms satisfying the braid relations, we see that (a) in the general case is a consequence of (a) in the special case. To prove (in the general case) that the objects defined in (b),(c) in terms of $\mathbf{h}, \mathbf{h}' \in \mathbf{H}$ coincide, we may assume, in view of the connectedness of the graph $\mathbf{H}$, that $\mathbf{h}, \mathbf{h}'$ are joined in $\mathbf{H}$, in which case the desired statements follow immediately from (a).

Thus, we may assume that we are in the special case above. In the case where $i \cdot j = 0$, the result is trivial. Hence we may assume that $i \cdot j = j \cdot i = -1$. Now $\mathbf{H}$ consists of two elements: $\mathbf{h} = (i, j, i)$, $\mathbf{h}' = (j, i, j)$. Besides $\mathcal{L}_{\mathbf{h}}, \mathcal{L}_{\mathbf{h}'}$, we introduce the $\mathbf{Z}[v^{-1}]$-submodule $\mathcal{L}$ of $\mathbf{U}^+$ generated by the set

$$B' = \{E_i^{(p)} E_j^{(q)} E_i^{(r)} | q \geq p + r\} \cup \{E_j^{(p)} E_i^{(q)} E_j^{(r)} | q \geq p + r\}$$

in which we identify $E_i^{(p)} E_j^{(q)} E_i^{(r)} = E_j^{(r)} E_i^{(q)} E_j^{(r)}$ for $q = p + r$.

By definition, we have $T'_{i,-1}(E_j) = E_j E_i - v^{-1} E_i E_j = T''_{j,1}(E_i)$ and $T'_{j,-1}(E_i) = E_i E_j - v^{-1} E_j E_i = T''_{i,1}(E_j)$. It follows that $T'_{i,-1} T'_{j,-1} E_i = E_j$ and $T'_{j,-1} T'_{i,-1} E_j = E_i$. Hence, if $\mathbf{c} = (c_1, c_2, c_3) \in \mathbf{N}^3$ and $\mathbf{c}' = (c'_1, c'_2, c'_3) \in \mathbf{N}^3$, we have

$$E_{\mathbf{h}}^{\mathbf{c}} = E_i^{(c_1)} (E_j E_i - v^{-1} E_i E_j)^{(c_2)} E_j^{(c_3)}$$

and

$$E_{\mathbf{h}'}^{\mathbf{c}'} = E_j^{(c'_1)} (E_i E_j - v^{-1} E_j E_i)^{(c'_2)} E_i^{(c'_3)},$$

where the notation $x^{(c)}$ is as in Lemma 42.1.2. Let $(p, q, r) \in \mathbf{N}^3$ be such that $q \geq p + r$. From 42.1.2(f), we have

$$E_i^{(p)} E_j^{(q)} E_i^{(r)} = \sum_{n=0}^{p} v^{-(p-n)(q-n)} \begin{bmatrix} p - n + r \\ p - n \end{bmatrix} E_{\mathbf{h}'}^{q-n,n,p-n+r}$$

where

$$v^{-(p-n)(q-n)}\begin{bmatrix} p-n+r \\ p-n \end{bmatrix} \in v^{-(p-n)(q-n-r)}(1+v^{-1}\mathbf{Z}[v^{-1}])$$

is in $v^{-1}\mathbf{Z}[v^{-1}]$, if $n<p$ and it equals 1 if $n=p$. Similarly,

$$E_j^{(p)}E_i^{(q)}E_j^{(r)} = \sum_{n=0}^{r} v^{-(q-n)(r-n)}\begin{bmatrix} r-n+p \\ r-n \end{bmatrix} E_{\mathbf{h}'}^{r-n+p,n,q-n}$$

where

$$v^{-(q-n)(r-n)}\begin{bmatrix} r-n+p \\ r-n \end{bmatrix} \in v^{-(r-n)(q-p-n)}(1+v^{-1}\mathbf{Z}[v^{-1}])$$

is in $v^{-1}\mathbf{Z}[v^{-1}]$, if $n<r$ and it equals 1 if $n=r$. These formulas show that $\mathcal{L}_{\mathbf{h}'}=\mathcal{L}$ and, if $\pi:\mathcal{L}\to\mathcal{L}/v^{-1}\mathcal{L}$ is the canonical map, we have $\pi(B')=\pi(B_{\mathbf{h}'})$; moreover, π maps B' onto $\pi(B')$ bijectively.

By the symmetry between i and j, there is an analogous statement for $\mathbf{h}$ (note that $\mathcal{L},B'$ are symmetric in i,j). Thus, we have $\mathcal{L}_{\mathbf{h}}=\mathcal{L}$ and $\pi(B')=\pi(B_{\mathbf{h}})$. It follows that (b),(c) hold. The formulas above show also that (a) holds. The proposition is proved.

Corollary 42.1.6. *The $\mathcal{A}$-subalgebra ${}_{\mathcal{A}}\mathbf{U}^+$ of $\mathbf{U}^+$ coincides with the $\mathcal{A}$-submodule ${}_{\mathcal{A}}\mathcal{L}$ of $\mathbf{U}^+$ generated by $\mathcal{L}$.*

The fact that ${}_{\mathcal{A}}\mathcal{L}\subset{}_{\mathcal{A}}\mathbf{U}^+$ has been noted in 42.1.4. To prove the reverse inclusion, it suffices to show that for any $i\in I$ and any $s\in\mathbf{N}$, ${}_{\mathcal{A}}\mathcal{L}$ is stable under x multiplication by $E_i^{(s)}$. Now ${}_{\mathcal{A}}\mathcal{L}$ has an $\mathcal{A}$-basis formed by the elements $E_{\mathbf{h}}^{\mathbf{c}}$ where $\mathbf{h}$ is a fixed element of $\mathbf{H}$ which starts with i and $\mathbf{c}$ runs through $\mathbf{N}^n$. Multiplication by $E_i^{(s)}$ takes each element of this basis to an $\mathcal{A}$-multiple of another element of this basis. The corollary follows.

42.1.7. From the definitions it is clear that $\mathcal{L}=\oplus_\nu\mathcal{L}_\nu$ where $\mathcal{L}_\nu=\mathcal{L}\cap\mathbf{U}_\nu^+$ for any $\nu\in\mathbf{N}[I]$. This induces a direct sum decomposition $\mathcal{L}/v^{-1}\mathcal{L}=\oplus_\nu\mathcal{L}_\nu/v^{-1}\mathcal{L}_\nu$. It is clear that B is compatible with this decomposition; in other words, we have $B=\sqcup_\nu B(\nu)$ where $B(\nu)$ is the intersection of B with the summand $\mathcal{L}_\nu/v^{-1}\mathcal{L}_\nu$ of $\mathcal{L}/v^{-1}\mathcal{L}$.

42.1.8. We consider the equivalence relation on $\mathbf{H}\times\mathbf{N}^n$ generated by $(\mathbf{h},\mathbf{c})\sim(\mathbf{h}',\mathbf{c}')$ whenever $\mathbf{h},\mathbf{h}'$ are joined in $\mathbf{H}$ and $R_{\mathbf{h}}^{\mathbf{h}'}(\mathbf{c})=\mathbf{c}'$. Let $\hat{\mathbf{H}}$ be the set of equivalence classes.

Lemma 42.1.9. *For any $\mathbf{h} \in \mathbf{H}$, the map $f : \mathbf{N}^n \to \hat{\mathbf{H}}$, which takes any $\mathbf{c}$ to the equivalence class of $(\mathbf{h}, \mathbf{c})$, is a bijection.*

From 42.1.5(a), we see that the (surjective) map $\mathbf{H} \times \mathbf{N}^n \to B$ given by $(\mathbf{h}, \mathbf{c}) \mapsto \pi(E_{\mathbf{h}}^{\mathbf{c}})$ is constant on equivalence classes; hence it factors through a (surjective) map $\hat{\mathbf{H}} \to B$. On the other hand, for any $\mathbf{h} \in \mathbf{H}$, the composition $\mathbf{N}^n \xrightarrow{f} \hat{\mathbf{H}} \to B$ is a bijection (again by 42.1.5). The lemma follows.

We have the following result.

Theorem 42.1.10. (a) *For any $b \in B$ there is a unique element $\tilde{b} \in \mathcal{L}$ such that $\pi(\tilde{b}) = b$ and $\bar{\tilde{b}} = \tilde{b}$.*

(b) *The set $\{\tilde{b} | b \in B\}$ is a $\mathbf{Z}[v^{-1}]$-basis of $\mathcal{L}$ and a $\mathbf{Q}(v)$-basis of $\mathbf{U}^+$.*

We shall regard the pairing $(,)$ on $\mathbf{f}$ as a pairing on $\mathbf{U}^+$ via the isomorphism $\mathbf{f} \to \mathbf{U}^+$ given by $x \mapsto x^+$. Let $\mathbf{h} \in \mathbf{H}$. By 38.2.3, the basis $E_{\mathbf{h}}^{\mathbf{c}}$ of $\mathbf{U}^+$, where $\mathbf{c}$ is running through $\mathbf{N}^n$, is almost orthonormal for $(,)$. Applying 14.2.2(b) to this basis, we see that any element $\beta \in \mathbf{B}$ satisfies $\beta^+ \in \mathcal{L}$ and $\pi(\beta^+) \in \pm B$. In particular, we have $\mathcal{L}(\mathbf{f}) \subset \mathcal{L}$. Applying 14.2.2(b) to the canonical basis $\mathbf{B}$ of $\mathbf{f} = \mathbf{U}^+$ and to $x = E_{\mathbf{h}}^{\mathbf{c}}$, which satisfies $(x, x) \in 1 + v^{-1}\mathbf{Z}[[v^{-1}]] \cap \mathbf{Q}(v)$ by 38.2.3, we see that $x \in \mathcal{L}(\mathbf{f})$; hence, by the previous sentence, $\mathcal{L} = \mathcal{L}(\mathbf{f})$ and $\pi(x) = \pm\pi(\beta^+)$ for some $\beta \in \mathbf{B}$. Since $\beta^+ \subset \mathcal{L}(\mathbf{f}) = \mathcal{L}$, we see that the existence statement in (a) holds. The uniqueness in (a), as well as statement (b) now follow from the known properties of $\mathbf{B}$. The theorem is proved.

42.1.11. We keep the notation from the proof of Theorem 42.1.10. We fix $i \in I$. Assume that $\mathbf{h} \in \mathbf{H}$ starts with i. Let $b \in B$ be such that $b = \pi(E_{\mathbf{h}}^{\mathbf{c}})$ where the first coordinate of $\mathbf{c}$ is 0. Let $\mathbf{c}' \in \mathbf{N}^n$ be such that $\mathbf{c}'$ has the same coordinates as $\mathbf{c}$ except for the first coordinate, which is $s \in \mathbf{N}$. Let $b' = \pi(E_{\mathbf{h}}^{\mathbf{c}'}) \in B$. We shall use the following notation. For $b \in B$, we define $\beta_b \in \mathbf{B}$ by $\beta_b^+ = \tilde{b}$ (see the proof of 42.1.10).

Lemma 42.1.12. (a) *Write $E_{\mathbf{h}}^{\mathbf{c}} = z^+$ where $z \in \mathbf{f}$. Then $z \in \mathbf{f}[i]$ and $E_{\mathbf{h}}^{\mathbf{c}'} = (\tilde{\phi}_i^s z)^+$.*

(b) *We have $\beta_{b'} = \tilde{\phi}_i^s \beta_b \mod v^{-1}\mathcal{L}(\mathbf{f})$.*

(c) *We have $\beta_b \in \mathbf{B}_{i;0}$.*

(d) *We have $\beta_{b'} = \pi_{i,s}(\beta_b)$.*

Since $\mathbf{c}$ starts with 0, the element $E_{\mathbf{h}}^{\mathbf{c}} \in \mathbf{U}^+$ is in $T'_{i,-1}\mathbf{U}^+$, hence by 38.1.6, z is in $\mathbf{f}[i]$, and ${}_ir(z) = 0$; hence $\tilde{\phi}_i^s(z) = \theta_i^{(s)}z$. It follows that $\tilde{\phi}_i^s(z)^+ = E_i^{(s)}E_{\mathbf{h}}^{\mathbf{c}} = E_{\mathbf{h}}^{\mathbf{c}'}$. This proves (a).

We prove (b). Using (a) and the definitions, we have $z = \beta_b \mod v^{-1}\mathcal{L}(\mathbf{f})$ and $\tilde{\phi}_i^s z = \beta_{b'} \mod v^{-1}\mathcal{L}(\mathbf{f})$. Since $\tilde{\phi}_i^s$ maps $v^{-1}\mathcal{L}(\mathbf{f})$ into itself it follows that $\tilde{\phi}_i^s z = \tilde{\phi}_i^s(\beta_b) \mod v^{-1}\mathcal{L}(\mathbf{f})$, hence $\tilde{\phi}_i^s(\beta_b) = \beta_{b'} \mod v^{-1}\mathcal{L}(\mathbf{f})$. This proves (b).

From the fact that the elements $E_{\mathbf{h}}^{\mathbf{c}''}$, where $\mathbf{c}''$ runs through $\mathbf{N}^n$, form a basis of $\mathbf{U}^+$, it follows immediately that

(e) for any $t \geq 0$, the elements $E_{\mathbf{h}}^{\mathbf{c}''}$ where $\mathbf{c}''$ runs through the elements of $\mathbf{N}^n$ with first coordinate $\geq t$, form a $\mathbf{Q}(v)$-basis of $E_i^t\mathbf{U}^+$.

We prove (c). Assume that $\beta_b \in \mathcal{B}_{i;t}$ with $t > 0$. Then $\beta_b^+ \in E_i^t\mathbf{U}^+$, hence it is a linear combination of elements as in (e); in particular, $E_{\mathbf{h}}^{\mathbf{c}}$ appears with coefficient 0, contradicting the definition of β_b. This proves (c). Since $\beta_b, \beta_{b'} \in \mathcal{B}$, we see from 17.3.7 that (d) follows from (b) and (c). This completes the proof.

Corollary 42.1.13. *We have* $\mathbf{B} = \{\tilde{b} | b \in B\}$. *(We identify* $\mathbf{f} = \mathbf{U}^+$ *as above.)*

From the proof of 42.1.10, we have that $\{\tilde{b} | b \in B\} \subset \mathcal{B}$. We show by induction on $N = \operatorname{tr} \nu$ that $\tilde{b} \in \mathbf{B}$, if $b \in B_\nu$. If $N = 0$, this is clear. Assume that $N \geq 1$. By 14.3.3, we can find $i \in I$ and $s > 0$ such that $\tilde{b} \in \mathcal{B}_{i;s}$. We then have $\tilde{b} = \pi_{i,s}\beta$ where $\beta \in \mathcal{B}_{i;0}$ (see 14.3.2). We have $\beta = \pm\tilde{b}_1$ where $b_1 \in B$. By the argument in the previous lemma we have that the sign is $+$, hence $\tilde{b} = \pi_{i,s}\tilde{b}_1$. By the induction hypothesis, we have $\tilde{b}_1 \in \mathbf{B}$; the previous equality then implies that $\tilde{b} \in \mathbf{B}$.

42.1.14. The basis $\{\tilde{b} | b \in B\} = \{\beta^+ | \beta \in \mathbf{B}\}$ of $\mathbf{U}^+$ is in a natural bijection with the set B, which in turn is in a natural bijection with the set $\hat{\mathbf{H}}$ (see the proof of 42.1.9). We thus have a purely combinatorial parametrization of the canonical basis $\mathbf{B}$.

The structure of left colored graph on $\mathbf{B}$ (see 14.4.7) corresponds to a structure of colored graph on $\hat{\mathbf{H}}$, which we will now describe in a purely combinatorial way.

For any $i \in I$, we define a function $g_i : \hat{\mathbf{H}} \to \mathbf{N}$ as follows. Let $c \in \hat{\mathbf{H}}$; we choose $\mathbf{h} \in \mathbf{H}$ such that the sequence $\mathbf{h}$ starts with i. By 42.1.9, c is the class of $(\mathbf{h}, \mathbf{c})$ for a unique $\mathbf{c} \in \mathbf{N}^n$. We set $g_i(c) = c_1$ where c_1 is the first coordinate of $\mathbf{c}$. To show that this is well-defined, we consider $\mathbf{h}' \in \mathbf{H}$ such that the sequence $\mathbf{h}$ starts with i. Let $\mathbf{c}' \in \mathbf{N}^n$ be such that c is the class of $(\mathbf{h}', \mathbf{c}')$ and let c'_1 be the first coordinate of $\mathbf{c}'$. We must show that $c_1 = c'_1$. Now the set $\mathbf{H}_i$ of all sequences in $\mathbf{H}$ which start with i can be naturally identified with the set of reduced expressions for $s_i w_0$; applying 2.1.2, we see that $\mathbf{H}_i$, regarded as a full subgraph of $\mathbf{H}$ is connected. Hence to prove that $c_1 = c'_1$, we may assume that $\mathbf{h}, \mathbf{h}'$ are joined in the graph. Then $\mathbf{c}$ and $\mathbf{c}'$ are related by an elementary move as in 42.1.3(a) or (b). This elementary move operates on coordinates other than the first, since $\mathbf{h}, \mathbf{h}'$ start with the same element i. Thus, we have $c_1 = c'_1$, as desired.

42.1.15. For any $i \in I$, we define a partition $\hat{\mathbf{H}} = \sqcup_{t \geq 0} \hat{\mathbf{H}}_{i,t}$ by setting $\hat{\mathbf{H}}_{i,t} = g_i^{-1}(t)$. We define a bijection $\pi_{i,t} : \hat{\mathbf{H}}_{i,0} \cong \hat{\mathbf{H}}_{i,t}$ as follows. Let $(\mathbf{h}, \mathbf{c})$ be a representative for an element of $\hat{\mathbf{H}}_{i,0}$. Then $\mathbf{c}$ starts with 0; let $\mathbf{c}'$ be the element of $\mathbf{N}^n$ which starts with t and has the same subsequent coordinates as those of $\mathbf{c}$. By definition, $\pi_{i,t}(\mathbf{h}, \mathbf{c}) = (\mathbf{h}, \mathbf{c}')$. One checks that this map is well-defined. From our earlier discussion, it is clear that the partitions of $\hat{\mathbf{H}}$ just described, together with the bijections $\pi_{i,t}$, correspond to the analogous objects for $\mathbf{B}$ which are the ingredients in the definition of the left colored graph.

42.1.16. We can also describe in purely combinatorial terms the left colored graph for not necessarily simply laced Cartan data, by reduction to the simply laced case, using 14.4.9 and 14.1.6.

42.2. Remarks on the Piecewise Linear Bijections $R_{\mathbf{h}}^{\mathbf{h}'} : \mathbf{N}^n \cong \mathbf{N}^n$

42.2.1. Let $\mathbf{h}, \mathbf{h}' \in \mathbf{H}$. We define a bijection $R_{\mathbf{h}}^{\mathbf{h}'} : \mathbf{N}^n \cong \mathbf{N}^n$ as a composition

(a) $R_{\mathbf{h}}^{\mathbf{h}'} = R_{\mathbf{h}(0)}^{\mathbf{h}(1)} R_{\mathbf{h}(1)}^{\mathbf{h}(2)} \cdots R_{\mathbf{h}(t-1)}^{\mathbf{h}(t)}$

where $\mathbf{h}(0), \mathbf{h}(1), \dots, \mathbf{h}(t)$ is a sequence of vertices of the graph $\mathbf{H}$ such that $\mathbf{h}(0) = \mathbf{h}, \mathbf{h}(t) = \mathbf{h}'$ and such that $\mathbf{h}(s), \mathbf{h}(s+1)$ is an edge of the graph $\mathbf{H}$ for $s = 0, 1, \dots, t-1$; the factors on the right hand side of (a) are the bijections defined in 42.1.3. (A sequence as above can always be found, by 2.1.2.) From 42.1.5, it follows that the definition of $R_{\mathbf{h}}^{\mathbf{h}'}$ is correct, that is, it does not depend on the choices made. Indeed, 42.1.5 gives us an intrinsic definition of this bijection: with the notation in 42.1.5, we have $R_{\mathbf{h}}^{\mathbf{h}'}(\mathbf{c}) = \mathbf{c}'$ if and only if $\pi(E_{\mathbf{h}}^{\mathbf{c}}) = \pi(E_{\mathbf{h}'}^{\mathbf{c}'})$. The bijections $R_{\mathbf{h}}^{\mathbf{h}'}$ are piecewise linear, since they are products of factors which are piecewise linear.

42.2.2. In this section we will show that the bijections $R_{\mathbf{h}}^{\mathbf{h}'}$ also appear in a completely different context.

Let K be a field with a given subset $K_0 \subset K - \{0\}$ containing 1, and such that the following holds:

(a) if $f, f' \in K_0$, then $f + f' \in K_0$, $ff' \in K_0$, $\frac{f}{f+f'} \in K_0$.

For example, we could take

(b) $K = \mathbf{R}, K_0 = \mathbf{R}_{>0}$ or

(c) $K = \mathbf{R}((\epsilon))$ where ϵ is an indeterminate and K_0 is the subset of K consisting of power series of the form $f = a_s\epsilon^s + a_{s+1}\epsilon^{s+1} + \cdots$ such that $s \geq 0$ and $a_s > 0$; we then set $|f| = s$.

42.2.3. We consider a split semisimple algebraic group $\mathcal{G}$ over K, corresponding to the root datum, with a fixed maximal unipotent subgroup $\mathcal{U}^+$ and a fixed maximal torus $\mathcal{T}$ normalizing $\mathcal{U}^+$, both defined over K. For each $i \in I$, we denote by $\mathcal{U}_i^+$ the simple root subgroup of $\mathcal{U}^+$ corresponding to i; we assume that we are given an isomorphism x_i of the additive group with $\mathcal{U}_i^+$, defined over K. Let B^- be the Borel subgroup opposed to $\mathcal{U}^+$ and containing $\mathcal{T}$. We shall identify $\mathcal{G}, \mathcal{U}^+, \mathcal{T}, \mathcal{U}_i^+, B^-$ with their groups of K-rational points. We shall regard x_i as an isomorphism of K onto $\mathcal{U}_i^+$.

Proposition 42.2.4. *Let $w \in W$. Let $\mathbf{h} = (i_1, i_2, \dots, i_n)$ be a sequence in I such that $s_{i_1}s_{i_2}\cdots s_{i_n}$ is a reduced expression for w.*

(a) *The map $K_0^n \to \mathcal{U}^+$ given by*

$$(p_1, p_2, \dots, p_n) \mapsto x_{i_1}(p_1)x_{i_2}(p_2)\cdots x_{i_n}(p_n)$$

is injective.

(b) *The image of the map (a) is a subset $\mathcal{U}^+(w)$ of $\mathcal{U}^+$ which does not depend on* $\mathbf{h}$.

(c) *If $w' \in W$ is distinct from w, then $\mathcal{U}^+(w) \cap \mathcal{U}^+(w') = \emptyset$.*

Let $\mathcal{U}^+(\mathbf{h})$ be the image of the map in (a). To prove (b), it suffices, by 2.1.2, to check the following statement: if $\mathbf{h}'$ is obtained from $\mathbf{h}$ by replacing h consecutive indices $i, j, i, \dots$ in $\mathbf{h}$ by the h indices $j, i, j, \dots$ (for some $i \neq j$ with $h = h(i,j)$), then $\mathcal{U}^+(\mathbf{h}) = \mathbf{U}^+(\mathbf{h}')$.

To prove this statement, we may clearly assume that I consists of two elements i, j. In the case where $i \cdot j = 0$, we have $x_i(p)x_j(p') = x_j(p')x_i(p)$ for any $p, p' \in K$. Assume now that $i \cdot j = -1$. We have the following identity, by a computation in SL_3:

$$x_i(t)x_j(s)x_i(r) = x_j(t')x_i(s')x_j(r')$$

where

(d) $t' = \frac{sr}{t+r}, \quad s' = t + r, \quad r' = \frac{st}{t+r}$

or equivalently,

(e) $t = \frac{s'r'}{t'+r'}, \quad s = t' + r', \quad r = \frac{s't'}{t'+r'}$.

By the definition of K_0, we have $s, t, r \in K_0$ if and only if $s', t', r' \in K_0$. This proves (b). We prove (c). Let $\dot{s}_i$ be an element of the normalizer of $\mathcal{T}$ in $\mathcal{G}$ which represents $s_i \in W$. If $p \in K - \{0\}$, we have $x_i(p) \in B\dot{s}_iB$. Hence if $p_1, p_2, \dots, p_n$ are in $K - \{0\}$, then

$$x_{i_1}(p_1)x_{i_2}(p_2)\cdots x_{i_n}(p_n) \in B\dot{s}_{i_1}B\dot{s}_{i_2}B\cdots\dot{s}_{i_n}B \subset B\dot{s}_{i_1}\dot{s}_{i_2}\cdots\dot{s}_{i_n}B$$

by properties of the Bruhat decomposition. Thus, $\mathcal{U}^+(w) \subset B\dot{s}_{i_1}\dot{s}_{i_2}\cdots\dot{s}_{i_n}B$ so that (c) follows from the Bruhat decomposition.

We prove (a). Assume that

$$x_{i_1}(p_1)x_{i_2}(p_2)\cdots x_{i_n}(p_n) = x_{i_1}(p_1')x_{i_2}(p_2')\cdots x_{i_n}(p_n')$$

where $p_1, \dots, p_n$ and $p_1', \dots, p_n'$ are in K_0. We prove that $p_l = p_l'$ for all l by induction on n. This assumption implies

$$x_{i_1}(p_1 - p_1')x_{i_2}(p_2)\cdots x_{i_n}(p_n) = x_{i_2}(p_2')\cdots x_{i_n}(p_n').$$

If $p_1 - p_1' \neq 0$, the two sides of the last equality are in

$$B\dot{s}_{i_1}\dot{s}_{i_2}\cdots\dot{s}_{i_n}B, B\dot{s}_{i_2}\dot{s}_{i_3}\cdots\dot{s}_{i_n}B,$$

by the argument above. This is a contradiction. Thus, we must have $p_1 = p_1'$. Then we have

$$x_{i_2}(p_2)\cdots x_{i_n}(p_n) = x_{i_2}(p_2')\cdots x_{i_n}(p_n')$$

and the induction hypothesis shows that $p_2 = p_2', \dots, p_n = p_n'$.

Corollary 42.2.5. *The subset $\cup_{w\in W}\mathcal{U}^+(w)$ of $\mathcal{U}^+$ is closed under multiplication. It coincides with the submonoid (with 1) of $\mathcal{U}^+$ generated by the elements $x_i(p)$, for various $i \in I$ and $p \in K_0$.*

Let $i \in I$ and $p \in K_0$. Let $\mathbf{h} = (i_1, i_2, \dots, i_n)$ be as in 42.2.4. If $s_is_{i_1}s_{i_2}\cdots s_{i_n}$ is a reduced expression in W, then $x_i(p)\mathcal{U}^+(\mathbf{h}) \subset \mathcal{U}^+(\mathbf{h}')$ where $\mathbf{h}' = (i, i_1, i_2, \dots, i_n)$. If $s_is_{i_1}s_{i_2}\cdots s_{i_n}$ is not a reduced expression in W, then we have $s_{i_1}s_{i_2}\cdots s_{i_n} = s_is_{j_1}s_{j_2}\cdots s_{j_{n-1}}$. for some $j_1, j_2, \dots, j_{n-1}$. Set $\mathbf{h}' = (i, j_1, j_2, \dots, j_{n-1})$. Clearly, $x_i(p)\mathcal{U}^+(\mathbf{h}') \subset \mathcal{U}^+(\mathbf{h}')$ and, by 42.2.4(b), we have $\mathcal{U}^+(\mathbf{h}') = \mathcal{U}^+(\mathbf{h})$. It follows that $x_i(p)\mathcal{U}^+(\mathbf{h}) \subset \mathcal{U}^+(\mathbf{h}')$.

We have thus proved that the set $\cup_{w\in W}\mathcal{U}^+(w)$ is stable under left multiplication by elements of the form $x_i(p)$ as above. The corollary follows.

42.2.6. From now on we assume that K, K_0 are as in 42.2.2(c). Recall that we then have a well-defined map $f \mapsto |f|$ from K_0 to $\mathbf{N}$.

For any $\mathbf{h} = (i_1, i_2, \dots, i_n)$ in $\mathbf{H}$ and any $\mathbf{c} = (c_1, \dots, c_n) \in \mathbf{N}^n$, we define a subset $\mathcal{U}^+(\mathbf{h}, \mathbf{c})$ of $\mathcal{U}^+$ as follows. By definition, $\mathcal{U}^+(\mathbf{h}, \mathbf{c})$ consists of all elements of $\mathcal{U}^+$ which are of the form $x_{i_1}(p_1)x_{i_2}(p_2)\cdots x_{i_n}(p_n)$ where $p_1, p_2, \dots, p_n$ are elements of K_0 such that $|p_1| = c_1, |p_2| = c_2, \dots, |p_n| = c_n$. From 42.2.4, we see that we have a partition

(a) $\mathcal{U}^+(w_0) = \sqcup_{\mathbf{c}} \mathcal{U}^+(\mathbf{h}, \mathbf{c})$.

Proposition 42.2.7. *Let* $\mathbf{h}, \mathbf{h}'$ *be elements of* $\mathbf{H}$ *and let* $\mathbf{c}, \mathbf{c}'$ *be elements of* $\mathbf{N}^n$ *such that* $R_{\mathbf{h}}^{\mathbf{h}'}(\mathbf{c}) = \mathbf{c}'$. *We have* $\mathcal{U}^+(\mathbf{h}, \mathbf{c}) = \mathcal{U}^+(\mathbf{h}', \mathbf{c}')$. *In particular, the partition 42.2.6(a) of* $\mathcal{U}^+(w_0)$ *is independent of* $\mathbf{h}$.

We may clearly assume that $\mathbf{h}, \mathbf{h}'$ are joined in the graph $\mathbf{H}$. That case reduces immediately to the case where I consists of two elements i, j. The case where $i \cdot j = 0$ is trivial.

Assume now that $i \cdot j = -1$. Using the identities (d),(e) in the proof of 42.2.4, we see that it is enough to verify the following statement. Let $t, s, r, t', s', r' \in K_0$ be such that $t' = \frac{sr}{t+r}, s' = t + r, r' = \frac{st}{t+r}$. Then

$$|t'| = |s| + |r| - \min(|t|, |r|), \quad |s'| = \min(|t|, |r|), \quad |r'| = |t| + |s| - \min(|t|, |r|).$$

This is immediate. The proposition is proved.

42.2.8. We now see that the set of subsets in the partition 42.2.6(a) of $\mathcal{U}^+(w_0)$ (which is intrinsic, by 42.2.7) is in natural $1-1$ correspondence with the set $\hat{\mathbf{H}}$, hence also with the canonical basis $\mathbf{B}$. At the same time we have obtained a new interpretation of the piecewise linear bijections $R_{\mathbf{h}}^{\mathbf{h}'}: \mathbf{N}^n \cong \mathbf{N}^n$ in terms of the geometry of the group $\mathcal{G}$.

Notes on Part VI

1. The braid group action on **U** has been introduced (with a different normalization) in [5], in the simply laced case, and in [6], for arbitrary Cartan data of finite type. Another approach (for Cartan data of finite type) to the braid group action has been found by Soibelman [8]. The general case has not been treated before in the literature. The fact that the braid group acts naturally on integrable modules over arbitrary ground rings (see 41.2) is also new.
2. The paper[3] of Levendorskii and Soibelman contains several results relating braid group actions (for finite type) with comultiplication and with the inner product. In particular, an identity like 37.3.2(a) appears (for finite type) in [3]. Our lemma 38.1.8 is closely related to [3, 2.4.2]; however, neither of these two results implies the other. Propositions 38.2.3 and 40.2.4 are generalizations of [3, 3.2].
3. Corollary 40.2.2 and Proposition 41.1.7 appeared in [6] and [2].
4. Most results in 42.1 appeared in [7]. The results in 42.2 are new; in obtaining them, I have been stimulated by a question of B. Kostant.

References

1. I. Damiani, *A basis of type Poincaré-Birkhoff-Witt for the quantum algebra of* $SL(2)$, J. of Algebra **161** (1993), 291–310.
2. M. Dyer and G. Lusztig, *Appendix to* [6], Geom. Dedicata **35** (1990), 113–114.
3. S. Levendorskii and I. Soibelman, *Some applications of quantum Weyl groups*, J. Geom. and Phys. **7** (1990), 241–254.
4. S. Levendorskii, I. Soibelman and V. Stukopin, *Quantum Weyl group and universal quantum R-matrix for affine Lie algebra A1*, Lett. in Math. Phys. **27** (1993), 263–264.
5. G. Lusztig, *Quantum deformations of certain simple modules over enveloping algebras*, Adv. Math. **70** (1988), 237–249.
6. ———, *Quantum groups at roots of 1*, Geom. Dedicata **35** (1990), 89–114.
7. ———, *Canonical bases arising from quantized enveloping algebras*, J. Amer. Math. Soc. **3** (1990), 447–498.
8. I. Soibelman, *Algebra of functions on a compact quantum group and its representations*, (in Russian), Algebra and Analysis **2** (1990), 190–212.
9. J. Tits, *Normalisateurs de tores, I. Groupes de Coxeter étendues*, J. Algebra **4** (1966), 96–116.
10. D. N. Verma, *Structure of certain induced representations of complex semisimple Lie algebras*, Bull. Amer. Math. Soc. **74** (1968), 160–166.

Index of Notation

G. Lusztig, *Introduction to Quantum Groups*, Modern Birkhäuser Classics,
DOI 10.1007/978-0-8176-4717-9, © Springer Science+Business Media, LLC 2010

Index of Terminology

Comments added in the second printing

1. Let $M \in \mathcal{C}'$ and let $i \in I, e = \pm 1$. We define two $\mathbf{Q}(v)$-linear maps $S'_{i,e}, S''_{i,e} : M \to M$ by

$$S'_{i,e} = \sum_{a,b,c} (-1)^b v_i^{e(c^2-a^2-ac+ab-bc+a+c)} F_i^{(a)} E_i^{(b)} F_i^{(c)} \tilde{K}_i^{e(a-c)},$$

$$S''_{i,e} = \sum_{a,b,c} (-1)^b v_i^{e(c^2-a^2-ac+ab-bc+a+c)} E_i^{(a)} F_i^{(b)} E_i^{(c)} \tilde{K}_i^{e(c-a)}$$

where a, b, c run over $\mathbf{N}$; although the sums are infinite, on any given vector in M, all but finitely many terms in either sum act as zero.

Several readers have asked me about the relationship between these operators (which appeared in [3, 3.1]) and the operators $T'_{i,e}, T''_{i,e} : M \to M$ in 5.2.1. The relationship is as follows:

$$S'_{i,e} = T'_{i,e} \tilde{K}_i^e, \quad S''_{i,e} = T''_{i,e} \tilde{K}_i^{-e}.$$

To prove this, we may replace $\mathcal{C}'$ by $\mathcal{C}'_i$ and we may assume that M is a simple object of $\mathcal{C}'_i$. Then the desired identities are checked by calculations similar to those in 5.2.2.

It follows that the braid group relations 39.4.3(c),(d) remain valid if T', T'' are replaced throughout by S', S''.

2. In the last sentence on p. 183, it was stated that it should be possible to remove the signs $\pm$ in Corollary 26.3.2. This is indeed true, as was shown by Kashiwara [2].

3. The question raised in the last sentence of 40.2.5 has been answered by J. Beck [1] in the untwisted case.

References

[1] J. Beck, *Convex bases of PBW type for quantum affine algebras*, Communications in Math. Phys. (to appear).

[2] M. Kashiwara, *Crystal base of modified enveloping algebras*, Duke Math. J. **73** (1994), 383-414.

[3] G.Lusztig, *Problems on canonical bases*, Amer. Math. Soc., Proc. Symp. Pure Math. (1994).

Errata to Introduction to Quantum Groups (1994 printing)

p. 3 line 13: after "homomorphism," and before "then" insert:
preserving $\mathbf{Z}[I]$-gradings

p. 9, line –5: inside the last bracket [ ], replace a' by a'' and t' by t''.

p. 11, in the numerator of the first fraction on line -4, insert a factor $v_i^n v_j$

p. 12, in the numerator of the fraction on line -8, insert a factor v_j

p. 13, just before 1.4.7, add the following text.
Here is another proof of Proposition 1.4.3 which is not based on 1.4.5 but is based, instead, on the following statement which is easily verified.
(a) For any $k \in I$ the left-hand side of the equality in 1.4.3 is in the kernel of $r_k : \mathbf{f} \to \mathbf{f}$.
Note that 1.4.3 follows from (a) in view of 1.2.15(a).

p. 31, in 3.4.6 line 6: delete (a) (twice) and in 3.4.6 line -2, replace
$E_i^{(a)}, F_i^{(a)}$ by E_i, F_j.

p. 32 line 3.5.3(b): replace
$E_j F_i^N + \mathbf{Q}(v) F_i^{N-1}$ by $E_j F_i^N + \mathbf{Q}(v) K_i F_i^{N-1} + \mathbf{Q}(v) K_i^{-1} F_i^{N-1}$

p. 32 line –1: add: See also 23.3.11.

p. 33, line 13, replace M_λ by Λ_λ.

p. 40, line –3: replace $n + 2m$ by $n + 2m - 2$.

p. 40, line –1: replace $n - 2m$ by $n - 2m + 2$.

p. 41 lines 5,6: replace If $x \in M^n(0)$, then by definition, $(c - s_n)x = 0$ by
We have $M^n(0) = 0$.

p. 41 lines 9, 10, 11, 12: replace (four times) $n + 2m$ by $n + 2m - 2$.

p. 42: delete lines 1,2.

p. 42 line 6: replace roman m,n by italic m,n.

p. 45: the equalities in line –2 of 5.2.6 should read
$v^{\langle \mu, \lambda \rangle - \langle i, \lambda \rangle \langle \mu, i' \rangle} = v^{\langle \mu, \langle \mu, i' \rangle i, \lambda \rangle} = v^{\langle \mu', \lambda \rangle}$.

p. 45: the line "This follows immediately from the previous lemma" in 5.2.7 should read: This has been stated in the proof of the previous lemma.

p. 49, line –5: insert 2 in front of the second Σ.

p. 49, line –7: insert 2 in front of the second Σ.

p. 50, line 4: replace 2 by 4 in front of the second Σ.

p. 56, Lemma 7.1.4 should read:

Assume that $m = \alpha n + 1$. We have

$$F_j f^+_{n,m;e} - f^+_{n,m;e} F_j = \tilde{K}_{-j} \frac{v_j^{n-1}}{v_j - v_j^{-1}} f^+_{n-1,m;1} - \tilde{K}_j \frac{v_j^{-n+1}}{v_j - v_j^{-1}} f^+_{n-1,m;-1}.$$

p. 57 line 3 and 4: the factors $v_i^{er(\alpha n - m+1)}$ should be be deleted.

p. 68 at the end of 9.1.2: insert the line:

"Here $g_{\mathrm{i}} : \mathbf{V}_{\mathrm{i}} \longrightarrow \mathbf{V}_{\mathrm{i}}$ is the restriction of g."

p. 71 line –8: replace $G \times_P F$ by $G \times_Q F$.

p. 72, line 7: replace $\mathbf{T}'_{\mathrm{i}}$ by $\mathbf{T}_{\mathrm{i}}$.

p. 77 in 9.3.1 line –8: replace

$\gamma'(\mathrm{i}) \geq \gamma(\mathrm{i})$ by $\gamma'_{\mathrm{i}} \geq \gamma_{\mathrm{i}}$

and $\gamma'(\mathrm{i}) > \gamma(\mathrm{i})$ by $\gamma'_{\mathrm{i}} > \gamma_{\mathrm{i}}$.

p. 85, line 11: insert) before =.

p. 87, lines 2,3 of 10.3.2: delete "and $\mathbf{T}_{\mathrm{i}'} = 0$ for all $\mathbf{i}' \in \mathbf{I} - \mathbf{I}'$ "

p. 91, lines 7, 8, 12, 18: replace $\ldots (a^*)$ by $\cdots \oplus (a^*)$ (in five places).

p. 113, lines –13, –12: replace "a-orbit" by "orbit".

p. 118, line 8 of 14.2.2: replace $p_b \in \mathbf{Z}$ by $p_b \in \mathbf{Q}$.

p. 124, line 2 of 14.4.11: replace $\theta^{\langle i,\lambda \rangle + 1}$ by $\theta_i^{\langle i,\lambda \rangle + 1}$.

p. 128: Ref. 6, add: I.H.E.S. before 67.

p. 132, line 1 of 16.1.2(b): replace $v^{-t(t-1)/2}$ by $v_i^{-t(t-1)/2}$.

p. 138 line –7, –5, –3: replace last $\mathcal{L}(M))$ by $\mathcal{L}(M)$.

p. 138 line –2: replace all $\mathcal{L}(M))$ by $\mathcal{L}(M)$ (three times).

p. 164, line 6 of 19.1.1, replace $S\rho_1$ by $\rho_1 S$.

p. 171, line 4: replace Let $b \in \Lambda_\lambda$ by Let $b \in \Lambda_\lambda$, $b \neq 0$.

p. 177 line –2: replace $\overline{\omega(u)\chi(\eta)}$ (following the second =) by $\overline{\omega(u)} \cdot \overline{\chi(\eta)}$.

p. 181 in lines 2, 5, 6: replace $d_{b,\theta_i,b',n}$ by $d_{\beta,\theta_i,\beta',n}$.

p. 181 in lines 2, 6: replace $|b'|$ by $|\beta'|$.

p. 185 in line –8, –9: replace

$\lambda'_1, \lambda''_1, \lambda'_2, \lambda''_2$ and any $t \in \mathbf{U}(\lambda'_1 - \lambda''_1)$, $s \in \mathbf{U}(\lambda'_1 - \lambda''_1)$,

by the following:

$\lambda'_1, \lambda''_1, \lambda'_2, \lambda''_2$ in X, any ν_1, ν_2 in $\mathbf{Z}[I]$ such that

$\lambda'_1 - \lambda''_1 = \nu_1$, $\lambda'_1 - \lambda''_1 = \nu_2$ (in X) and any $t \in \mathbf{U}(\nu_1)$, $s \in \mathbf{U}(\nu_2)$,

p. 190 line –1: replace $M_\lambda \otimes T'$ by ${}^\omega M_\lambda \otimes T'$.

At the end of p. 191 (after "...from 23.3.6") add a new subsection:

23.3.11.

We give a more detailed proof of Proposition 3.5.4 based on results in this chapter. Let $\pi_{\lambda',\lambda''}$ be as in 23.1.1. From the definition we see that if $u \in \mathbf{U}$ satisfies $\pi_{\lambda_1,\lambda_2}(u) = 0$ for any λ_1, λ_2 in X then $u = 0$.

Let $a = \sum_i a_i i \in \mathbf{N}[I]$, $b = \sum_i b_i i \in \mathbf{N}[I]$, $\lambda \in X$, $J(\lambda, a, b) = \sum_i \mathbf{U}F_i^{a_i+1} + \sum_i \mathbf{U}E_i^{b_i+1} + \sum_{\mu \in Y} \mathbf{U}(K_\mu - v^{\langle \mu,\lambda \rangle})$. Let $u \in \cap_{a,b,\lambda} J(\lambda, a, b)$. By 3.5.3 it is enough to show that $u = 0$. Hence it is enough to show that $\pi_{\lambda_1,\lambda_2}(u) = 0$ for any λ_1, λ_2 in X. If $\langle i, \lambda_2 \rangle = a_i - b_i$ for all $i \in I$ we have $\pi_{\lambda_1,\lambda_2}(J(\lambda_2, a, b)) \subset P(\lambda_2, a, b)$ (notation of 23.3.3). Hence it is enough to show that for any $\zeta \in X$ we have $\cap P(\zeta, a, b) = 0$ where the intersection is taken over all a, b such that $\langle i, \zeta \rangle = a_i - b_i$ for all i. Using the isomorphism $\dot{\phi}$ in 23.2.5 we see that we may assume that our root datum is simply connected. We identify X^+ with $\mathbf{N}[I]$ by $\lambda \mapsto \underline{\lambda}$ where $\underline{\lambda} = \sum_i \langle i, \lambda \rangle i$. It is enough to show that for any $\zeta \in X$ we have $\cap_{\lambda',\lambda \in X^+; \zeta = \lambda' - \lambda} P(\zeta, \underline{\lambda}', \underline{\lambda}) = 0$. For $\lambda', \lambda \in X^+$ such that $\zeta = \lambda' - \lambda$ we have an isomorphism $\dot{\mathbf{U}} 1_\zeta \to \mathbf{f} \otimes \mathbf{f} = {}^\omega M_\lambda \otimes M_{\lambda'}$ (see 23.3.1(c)) which by 23.3.5 carries $P(\zeta, \underline{\lambda}', \underline{\lambda})$ onto $\mathcal{T}_\lambda \otimes \mathbf{f} + \mathbf{f} \otimes \mathcal{T}_{\lambda'}$ where $\mathcal{T}_\lambda$ (resp. $\mathcal{T}_{\lambda'}$) is the kernel of the canonical homomorphism $\mathbf{f} = {}^\omega M_\lambda \to {}^\omega \Lambda_\lambda$ (resp. $\mathbf{f} = M_{\lambda'} \to \Lambda_{\lambda'}$). It is then enough to show that $\cap_{\lambda,\lambda' \in X^+; \zeta = \lambda' - \lambda} (\mathcal{T}_\lambda \otimes \mathbf{f} + \mathbf{f} \otimes \mathcal{T}_{\lambda'}) = 0$.

Now assume that x belongs to the last intersection. We have $x = \sum_{\nu \in F, \nu' \in F'} \mathbf{f}_\nu \otimes \mathbf{f}_{\nu'}$ where F, F' are finite subsets of $\mathbf{N}[I]$. We can find $\lambda \in X^+$ such that $\lambda + \zeta \in X^+$, $\sum_i \langle i, \lambda \rangle > \mathrm{tr}(\nu) + \mathrm{tr}(\nu')$, $\sum_i \langle i, \lambda + \zeta \rangle > \mathrm{tr}(\nu) + \mathrm{tr}(\nu')$ for any $\nu \in F, \nu' \in F'$. If $x \neq 0$ then for such λ we have $x \notin \mathcal{T}_\lambda \otimes \mathbf{f} + \mathbf{f} \otimes \mathcal{T}_{\lambda+\zeta}$. This shows that $x = 0$. This proves 3.5.4.

p. 200 line –5: replace last λ' by λ

p. 220: part (d) of Theorem 27.3.2 and the first two lines of the proof are missing. Thus, at the bottom of p. 220 (after 27.3.2(c)) add:

(d) The natural homomorphism $\mathcal{L} \cap \Psi(\mathcal{L}) \to \mathcal{L}/v^{-1}\mathcal{L}$ is an isomorphism.

The element $b_1 \diamond b'_1 = \sum_{b_2,b'_2} \pi_{b_1,b'_1,b_2,b'_2} b_2 \otimes b'_2$ satisfies the requirements of (a). This shows the existence in (a). It is also clear that the elements

p. 228: the diamond in line 2 of 28.2.6 should be bigger (of the same size as the diamond in line 4 of 28.2.6).

p. 237 in line 2 of 29.5.1: replace: on $\dot{\mathbf{U}}[\geq \lambda]$ for some $\lambda \in X^+$ by on some two-sided ideal of finite codimension of $\dot{\mathbf{U}}$.

p. 240 first line of 30.1.7: remove one dot after "Example".

p. 246 line 8: replace $x \in \mathbf{f}_\nu$ by $x \in {}_R\mathbf{f}_\nu$

p. 320 in line 5 of 40.1.2: replace $l(w) = 1$ by $l(w) = 0$.

p. 320 in line 6 of 40.1.2: replace $l(w) > 1$ by $l(w) > 0$.

p. 325 line –10: replace [7] by [6].